Nature Management by Grazing and Cutting

Geobotany 14

Series Editor

M.J.A. WERGER

Nature Management by Grazing and Cutting

On the ecological significance of grazing and cutting regimes applied to restore former species-rich grassland communities in the Netherlands

by

JAN POUWEL BAKKER

KLUWER ACADEMIC PUBLISHERS
DORDRECHT / BOSTON / LONDON

Library of Congress Cataloging in Publication Data

Bakker, Jan Pouwel.
Nature management by grazing and cutting : on the ecological significance of grazing and cutting regimes applied to restore former species-rich grassland communities in the Netherlands.
p. cm. -- (Geobotany ; 14)
Thesis (doctoral)--Rijksuniversiteit te Groningen, 1989.
Bibliography: p.
ISBN 0-7923-0068-8
1. Range management--Netherlands. 2. Rangelands--Netherlands. 3. Range plants--Netherlands--Ecology. 4. Restoration ecology--Netherlands. I. Title. II. Series.
SB193.3.N4B35 1989
639.9'9--dc19 88-31951

ISBN 0-7923-0068-8

Published by Kluwer Academic Publishers,
P.O. Box 17, 3300 AA Dordrecht, The Netherlands

Kluwer Academic Publishers incorporates
the publishing programmes of
D. Reidel, Martinus Nijhoff, Dr W. Junk and MTP Press.

Sold and distributed in the U.S.A. and Canada
by Kluwer Academic Publishers,
101 Philip Drive, Norwell, MA 02061, U.S.A.

In all other countries, sold and distributed
by Kluwer Academic Publishers Group,
P.O. Box 322, 3300 AH Dordrecht, The Netherlands.

printed on acid free paper

Printed in the Netherlands

To Prof. Dr. Dingeman Bakker
Prof. Dr. Dik Thalen†

Acknowledgements

This study is part of the research programme on grasslands carried out by the Department of Plant Ecology of the University Groningen in the Netherlands. In 1970 the department decided that some fundamental work on nature management should take place. I started this work in 1971 on the salt marsh of the island of Schiermonnikoog. I would like to thank the authorities in charge of the management (Dienst der Domeinen, recently the State Forestry Commission) P. Harthoorn, R. Nieuwenhuis, A. Talsma and colleagues for their cooperation and the fact that they allowed me to work in the nature reserve. I also wish to express my gratitude to the farmers of the island for their kind cooperation.

The study was extended to the inland area of the nature reserve 'Stroomdallandschap Drentsche A' in 1972. I would like to thank the authorities in charge of the management (State Forestry Commission) Alex Ernst, Hendrik Lanjouw, Henk Post, Bouwe Riemersma and colleagues for both their cooperation and allowing me to work in their domain. J. Nijwening and J. Reinders kindly provided the sheep for the grazing experiment.

I am much indebted to my teacher in plant ecology Professor Dingeman Bakker. The memories of the long excursion we made along the French Atlantic coast in my student period and the many trips we later took in the Drentsche A area and on Schiermonnikoog will always remain with me. We discussed the plans for the management experiments and intended to follow them for a period of five years. It soon transpired that such a period was too short for new equilibria to establish after the imposition of the regimes. The publication of this study does not mean that such equilibria have now been reached, but it gives the state of the art after 10 to 15 years. Professor Dingeman Bakker did unfortunately not live to see the later phases of it and I, therefore, dedicate this book to his memory.

Two weeks after the start of my job a letter came from the Research Institute of Nature Management. This institute started research on nature management by grazing with large herbivores and requested the cooperation of Dutch universities. I agreed to their proposals and I am much obliged to Piet Oosterveld who introduced me into the world of grazing.

Yzaak De Vries became my assistent in 1973 and it is certain that but for him only a small part of the work would have been done. He worked in the field, in the greenhouse and on the computer, he instructed students and arranged the carrying out of management practices and never lost his temper. Thank you Yzaak, I hope we will cooperate for many more years.

I thank Jos De Wiljes, Nellie Eck, Willem Van Hal and Lena Rusthoven, for the chemical analyses, Kor Beukema, Jouke Franke, Jacob Hogendorf, Sieze Nijdam, the late Menno Robertus and Ebel Top for technical assistance and Clara Biewenga for her administrative assistance. I also thank Gerda Allersma, Leendert Brand, Theo De Jong, Marga Dekker, Jan De Leeuw, Jan Leemburg, Ab Masselink, Janke Schoonewille, Ans Schoorlemmer, Gerard Ter Heerdt, Lodewijk Van Duuren, Arend Van Dijk and Rudi Zwiers who were appointed to our department for short projects.

Many students were involved in studies on nature management practices from 1971 up to and including 1987. The results of many projects and ideas which were generated, form an important part of this book and

I, therefore, wish to thank chronologically Feiko Prins, Margje Van Der Duim, Joke De Boer, Niels Schotsman, Jan Leemburg, Jannie Schokker, Jan Klooster, Rudolf Schwab, Tineke Schwab-Vos, Kees Canters, Ada Zeevalking-Van Yperen, Ab Grootjans, Steven De Bie, Gerda Allersma, Gretha Hoekstra, Piet Swierstra, Bram Vreugdenhil, Fons Vos, Sip Van Wieren, Ko De Witt, Ger Boedeltje, Jacob Ruyter, Jelle Norder, Herman Hofs, Lilli-jan Rohof, Paul Struyck, Hans Dallinga, Peter Tjaden, Jan De Leeuw, Henk Van Netten, Anneke Pon, Margreet Zijlstra, Mirjam Huisman, Jacob De Vries, Edwin Hoekstra, Rika Brands, Henk Kasteleijn, Rolf Boersma, Chris Brouwer, Bertus Van Den Hof, Hans Zomer, Rick Looijen, Joan Looijen, Dick Veltman, Belia Van Den Berg, Ron Fijn, Karin Sjoukes, Aly Schutter, Gerard Ter Heerdt, Arjan Braaksma, Peter Russchen, Jos Hoogveld, André Hendriks, Obe Brandsma, Menno Dijkstra, Regien Schoemaker, Brigit Van Berkel, Hendrik Jan Muller, Erik Quené, Johannes Tonckens, Marc 't Hart, Marion Brongers, Jacob Vogel, Janine Van Den Bos, Dick Bekker, Robert Hofstede and Han Olff. It is a pleasure to know that several of them now have jobs in nature conservation.

Professors Jelte Van Andel and Jan Barkman became involved in a late phase of this study, namely, the setting up of the book and its writing. I wish to thank them for their constructive criticism. Jelte, thank you for the opportunity of being able to go my own way during the writing of the book and other activities in the department. I would like to thank Professor H.T. Waterbolk for his comments on the chapter concerning the Drenthe Plateau. The promotor Professor Dik Thalen and the referent Dr. Ger Londo were already involved in the early stages of the study. I thank them for the discussions on the plans of the experiments and in the field and their comments on the manuscript. Unfortunately Professor Dik Thalen died when I was correcting the proofs, and I, therefore, dedicate this book to his memory.

I learned a lot about toponymes and former land utilization in Drenthe from Bart Van Heuveln and Jan Wieringa and in Schiermonnikoog from Durk Reitsma. The literature study, the personal communications with colleagues, during symposia and field visits in the Netherlands and other countries, working on similar problems have been of great importance for the analysis, interpretation and final presentation of the data. I gratefully acknowledge the different contributions of my present and former colleagues in the Department of Plant Ecology.

I would like to thank those involved in producing the book, namely, Evert Leeuwinga, Chris Steenmeijer and Dik Visser who made the figures, Clara Biewenga who typed the tables, Joan Purdell-Lewis for the correction of the English text and Wil Peters and Professor Marinus Werger for their activities on behalf of Kluwer Academic Publishers. Suus Geluk, my wife, typed various versions of the manuscript and our children Liesbeth and Chris often accompanied me in the field and did some boring jobs in the laboratory. Thank you all.

I had initially intended to present the results of this study in a Dutch book, since the Dutch authorities in charge of the management should be able to read it. Later I changed my mind for two reasons. The first was the publication of many articles in Dutch journals and the opportunity of writing in the series of Dutch books on nature conservation 'Handboek Natuurbeheer'. The second was the increasing pressure for scientists to publish on the international scene. This has the advantage of demonstrating some of the Dutch knowledge on the conversion of grassland from agricultural production to nature reserves, which is much needed in the EEC. My wish is that this book contributes to the increase of such knowledge and its extension in other countries.

Table of contents

PART ONE

Introduction

CHAPTER I

Introduction

I.A. Background and scope of the study

The degradation of flora and vegetation in natural and semi-natural landscapes has become a matter of great concern. The problem affects the whole of Western Europe as well as many other parts of the temperate world, but the Netherlands seem depressingly to occupy the top position. This may be attributed to the dense population and the high level of technological development resulting in expanding towns, villages and industrial areas connected by a vast network of roads. The rural areas outside the population concentrations have a predominantly agricultural land use. Agricultural practices have, however, had the greatest detrimental influence on the country side, at least with respect to the area under study. These practices included reclamation of 'waste' areas during the 1920's and 1930's, and reallotment programmes after the Second World War. The increase in agricultural production and quality of the crop was the result of drainage and the application of fertilizers (Fig. I.1). The reduction of species diversity in trials, as a result of the application of fertilizers as compared with zero-plots, has been described in the United Kingdom (Williams, 1978). Germany (Klapp, 1965), Belgium (Van Hecke, Impens & Behaeghe, 1981) and the Netherlands (Oomes & Mooi, 1981; Elberse, Van Den Bergh & Dirven, 1983; Willems, 1983A).

A 'feed-back' against the degradation of flora and vegetation has been developed. Again the Netherlands seem to occupy a top position. From 1906 onwards, areas have been acquired by private organizations for landscape and nature conservation purposes, more recently also by the State. Due to the low intensity of human impact early this century the main conservation practices implied complete non-interference, effectuated by fencing-in the nature reserves. During the Second World War both British (Anon., 1943; Gabrielson, 1947, in Westhoff, 1983) and Dutch conservation workers (Westhoff, 1945; Westhoff & Van Dijk, 1952) developed the idea of necessary intervention in semi-natural landscapes to preserve communities of nature conservation interest. Reduction of species diversity in abandoned grassland is well known from the comparison of managed and unmanaged grassland, cf. Wells (1980) for the United Kingdom, Schiefer (1981) for Germany, Zoller et al. (1984) for Switzerland, Persson (1984) and Tyler (1984) for Sweden, Willems (1983A) and During &

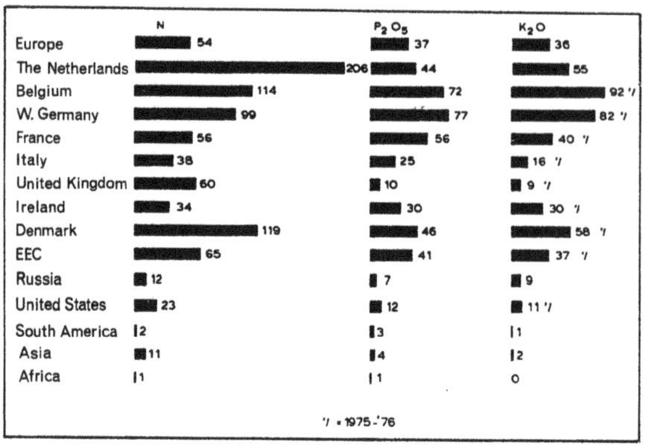

Fig. I.1. The application ($kg \cdot ha^{-1}$) of artificial fertilizers N, P_2O_5 and K_2O in various countries in 1976/1977 (after Rinsema, 1979).

Willems (1986) for the Netherlands. The organisations owning nature reserves were, therefore, transformed into authorities in charge of active management.

Since most semi-natural grasslands are marginal areas, from an agricultural point of view, these areas tend to be the first to be neglected or abandoned. This was done earlier, but recently it was enforced by the EEC agricultural policy which facilitated highly productive farms and closed down less productive ones.

Nature management by the continuation of the agricultural practices existing before acquirement is the most logical option in maintaining species-rich grassland communities. During the last 10 to 20 years the scope of nature management has, however, been extended to the restoration of formerly species-rich plant communities that had been degraded by abandoning or the application of fertilizers. The question arises: *Is there a way back from degraded grassland to species-rich grassland communities?* The present study focusses on this aspect of nature management practices.

Small-scale experiments including grazing, cutting, or burning revealed a successful restoration of abandoned grasslands in the United Kingdom (Green, 1980; Wells, 1980), in France (Poissonet et al., 1978), in Denmark (Bülow-Olsen, 1980B; Buttenschøn & Buttenschøn, 1982A) and in Germany (Gerth, 1978; Schiefer, 1981). Most results have not, however, been systematically analyzed or documented, but exist as a vast fund of practical experience in the minds of managers (Green, 1980).

The Dutch authorities in charge of the management did a lot of work based on practical experience in reducing the residual effects of high fertilizer applications (impoverishing the soil, 'Aushagerung' in German, 'verschraling' in Dutch) (Research Institute for Nature Management, 1979). This aim has been persued because a nutrient-poor soil was considered a prerequisite for the maintenance and/or development of species-rich plant communities. The management results were interesting, but, unfortunately, they have again not been documented. If temporal results are available these do not go beyond a description of successional changes in plant communities observed in permanent plot studies. Moreover, only a few comparisons between different management practices starting from an initially homogeneous grassland have been made (Van Den Berg, 1979; Oomes & Mooi, 1981; Willems, 1983A). In many cases changes in permanent plots do not fully represent the vegetation processes on a larger scale (Londo, 1974). Also, virtually no attempt has been made to use permanent plot data as a starting point for population dynamical studies for a better understanding of the observed vegetation dynamics.

The aim of the present study may be formulated as follows:

- to describe the longlasting human impact on the landscapes under study and
- to analyse the ecological significance of (a) the agricultural practices fitting in the previous land use system and of (b) the nature management practices hay-making and grazing by comparing them in several former species-rich grassland communities that have been degraded by abandoning or the application of fertilizers.

The analysis of ecological significance was focused on:

1. Land use dynamics:

- How did land use practices change until they became nature management practices?
- How is the acquisition of nature reserves practised?
- How are the nature management practices carried out?

2. Vegetation dynamics:
- How does species composition change under the management regimes?
- How is the stand structure (vegetation horizontal pattern and vertical structure of the sward) influenced?
- How are production ecological variables (standing crop, production, consumption, decomposition) influenced?
3. Population dynamics:
- How does the seed dispersal of existing and expected plant species occur?
- What is the role of the seed bank?
- How is the germination ecology of existing and expected plant species?
- What is the fate of seedlings in the sward?
- To what extent can the re-establishment of new species be enhanced by re-introduction?

Vegetation dynamics and population dynamics may result from direct and indirect effects of management practices. Attention was, therefore, also paid to the abiotic factors which are modified by management regimes:
- What is the impact of grazing and hay-making ('internal' management) on soil nutrients, soil salinity, soil compaction?
- Although no experiments were performed with respect to groundwater regimes, some attention is paid to the influence of groundwater on the vegetation since it is the most important 'external' abiotic factor for nature reserves.

The information presented is mainly relevant to the Netherlands. Abandoned and marginal agricultural grasslands, however, occupy vast areas of Western Europe as will be shown below and most processes studied are not limited to the Netherlands. Some of the findings presented here will, therefore, also have an application value elsewhere.

I.B. Abandoned and marginal agricultural grasslands

I.B.1. Definition

The grasslands dealt with in the present study cannot merely be called amenity grasslands. The Natural Environmental Research Council in the United Kingdom defined amenity grassland as 'all grass with recreational, functional or aesthetic value and of which agricultural productivity is not the primary aim' (NERC 1977 in Rorison, 1980). Vast areas of amenity grasslands are managed intensively and can be considered as cultural landscapes, e.g. lawns, urban parks and open spaces, urban road verges, golf fairways. Cultural landscapes are characterized by flora and fauna which are essentially influenced by man, in such a way that the dominant natural species may have been replaced by other ones (Westhoff, 1983). Semi-natural landscapes are landscapes in which the flora and fauna are largely spontaneous, but where the vegetation structure has altered in such a way that it now belongs to a different formation type (Westhoff, 1983; see also Dierschke, 1984). They include rural road verges, common land, golf course, National Trust land and nature reserves. Only the latter two are interesting for our purpose, because they will be managed primarily for nature conservation purposes. The marginal agricultural grassland types, which were studied, therefore form part of the semi-natural amenity grasslands and they are restricted to nature reserves (Photo I.1). Abandoned grasslands are characterized by non-interference. This can occur in agricultural areas when farmers loose interest, or in nature reserves where a shortage of money prevents the application of nature management practices. Marginal agricultural areas are not an absolute conception. These have relatively low yields and the quality of the forage is poor compared to the areas with intensive agricultural practices. In general the former consist of grasslands in remote valleys, in rough mountain areas, but also in lowland areas where sufficient drainage is too expensive to be economically feasible in view of the low yield and/or quality of the products. Their character can, therefore, depend on abiotic environmental factors; too dry, too wet, too steep or by the distance from the farm, or the obstacles (hedgerow, wooded hedgebanks, small woods, ditches) in small-scale landscapes.

The differences between countries should be considered with respect to the local definition of

Photo I.1. Hay-making by scythe in the salt marsh study area.

marginal agricultural areas. Relatively low yields in some countries are regarded as being relatively high in other countries with an overall lower agricultural production. In other words, 'marginal' areas taken out of the agricultural system in, for instance, the Netherlands or Denmark are still exploited agriculturally in, for example, the United Kingdom or France. That is the reason why even artificially fertilized grasslands can be taken out of the agricultural system in the Netherlands with the aim of restoring species-rich grassland communities (Photo I.2) or heathland, vegetation types still widespread in other countries. Most striking in this respect, are the attempts to restore former hayfield communities in the Netherlands, while comparable existing plant communities in northeastern Poland have been damaged due to the recent introduction of artificial fertilizers (Bakker & De Vries, 1982). This, in fact, has a double negative influence. The productivity of the grasslands adjacent to the farm increases in such a way that the farmer can produce the same amount from a smaller area than he did before the introduction of fertilizers, while the second effect is the abandoning of vast grassland areas situated further away.

One should not misinterpret the above. A possible interchange of plant communities might be suggested between the Netherlands and Poland. This is impossible since grassland communities on comparable abiotic substrates show local variations due to geographical differences. The above reasoning is a plea to maintain still valuable semi-natural grassland communities including the geographical differences. The present study will show that it takes a great deal of time and money to restore former species-rich communities, if restoration is at all possible.

I.B.2. Inland areas in western Europe

The agricultural EEC Committee (1979) drew up an inventory of the extent of areas taken out of the agricultural system. The northwest European countries will be dealt with in turn. The situation of these areas is shown in Fig. I.2.

In the United Kingdom about 210 000 ha of aban-

Photo I.2. Hay-making in previously artificially fertilized grassland in the inland study area.

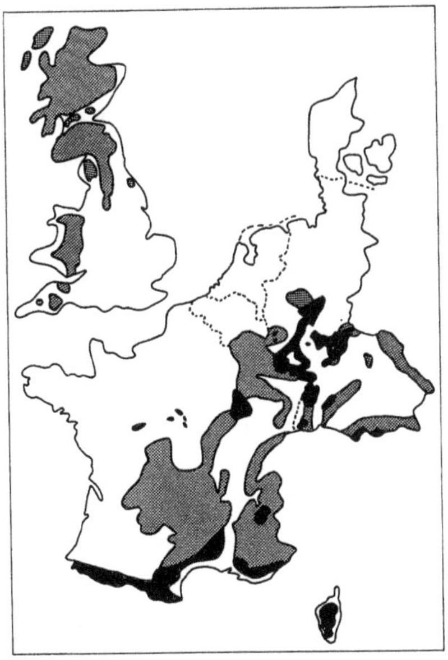

Fig. I.2. Location of regions with large abandoned agricultural areas (black) and small abandoned agricultural areas, but threatened in the near future (dotted) in northwest Europe (after EEC Committee, 1979).

doned land is found. This is expected to increase greatly, because large areas are under rough grazing with very low stocking rates. Germany has 300 000 ha abandoned land (Schreiber, 1980).

The definition of land taken out of the agricultural system is not clear in France, but at a rough estimate the area of abandoned land amounts to 1 500 000 ha. It is expected to increase greatly in the near future.

Denmark has hardly any land which has been taken out of the agricultural system. It amounts to between 5 000 and 10 000 ha and it is not expected to increase.

Belgium only has 3 200 ha of land taken out of the agricultural system and this area is not expected to extend.

In the Netherlands only a few fragments of abandoned land are found in nature reserves and this is likely to remain the case in the future.

At the end of the 1970's there were approximately 2 000 000 ha of abandoned land in northwest Europe (Table I.1). Mansholt (1969) estimated that about 5 000 000 ha of marginal land existed in the

EEC countries just before the EEC improvements (without the United Kingdom at that time).

The United Kingdom, Germany and France have large areas of abandoned land and the area will increase. Its former agricultural use was often rough grazing and hay-making with only a little application of fertilizers. In these countries, restoration management, therefore, often involves the removal of scrub and litter, but not a reduction in the residual effect of the former application of fertilizers. This latter type of restoration practice will particularly be carried out in nature reserves in countries with a generally more intensive agricultural system like Belgium, Denmark and the Netherlands. Owing to a shortage of money this is only done in the Netherlands and not in Belgium and Denmark, although certain areas of abandoned land do occur in these countries.

Prillevitz (1969) expected that approximately 160 000 ha grassland might be taken out of the agricultural system in the Netherlands during the 1970's and 1980's. These areas include dry sandy soils, wet sandy and peaty sand-soils, weak peat and clay-on-peat soils. When areas with other uses e.g. recreation, roads, lawns are excluded, 60 000 ha might be considered for the extension of nature reserves. Which part of this area might come under restoration management or non interference will depend on the management options of the future authorities in charge. At the present time there is very little abandoned land in the Netherlands.

Table I.1. Abandoned land in Northwest Europe inland sites (after EEC Committee 1979).

Country	Area ha
United Kingdom	210 000
Germany	300 000
France	1 500 000
Belgium	3 200
Denmark	c. 7 000
The Netherlands	p.m.
Total	c. 2 020 000

I.B.3. Coastal areas in western Europe

Salt marshes are a source of plant materials which can be utilized for agricultural purposes. They are best suited for grazing and hay-making. Hay-making has gradually decreased and nowadays the most widespread use is open-range grazing for domestic animals (Photo I.3), especially in northern- and western Europe including northwest France. In his study on European salt marshes and salt steppes for the Council of Europe, Dijkema (1984) made a provisional inventory of salt marshes. His results for the northwest European coastal areas have been listed in Table I.2. Grazing and hay-making were not always documented separately. On average these practices are carried out on 63% of the salt marsh area, with the lowest percentage for the United Kingdom (53%) and a very high percentage in Germany (95%). Recently Kempf et al. (1987) compiled the land use in the whole Waddensea area. They found that the overall intensity of exploitation decreased in the sequence Schleswig-Holstein>Denmark>The Netherlands>Niedersachsen. Moreover, a remarkable difference was revealed between the sandy island salt marshes (48% was abandoned, 52% was grazed) and the clay containing mainland salt marshes (27% was abandoned, 65% was grazed and 8% was cut for hay).

Along the Baltic shore marshes the stocking rate of domestic animals decreased over the last decades and many marshes are no longer part of the agricultural system (Tyler, 1969). The Dutch, German and Danish mainland salt marshes and the larger Halligen in the Waddensea area are heavily grazed. In general, the sandy island salt marshes are grazed more lightly than the mainland salt marshes (Kempf et al., 1987). Some Wadden island salt marshes have never been grazed or have only been done so very lightly. In the southwest of the Netherlands (Kogel & De Jong, 1983) and Belgium the grazing of salt marshes has strongly diminished. In the south and southeast of England stock grazing on salt marshes is very limited. The only major area where grazing still takes place is in the Wash. Many of the marshes in north Norfolk, Essex and Kent had grazing up to 40 or 50 years ago. Most salt

Photo I.3. Cattle-grazing on the salt marsh study area.

marshes on the west coast of England, Wales and Scotland are heavily grazed (Adam, 1978). In France 78% of the salt marsh area along the Channel coast, is grazed, whereas more to the south only 26% is grazed. Approximately 1500 ha has been taken out of agricultural practices along the Channel. It is not known whether the French non-interfered salt marshes have been taken out of the agricultural system recently or have never been grazed.

Apparently the costs of hay-making and sheep and cattle breeding have become and are still becoming too high in many European countries to be economically attractive. Consequently, the area of salt marsh pasture taken out of the agricultural system is gradually increasing nowadays and amounts to approximately 10 000 ha for the past 40 or 50 years.

The cessation of grazing often led to a dominance of one or a few plant species and a subsequent decrease in species diversity. Depending on the tidal level and the soil salinity, different

Table I.2. Salt marshes (ha) in northwest Europe (after Dijkema, 1984).

Country	Area	Nature reserve also in prep.	Agricultural exploitation			
			Grazing, mowing ha	Non interference	Unknown	Recently abandoned
Denmark	19 500	9 300	12 800 (66%)	200 (2%)	6 500	200
Germany	19 000	7 400	18 100 (95%)	500 (2%)	600	200
The Netherlands	12 800	10 500	9 500 (74%)	3 300 (26%)	0	1 300
Belgium	400	200	200 (50%)	200 (50%)	0	200
United Kingdom	37 500	17 000	19 800 (53%)	15 200 (40%)	2 500	6 600
France	14 400	1 000	8 100 (56%)	6 300 (44%)	0	c. 1 500
Total ha	103 600	45 400	68 500 (66%)	25 700 (25%)	9 600	c. 10 000

species can become dominant, e.g. *Puccinellia maritima* (Ranwell in Beeftink, 1977A), *Festuca rubra, Elymus repens* (Schmeisky, 1977), *Phragmites australis* (Siira, 1970; Jensen, 1978).

I.C. Nature management practices

To achieve the aim of nature conservation in grassland nature reserves (shortly: high species and community diversity including the occurrence of rare and endangered species) several notions have been used with respect to nature management practices. In the United Kingdom Wells (1980) distinguished between reclamation management, performed only once, and regular maintenance management. When grasslands are taken out of the agricultural system, coarse grasses, sedges and scrub take over. When such an abandoned area is required for some form of amenity use, it first has to be reclaimed before regular, or maintenance management can proceed. For example, a great deal of scrub removal and burning of grass litter took place in many public open spaces, which had been neglected during the Second World War before the land was made available for amenity purposes.

In the Netherlands two types of nature management practices have been distinguished (Londo, 1982A). Both aim at creating conditions for as many types of organisms as possible. The difference is the situation at the start of the management practices. In one situation the nature conservation interest is already high, whereas in the other it is low. Management practices aimed at maintaining the already high nature conservation interest of an area (Photo I.4) have been called 'natuurbeheer' and seem to be comparable to maintenance management. These imply the continuation of the existing management practices e.g. hay-making, chopping, non-interference, cutting sods with a certain regularity. Management practices aiming at the increase of the nature conservation interest of an area have been called 'natuurbouw' and may be distinguished in restoration management practices and the creation of entirely new habitats. These practices create the environmental conditions for the desired plant communities by affecting the soil or the existing plant communities. Restoration management includes both practices carried out only once (equivalent with reclamation management), and regular management practices. Restoration management always implies a change in the existing management practices e.g. cessation of the application of fertilizers followed by regular hay-making or grazing with a low stocking rate, removal of the eutrophic topsoil by cutting sods or deep ploughing followed by regular hay-making.

Experimental research in small-scaled plots is going on in the United Kingdom, France, Denmark, Switzerland and Germany, to study the effects of grazing, burning and cutting regimes. Questions of burning or grazing versus cutting and return of the cuttings versus removal can be evaluated for large-scaled practical management. At present, in the absence of the necessary knowledge and conceptual frame work, management is often carried out by trial and error, precedent or intuition, rather than by rational decisions (Snaydon, 1980). With or without a scientific base, the maintenance of high species diversity in the sward and the management costs of cutting give rise to major dilemma's (Green, 1980). Restoration management practices outside the Netherlands seem to be restricted to the reclamation of abandoned grasslands (Wells, 1980), but do not form an important part of the activities of authorities in charge of the management.

In the Netherlands, on the contrary, restoration management practices are performed manifold in nature reserves. A minor part consists of practices in abandoned grasslands, whereas the major part consists of reducing the residual effects of the application of fertilizers. The tradition of rough grazing almost disappeared with the decline in heathland area in the Netherlands. Controlled hay-making practices were, therefore, started a long time ago in nature reserves. Like in other countries the costs of cutting became very high. Another disadvantage is the uniformity of the practice, whereas grazing with low stocking rates might enhance a higher community diversity than cutting regimes. These insights have been gained a.o. during an excursion of several members of the Netherlands' Research

Photo I.4. Sheep-grazing on the heathland part of the inland study area.

Institute for Nature Management (Van Leeuwen et al., 1971) to Gotland and Oeland in Sweden. On these islands grasslands had been exploited for centuries by a.o. grazing with low stocking rates, resulting in an optimal differentiation with respect to grassland, scrub and woodland in park meadows. Moreover, on these islands grazing is associated with ant-hills which further enhance small scale botanical differentiation.

I.D. Theoretical considerations on nature management with special reference to diversity

I.D.1. General

In general nature conservation aims at a high species diversity and, as a consequence, the maintenance, or the creation, of conditions for as many species as possible (Westhoff, 1971). This aim should be extended to taxa below the species level such as ecotypes but for convenience sake the species is used. Diversity should be defined in order to evaluate the aim and achievements of nature conservation. Diversity has been subdivided into several levels (Whittaker, 1972). Alpha diversity refers to species diversity within a community, bèta diversity to the community diversity within a landscape, and gamma diversity to the total diversity of the landscape (incorporating both alpha and bèta diversity). Since both alpha and gamma diversity refer to the numbers of species (per community stand, respectively per landscape), both will be referred to here as species diversity which contradicts the proposal made by Peet et al. (1983). They wanted to distinguish between species number and species density.

The implications for nature management that can be derived from the theoretical considerations will not be enumerated in each of the following sections. They will be summarized under section I.D.7.

I.D.2. Some equilibrium models explaining species diversity

Since 1945 Van Leeuwen and Westhoff have regularly studied approximately 500 permanent plots in nature reserves all over the Netherlands. Based on these data they formulated some general ideas about species diversity (Van Leeuwen, 1966; Westhoff, 1971). The high species diversity in semi-natural landscapes seemed to be the result of the previous agricultural system: (i) for centuries the methods did not change, (ii) isolation by distance gave a gradually increasing restriction to human influence on the landscape: minimal restraint near the dwelling places and maximal at the outside of the action radius with all possible degrees of impact between the extremes, and (iii) our ancestors worked on a small scale. In other words: a high species and community diversity seem to be positively correlated to a high degree of constancy of environmental factors, which suggests that the variation around the average values of environmental factors should not surpass certain limits (Slobodkin & Saunders, 1969).

Grime (1973; 1979) collected numerous samples over a wide range of herbaceous vegetation types near Sheffield in one particular period and compared them (Fig. I.3). He observed that species diversity can be plotted as bell-shaped curves along gradients of maximum standing crop including litter. On the one hand small maximum standing crop and small species diversity might be correlated with a high 'environmental stress' and/or high 'disturbance'. Stress is defined as the sum of external constraints which limit the rate of dry matter production of all or part of the vegetation: e.g. very dry, dark, salt or nutrient poor conditions. Disturbance is defined by Grime as the sum of mechanisms which limit the plant biomass by causing its partial or total destruction: e.g. burning, grazing, cutting, and thus can be considered as an intensity of management. On the other hand a high standing crop, including litter, means a high productivity of at least some species which might limit species diversity by competitive exclusion. This is consistent with the observations on the application of fertilizers in section I.A. as fertilizers increase productivity (Peet et al., 1983). Grime's model implies that a moderate level of stress is necessary to prevent competitive exclusion.

Grime's original model (1973) interpreted the axis of maximum standing crop (including litter) as environmental stress and as disturbance (which

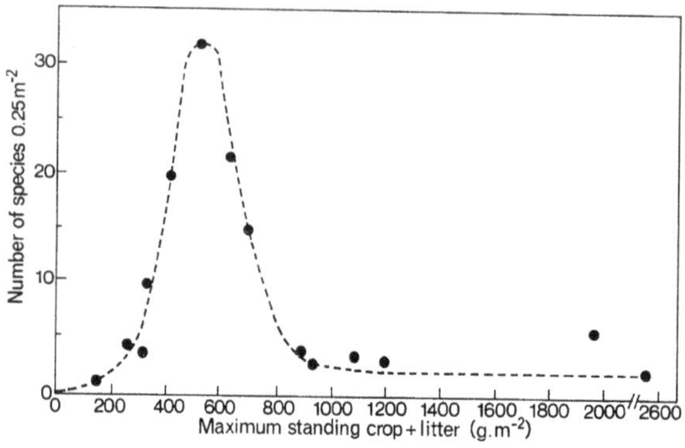

Fig. I.3. The number of species per unit area as related to the maximum standing crop including litter (after Al-Mufti et al., 1977).

equals intensity of management). Huston (1979) considered their effects simultaneously (Fig. I.4). He replaced Grime's stress gradient by a 'rate of displacement' and interpreted it as a gradient of growth rate. This interpretation is, of course, again consistent with observations on the application of fertilizers. Huston replaced Grime's disturbance gradient by a 'frequency of reduction'. Huston's model based on computer simulation, shows, therefore, that moderate levels of both disturbance and environmental stress are prerequisites for a high species diversity.

Grubb (1985) further elaborated the concepts of stress and disturbance. He is of the opinion that Grime's (1979) concept of stress is only a subset of the Levitt's 'stresses' (1972) i.e. those associated with a reduction in growth rate in the field. It would, therefore, be better if the stress axis in Grime's scheme was renamed 'productivity' axis, since that is what it really is. Grubb (1985) divided disturbance into various types, varying in frequency, extent and intensity. A basic sub-division is into continual and periodic disturbance. Examples of continual disturbance are grazing and trampling, mowing, destruction by wave action or flood, burial by sand or mud producing distinctive, long-persistent types of vegetation. Periodic disturbances such as storms and volcanic eruptions produce transient gaps in which regeneration occurs. There is also a great deal of variation in the types of disturbance within this category.

Grubb (1985) also mentioned the scale of obser-

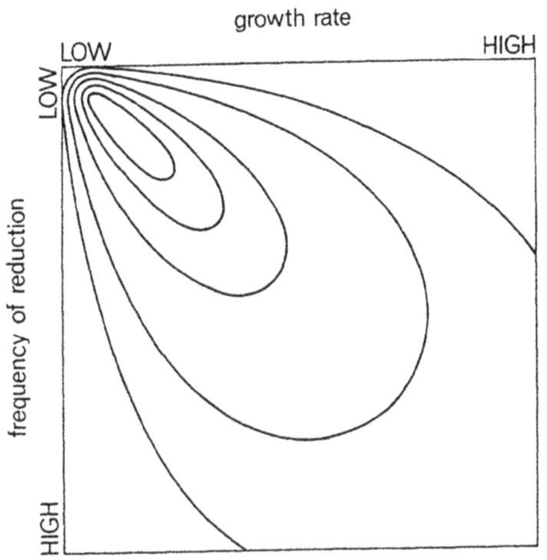

Fig. I.4. A trend surface of the dynamic equilibria between the rate of competitive displacement and the frequency of population reduction; the highest diversity is within the inner ellipsoid in the upper left corner (after Huston, 1979).

vation when he discussed two disturbance types. Both disturbance types can occur in the same area e.g. grazed or mown grasslands often suffer not only continual disturbance all over, but also periodic disturbance locally through the formation of heaps of soil by moles, or the deposition of faeces by cattle.

Using the term disturbance one should make a distinction between frequency, intensity and scale. According to Grubb this applies to continuous as well as periodical interference. Although consistent grazing and mowing are inherent to a system, Grubb considers these environmental factors to be a continual disturbance. Storms and volcanic eruptions are a couple of examples of periodical disturbances, not taking into account whether it as an isolated event or will re-occur (irregularly). If one adopts Van Andel & Van Den Bergh's (1987) definition, as is done in this study, an end can be put to the discussions about the various types of disturbance. In their opinion disturbance is 'a change in conditions which interferes with the normal functioning of a particular biological system'. They only use the term 'stress', in an eco-physiological sense, to indicate a specific negative reaction of individuals to a disturbance. 'Disturbance' and 'stress' are, in these definitions, related to an existing system and a disturbance is only a sudden change in for example a grazing or mowing regime. If the new management regime, for instance restoration management, is carried out consistently, a new system wil arise after some time and this will be dependent on that particular grazing or mowing regime.

To summarize, one can expect a high species diversity in mesotrophic environments where the above-ground biomass is removed frequently. The species diversity can also be high in poor environmental circumstances, where the above-ground biomass is removed occasionally.

Whittaker found surprisingly high species diversities in overgrazed pasture land i.e. under a high environmental stress (Naveh & Whittaker, 1980). Like Grime, Whittaker suggested that grazing can prevent competitive exclusion. He added to Grime's model by suggesting that over long periods of heavy grazing plant communities can adjust through the immigration of new species and evolutionary changes in those already present (Fig. I.5).

A quite different approach to explaining the maintenance of species-richness was followed by Grubb (1977). He introduced the concept of the regeneration niche by suggesting that the most important factor, making possible the long-term coexistence of species with the same habitat tolerance and life form, is differentiation in the requirements for regeneration, i.e. the sum total of processes leading to perpetuation of a species despite death of individuals or large parts of individuals. Some useful and easily quantifiable measures of the regeneration niche seemed, for example, seed-size, time of flowering, means of pollination and means of dispersal in space. A study on short-lived plants in chalk grassland in southern England showed, however, that the sensitivity to water supply, canopy height, predation by molluscs, dormant seed bank and conditions required to induce germination, control the population size of these species (Grubb, 1984). These factors were difficult to measure. So we are still far from being able to assess the importance of differences in the regeneration niche in the maintenance of species-richness in many communities.

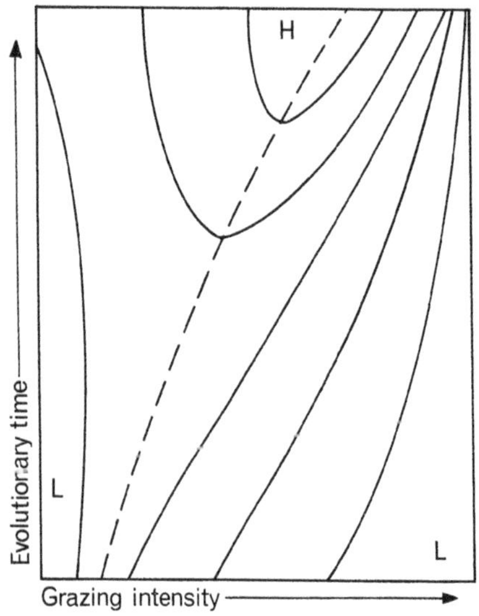

Fig. I.5. A trend surface of the relationship between species diversity and grazing intensity; L refers to low species diversity and H to high diversity (after Naveh & Whittaker, 1980).

I.D.3. Change of the steady state

The afore-mentioned models are, in fact, equilibrium models in describing species diversity under constant conditions (Peet et al., 1983). Management practices like burning, grazing and cutting with a regular frequency and constant intensity are essential in maintaining an equilibrium and are, therefore, considered 'system-inherent' environmental factors. The models do not describe what happens following a change in the environmental conditions, i.e. in the management regime ('system-non inherent' environmental factors). Grime's and Huston's (1979) models both imply that the change in species diversity can be predicted by a simple shift in position on the graphical model. This may be true if species equilibrium has been re-established, which does not hold for short-term changes (Peet et al., 1983).

In semi-natural landscapes a sudden change in abiotic environmental conditions or management practice brings about a succession which leads to a decrease in species diversity (Westhoff, 1971). Van Leeuwen (1966) extends this to suggesting that any change in environment or management regime will lead to a decrease in species diversity.

I.D.4. Dispersal of plant species

Restoration of degraded grassland supposes the (re-)establishment of plant species. Maintenance of species in a sward also implies a continuous process of establishment of individual seedlings, because a lot of perennial species have a turn-over period of just a few years e.g. *Ranunculus repens* (Sarukhán & Harper, 1973) and *Anthoxanthum odoratum* (Davies & Snaydon, 1976). Establishment is often the critical phase in plant invasion (Grubb, 1977). Disturbance in grasslands provides invasion opportunities as well as favourable conditions for early growth.

Newly appearing species may have two sources of origin. They can originate from a persistent seed bank (Thompson & Grime, 1979) as an ecological record of former plant communities (dispersal in time). This concerns species with long-lived seeds which have some type of dormancy. A lot of species, however, have short living seeds and these species will not contribute to a persistent seed bank. They have a transient seed bank sensu Thompson & Grime (1979). In order to become re-established this category of species has to come from elsewhere (dispersal in space). Dispersal possibilities of such species are, therefore, important.

Mac Arthur & Wilson (1976) formulated the equilibrium theory of island biogeography. It describes the species diversity of islands as the equilibrium point between the immigration and extinction of species. Larger islands have larger populations and thus lower extinction rates than small islands with consequently small species diversity. More isolated islands have lower immigration rates due to their distance, with consequently small species diversity, than islands closer to a source area. The equilibrium theory seems to predict the species diversity of nature reserves as islands in the sea of the cultural landscape. The extreme degree to which semi-natural landscapes have been reduced to small islands was well documented in case studies of northwest European heathland by De Smidt (1979). The accessibility sensu Heimans (1954) i.e. the possibility for diaspores of species to reach such islands became smaller and smaller. Though experimental testing of the equilibrium theory, particularly with respect to terrestrial plants, has not been done, it seems obvious that nature managers should take it into account.

I.D.5. 'Ecotypic' diversity

The extinction of species due to increasing fragmentation and isolation of semi-natural landscapes in the sea of cultural landscapes might have a continuation at a taxonomic level below the species. Populations of remaining species will decrease and the exchange of genetic material will be hampered due to isolation. Eventually this process may lead to the origin of ecotypes due to disruptive selection (Bradshaw, 1965).

Population differentiation in response to disruptive selection pressures has been recorded in several naturally occurring plant species. Selection pressure may be caused by drought (Quarrie, 1981), temperature (Graham & Petterson, 1982), light intensity (Wild, 1979), heavy metals (Ernst, 1982), mineral nutrition (Chapin, 1980), salt (Albert, 1982). Population differentiation as a response to human interference viz. mowing and grazing of grassland communities has been recorded in several species: e.g. *Rhinanthus angustifolius* (Ter Borg, 1972), *Achillea millefolium, Bellis perennis, Plantago lanceolata, P. major, Prunella vulgaris* (Warwick & Briggs, 1979). Disruptive selection pressures operate in the long run, but may begin to operate also after relatively short periods.

Snaydon & Davies (1972) found that populations of *Anthoxanthum odoratum* from limed and unlimed plots with hay-making in the Rothamsted Park Grass Experiment in the United Kingdom, appeared to be genetically adapted to the specific environmental conditions of their source site within 50 years. Snaydon & Davies (1976) proved that population differentiation had evolved at sharp boundaries, the sharpest differences occurred between populations collected only 0.1 m apart at each boundary. The steepness of the cline depended on the duration and constancy of the selection pressure involved. Conditions at the boundary that had been imposed for 60 years and then only at 4-year intervals were regarded less constant than those that had been imposed annually for 112 years. Genetic variation within populations was smaller and differences between populations were more distinct at the older boundary.

Gray et al. (1979) studied an ungrazed and a 120 year grazed salt marsh in the United Kingdom. *Puccinellia maritima* from the ungrazed mature salt marsh showed greater length of the flowering tillers, more allocation of resources to the production of flowering tillers and heavier plants than at the grazed salt marsh. Apparently grazing favours the selection of small, prostrate, short-leaved biotypes which allocate a relatively small proportion of their resources to flowering.

I.D.6. Patterns in plant communities

Different plant communities can exist side by side on slightly different substrates. Following the application of fertilizers or after the cessation of a management regime a few species tend to crowd out the species making up the differences between the communities and hence different plant communities converge. The result is a decrease in community diversity. In general the reverse would be expected sooner or later under restoration management practices. Substrate differences are again reflected in the distribution patterns of plant communities.

Micro-patterns with large abiotic differences and different plant communities between heavily grazed areas and lightly grazed patches were mentioned by Pigott (1956) in Upper Teesdale, United Kingdom and Regnéll (1980) in southern Sweden. They described peaty tussocks on a calcareous and loamy subsoil. The development of tussocks was considered to be the residual part of initially continuous turf, split by trampling cattle and subsequent water erosion. Important abiotic differences are found between tussocks and depressions. This finding, in combination with the age of the tussocks, explains their diverging plant composition. Once tussocks are formed they perpetuate and govern the trampling pattern of the cattle. Out of 117 'steps', observed on freely grazing cattle, 106 were placed in depressions, 5 in an intermediate position and only 6 on tussocks (Regnéll, 1980).

The presence of a fold can be a point from which

trampling and manuring gradients originate. Rosén (1973) described a macro-gradient in plant communities in a 400 m transect on the alvar (limestone heathland) of Oeland, Sweden. A few ruderals occurred close to the fold, mosses were found at 70 m, whereas lichens occurred 200 m from the fold. Total species diversity reached an optimum at 100 m from the fold.

In the preceding paragraphs the fact was mentioned that the distinctions of boundaries between plant communities may depend on the scale of the survey. Two boundary types can be discerned. The limes convergens or ecotone (Van Der Maarel, 1971) is characterized by a coarse pattern and sharp, often straight borderlines, while the vegetation is species-poor. It has strong and often irregular, temporal variations in the habitat. These often sudden and marked changes may result in soil compaction (Van Leeuwen, 1966; Westhoff, 1971). The limes divergens or ecocline shows a fine pattern with faint demarcation lines. It has little temporal variation in the habitat. It has attracted attention because of its species-richness and the often very small populations of its species. The former fact is in accordance with the general law formulated by Odum (1959) which states that the greatest diversity occurs in the middle range of a gentle physical gradient.

The floristic maximum of the limes divergens is found in the gradual transition between woodland and grassland, particularly represented by scrub and tall forb fringe communities (Westhoff, 1971). Dierschke (1974) gave an example of the transition from a hayfield into a beech forest in Germany. The hayfield had 20–25 species $\cdot$ m^{-2}, the forest 6–10 species $\cdot$ m^{-2} and the fringe 32 species $\cdot$ m^{-2}. Moreover, 8 species were restricted to the fringe zone.

In a dune slack transect in the Netherlands, with a height difference of 14 cm and a pH range from 4.2–7.0, species diversity was highest at the middle of the gradient: 36–43 phanerogamic species $\cdot$ m^{-2}. Thalen (1971) analyzed this species-richness by detailed mapping of the distribution of individual species. Four types of distribution could be distinguished. Species like *Salix repens* occurred all over the transect. *Mentha aquatica* occurred in the lower and middle part of the transect, *Potentilla erecta* in the higher and middle part. *Linum catharticum* was confined to the mesic environment of the middle of the gradient.

Grubb (1985) found that the gradient-species (such as *Linum catharticum* in the above study), which he calls edge-species, are not always restricted to gradients. Some are known to be common in extensive patches of one vegetation type, while being gradient-species in another situation. Little is known about the characteristics which separate gradient-plants from plants of gaps in nearby homogeneous vegetation stands.

Gradients can be a prerequisite for the maintenance of populations in irregularly fluctuating habitats. The presence of small height gradients, for example, effectively prevents the extinction of dune slack species from a whole dune valley, because the factors which threaten the maintenance of the population differs between height zones (Schat, 1983).

I.D.7. Implications for management

The implications for nature management that can be derived from the findings and theoretical considerations mentioned above are summarized below. The importance of community diversity may imply efforts to maintain species-poor plant communities provided these plant communities are rare and threatened. A high species diversity is not, therefore, the only true aim of nature conservation. Quantification of (species)diversity should be related to the area concerned or to the scale of observation. The occurrence of both macro- and micro-gradients complicates the quantification of (species)diversity. Gradients are very important in themselves since they are often species-rich and can have a species-rich flora including rare and threatened species (Westhoff, 1971). Since burning and cutting are uniform in their spatial effects, grazing with a low stocking rate gives rise to extra diversity with respect to patterns and gradients on different scales. Finally, gradients can be a prerequisite for the maintenance of populations in irregularly fluctuating environments (Schat, 1983).

Semi-natural landscapes and communities are

dependent on management regimes like burning, grazing, mowing. Each ecosystem in equilibrium, depends on its own system-inherent environmental factors including the ones governed by management practices. The best management practice for newly acquired nature reserves with a high nature conservation interest is, therefore, the continuation of the existing agricultural or other human practices (Van Leeuwen, 1966). This is in accordance with the idea that communities with a long management history have a higher species diversity than those subject to a recent management regime (Naveh & Whittaker, 1980). At present the detrimentous effects of changed hydrological conditions and deposition of nutrients in the 'acid rain' should be taken into account. Frequency, intensity and extent of a management regime should not exceed certain limits (Slobodkin & Saunders, 1969), but the limits depend on the origin of the managed ecosystem. This should be interpreted in two ways, e.g. the frequency of mowing can be both too high and too low (Van Leeuwen, 1973; Huston, 1979). The species diversity also shows an optimum in relation to the productivity of the substrate, with the caveat that the accumulation of litter is prevented, since this prevents the regeneration of species (Grime, 1981; Grubb, 1985).

Restoration management aims at increasing species diversity on fertilized or abandoned areas. The initiation of restoration management implies, however, a disturbance which in itself leads to a decrease in species diversity (Van Leeuwen, 1966). The observed effects of such a disturbance will depend on the plant community, the type of disturbance and the length of the period since the disturbance occurred (Peet et al., 1983). New species can either establish themselves from a persistent seed bank or from dispersal of seeds with a transient seedbank. (Re-)establishment of plant species with short-lived seeds requires either a small distance to the nearest population or a long action radius. Otherwise, transplanting or deliberate sowing may be necessary (Grime, 1981). The increasing isolation of semi-natural landscapes in the sea of cultural landscapes may lead to the creation of ecotypes. Due to diverging adaptations to the local environment the extinction of an ecotype cannot be compensated by the immigration of another ecotype, even if dispersal is not limited.

CHAPTER II

Research approach and methods

II.A. Organization

The fieldwork for this study started on the salt marsh of the island of Schiermonnikoog in 1971 and was extended to the valley grasslands and reclaimed heathland of the Drentsche A river valley in 1972. It included mapping of the vegetation, recording of permanent plots, estimating of above- and below-ground biomass, soil chemical analyses and the measurement of terrain-use by grazing cattle and sheep. Changes in the vegetation are still occurring, 10 to 15 years after the start of the restoration management practices. This report can, therefore, only give preliminary conclusions on trends. The recording of changes in species composition and structural characteristics of the vegetation still continues.

The work on the seed bank in the field and in the greenhouse, on germination ecology and the introduction of seeds was carried out mainly between 1977 and 1983. Estimates of the utilization factor took place between 1981 and 1983.

Experimental work was undertaken in the field as well as in the laboratory. The procedures adopted vary widely with the objectives and are, therefore, discussed in the sections concerned.

Part of the results of studies on selected topics were reported in Dutch or international journals.

II.B. Land use dynamics

The geomorphological, palynological and archaeological information was derived from the literature. The toponyme data and their interpretation are part of a research project by Mr. J. Wieringa from the Nedersaksisch Institute, University of Groningen. It covers the whole province of Drenthe and has been published in parts. The historical data on the island of Schiermonnikoog could be provided thanks to a private archive study by Mr. D.T. Reitsma of Stiens in the province of Friesland.

II.C. Vegetation dynamics

II.C.1. Nomenclature of taxa and syntaxa

The nomenclature of taxa follows the 'Flora van Nederland' (Van Der Meijden et al., 1983A), which, with the exception of a few details, is based on the 'Flora Europaea' (Tutin et al., 1964–1980).

Although we did not discern at the subspecies level, it is likely that several species which were recorded were represented by particular subspecies, viz. *Festuca ovina* ssp. *tenuifolia, Bromus hordeaceus* ssp. *hordeaceus, Phleum pratense* ssp. *pratense, Eleocharis palustris* ssp. *palustris, Galium palustre,* ssp. *palustre, Luzula multiflora* ssp. *multiflora, Plantago major* ssp. *major,* and *Festuca rubra* ssp. *commutata* in the inland study area. In salt marsh areas *Festuca rubra* ssp. *arenaria* and spp. *commutata* cannot be separated when grazed.

Salicornia brachystachya and *S. dolichostachya* were not always discerned and were, therefore, classified as *Salicornia europaea. Juncus ambiguus* and *J. bufonius* were classified as *Juncus bufonius. Lotus corniculatus* and *L. tenuis* both were labeled *Lotus corniculatus.* It was sometimes difficult to distinguish between *Agrostis capillaris* and *A. stolonifera.* Hybridization was an added complication. This was also the case for *Juncus acutiflorus* and *J. articulatus,* for *Rhinanthus angustifolius* and *R. minor* and for *Elymus pycnanthus* and *E. farctus.* The latter cannot be separately identified in grazed areas. Recently Van Glahn (1987) published, however, a key for the determination of *Elymus* taxa including hybrids according to vegetative characters.

The nomenclature of syntaxa mainly follows 'Plantengemeenschappen in Nederland' (Westhoff & Den Held, 1969). Some details are derived from Beeftink (1965), Sykora (1982; 1983), and Everts et al. (1980; 1984). The latter made a local classification of the Drentsche A communities taking the views of Ellenberg (1978) and Oberdorfer (1983) into account. The authors of syntaxa are mentioned only when the syntaxa are not included by Westhoff & Den Held (1969).

II.C.2. Vegetation mapping and recording of permanent plots

Vegetational succession can be described by both sequential mapping and the recording of permanent plots. Küchler (1967) considered the successive mapping method to be the most accurate and reliable for the study of succession in vegetation complexes.

According to Londo (1974) several criteria must be satisfied if vegetation maps are to provide a maximum amount of information about succession. Firstly, a suitable scale is necessary. The scale should be large enough to show the population patterns of important i.e. dominant species. In the present study the scale 1 : 250 was used for the field map and this was later reduced to 1 : 1000. Mapping of vegetation complexes should be avoided. Nevertheless, small-scale patterns were sometimes mapped as a mosaic for practical reasons. Secondly, each legend unit needs to be described consistently to allow a comparison between the subsequent maps. This problem was partly overcome by a check of the authors of the previous mapping of the adopted typology and legend for each subsequent mapping. For the different communities included in the legend units various criteria for classification were applied. It was found that the distribution of certain communities could best be mapped on the basis of the dominance of a single species or a few species, whereas for other communities the entire species composition played an important role. Thirdly, the communities constituting the legend units should be well-documented in phytosociological tables, so that a reliable comparison of the subsequent maps is possible. Fourthly, it may be necessary to choose particular areas to be mapped in detail. Mapping an entire area at the desired scale provides the most complete information about patterns and processes, but this is often difficult or impossible to achieve in practice. Detailed, large-scale, maps of parts of a study area are, therefore, often made. The choice of the parts to be mapped should guarantee a representative picture of the whole study area. In his study area, Londo (1974) always included gradients in detailed maps in order to gain as much information as possible with limited means. The possibility of annual recording seems an advantage over periodical (e.g. five years) mapping, since year-to-year fluctuations can be studied separately from vegetational succession. All field ecologists know that events may occur outside their 'representative' part of the study area. Londo (1974) analyzed the information gained by sequential mapping and by permanent plot recording. He concluded that a limited number of permanent plots gave a relatively large amount of information about the succession of a dune slack vegetation. A large increase in the number of permanent plots resulted in a relatively small increase in the information obtained.

The best method for the investigation of vegetational succession proved to be a combination of a limited number of permanent plots and sequential mapping of the entire study area (or representative parts of it). This method has been adopted in the present study.

II.C.3. Processing of vegetation data

In order to make a comparison of vegetation maps, transparant maps (scale 1 : 1000) of two years were overlaid with a dot grid. The distribution of the legend units was worked out for each separate point by means of the dot grid with a dot spacing of 0.5 cm. This enabled the changes in communities and the areas of the various communities to be calculated (400 points equals 1 ha at the scale 1:1000).

The vegetation was described through relevés $(2 \times 2)\,m^2$. Initially the Braun-Blanquet scale was used, as modified by Barkman et al. (1964), but the decimal scale (Londo, 1976) was used from 1976 onwards. The relevés recorded with the Braun-Blanquet scale were transformed to the real cover percentages in accordance with the middle parts of the Braun-Blanquet scale: r = 1, + = 2, 1 = 4, 2m = 4, 2a = 10, 2b = 10, 3 = 40, 4 = 60 and 5 = 90%.

Dissimilarities between relevés can be calculated taking only the presence or also the cover percentage into account. Londo (1971) found that the changes of both types of percentages dissimilarity during a succession study revealed similar trends, nevertheless large differences were found in some years. Although it seems not necessary to calculate both types of percentage dissimilarity some intermediate between presence and cover percentage has been looked for by different authors. In this study dissimilarities were calculated by means of the percentage dissimilarity according to Gauch (1982): $PD_{ab} = 100 - \frac{200\sum \min(a_i b_i)}{\sum(a_i + b_i)}$ in which a_i = the transformed value of species i in relevé a. The cover percentage of species i is subjected to a square root transformation in order not to overestimate high cover percentages. The 'Euclidian distance' values of other commonly applied dissimilarity measures are determined mainly by the dominant species, whereas the 'complemented coefficient of community' gives minor and major species the same emphasis. The intermediate weighing of percentage dissimilarity applied in this study, therefore, seems best since it emphasizes dominant species to a certain degree, but still considers minor species (see also Whittaker & Gauch, 1973). An agglomerative group average clustering (Everitt, 1974) based on the quantitative dissimilarities was carried out in order to obtain local coena. The average dissimilarity to a newly formed cluster was defined with the formula $D(a + b, c) = \frac{1}{2}(D(a, c) + D(b, c))$ in which a = the dissimilarity of relevé a, b = the dissimilarity of relevé b, and c = the dissimilarity of the newly formed cluster.

Each annual relevé of each permanent plot was ascribed to one of the (sub-clusters resulting from the clustering. According to this analysis the vegetational change was described schematically in terms of (sub)cluster transitions (e.g. Van Noordwijk-Puijk et al., 1979) in the permanent plots.

Vegetational changes in the permanent plots were quantified by comparing the dissimilarity for each year with the initial year for each permanent plot. The increase, decrease or fluctuations of dissimilarities were determined as the correlation between their values in each year and the matching year number during the study period. Vegetational changes were considered as trends or gradual deviations from the initial stand when the rang correlation coefficient proved to be significant ($P<0.05$). The occurrence of a trend does not imply that the original value is significantly higher or lower than that in the last year of the study period.

The successional response of each species was analyzed by comparing the observed changes in cover percentages to n (= as much as number of years of the experiment) standardized time series, having transformed these in an average relevé of the permanent plots which occurred in a subset (management regime per community). Incorrect conclusions with respect to increase or decrease are only made in a minority of cases when lumped permanent plots are compared to separate permanent plots (see Van Duuren et al., 1981). In each of these standardized series, maximal coverage occurred in a different year (in the first series the maximum in the first year etc.). The comparison was carried out by adding rank numbers to cover values of both actual observations and standardized series. The correlation of these series was expressed by Spearman's coefficient, and a significance level of 5% was used. If correlations with more than one series proved to be significant, the species in the permanent plot was ascribed to the standard series to which it showed the highest correlation (Van Duuren et al., 1981). If the maximal cover percentage was reached in one of the first three years the species was considered as a decreaser. If maximal coverage was reached in one of the last three years the species was considered as an increaser. If maximal coverage was reached in one of the intermediate years the species was considered as a temporarily increaser and if no maximal cover percentage was found the species was regarded as constant.

The diversity was measured as the number of species per relevé of 4 m^2. Since this attribute does not take the coverage of the species into account the equitability was also measured. The Shannon/Weaver equitability according to Whittaker (1965) depends on the number of species, s. If the latter is taken into account, as in this study, the relative equitability is measured

$$H^1 = - \frac{\sum_{i=1}^{s} p_i \cdot {}^2\log p_i}{{}^2\log s}$$

in which P_i = cover percentage of species i.

Most calculations mentioned above were done using computer programmes developed by L.F.M. Fresco in the Department of Plant Ecology at the University of Groningen. Dissimilarity matrices were processed with the programme DIMAT, clustering with SIMCLUS, graphical reproduction with DENDROS, trends with RECO and successional response of species with OPTI.

II.C.4. Indicator values for soil fertility

The effects of restoration management on the vegetation are measured here by the changes in several attributes.

The first attribute is soil nutrient content and simple parameters were adopted: N_{total}, P_{total}, K_{total}, C/N quotient (section II.E.9). The availability of macro-nutrients was not measured. These analyses, although they are comparable with the results of other authors, are of a very rough character.

An indirect way of quantifing the effects of management could be a species qualification with respect to the nutrient status of the soil. The second attribute is, therefore, the indicator value of the species present. Other authors recognize the need to describe environmental changes related to vegetational succession using indicator values (Loopstra & Van der Maarel, 1984; Melman et al., 1985).

Klapp (1965) and Ellenberg (1979) based nitrogen values on field resp. greenhouse experiments including the application of fertilizer. Nitrogen values are figures giving a rough, qualitative indication of the nitrogen preference of the species concerned. Kruijne et al. (1976) correlated the occurrence of species in the field with soil chemical analyses of phosphate. Ellenberg (1953) also measured response at a nutrient gradient level as a proportion of the sum of yields for all levels. He ascribed the discrepancy between physiological and ecological response to interspecific competition. Ernst (1978) only considered absolute yields and he attributed discrepancies to misinterpretation and insufficient knowledge of interacting environmental factors and ignorance of the phenomenon of ecotypic differentiation. Austin (1982) introduced the relative physiological response as being the yield of a species grown in monoculture in a particular environment divided by the yield for a monoculture of the most productive species (indicating the potential productivity) in that environment. The normalized ecological performance value for multispecies mixtures was defined analogically to the relative physiological response by replacing yields in monocultures by those reached in mixtures. He found that the maximal values of individual species in multispecies mixtures correspond in position along a nutrient gradient within the species maximum values for the relative physiological performance. They do not, however, correspond with the physiological optima measured using absolute yield. In the opinion of Austin this contradicts both Ernst's and Ellenberg's views of the relationship between ecological and physiological optima.

Other points of discussion on indicator values concern the soil chemical analyses: (i) the extraction method, and (ii) the seasonal fluctuations. Fresco et al. (1984) demonstrated that different methods for the determination of available phosphate reveal different correlations with the ordination axes in grassland vegetation. Andic (1971) showed that NO_2, P_2O_5 and K_2O decreased during the growing season in the subsoil of a *Senecioni-Brometum* and of a *Lolio-Cynosuretum*. Moreover, annual variation was found. In 1968 the correlation between NO_2 level and production was positive, whereas it proved negative in 1969 at the same time of the year (Andic, 1971). Seasonal patterns in nitrogen availability were measured in acidic soil supporting *Deschampsia flexuosa* and a calcareous soil supporting *Bromus erectus* (Taylor et al., 1982). The same was measured in chalk soil, calcimorphic brown earth and acid mull soil, all supporting *Deschampsia cespitosa* (Davy & Taylor, 1974). Gupta & Rorison (1975) found seasonal patterns in the availability of phosphorus and nitrogen of a podzolic profile using *Rumex acetosa* as a bio assay. Pegtel (1987) gives a survey of problems in the measurement and interpretation of soil chemical analyses.

Anticipating the theoretical discussion, indicator values are often used. Ellenberg's nitrogen values are, in particular, often applied when describing the synecology of plant communities in combination with the stand ordination (Werger et al., 1978; Persson, 1981). Although the list of nitrogen values for species, compiled by Ellenberg (1979) referred to Central Europe, it was also used for southern Sweden (Persson, 1981) and the Netherlands (Wind, 1980; Sterk et al., 1983). Loopstra & Van der Maarel (1984) concluded that Ellenberg's figures can be used in the Netherlands, if no absolute values, but only ranking values, are attached to plant species.

Some authors did not want to rely on the indicator values given by one author. They, therefore, compared the values of several authors. Sterk et al. (1983) made separate indicator value spectra according to Ellenberg (1979) and Kruijne et al. (1967). Those spectra appeared to be fairly similar. Wind (1980) combined Ellenberg's, Kruijne's and Klapp's (1965) indicator values by calculating their mean after having them transformed into a decimal scale. Schiefer (1983) did not rely on the detailed Ellenberg scale because of pseudo-exactness and simplified it into nutrient-rich ('Nitrophyten') and nutrient-poor ('Magerkeitszeiger') groups. This simple subdivision was adopted in this study. All species occurring in the study area were subdivided into three groups ('poor', 'intermediate' and 'rich' species), indicating the nutrient status of the soils in which they normally occur. Another well known subdivision is that in 'oligotrafent', 'mesotrafent' and 'eutrafent' species, but it is not indicated how this refers to a particular nutrient or groups of nitrogen or phosphate values as is done in this study. Species with nitrogen values of 1 to 3, according to Ellenberg (1979), were grouped as 'poor' species and values of 7 to 9 as 'rich'. Species with nitrogen values of 1 and 2, according to Klapp (1965), were grouped as 'poor' species and those of 4 and 5 as 'rich'. The phosphate values according to Kruijne et al. (1967) range from −100 to 100, though only some salt-marsh species show values of over 50. Species with phosphate values of −100 to −50 were, therefore, grouped as 'poor' species and values of 1 to 50 as 'rich'.

In the inland study area 145 phanerogamic species were recorded. Two or three authors agreed on 67 species (46%). If the authors differed in their assessment of the species by one grouping, the majority view was accepted (36 species, 25%); if they differed by two groupings, the species was categorized as unknown. The same was the case if only two authors classified a species and disagreed (20 species, 14%). If only one author classified a species this classification was used (22 species, 15%). Of the 36 species where the majority view was accepted, 7 species could be attributed to Ellenberg (all classified poorer than the other authors), 11 species to Klapp (most of them classified richer than the other authors) and 18 species to Kruijne (9 of which were richer and 9 of which were poorer than those of the other authors). It can be concluded that Kruijne's 'phosphate values' do not deviate systematically from Ellenberg's and Klapp's 'nitrogen values' (Table II.1).

The ratio

$$\frac{\text{number of species indicating nutrient poor soils}}{\text{number of species indicating nutrient rich soils}}$$

was used as a quantitative index of soil fertility. A quantification index was calculated by taking the cover percentage of the various species into account.

Ideally, if one doesn't want to go round in circles, the species indicator values should be compared with independent parameters. Melman et al. (1985), for example, considered the above-ground standing crop as a good measure of nutrient availability (cf. Grime, 1979) and they found very good correlations (r = 0.90) between the above-ground standing crop and the indicator values of both

Table II.1. Frequency of occurrence (%) of 145 plant species found in the inland study areas in the groups indicating nutrient poor soil conditions, intermediate- and nutrient rich soil conditions according to assessments in the literature.

	Kruijne et al. (1967) (n = 113)	Ellenberg (1979) (n = 102)	Klapp (1965) (n = 120)
Nutrient poor soil conditions	33%	39%	39%
Nutrient rich soil conditions	22%	25%	29%
Intermediate soil conditions	45%	36%	32%

Kruijne et al. (1967) and Ellenberg (1979). In the present study there was also a good correlation between the chemical characteristics of the soil and the indicator values of the species present except for the *Poo-Lolietum* (Table II.2). The indicator values of the species, therefore, seems to be satisfactory.

II.C.5. Species classification with respect to Mean High Water

Every field ecologist has an ecological classification of *species* in the back of his mind which includes lower salt marsh-, mid salt marsh-, upper salt marsh- and dune species. Such a classification is not, however, found in the literature. Several authors have described the zonation of plant *communities* down a salt marsh gradient and correlated it to critical sea water levels (Beeftink, 1977A; Dijkema, 1983) or to soil salinity (Waisel, 1972). The zonation of the species was also related to water salinity (Gilham, 1957A) or soil salinity (Gillham, 1957B; Beeftink, 1977A), but the species were not labelled with respect to their occurrence in the zonation or their salt tolerance. The latter was probably due to their wide amplitudes, which hindered a classification. Waisel (1972), therefore, classified nearly all salt marsh species into the category euhalophytes which is not informative. Westhoff (1947) related the plants' vitality to late spring and summer soil salinity and classified most species in the Boschplaat salt marsh on the Dutch Waddensea Island of Terschelling into four categories. The highest salinity category, however, proved to be heterogeneous, because it contained species related to a high mean soil salinity as well as species with high peak values during the dry summer. These difficulties could only be solved by very frequent salinity measurements. The problem still remains that any local classification cannot be generalized for all salt marsh systems (Westhoff, 1947). The then Boschplaat salt marsh was only 20 years old with a thin clay layer on top of the sandflats, whereas the Oosterkwelder salt marsh had been developing for a 100 years. A comparison of species classification in different areas only seems applicable with respect to MHW and not to a fixed Ordnance Date. Mean High Water varies from 80 cm + NAP (Amsterdam Ordnance Date) on Terschelling (Galle & Bontemps, 1975) to 95 cm + NAP on Schiermonnikoog (Fig. IV.10) and 110 cm + NAP on the Groningen mainland salt marsh (Dijkema, 1975).

Rozema et al. (1985) studied the growth response of species occurring in the lower and upper salt marsh to flooded and saline conditions and found that seawater salinity governs salt marsh zonation and inundation is superimposed on it. Armstrong et al. (1985) emphasized the relationships between species distribution, tidal activity and subsequent soil aeration patterns. One, therefore, assumes that species classification should be related to their position with respect to MHW. Based upon their abundance in the different main clusters, in the Schiermonnikoog situation most species could be classified into the categories lower salt marsh, mid- and upper salt marsh, and dunes respectively, some species being assigned to two categories (see Chapter VII). Since it is very difficult to draw lines between plant communities assigned to the mid-marsh *Amerion maritimae* and the upper-marsh *Lolio-Potentillion anserinae* these categories were lumped together (cf. Fresco, 1971).

II.C.6. Groundwater table and precipitation

Groundwater tables were measured every two or three weeks for different periods of time, viz. 1976–1977 in the Anlo and Taarlo valley grasslands, 1977–1978 in the Westerholt mixed grassland,

Table II.2. The nutrient content in autumn of the surface soil (0–5 cm) beneath some plant communities in the Netherlands. The number of species indicating nutrient-poor and nutrient-rich soil conditions in each relevé (2 × 2) m² is also shown. The plant communities chosen are supposed to succeed each other in the given sequence during the process of hay-making without fertilizer application (Everts et al., 1984).

Community	mg · 100 g dry soil^{-1}			C/N	Number of species		Poor/Rich species ratio
	N_{total}	P_{total}	K_{total}		Poor	Rich	
Poo-Lolietum	1880	510	71	14	1	7	0.1
Poo-Lolietum/Lolio-Cynosuretum	2320	725	65	15	2	6	0.3
Lolio-Cynosuretum	1930	575	52	14	6	5	1.2
Senecioni-Brometum	1070	189	47	16	8	5	1.6
Juncetum acutiflori	390	80	41	15	12	4	3.0

1974–1979 in the Lammertshem valley grassland and 1974–1985 in the Loefvledder valley grassland. Groundwater dynamics have been summarized by the transformation of the groundwater table fluctuations into duration lines i.e. graphs representing the cumulative number of days with a certain groundwater group (5 cm). The graph represents the length of time the groundwater is in the soil layers (Niemann, 1963; 1973).

Climatological data were derived from meteorological stations in Eelde and on Schiermonnikoog. Total monthly rainfall and transpiration were transformed into one figure, the rainfall surplus (mm) for hydrological years (1 April to 31 March), hydrological winter half-years (1 October to 31 March) and hydrological summer half-years (1 April to 30 September). This means total monthly rainfall minus total monthly transpiration of an open watersurface multiplied by a constant factor, representing the open water/grassland transpiration ratio. This factor is 0.81 from May to September, 0.72 in March and April, and September and October and 0.65 from November to February (Penman, 1948). The average rainfall surplus over the study period and its standard deviation were calculated in order to recognize dryer or wetter than average (half)years (Both & Van Wirdum, 1981).

II.D. Ecosystem dynamics

II.D.1. Terrain-use or occupancy

Direct observations of animal presence and behaviour were carried out every ten minutes in ten plots (25 × 25) m^2 laid down systematically in each of four plant communities in the salt marsh study area. The plots were marked with plastic sticks to enable direct observation from a watch-tower. Dung patches were counted every three weeks, which is shorter than the disappearance period (Fig. II.1) (De Boer, 1974). Double counting was prevented by paint-marking (Allersma, 1977).

The mixed grassland of the Westerholt area had been subdivided by former agricultural practices and by small ditches, which enabled the subdivision of the study area into twenty sections, each initially containing a relatively homogeneous vegetation. Direct observation by counting the number of animals and their behaviour per section every ten minutes and by following individual animals for a period of 24 hours, was easy in the initial period, but it became more and more difficult in the fine grained vegetation pattern which gradually emerged. With the changing pattern the boundaries between the vegetation types could not be distinguished from a distance. Ten plots were, therefore, laid down systematically in each section. The dung voided in these plots was quantified by measuring the volume of water displaced in a 1000 ml cylinder (Rawes & Welch, 1969). To obtain representative samples, minimal dung plot areas were determined for sections with different amounts of dung, resulting in 4 m^2 for heavily dunged sections, 32 m^2 for lightly dunged sections and 16 m^2 for intermediate sections (De Bie, 1976). The frequency of sampling was made conditional to the rate of decay of the pellets, which turned out to be greater in pasture than in heathland (Fig. II.2). Hence, the dung plots were sampled two weekly in the pasture sections and monthly in the heathland sections.

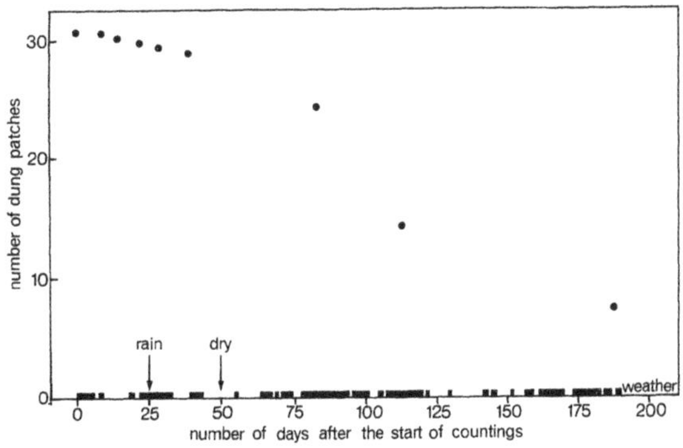

Fig. II.1. Mean disappearance rate of cattle dung patches in two plots (10 × 10)m^2 on the salt marsh (after De Boer, 1974).

Dung counting techniques to quantify occupancy can include sources of inaccuracy (see also De Bie, 1976; Allersma, 1977; Bailey & Putman, 1981). Potential errors are:

1. Non-representative sample plots. This can be countered by determining minimal dung plot areas in preliminary investigations, thus, resulting in different areas for different plant communities.

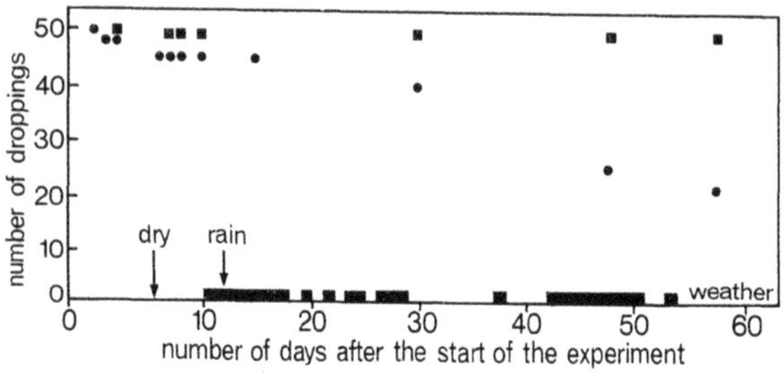

Fig. II.2. Disappearance rate of sheep dung in pasture (circles) and in heathland (squares) (after Bakker et al., 1983).

2. Differential dung decay rates. The frequency of sampling should be made conditional to the rate of decay of the dung, which can, in turn, depend on the vegetation type.
3. Differential search ability. The recovery of dung may not be equally easy in all communities.
4. Irregular defecation by the animals. Both cattle and sheep void large amounts of dung after a period of rest. Moreover, the time between two subsequent defecations can vary between 30 and 270 minutes in the case of sheep. Large individual variations were measured for four sheep during a 24 h period (131 ± 33, 90 ± 22, 80 ± 21, 50 ± 13 min). Irregular defecation can easily result in descrepancies with respect to communities in which the animals forage and in which they defecate. This is also the case for the next item.
5. Differential mobility of animals at different times of the day or season or in different communities resulting in non-random use of the animal's range. It is known from direct observation that resting and ruminating takes place during the night and during several periods in the day-time, and that the time spent resting and foraging varies amongst communities.

Despite these problems, measurements of dunging intensity are the only technique that can be used extensively to estimate the occupancy of grazing animals.

II.D.2. Utilization factor

The utilization factor (Esselink et al., 1989) is the herbage utilization ('t Mannetje, 1978) per unit area per unit time expressed as a proportion of the net herbage accumulation (Hodgson, 1979) per unit area per unit time. Herbage utilization is the sum of the herbage consumed by the animals and subsidiary losses of herbage due to spoiling by trampling and fouling and decomposition. Hence, the term 'utilization factor' is analogous to Hodgson's 'efficiency of grazing' (Hodgson, 1979), being the herbage consumed expressed as a proportion of the net herbage accumulation. Both herbage accumulation and herbage removal were measured using removable exclosures (Photo II.1) of (1×1) m^2 and periodically sampling the above-ground standing crop inside and outside (Milner & Hughes, 1968), also known as the 'difference' method. A large exclosure of (7×5) m^2 was erected in the short turf of a *Juncus gerardii* community on the Schiermonnikoog salt marsh in order to study the period of continual regrowth in exclusion cages. It appeared that constant regrowth occurred for a period of six weeks in July and August, when the sward was clipped every four days to 2 cm above ground-level in different places in the exclosure (Fig. II.3). This means that the 4 to 6 weeks period of protection from grazing in exclusion cages (Milner & Hughes, 1968) did not lead to underestimations of herbage accumulation and utilization. It is not known whether this is also the case in swards with a tall canopy in this study area, because it is very difficult to only harvest the upper layer from a tall canopy.

The cages were relocated each month to randomly chosen sites in similar communities on the salt marsh. Three circular plots of 1.8 dm^2 were handclipped to ground level both inside and outside each cage. These included all the litter. A sub-sample of the mixture of the three plots was separated manually into living (green) and dead fractions. Dry weights (24 h at 70° C) of these fractions were converted to the total weight of the three plots. Therefore, the following was defined for each sampling time:

living herbage mass inside : CL (C from caged)
dead herbage mass inside : CD
living herbage mass outside : GL (G from grazed)
dead herbage mass outside : GD

Net herbage accumulation was computed according to Smalley's method (Smalley, 1959; see also Job & Taylor, 1978), which is based on the accumulation of living and dead herbage. Let $\triangle CL = CL(t_1) - L(t_0)$ be the change in living herbage mass over each sampling interval, $\triangle CD = CD(t_1) - D(t_0)$ be the change in dead herbage mass, then $\triangle CT = \triangle CL + \triangle CD$ is the change in total herbage mass. Net herbage accumulation over each interval (A_i) was then estimated as

Photo II.1. Removable exclosure at the Oosterkwelder salt marsh for sampling the above-ground standing crop inside and outside.

shown in Fig. II.4. Total net herbage accumulation over the grazing period or over the grazing season is then computed as the sum of respective A_i's.

The mass of herbage utilized was computed from the difference in living and dead herbage inside and outside the cages at the end of the time interval between each sampling. Let $\triangle GL = CL(t_1) - GL(t_1)$ be the difference in living herbage mass at the end of each interval, $\triangle GD = CD(t_1) - GD(t_1)$ be the difference in dead herbage mass, then $\triangle GT = \triangle GL + \triangle GD$ is the difference in total herbage mass. Herbage utilized per sampling period (U_i) is then estimated as shown in Fig. II.4. Total herbage utilized over the grazing period is then computed as the sum of respective U_i's.

A more simple method was initially used in the mixed grassland area. Only one (40×40) cm^2 plot was clipped inside and outside each cage and no subdivision was made into living and dead biomass, because of lack of time.

Since it may be assumed that the animals do not only utilize the herbage accumulated in the sampling period, but also herbage which was accumulated earlier, the initial standing crop, i.e. the standing crop present at the start of the interval, can be taken into account in the estimations.

An almost identical procedure as that on the salt marsh was adopted for estimating the utilization factor in the heavily grazed areas and the lightly grazed patches in the micro-pattern of the mixed grassland. The only modifications were the area of circular plots (2.8 dm^2) and the fact that samples from inside the cages were not separated into living and dead fractions because of lack of time. The amount of living material inside the cages was, therefore, estimated assuming that the ratios of living : dead material were similar to those outside the cages. Because of the low stocking rate it was expected that the sheep would select, particularly, on live herbage, and therefore the relative herbage accumulation was estimated, i.e. the accumulated live herbage related to the live herbage present at

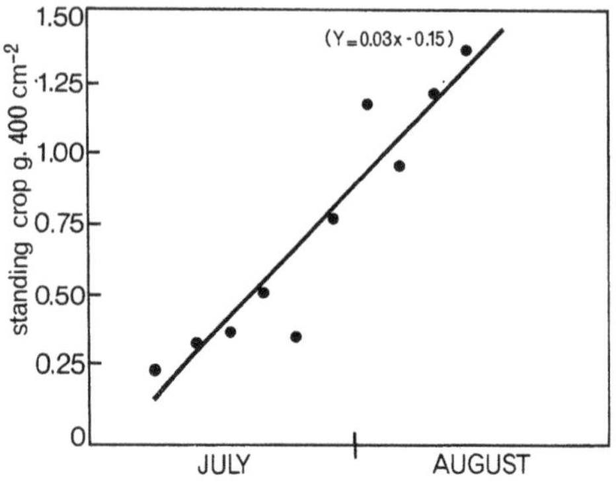

Fig. II.3. The increase of the above-ground standing crop in the *Juncus gerardii* community protected from grazing, by clipping every four days to 2 cm above ground-level in different places (after Looijen et al., 1984).

the start of the interval and thus resembling a measure for relative growth rate. The relative herbage utilization was estimated analogously.

II.D.3. Above-ground standing crop in cut areas

The cut areas on the Schiermonnikoog salt marsh varied between 20 and 40 m^2. In order to determine the dry weight of the biomass of the whole cut area, the fresh weight was measured and the ratio dry: fresh was calculated for a small sample. The samples were dried at 70° C for a period of 24 hours and not at 105° C because some samples had to be subjected to chemical analyses.

The cut areas in the river valley and mixed grassland amounted to c. 500 m^2 in each management regime. Aerial biomass was estimated by clipping a number of small plots. First the number and the area of plots was determined by clipping 3, 6, 9, 12, ..., 30 randomly chosen plots (see Milner & Hughes, 1968) of (10×10) cm^2, (20×20) cm^2 and (40×40) cm^2 to ground level in a *Holcus lanatus* stand and in a tall *Phalaris arundinacea* stand in July 1975 (Fig. II.5A–B). No subdivision was made into green and dead biomass. It is striking that the (10×10) cm^2 plots showed condiderably higher dry weights than the other plots, apparently because in doing the clippings the person unconsciously tended to add some material to the very low amount resulting from the small plots. Ten (20×20) cm^2 plots were clipped to ground level in each management regime just before cutting, thus accepting an inaccuracy of c. 9% (standard error related to mean weight).

II.D.4. Below-ground biomass

The below-ground biomass was estimated in each management regime, amounting to c. 500 m^2, in the river valley and mixed grassland areas by sampling a number of small plots with the auger method. The number of these plots was determined by sampling 3, 6, 9, 12, ..., 30 randomly chosen cores of 38.5 cm^2 to a depth of 15 cm in a *Holcus lanatus* stand and in a *Phalaris arundinacea* stand in July 1975 (Fig. II.6 A–B). The roots were subdivided into living and dead parts. Dead roots are far less elastic, the colour is often greyer, the cortex is absent and the lateral roots have often already broken off, leaving stumps with ravelled ends. The combined evaluation of these four features determines whether the root is to be regarded as living or dead (Schuurman & Goedewaagen, 1965). Ten plots were cored in each management regime, thus accepting an inaccuracy of c. 18% for living material and c. 10% for dead material (standard error related to mean weight).

II.D.5. Decomposition

The decomposition rate was measured by means of litter-bags. Nylon bags (6×29) cm^2 were used with a mesh width of 2 mm and some 6 mm perforations to allow microbes, micro-, meio- and macrofauna <6 mm to enter. The bags were loosely filled with approximately 8 g of oven-dried (70° C) green plant tissue, but some standing dead material was inevitably also included during clipping of the samples in each community. The material was not re-wetted before it was returned to the field. Special care was taken to avoid spillage during transport. Bags containing tissue from the 'own' community were placed on the soil surface in each community.

Bags were also filled with dry cellulose strips of (4×21.5) cm^2 weighing 2.111 g and placed in each community. These were used as a standard for comparing the effect on the decomposition rate under the given abiotic conditions.

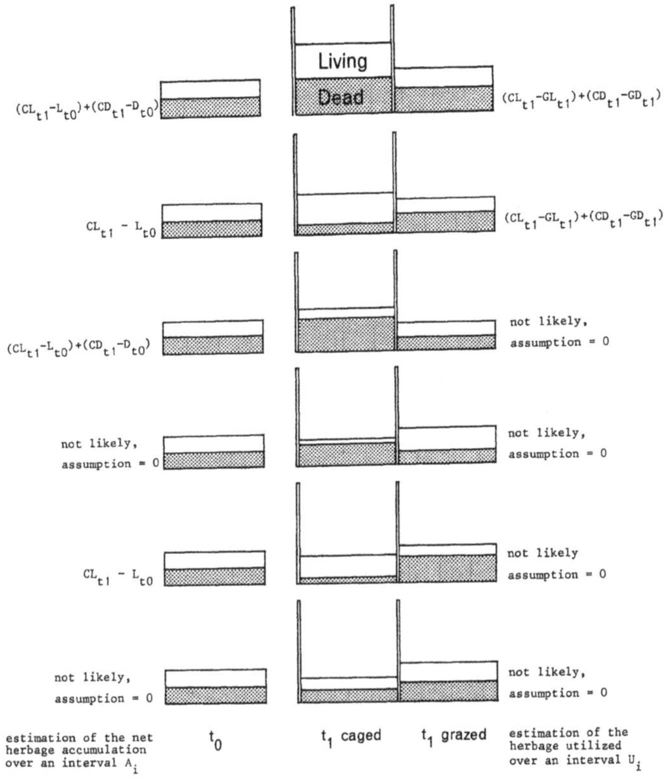

Fig. II.4. Estimation of the net herbage accumulation and herbage utilization. The initial above-ground standing crop in the grazed area is measured at t_0 and again after caging at t_1. The different possibilities with respect to the change of living (white) and dead (dotted) material and the estimation of the net herbage accumulation over the period $t_1 - t_0$ are indicated at the left. The above-ground standing crop in the caged and grazed area is measured at t_1. The various possibilities with respect to differences in living and dead material and the estimation of the herbage utilization over the period $t_1 - t_0$ are indicated at the right.

Four or five replicates of bags containing herbage or cellulose strips were collected on sampling dates with intervals of c. 6 weeks. The organic matter weight was determined after ignition (3 h at 950° C), in order to prevent over-estimation of the decomposition rates because of the presence of sand.

II.D.6. Vegetation patterns and canopy structure

Heavily grazed areas and lightly grazed patches could be distinguished by their stand height (<5 and >10 cm, respectively) and by the absence and/or presence of litter. Patterns could easily be detected because of the sharp boundaries.

The height of the canopy was measured by dropping a foam-plastic disc (diameter 15 cm, 19g) along a calibrated stick until it rested on the canopy. The thickness of the litter was measured with the same stick.

The density of the stand was estimated by measuring the number of visible dots out of 100 dots on a calibrated stick of 1 m at 0, 5, 10 and 15 cm above soil level at ten randomly chosen plots. Light was measured 4 cm above the soil (five replicate readings) using a Lux-meter.

II.D.7. Grazing frequency

The grazing frequencies to which salt marsh species were subjected, were measured by repeated control of individual plants at distances of 25 cm in two 6 m transects. The bottom and top of all stems and/or leaves were marked with a small dot of paint. The removal of marked herbage was recorded every week from the end of May to the end of September.

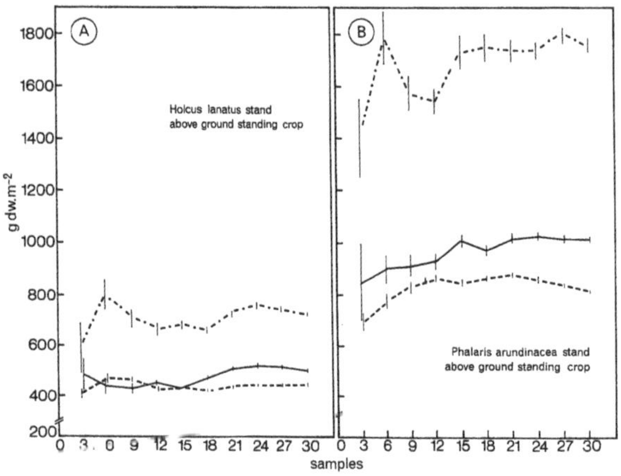

Fig. II.5A–B. Above-ground standing crop (± S.E.) in A: a *Holcus lanatus* stand in the Loefvledder area and B: a *Phalaris arundinacea* stand in the Lammertshem area estimated by different sample numbers with areas of (10 × 10) cm^2 (upper line), (20 × 20) cm^2 (solid line) and (40 × 40) cm^2 (broken line).

The grazing frequency of plant species in the mixed grassland was measured in a different way. In May, June and September, during point quadrat analysis of the vegetation in the heavily grazed areas and the lightly grazed patches, it was noted whether tillers or leaves were grazed. The pins were placed with a 30° inclination from the zenith (Hodgson, 1981) and 1000 scores were made in both the heavily grazed areas and lightly grazed patches.

Selective feeding is most often defined as forage- or preference ratio, which is the proportion r of the food in the diet divided by the proportion p of the food in the habitat: values greater than 1 show preference, values less than 1 avoidance (Crawley, 1983). The electivity index E, according to Ivlev (1961), is also often used. E is defined as the relative difference between the r and p: $E = \frac{r - p}{r + p}$. E varies from −1 to 0 for negative selection and from 0 to +1 for positive selection. A quantitative comparison of the selection between different food types with the aid of preference ratio or electivity index, can, however, only be made if the food types have the same relative abundance. Jacobs (1974), therefore, proposed to use the index

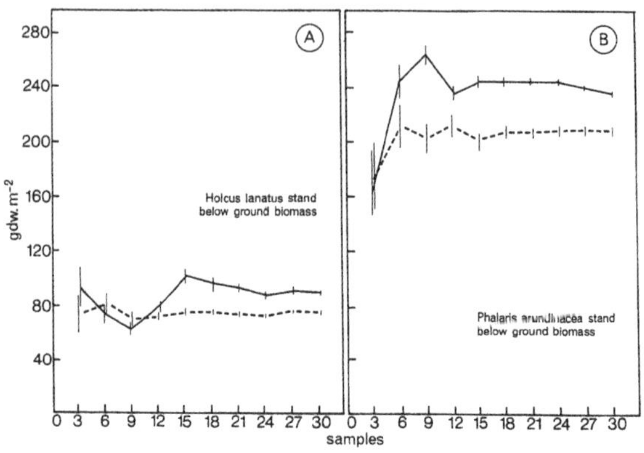

Fig. II. 6A–B. Dead (solid line) and living (broken line) below-ground biomass (± S.E.) in A: a *Holcus lanatus* stand in the Loefvledder area and B: a *Phalaris arundinacea* stand in the Lammertshem area at 0–15 cm depth estimated by different sample numbers with an area of 38.5 cm^2.

$$D = \frac{r - p}{r + p - 2rp}.$$

This has been adopted in the present study.

II.E.8. Chemical analysis of plant material

Since the grazing animals showed little preference for individual species (see Chapter VIII), samples were taken as a random mixture from all the occurring species in the cut (green and dead material mixed) and grazed (only green material) areas with a few exceptions. Since *Ammophila arenaria* and *Juncus maritimus* were hardly grazed only the vegetation between the tussocks of those species was sampled. Samples of only the very dominating species *Artemisia maritima* and *Elymus pycnanthus* were taken in those communities.

All dried samples were ground. Depending on the amounts necessary for the respective analyses, each sample was analysed or some samples were pooled. All analyses were carried out in duplo. Only a deviation of less than 5% between the duplo's was accepted, otherwise a third analysis was carried out. Since cattle on the salt marsh traverse bare sandy and clay soil, they can easily foul the herbage. The chemical analyses of the salt marsh samples were, therefore, based on organic matter and not on dry matter like those of the mixed grassland. Organic matter weight was determined after ignition (3 h, at 950°). It appeared that the percentages of crude protein were 1 to 2% higher, percentages cell wall were 4 to 5% higher and the in-vitro digestion was 1 to 8% higher, based on organic matter compared to dry matter analysis (see also Looyen et al., 1984).

The neutral detergent solution dissolves the cell content (CC) and hence leaves the cell wall fraction (NDF). The in-vitro digestion of the cell wall fraction and of all organic material was measured in cattle rumen fluid according to Van Soest (1975). The cell wall digestibility (DCW) is thus measured according to Deinum & Van Soest (1969):

$$DCW = (1 - \frac{\text{NDF after in-vitro digestion}}{\text{NDF before in-vitro digestion}}) \times 100.$$

The digestion of all organic matter of the samples (DOM) includes the digestion of cell wall and cell content and is thus measured:

DOM = DCW × NDF before in-vitro digestion + 0.98 (100 − NDF) before in-vitro digestion. The cell contents are digested for 98% (Deinum, 1976). Because the mixed grasland in this study was grazed by sheep the digestibility values according to Van Soest must be considered with caution. The difference in organic matter digestibility between sheep and cattle, however, does not exceed 10% and is much smaller than differences in digestibility between plant parts (Hacker & Minson, 1981).

Both the measurements of the cell wall fraction and the in-vitro digestion were carried out in the laboratory of the Department of Nature Management of the Agricultural University at Wageningen. All other measurements were done in the laboratory of the Department of Plant Ecology of the University at Groningen (see Pegtel, 1976, for details).

The crude protein percentage was estimated as 6.25 × N_{total}. Total nitrogen was determined with a modified Kjeldahl-technique (Wieniger, 1936).

Total H_2PO_4 was measured according to Murphy & Riley (1962) and potassium by flame photometry.

II.D.9. Soil sampling and soil analysis

Samples were taken as a mixed sample of 12 subsamples from 1 m^2 on the salt marsh in various management regimes. The subsamples (diameter 2 cm, 0–5 and 5–20 cm depth) were taken randomly, sometimes under or between plants, sometimes on bare soil, particularly in grazed areas.

The samples were taken from an area of (10 × 10) m^2 in the valley grassland and mixed grassland areas. in 1977 twelve areas of 1 m^2 within the (10 × 10) m^2 areas were sampled with 12 subsamples each in order to quantify the variation of the samples which were normally taken. It transpired that the standard error of the pH values varied between 0.03 and 0.08 and that an inaccuracy of mostly below 10% had to be reckoned with for the other soil chemical parameters in the study areas on 'Beek' earth soil (*Holcus lanatus* stand), on 'Made' peat soil (*Phalaris arundinacea* stand), on ploughed 'Veld' podsol (*Holcus lanatus* stand), and on untouched 'Veld' podsol soil (*Erica tetralix* stand) (Table II.3).

All dried samples, including small roots, were ground to pass through a 1 mm sieve. All analyses were carried out in duplo. Only a deviation of 5% from the average of the duplo's was accepted otherwise a third analysis was carried out.

All analyses were carried out in the laboratory of the Department of Plant Ecology of the University at Groningen (see Pegtel, 1976, for details).

The moisture content was measured several times a year as weight loss after drying at 105° C for 24 h, expressed on a dry weight base × 100%. Pore volume of the soils was calculated from the particle density and bulk density of undisturbed soil cores of a known volume (water saturated soil minus soil dried at 105° C). Since roots are also included and contain 90% water, ten times the dry weight of roots present were subtracted from the saturated soil weight in order to get a more realistic pore volume (Don et al., 1980).

Salinity was determined in g NaCl · 100 g dry $soil^{-1}$, and g NaCl · l soil $solution^{-1}$, respectively, after the titrimetrically measurement of Cl^- by chlorocounter.

Total-nitrogen was determined with a modified Kjeldahl-technique (Wieniger 1936) and potassium by flame photometry. The total H_2PO_4 was measured according to Murphy & Riley (1962) and P-water 1 : 60 according to Sissingh (1969).

II.E. Population dynamics

II.E.1. Seed dispersal

The dispersal of seeds by means of several agents was studied by collecting samples of cattle dung patches, geese pellets, drift line material on the salt marsh, sheep pellets and wool fragments from the grazed inland communities and material from the body work of tractors working in the hayfields.

These samples were treated in a similar way to the soil samples when measuring the seed bank (see II.E.2). Because most of the samples were collected in the autumn they were put into the experimental garden for two months in order to meet the stratification demand of some species. A very fine wire-mesh prevented contamination with fresh seed rain.

II.E.2. Seed bank

The seed bank was measured as germinable seeds (Thompson & Grime, 1979) kept for six months in a greenhouse providing day and night intervals at 20° C. The presence of seeds in the soil was recorded in several 2 cm thick layers of five replicates each of 5 or 7 cm diameter. The seed-containing soil was spread in a plastic tray filled with sterilized (in order to prevent the germination of not sampled seeds) potting ground and watered daily. Roots and tillers were removed. Emerged seedlings were removed after counting. Seedlings which could not easily be identified were grown separately. Watering was ceased after a few months and, after drying, the top soil was crumbled resulting in the emergence of new seedlings.

In order to meet the stratification demand of several species samples were taken in both the autumn and early spring.

II.E.3. Germination and early growth

Seeds were collected in the summer or autumn, dried at 25° C and then stored dry at 5° C. Fresh seeds were used in some experiments, whereas in others the germination response was measured after stratification at 5° C in fresh water for one month. Seeds were also sometimes stratified in seawater and for three months.

Germination response was measured in germination chambers with alternating 12 h light/dark regimes. Fifty seeds were tested for germination (appearance of the radicula) on filter paper in each of two Petri-dishes under various temperature regimes, with either a constant temperature or diurnal temperature fluctuations. The final germination percentage was usually recorded after 28 days.

In the case of some marsh species the dry matter production of the young plants in different NaCl concentrations was measured in a greenhouse supplied with natural daylight at a constant temperature of 20° C. An undefined number of not strat-

Table II.3. Mean value ± S.E. of pH and percentages inaccuracy (standard error related to the mean values) of several soil chemical parameters (0–5 cm and 5–20 cm) from 12 randomly chosen 1 m^2 areas within 100 m^2 in October 1977 in 'Beek' earth soil (Loefvledder), 'Made' peat soil (Lammertshem), 'Veld' podsol soils (Westerholt).

		pH (KC1)	Humus %	C/N	Ntotal %	100 g dry soil		1 cm^3 dry soil
						$K_{exchange}$ mg K_2O	P_{total} mg P_2O_5	P_{water} μg P_2O_5 1 : 60
'Beek'	0–5 cm	4.11 ± 0.03	5.6	2.1	4.5	6.3	4.3	7.7
earth soil	5–20 cm	4.08 ± 0.03	9.6	4.7	6.0	8.9	5.2	9.3
'Made'	0–5 cm	5.12 ± 0.05	0.7	1.4	1.3	4.1	2.2	8.1
peat soil	5–20 cm	5.19 ± 0.05	0.9	1.3	1.7	4.3	7.7	15.6
'Veld'	0–5 cm	4.29 ± 0.08	5.9	2.9	6.3	9.8	10.2	10.1
podsol soil	5–20 cm	3.91 ± 0.08	9.7	0.9	11.8	10.5	12.7	13.3
'Veld'	0–5 cm	3.11 ± 0.06	7.6	2.7	8.7	10.4	6.1	14.5
podsol soil ploughed	5–20 cm	3.18 ± 0.03	10.6	2.6	10.0	12.9	9.7	9.1

ified seeds were sown in plastic trays 30 × 30 × 15 cm filled with 7 kg of coarse sand. Each tray was watered with 0.7 l nutrient solution (50% Hoagland) with the addition of a range of NaCl concentrations 0, 5, 10, 25 and 35 $g.l^{-1}$. After germination (seedling emergence) seedlings were removed so that 15 were left in each tray arranged equidistantly. The mean dry weight, including roots, was measured after two months. In a separate but parallel experiment, the final germination percentages, after two months, were recorded in samples sown in Petri-dishes with the same range of NaCl concentrations.

In other experiments with salt marsh species the response of seedlings to different light intensities was measured in the greenhouse in order to simulate the effect of different canopy structures. Ten seedlings (obtained by the germination of unstratified seeds) were grown in plastic trays, prepared as has previously been described for the NaCl response experiments, and were given 13, 38 or 100% of full outdoor light intensity. The mean dry weight (including roots) of the individuals was measured after two months and the final germination percentages were also determined after the same period of time.

The viability of seeds to be sown in the valley grasslands was tested in the experimental garden by sowing 100 unstratified seeds into plastic trays with sterilized soil from the valley grasslands. The final germination (seedling emergence) percentages were determined after eight months.

II.E.4. Introduction of seeds

Seeds of species of the same community type, but not present in the local vegetation or seed bank were sown in various sites in the undisturbed vegetation in order to find out whether the habitat provided suitable germination conditions, or whether a lack of dispersal caused their absence. Unstratified seeds (dried at 25° C and then stored dry at 5° C) were used in sowing experiments performed in the autumn and early winter. Sowing experiments in the early spring were performed with seeds stratified at 5° C in both fresh- and seawater for meadow and salt marsh species, respectively.

Seeds were sown in three or five replicate plots (10 × 10) cm^2. Seedling emergence and survival of each cohort of seedlings were recorded every six to eight weeks. This approach yields minimum numbers since seedlings may have emerged and disappeared in the intervening time between the two recordings. The recordings were done by inserting little rings or coloured pins next to each individual.

Naturally-emerging i.e. not introduced seedlings were recorded in five replicate plots (10 × 10) cm^2 at each site and in some cases sods of the same area were investigated closely in the laboratory (cf. Linkola, 1930).

PART TWO

Land use dynamics

CHAPTER III

The Drenthe Plateau

III.A. Introduction

The area in which the restoration experiments of former species-rich grassland took place is described in this chapter. It lies in the inland nature reserve 'Stroomdallandschap Drentsche A' (53° 05′ N, 6° 40′ E in the centre of the long narrow area along a brook system: 3500 ha). It forms part of the so-called 'esdorpen' landscape (Bijhouwer, 1971; Renes, 1981) or 'brinkdorpen' landscape (Lambert, 1971) featuring a previously characteristic combination of arable land (the 'es'), grassland, hedgebanks, heathland and woodland. The old moist meadows adjacent to the stream form one study area and recent grassland, which was reclaimed from heathland in the 1950's, forms the other study area.

The land use history and particularly the agricultural activities are described to emphasize the semi-natural character of the nature reserve. Drenthe has a land use history which goes back several millenia. The agricultural history largely explains the nature management practices carried out after the acquisition of the nature reserve.

The description of the abiotic environment, namely, geomorphological and hydrological conditions is followed by a discussion of the land use. The settlement history goes back to c. 4000 B.C. and is analysed by pollen analytical, archaeological and historical geographical data. The continuity of settlements and the origin of the 'esdorpen' landscape are also discussed and the changes in the 'esdorpen' landscape from 1800 onwards are quantified. A great deal of attention is paid to the development of anthropogenic valley grasslands and the relationship between plant communities and hydrological conditions. The species diversity is discussed in connection with spatial heterogeneity, gradients concerning both abiotic conditions and human impact and continuity of human impact in order to illustrate some theoretical considerations covered in chapter I.

The acquisition of the nature reserve and finally the carrying out of management practices from

1965 onwards is quantified. The involvement of the farmers in the nature management practices will be shown.

III.B. Land use of the Drenthe Plateau

III.B.1. Geomorphological and hydrological conditions

The Drenthe Plateau was covered by ice during the Saalien glacial period around 200,000 B.C. leaving a layer of boulder-clay. Large amounts of water, released by the retreating ice, eroded valleys causing the local disappearance of boulder-clay. The valley was partly re-filled with eroded material from elsewhere. Drenthe was not covered by ice during the last glacial period, the Weichselien. It was in this period that the main sediments and the present geomorphological conditions of the Drentsche A basin were formed. A deep and broad valley had been eroded during the wet early Weichselien (Fig. III.1). Due to the permanently frozen subsoil the water saturated topsoil could slide away in inclined areas (solifluction) and materials from the plateau, including boulder-clay, were transported into the valleys where they created less pervious layers (De Gans, 1981). The initially homogeneous boulder-clay, which was deposited during the third glacial period, broke up as a result of erosion.

Solifluction probably created new upper course valleys on the plateau (Van Der Hammen, 1951). Large amounts of aeolian material were transported during dry periods in the Weichselien. This process resulted in coversands filling some original valleys, whereas others remained intact (De Gans, 1981). The latter formed the primary drainage system of the plateau. Later a secondary drainage system evolved during wet periods and most of the narrow uppercourse valleys were a part of it. All the valleys were mainly filled by peat during the moist Atlanticum (De Gans, 1981).

Up to the present day the valleys are the natural drainage system of the Drenthe Plateau. Before reaching the valleys the infiltrating rainwater passes through the soil along different pathways (groundwater flows) as depicted in Fig. III.2. Grootjans (1980) described these groundwater flows and the quality of the groundwater.

1. In some places impervious strata, like boulder-clay or iron pans, prevent percolation and rainwater is impounded in e.g. heath pools.

In other places there is a run-off of various types:

2. relatively fast moving water underlain by impervious lenses of boulder-clay. This water is rich in oxygen and the levels of Ca^{2+} and Fe^{2+} ions are relatively low;
3. relatively fast moving water from beneath the boulder-clay. Since it has been in contact with geological strata for a longer time it contains more Ca^{2+} and Fe^{2+} ions than the previous type of groundwater;
4. deep groundwater flow travelling over long distances, containing even more Ca^{2+} ions (high total hardness);
5. deepest groundwater flow moving very slowly. It is probably less important in the valley systems of the Drenthe plateau;
6. in the lower course of the stream, nutrient-rich riverwater inundates the valley in wet seasons.

The overall consequence of the different groundwater flows is the occurrence of seepage water in the valleys and a longitudinal gradient in nutrient level, according to which the water course can be divided into source area, upper course, middle course and lower course. In addition, a gradient perpendicular to the river occurs at every place along the valley. This will be discussed in more detail in section III.B.11.

III.B.2. The natural vegetation in 4000 B.C.

To understand the effects of human impact upon the Drenthe Plateau, the natural vegetation will be described first.

The Drenthe Plateau lies in the province of Drenthe as well as in parts of the provinces of Groningen and Friesland. Streams drained radially into the bogs at its margins. These large bogs formed an almost impassable natural frontier. Smaller bogs were found on the plateau in places with impervious layers of boulder-clay. Marshes and swamping woodlands also extended far up the valleys of the main streams and their tributaries.

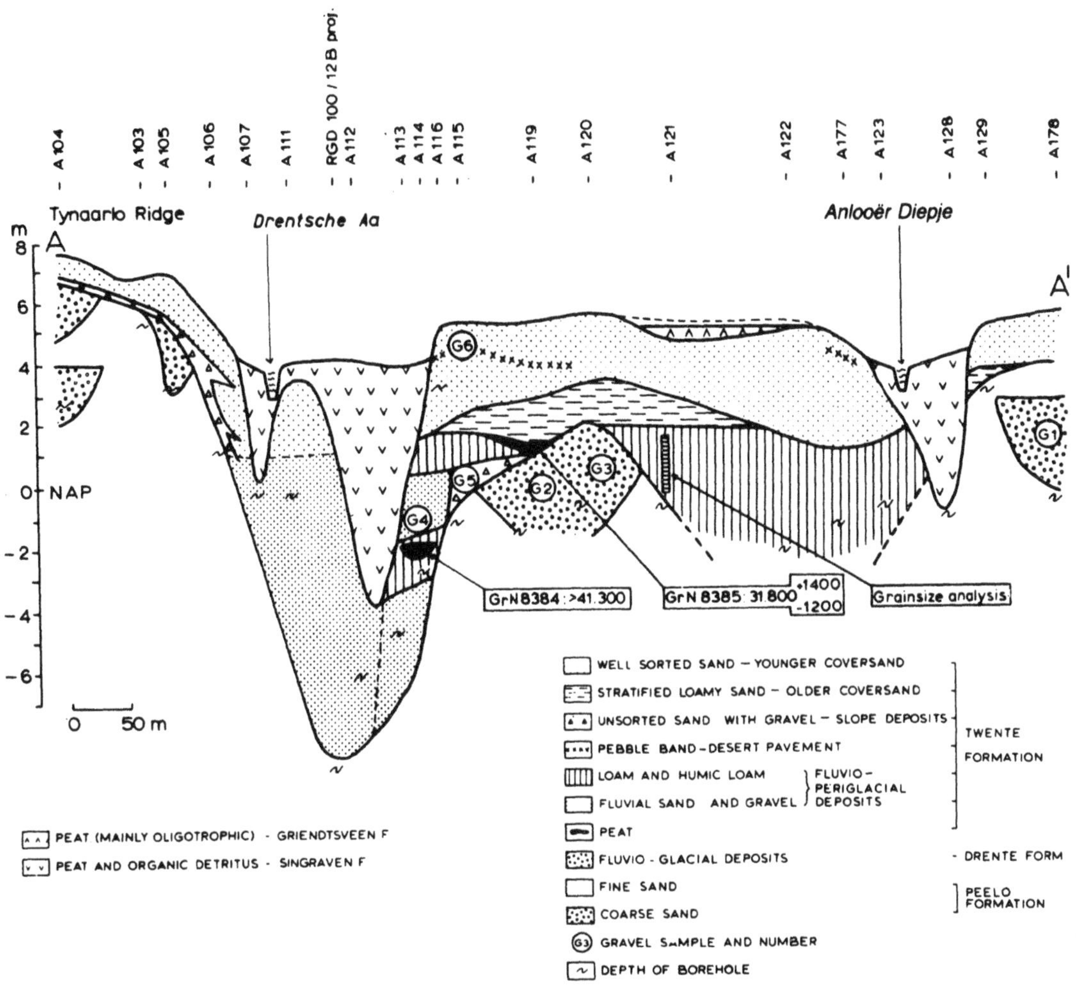

Fig. III.1. Detailed cross section showing the geological strata of the nature reserve Stroomdallandschap Drentsche A and the adjacent plateau's (after De Gans, 1981).

Most of the plateau was, therefore, subdivided into a series of units with their core on a drier sandy ridge, and separated from their neighbours by broad belts of swamp and bog (Fig. III.3). In time these units came to form the settlements. Since the units were generally large and the various types of substrate within them showed a more or less distinct arrangement, a very recognizable pattern of settlement and land utilization evolved.

The landscape and especially the vegetation which the first neolithic farmers found on the Drenthe Plateau about 4000 B.C. is described briefly below. This description was derived from the assumptions of Waterbolk (1954), Van Zeist (1959) and Casparie (1982). Although the situation is not quite comparable (*Fagus sylvatica* and *Carpinus betulus* had not yet reached the Netherlands, the sea level was c. 8 m lower and no human interference had taken place), the vegetation is interpreted according to the present syntaxonomic nomenclature according to Westhoff & Den Held (1969). A *Fago-Quercetum* was probably found on dry sandy, more acid soils, relatively poor in humus. On somewhat more loamy soils, containing a lot of humus, a *Stellario-Carpinetum* with *Tilia cordata* and *Ulmus minor* occurred. *Circaeo/Alnion* and *Alnion glutinosae* probably occupied the valleys, the first with a rather thin and the second with a thick layer of peat on the sandy subsoil. These situations might have been found in the narrow

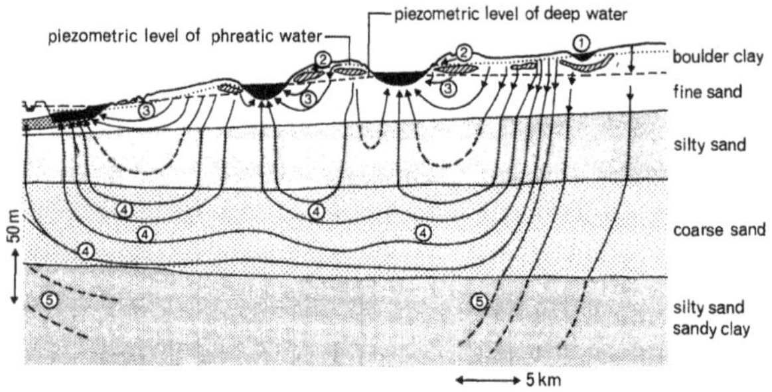

Fig. III.2. System of geological strata and direction of groundwater flows in the catchment of Drenthian brooks. 1: impounded water on boulder clay, 2: relatively fast moving water through sand aquifers, underlain by boulder clay, 3: relatively fast moving seepage water from beneath the boulder clay, 4: deep ground water flow through coarse sand aquifers, 5: deep groundwater flow, passing very slowly through fine marine sediments (after Grootjans, 1980).

upper course valleys and at the fringe of the broad middle course valleys. In the centre of the broad middle course valleys autonomous peat building may have occurred because of the continuously large supply of seepage water similar to that which is found nowadays in East-Germany and Poland (Succow, 1971). Elements of *Parvocaricetea* and *Oxycocco-Sphagnetea* may, therefore, have been found in the middle course. *Magnocaricion* and *Phragmition* probably occurred in the very wet and sometimes inundated lower course.

The relationship between hydrological conditions and alder swamps was demonstrated nicely by Schwaar (1976) in a calcareous fen near Osnabrück in Germany. An alder swamp (*Alnion glutinosae*) was succeeded by sedge communities (*Magnocaricion* and *Caricion curto-nigrae)* between 1600 B.C. and 400 A.D. These changes could be attributed to both a rise in the groundwater table and to human activities. The alder swamps recovered from 400 to 800. C. 800 plant communities preferring wet conditions like sedge and reed communities (*Phragmition*) succeeded the alder swamps which had developed under less wet conditions. This change coincided with the building of a watermill and a subsequent rise in the groundwater table.

III.B.3. Neolithic Age (3800–1800 B.C.) and Bronze Age (1800–800 B.C.) farmers

The neolithic farmers practiced a kind of shifting cultivation (Iversen, 1949, 1967; Waterbolk, 1954). For a time the original deciduous woodland (primeval forest) was slashed, burned and cultivated as arable land. Two types of reclamation might have been practised (i) the Troels-Smith (1960) type: small clearances and the livestock kept in byres or inside an enclosure, and (ii) the Iversen (1949) type: fairly large reclamation of forest with most of the land which has been reclaimed subsequently being used for grazing, together with the adjacent forest.

Iversen (1967) schematically distinguished three phases in each cultivation using pollen analytical data. The first phase is characterized by a decline in the proportion of pollen from the tall trees, *Ulmus, Tilia, Fraxinus, Quercus* and the sudden increase in that of grasses, composites and *Pteridium aquilinum,* and the first pollen of *Plantago lanceolata* (indicative of treading), *Rumex acetosella* and *Cerealia* appeared. Pollen from *Frangula alnus* and *Populus tremula* are found at the beginning of the second phase, followed by an increase in the proportion of pollen from *Betula*, which is, according to Iversen, an indication of burning. This phase is also characterized by minimal proportions of pollen of *Ulmus* and *Tilia* and maximal proportions of *Plantago lanceolata, Cerealia, Rumex acetosella* and grasses. This might be attributed to the influence of grazing cattle (Iversen, 1967). In the third phase the proportion of *Corylus avellana* pollen is high (it had already increased in the second phase), whereas the tall trees *Ulmus* and *Tilia* seem to recover and *Fraxinus* and *Quercus* reach even higher values than before the cultivation since, according to Iversen, they might be favoured by the change in light conditions. In this phase the pollen of *Cerealia,* grasses, *Rumex acetosella* and *Plantago lanceolata* nearly disappear.

The Funnel Beaker people (3800–2700 B.C. (chronology according to Waterbolk, (1985)) reclaimed the *Ulmus*-rich forest and the *Tilia*-rich forest on weathered boulder-clay (Van Zeist, 1959). Starting from the level of 3600 B.C. *Ulmus* and *Tilia* pollen declined considerably, possibly because the foliage was eaten by cattle. The causes of *Ulmus* decline are, however, in discussion since the decline seems not always to coincide with the beginning of agricultural activities (Kubitzki, 1961; Groenman-Van Waateringe, 1983). Perhaps the Dutch elm disease caused the decline. As the *Fraxinus* pollen did not decrease it is assumed that people avoided the wet soils (Casparie, 1982).

From 3100 B.C. the Protruding Foot Beaker (3100–2500 B.C.) and later the Bell Beaker people (2700–1800 B.C.) also cleared forests that had developed on cover sands. The more frequent occurrence of several types of herbs, the slightly higher values of *Corylus avellana* and *Pteridium aquilinum* may be related to a further extension of the cultivated area. The increase of *Prunus spinosa* and *Crataegus monogyna* might point to the development of shrub.

After the reclamation of the cover sands *Plantago lanceolata* could reach high values in a short time i.e. within 50 years (Casparie & Groenman-Van Waateringe, 1980). Transitions between different plant communities, however, can take place

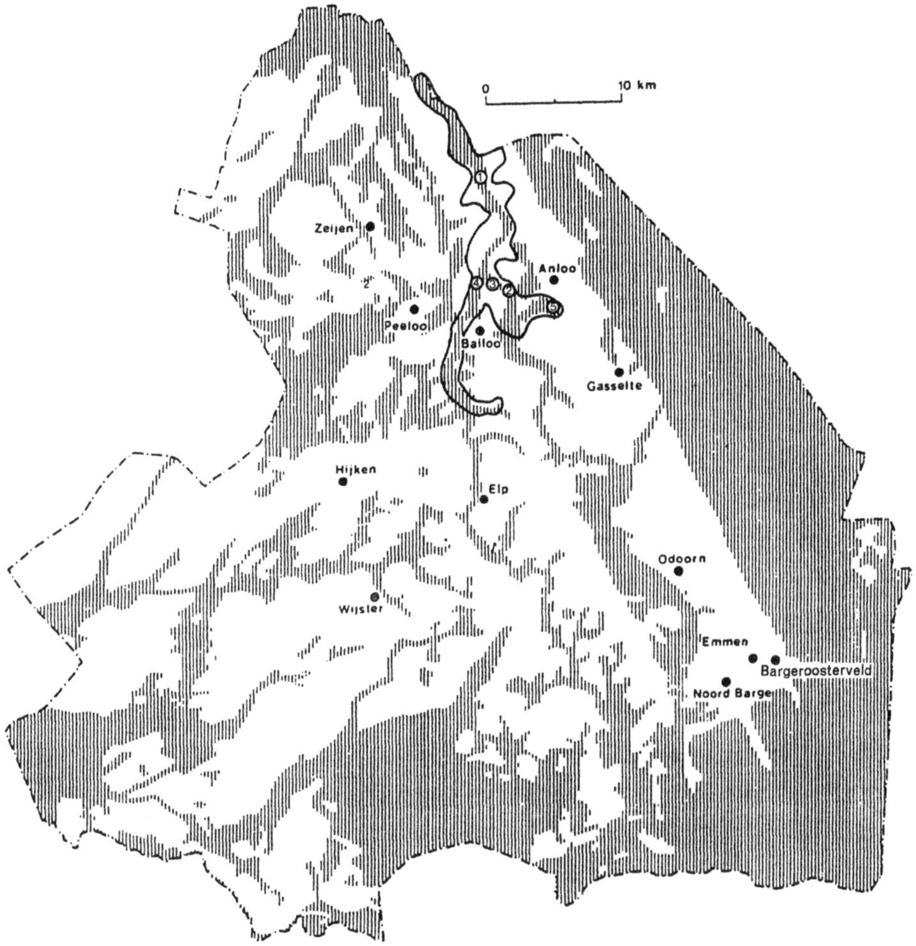

Fig. III.3. The dry sandy soils in Drenthe for settlements between 3000 B.C. and the Middle Ages with the settlements excavations mentioned in the text and the location of the reserve Stroomdallandschap Drentsche A and of the study areas 1: Lammertshem, 2: Anlo, 3: Loefvledder, 4: Taarlo, 5: Westerholt. The hatched area indicates river valleys and moors where no settlements were possible (after Harsema, 1980).

in a very short time and it is questionable whether those rapid changes can be traced in pollen diagrams. Pott (1986) described the recent rotational system of a specific land utilization in Germany in which slash and burn, arable land, utilization of broom (*Cytisus scoparius*), wood-pasture, bark-stripping on oaks and coppice-wood for charcoal production lasting only 20 years. Groenman-Van Waateringe (1986) studied pollen diagrams of the surface layer and different soil horizons up to 5 cm. These samples were taken in present reserves in which changes from heathland to *Deschampsia flexuosa* stands, to woodland or to reclaimed grassland took place within a few decades. She demonstrated that inconsistencies in the pollen spectra in soil layers with different depth can only be understood if the succession of the site is also known. The latter is of course not the case in studying pollen diagrams of old surface layers under prehistoric burial mounds, which are important sources of information. Similar rapid vegetational changes as in present times are likely to have taken place in the Neolithic, which can hamper the interpretation of those pollen diagrams.

The development from 3800–1700 B.C. is illustrated in the pollen diagram in Fig. III.4 of the locality of Bargeroosterveld in south-east Drenthe. Casparie (1982) concluded that the region had been inhabitated continually since c. 3000 B.C. A number of cultivations can be distinguished in the Neolithic Age for example c. 3000 B.C., c. 2500 B.C., c. 2200 B.C., c. 1900 B.C. and c. 1700 B.C. Later clearance activities were also recorded in the region such as c. 1500 B.C., c. 1300 B.C., c. 800 B.C. and c. 500 B.C. (Casparie, 1982).

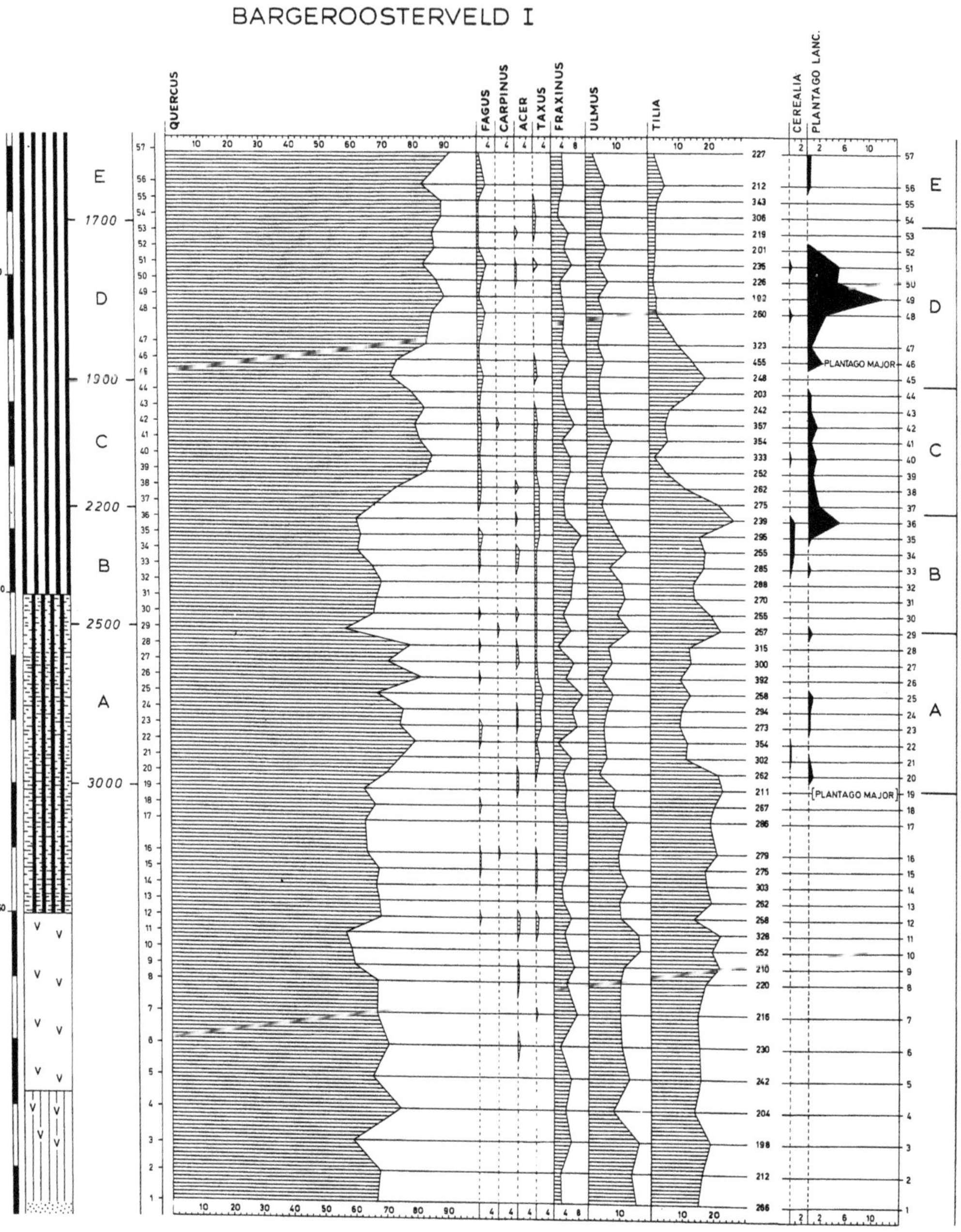

Fig. III.4. Pollen diagram from Bargeroosterveld during the Neolitic with several periods of cultivation A, B, C, D, E (after Casparie, 1982).

During the repeated clearing and regeneration process gradually more heathland developed on the most impoversished areas. Abandoned arable fields and pastures evidently became overgrown by heath fairly rapidly during the late Neolithic Age (Casparie & Groenman-Van Waateringe, 1980). This made it necessary to reclaim more and more forest because heathland was unattractive in the agricultural economy with cattle and swine grazing (Firbas, 1949; Ellenberg, 1954). The cultivated areas grew so large that sanddrifts could develop (Waterbolk, 1985).

During the Bronze Age (1800–800 B.C.) heathland development was further enhanced by the intensification of grazing. The resulting podzol profile probably, in its turn, delayed the woodland regeneration. Therefore, heathland areas alternated with degenerated secondary forest during the late Bronze Age, whereas most primeval forest had disappeared on the Drenthe Plateau and only continued on the wet soils of the Plateau and in the wet valleys (Harsema, 1980).

It is assumed that neolithic people lived in concentrated settlements. Palisaded enclosures found in Anlo support this assumption (Waterbolk, 1977). Excavations in Elp (Waterbolk, 1964), near Emmen (Van Der Waals & Butler, 1974), and near Hijken (Harsema, 1974) revealed Bronze Age settlements in the centuries around 1000 B.C. Harsema (1980) assumed that from the Bronze Age onwards each community obtained some territorial rights by settlements which were continuous but moved occasionally. Due to impoverishment of the soil such a location could have been abandoned for several decades, during which period the settlement was moved a few hundred meters. It might later return to the original place. Such a return took place at least four times in Elp. The territorial claim can be made by a burial mound (Harsema, 1984). Each settlement contained one or more complete agricultural units or farmsteads. The strength of a unit depended on the number of farmsteads in a settlement. If one unit was found it held between 12 and 16 people and if a settlement contained about four units each of them held between 6 and 10 people. Therefore, the Elp settlement might have held about 20 people, the settlement near Emmen between 40 and 50 people and the settlement near Hijken between 24 and 30 people. Each farmstead had stables for several dozen cattle (Harsema, 1984). With the development of heathland areas sheep were also kept (Clason, 1980A).

III.B.4. Iron Age (800 B.C.–0)

During the late Bronze Age and the Iron Age the celtic field method to grow crops was not only adopted on sandy soils in the Netherlands, but also in the United Kingdom, northwestern Germany and Denmark. The celtic fields were often situated along roads which sometimes went back to neolithic times. Large rectangular fields (e.g. 1.5 km^2 in Hijken) (Waterbolk, 1985) were reclaimed on loamy soils. These were later subdivided into smaller fields. This resulted in a chess-board like appearance. Each field was surrounded by levees containing the litter, tree stumps, stones and weeds which were deposited during reclamation. These levees had a width of four to five metres and widened during the cultivation of the arable field which measured c. 30 × 30 m (Brongers, 1976). Wooden stakes were placed on the levees and these levees were gradually overgrown. The fence which resulted might, therefore, have functioned against animals and wind erosion of the arable fields (Harsema, 1984). The latter has always been a major problem on the poor cover sands of Drenthe. Near Hijken each agricultural unit averaged 6 people and needed about 25 of these small arable fields (Harsema, 1980).

The celtic field agriculture introduced the more-course system in which fallow fields were grazed and consequently dunged periodically. After the harvest the stubble was grazed. However, cattle grazed mainly in the woodland. The number of cattle was smaller than in the Bronze Age. The stables provided shelter for 20 cattle per family. The Drenthian cover sands were so poor in nutrients that after a time simply fallowing and grazing were not sufficient to maintain soil fertility unlike elsewhere in Europe on more loamy soils. Replenishing of nutrients may have been practiced by applying organic material onto the arable fields: woodland litter, sods and/or waste from the settlement and cattle dung when the stubble was grazed (Brongers, 1976). The more intensive use of arable fields in the Iron Age compared to the Bronze Age is reflected by a relatively large amount of pollen of *Rumex acetosella, Plantago lanceolata, Artemisia, Chenopodiaceae, Cerealia* and summer corn weeds. These were not characteristic weed species but they must have been introduced with the organic material from elsewhere (Groenman-Van Waateringe, 1979).

One of the small fields was used as a dwelling-place. After some time the house was rebuilt on the same or on another field. The dwelling-place then became arable field again. Hence the residences moved around within the celtic field pattern and formed no real village (Waterbolk, 1973A). Hijken was already mentioned as an example of a celtic field settlement (Harsema, 1974). Comparable celtic fields were found near Zeijen, Ballo and Emmen. These settlements numbered between 15 and 30 people. From these figures Harsema (1980) calculated about 5000 inhabitants for the whole of Drenthe. Based on the number of interments in funeral fields, Kooi (1979) suggested between 10 and 20 people for each settlement.

III.B.5. The Roman period and the early middle ages (0–800)

During the Roman period settlements became semi-permanent probably influenced by the customs of the time. A distinct pattern developed featuring two parallel north-south roads intersected by several east-west roads which resulted in rectangular palisaded residence allotments. These settlements, with a systematic lay-out, might be the first villages (Waterbolk, 1979). The palisaded block-structure continued in the adjacent arable fields (Waterbolk, 1973A).

The development of semi-permanent villages can be traced in Noordbarge (Harsema, 1979), Wijster (Van Es, 1967), Peelo (Bardet et al., 1983), Gasselte (Waterbolk & Harsema, 1979) and especially in Odoorn (Waterbolk, 1973B).

The location of these villages proved to be permanent for at most, three or four centuries, but when the arable field suffered from impoverishment, rising groundwater table or formation of sand drifts the villages shifted their position within their territories (Waterbolk, 1979). Hence the exploitation of celtic fields ceased during the Roman period.

The grazing intensity changed as a consequence of the shifting settlement and arable field within the territory. Most former celtic fields were grazed and could become heathland under heavy grazing or sand drifts under overgrazing or woodland in the absence of grazing. Some grazing might have been practiced in the still unreclaimed valleys near the villages. During the seventh and eighth century the plateau bogs were at their largest. The inhabitable part of Drenthe has never been smaller or more isolated than during that period (Waterbolk, 1984).

III.B.6. The traditional view on the 'esdorpen' landscape (800–1800)

The transitions of the Roman period and the early middle age farming system to the 'esdorpen' system is not quite clear. Waterbolk (1973A) suggested that the palisaded block-structure of the early medieval period villages extended into the adjacent arable fields and can be considered as being the oldest part of the later 'es' (permanent arable field) in the 'esdorpen' agricultural system. Therefore, the village is often older than the 'es' since the former found its definite location in the ninth century (Waterbolk, 1982).

A number of fieldnames or toponymes refer to the characteristic features of the 'esdorpen' landscape. The meaning of the toponyme 'woert' is troublesome. It might be associated with the allotments of the original farms (Waterbolk, 1973A) or an arable field enclosed with a wooded bank (Wieringa, 1979). Glässer (1976) suggested that they even might represent the oldest, irregular arable fields from the period before the medieval establishment of regular formed lots. The woodland adjacent to the settlement was open to grazing and was called 'lo', the somewhat more distant and less influenced (oak) forest was called 'holt' (Wieringa, 1979). Also the word 'made' originated from this period, indicating grassland areas in the river valleys. Perhaps the first clearances in the *Alnus* forests were made to allow for hay-making and grazing of the aftermath. The heathland areas were given the toponyme 'veld'. The arrangement of toponymes will be discussed later in section III.B.9 (Fig. III.11).

In Drenthe the location of the 'es' was on somewhat higher areas with boulder-clay only several decimeters below the surface (De Roo, 1953). The arable field was reclaimed by a large number of parallel trenches which were filled with sods and litter from the whole area, whereas the sand from the trenches then covered the whole reclaimed area including the filled in trenches (Waterbolk, 1985). A mixture of manure and sods was added later. These so-called 'plaggen' soils occur from Antwerpen to Hamburg (Pape, 1970) coinciding with the area of pleistocene cover sands and the accompanying heath community *Genisto-Callunetum* (De Smidt, 1979).

The summer corns *Triticum dicoccum* and *Hordeum vulgare* played an important role in the Neolithic Age, Bronze Age and early Iron Age, but were largely or completely replaced by the winter corn *Secale cereale* in the middle ages in Drenthe as suggested by macrofossil findings near Gasselte (Van Zeist & Palfenier-Vegter, 1979) from the period of the middle of the 9th to the middle of the 12th century (Waterbolk & Harsema, 1979). Winter corn weeds such as *Scleranthus annuus, Spergula arvensis, Agrostemma githago* and *Centaurea cyanus* were also found. It is suggested that *Centaurea cyanus,* which became a characteristic species in winter grain fields on sandy soil did not arrive in the northern part of the Netherlands until c. 1000 (Van Zeist & Palfenier-Vegter, 1979; Van Zeist et al., 1986). Pollen analytical investigations in northern Germany indicate considerable amounts of pollen of *Secale cereale* together with those of the winter corn weeds mentioned from the 9th century onwards as shown in Fig. III.5 (Behre, 1976).

It is assumed that the growing of both winter and summer crops only left a two months resting period, therefore, dunging became necessary. In Central Europe, on relatively fertile loamy soil, the nutrient shortage after cropping was encountered by fallowing the arable field every third year (three-course system). On the pleistocene cover sands the nutrient shortage

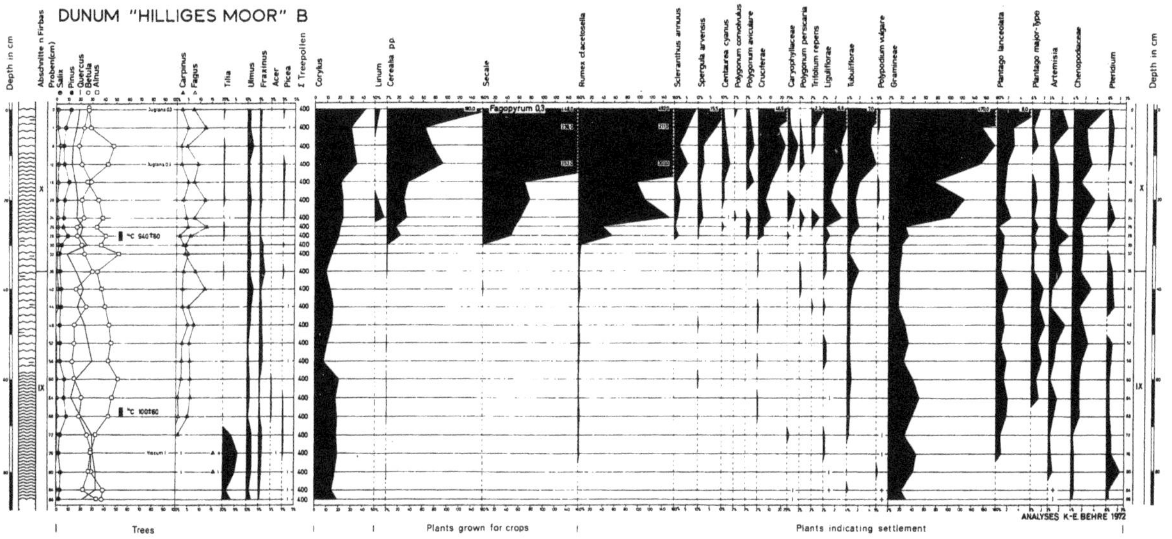

Fig. III.5. Pollen diagram from the arable field near Dunum (northwest Germany). Only the changes of tree pollen, grown crops and settlement indicators (the latter two related to tree pollen = 100%) are indicated. The start of the cultivation of rye lies at 28 cm (after Behre, 1976).

was compensated by sod manuring (Fastabend & Von Raupach, 1961; Pape, 1970; Behre, 1976) enabling the 'all-corn-system' of rye husbandry (in fact it seemed to have been alternated with *Hordeum* spp. and *Avena strigosa* or *A. sativa* (Behre, 1976; Van Zeist & Palfenier-Vegter, 1979; Van Zeist et al., 1986).

It is assumed that the number of livestock was increased and stabled indoor overnight in order to gather as much manure as possible. A great deal of litter was necessary to restrict the liquid proportion in order to obtain a tolerable material for the animals to tread. The litter used was mainly heather sods, grass sods, forest litter and peat litter, and it was mixed with sand (Pape, 1970). Most of it came from previous waste land on which the live stock also grazed. Heathland, therefore, became part of the agricultural system. Until the introduction of artificial fertilizers in the twentieth century a continuous transport of organic material from the 'waste land', via the stables, to the arable field took place, implying a considerable impoverishment of the waste land soil (Fig. III.6) (Vink, 1975).

It is assumed that cattle and particularly sheep grazed on the common heathland areas (Photo III.1) of the settlement. Because of the relatively rapidly increasing population after the twelfth century (Slicher Van Bath, 1963) more manure was necessary. As a result of heavier grazing the area outside the 'es' gradually transformed from woodland with heathland into an extensive treeless heathland with scrub, fringing the 'es' (Iversen, 1969). During the thirteenth and fourteenth centuries a decline of tree species pollen was found in Schleswig-Holstein (Tidelski, 1933) and at the Lüneburger Heide (Selle, 1941) in northern Germany. *Alnus* pollen reached relatively high values (Behre, 1976) towards the end of the middle ages, probably because the valley forests had not yet been destroyed to as great an extent as the plateau areas.

Hedges were used in order to protect crops from grazing animals (Photo III.2). These were raised around grassland lots in the neighbourhood of the villages. A particular technique, consisting of an earth bank with ditches at both sides and planted with shrubs and trees resulting in a hedge-bank, was adopted in the Netherlands and northern Germany (Dutch: 'hout-wal'; German: 'Wallhecke') (Klaasen, 1961; Wittig, 1979). Young twigs were bend back and wattled through the stems resulting in impenetrable fences. These fences were cut every 8–20 year (several authors in Wittig, 1979). Due to the increasing need for wood, chopping became

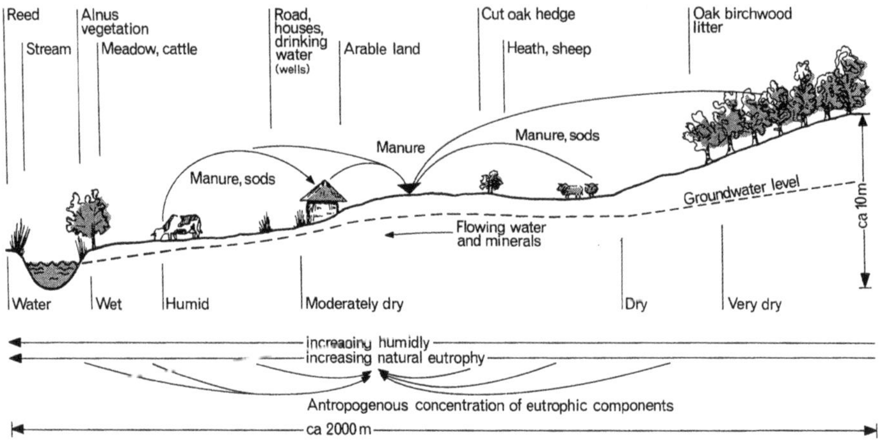

Fig. III.6. Schematic cross section through traditional land use system practised in the esdorpen landscape from c. 800 onwards (after Vink, 1975).

an important adventitious function of the hedge-banks. Although the age of the hedge-banks is not known, it is assumed that their construction started after the partition of the common grassland area in the river valleys of the village territories (Photo III.3), which was often carried out as early as the 16th and 17th century (Garming, 1955; Kocks, 1979; Heringa, 1982).

The origin and development of anthropogenic grasslands, which are the objects of our study, is considered in a separate section (III.B.11).

During the middle ages the characteristic lay-out of the 'esdorpen' landscape developed, featuring the permanent arable field or 'es' surrounded by remnants of the original forest or by secondary woodland; the valley meadows fringed by hedge-banks, and the extensive heathland area. The limiting factor of this agricultural system was the heathland area. In 1840 the heathland area in the municipalities of north Drenthe carried about 1 sheep · ha^{-1}, which is in agreement with De Smidt (1975). Contrary to the opinion of De Smidt, the heathland/'es' ratio varied considerably (Alstorphius Grevelink, 1844) (Table III.1). He also suggested that between 5 and 10 ha of heathland was needed to maintain the fertility of 1 ha of the 'es'. This was not found to be the case for northern Drenthe. Moreover, 1 ha was needed for cutting sods: 0.1 ha was cut annually in a ten year cycle. Although the meadow/pasture ratios varied considerably between the municipalities, the large herbivores/grassland ratios only fluctuated between 0.7 and 1.6 ha^{-1} thus indicating some balance (Table III.1). The municipalities of Norg and Zuidlaren had deviating ratios probably because of the large areas of reclaimed peat bog.

It is assumed that a well-balanced relationship existed between the increasing number of people, the size of the 'es' area and the heathland area and the number of livestock. Rigid rules for the village territories (Dutch: 'marken'; German: 'Gemarkungen') were developed increasingly from the second half of the thirteenth century onwards in order to regulate the utilization of areas in common (Slicher Van Bath, 1963). Only owners of full shares (Dutch: 'gewaardeelden' cf. Lambert, 1971, or 'vol bedrijf' cf. Bieleman, 1987) with large farms or part shares with smaller farms were entitled to the use of common areas. Cottars had no rights and could, therefore, not participate.

De Smidt (1975) suggested that these village territories were so well organized, due to their strict rules, that no agricultural crises arose in the 14th and 17th centuries. Elsewhere in Europe large areas had been abandoned (so-called 'Wüstungen' in German) during these periods (Slicher Van Bath, 1963). After the celtic field period the population of the whole of Drenthe only increased four or five fold reaching a 22,000 in 1630 (Heringa, 1977).

The population in the whole of Drenthe increased from 22,000 in 1630 to 40,000 in about 1800 (Scheltens, 1905). The total 'es' area, however, hardly increased between 1650 and 1800. This was probably because the yields increased, owing to more available manure. The latter was made possible by the increased grassland area, pasture for yearlings to graze whereas they were kept in byres overnight (Alstorphius Grevelink, 1844; Garming, 1955), and hay-fields for winter forage.

Photo III.1. Source area of the Drentsche A river valley featuring the Westerholt study area. Some remnants of the heathland are left (light areas at the background and the right), but are partly covered with spontaneous woodland after the ceasing of cattle and sheep-grazing in the beginning of the 20th century. The grasslands were reclaimed from heathland until 1950. The line from left to right somewhat above the middle of the photo is the limit between two village territories which may explain the different directions of reclamation.

Photo III.2. Upper course area of the Drentsche A river valley featuring part of the Anlo study area. The hedge parallel to the brook acted as a barrier for the grazing animals at the time that the higher soil (now reclaimed) was heathland. Some grassland area was abandoned this century resulting in spontaneous woodland (left).

Photo III.3. Upper course area of the Drentsche A river valley featuring part of the Anlo study area and the Loefvledder area at the background. The hedges perpendicular to the brook indicated ownerships of the valley grasslands.

III.B.7. A dynamic view on the 'esdorpen' landscape (1600–1900)

A great deal of the literature on the 'esdorpen' landscape, mentioned in the previous section, suggested a more or less consistent view on the agricultural system in the period between 1000 and 1850, until the introduction of artificial fertilizers. A picture, therefore, emerged of a very long lasting and constant agricultural system on the Drenthe Plateau. The system was said to be characterized by a number of features (Bieleman, 1987), namely (i) a

Table III.1. Areas of landscape elements, numbers of animals and derived large herbivores/grassland and sheep/heathland ratios for the municipalities of the northern part of Drenthe around 1840 (municipalities in italics have considerable areas in the Drentsche A river valley) (after Alstorphius Grevelink, 1844).

	Es (ha)	Meadow (ha)	Pasture (ha)	Grass-land (total ha)	Meadow/ pasture ratio	Cattle number	Horses number	Cattle horses total	Large herb-ivores/ grassland ratio	Heath-land (ha)	Sheep number	Sheep/ heath-land ratio	Heath-land/es ratio
Anlo	502	823	1580	2403	0.52	1865	641	2506	1.04	4570	5120	1.12	9.1
Assen	209	359	187	546	1.92	717	179	896	1.64	2561	2447	0.95	12.2
Eelde	179	850	216	1066	3.94	810	170	980	0.92	419	420	1.00	2.3
Norg	243	600	800	1400	0.75	864	168	1032	0.74	8749	4980	0.57	36.0
Peize	199	697	215	912	3.24	776	162	938	1.03	488	436	0.89	2.4
Roden	451	1500	1000	2500	1.50	1587	366	1953	0.78	2031	2320	1.14	4.5
Rolde	439	625	310	935	2.01	1037	250	1287	1.38	8168	6731	0.82	18.6
Vries	648	573	1275	1858	0.45	1492	369	1861	1.00	7580	6680	0.88	11.7
Zuidlaren	245	800	1300	2100	0.62	936	171	1107	0.53	1253	1236	0.98	5.1

separation of the arable field and grassland, (ii) rye was grown continuously on the arable field, (iii) the area of grassland was small as compared to the arable field area and as a result cattle livestock was small, (iv) the supposed shortage of manure for the arable field was supplied by sheep from which dung was collected in stables during the night time, (v) the arable field was sub-divided in private lots on which the sheep were kept for a few days after the rye harvest in order to provide some manure, (vi) the productivity of the arable field was enhanced by sods cut from the heathland which were mixed with the dung voided by the sheep in the stable, (vii) in order to provide sufficient manure for the arable field an extensive heathland area was necessary for the sheep to graze and for sod removal. A more or less fixed arable field:grassland:heathland ratio of 1:2:7 was, therefore, found.

Bieleman (1985, 1987) has studied archives which existed from 1600 onwards and he made clear that the existing view on the agricultural system was based on the extrapolation of the situation around 1850. He showed that the agricultural system changed considerably in the period between 1600 and 1900 and with it the 'esdorpen' landscape.

Farmers were continuously threatened by crop failures around 1600. Moreover the year-to-year fluctuations of the average yield were extremely large. The very infertile soil of the Drenthe Plateau added to the problems. Risk-minimizing was, therefore, one of the main principles that ruled farming in those days. It encouraged the farmers to run a diversified type of farming.

A farm with full rights (a so-called 'vol bedrijf') included nearly 9 ha of arable land. These 'essen' were used in an extensive way. Not all arable land (c. 15%) was actually ploughed and the marginal fringes of the arable fields were in use as 'out-field'. Part of this outfield usually lay fallow. Once every 10 to 15 years a farmer grew one or two crops of rye or buckwheat without using manure and labour. One of the advantages of this system was the accumulation of humus in the marginal parts of the arable field. The central part of the field, the 'infield' fell under a crop sequence based on a two-course rotation system. The infield was, therefore, divided into two equal areas, more or less connected. These areas were alternately sown with winter rye and summer rye. This rotation system enabled the community to use the arable field for pasturing, mainly cattle but also horses and sheep. In fact it meant that half the arable field was open for grazing almost all the year except for the growing season (May, June, July) of the corn crop. To enable the use of the arable field each farmer was bound by a set of regulations. Rules for grazing meant that the rights of the individual farmer were secondary to the rights of the whole community. The practice of pasturing on the stubble and weeds contributed considerably to the maintenance of the soil fertility. It was, therefore, probably sufficient to manure only those parts of the arable field which were to be sown with winter rye. The manure consisted of dung from the stables and was mixed with sods collected from the very poor valley grasslands and with cuttings from the heathland (not the sods). The consequence of this extensive treatment of the arable land was very low yields.

Twenty four cows were kept on farms with full rights. The remarkably large number of cattle on the 'esdorpen' farms included over 50% of young bullocks which grazed on the extensive poor commons. After some years these bullocks were driven to the much better pastures in the coastal provinces and then sold for slaughter. The most important grazing area for the cattle was the arable field for long periods of the year. Four adult horses were kept on farms with full rights. The horses were not only kept for the supply of draught power on the farm. Horse breeding was an important source of income.

Although the production of wool by Drenthian sheep was small, this part of farming was an important source of income at that time. Farms with full rights kept 40 sheep. The very wet heathland of the commons and bad weather frequently caused animals, especially sheep, to fall victim to liver fluke diseases. For example the wet summer of 1621 followed by a severe winter caused the death of 50% of sheep, 25% of cattle and 20% of horses on the Drenthe Plateau. This combination of bad weather conditions was repeated in 1632, 1650, 1699, 1710, 1740 and 1770, thus throwing farming out of balance.

The livestock grazed on the common heathland and grassland if the amount of forage on the arable field was too little, or for the periods in which grazing was not allowed. The ratio between common grassland and private grassland in the 17th century is not known. Approximately 10% of the private grassland consisted of pasture and nearly 90% of meadow areas. The latter were very important for hay-making in order to keep cattle and horses throughout the winter period. The meadows, therefore, became already private as early as the first half of the 17th century. Only grazing in early spring and after hay-making was common. The quality of the meadows was very poor since they were hardly drained.

At the end of the 17th century the prices of agricultural products declined and at the same time new and heavier taxes were imposed. The culmination of falling prices and rising costs was in the years 1690 to 1720. The only way to escape from these problems was trying to raise the production. This was done by intensification, since the area of arable land did not increase between 1650 and 1800. The farmers took the outfield part of the 'es' into more permanent use and the infield part was used more intensively. This was only possible using more manure. There are strong indications that the techniques used to maintain the fertility of the arable field were fundamentally changed after the beginning of the 18th century. The quantity of sods cut for the absorption of dung in the stables increased strongly and they were not longer taken from the grassland but from the heathland. At the same time the common grazing of the arable fields decreased. What remained was a system of holding sheep behind hurdles for some nights after the harvest. Farmers could now cultivate their land much better and could, therefore, control the weeds better. This resulted in increasing yields of profitable winter rye. The process of intensification resulted in a decrease in horse and cattle breeding, especially after heavy losses during a cattle plague in the 1740's. The number of sheep increased at the same time. Losses of sheep from liver fluke decreased throughout the 18th century as a result of draining the wet commons. A farm with full rights had 11 cattle, 2 horses and 50 sheep in 1800.

The cessation of cattle grazing on the arable field lead to an increase of the pasture area despite the declining number of livestock. It is not known to what extent the whole grassland area increased, but at the beginning of the 19th century 60% of it was pasture, whereas meadows only amounted to 40%.

The changes in agricultural practices can be summarized by concluding that there was, in fact, a process of intensification in which the earlier diversified type of farming with income from rye, wool, horses and bullocks was transformed by specialization into one which was based more exclusively on the production of rye. As a result the impact of man was not the same in the whole period of the 'esdorpen' landscape, but changed considerably.

Table III.2. Area (ha) and percentages of landscape types and hedge-banks (km) in the esdorpen landscape in the catchment area of the Drentsche A river system from 1810 onwards. The total area is smaller in 1970 than in previous years because of the extension of villages.

Year	1810		1850		1900		1930		1970	
	ha	%	ha	%	ha	%	ha	%	ha	%
Heathland/peatbog	12820	70	10990	60	9850	54	7260	40	940	5
Woodland	750	4	560	3	1450	8	790	4	2900	16
Arable field	1760	10	2480	14	3150	17	4950	27	5700	32
Grassland	3060	17	4370	24	3950	21	5400	29	8330	47
Hedge-bank (km)	?		266		295		354		160	
Total	18400		18400		18400		18400		17870	

III.B.8. Quantitative changes in the 'esdorpen' landscape of the Drentsche A area (1650-present)

The locations and areas of 'es', grassland and heathland were not mapped accurately until the beginning of the nineteenth century after the introduction of trigonometry (Van Der Linden, 1973). In order to quantify the landscape changes a comparison is made of maps of part of the northern Drenthe Plateau covering more or less the Drentsche A catchment area in c. 1810, 1850, 1900, 1930 and 1970. The changes in the landscape are depicted in Figures III.7 and III.8. The area changes are summarized in Table III.2 for the catchment area and in Table III.3 for the nature reserve, respectively.

Cultivated land (Fig. III.7A–F). Until 1900 the arable fields were mainly restricted to the 'es' and the grassland to the valleys. After that large extensions occurred at the cost of heathland areas. The application of artificial fertilizers and increasing drainage enabled an extension of cultivated grassland in the low-lying areas and of arable fields on elevated ground. The grassland area, in particular, increased after 1930, since it was more profitable to the farmer than arable fields.

Woodland (Fig. III.7A–F). Although remnants of the primeval forests were still present in 1810, these had nearly completely been reclaimed to grassland in 1850. Even more woodland occurred in 1900 than in 1850, a great deal of which had been planted (Photo III.4). The deciduous trees were mainly oaks the bark of which was used in tanneries. Conifers were planted for mine-supports in the southern part of the Netherlands. Up until 1930 many forests, which were planted around 1900, were cut down again in favour of agricultural areas. The woodland area had increased greatly again in 1970 as a result of coniferous plantations for both production and recreational purposes.

Heathland (Fig. III.7A–F). Heathland, peat bogs and sand dunes which developed in the 18th and 19th century (Waterbolk, 1985) are in the following referred to as heathland for convenience. The extensive heathland area decreased between 1810 and 1850 and became partly grassland, mainly because of drainage and the exploitation of the low and wet areas. The main decreases up until 1900 were due to woodland plantations. Reclamation for arable fields or grasslands was still going on on a small scale and involved the beginning of the dividing of the vast linked heathland areas for private purpose. This could be attributed to the subdivision of the common territorial heathland areas into private properties around 1850 (Kocks, 1979; Heringa, 1982). The heathland area lost its agricultural function because of the introduction of artificial fertilizers. Moreover, wool imported from Australia was cheaper than home produced wool. The reclamation process went on on a relatively small scale until 1930. It increased rapidly in the thirties and was stimulated by the State as unemployment relief work. By 1970 only some small heathland islands were left in an agricultural sea.

Hedge-banks (Fig. III.8A–E). In 1810 most hedge-banks occurred in the valleys. Hedge-banks were only found perpendicular to the brook locally and mainly in the middle course. This is remarkable since the subdivision of common grasslands had already taken place in the 17th century (Heringa, 1982). Perhaps private properties were marked by ditches in the peaty soils close to the brook. It seemed, however, that around 1800 large valley areas were not protected against cattle or sheep grazing on the common heathland by hedge-banks.

The hedge-banks on the 'es' probably prevented sand drifting and those at the fringe of the 'es' enclosed the so-called kampen. The hedge-bank length increased until 1930 due to a better fencing-in of the valley grasslands and a subdivision of private properties. A lot of new grassland was fenced-in by hedge-banks after heathland reclamation on moist soils because sheep were still kept on the remaining heathland. After 1930 the hedge-banks lost their agricultural function and many were removed.

A rough mapping of the heathland-, arable field-, woodland-, and grassland areas in the 'Stroomdallandschap Drentsche A' around 1650 was performed based on toponymes (see section III.B.9). It is enlightening to compare this map with the maps from 1810 onwards featuring the catchment area of the Drentsche A.

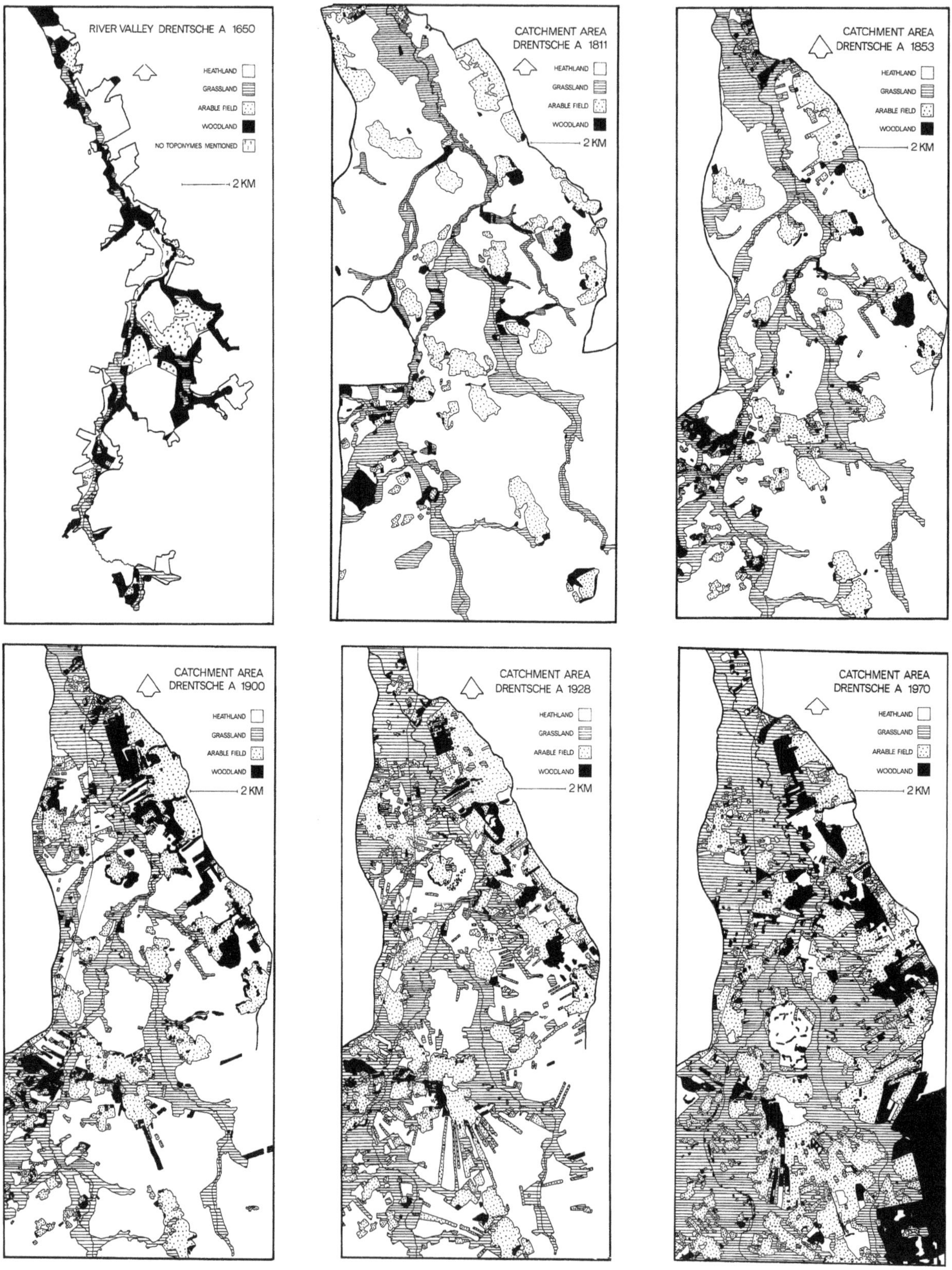

Fig. III.7A–F. Changes in the location of the main landscape elements heathland, grassland, arable field, woodland in the Drentsche A catchment area in the period between 1650 and 1970. For 1650 the location within the present nature reserve Stroomdallandschap Drentsche A is known, and has been derived from toponymes.

Photo III.4. Upper course area of the Drentsche A river valley featuring part of the Anlo study area. The woodland on the higher soil along the brook was planted on heathland at the beginning of the 20th century.

To study whether the 'esdorpen' landscape of the Drentsche A area is unique, it is compared to that of the 'Lüneburger Heide' in north-west Germany. The latter featured 85% heathland, 5% woodland, 8% arable field and 1% grassland in 1775 in an area of 19400 ha including the present reserve (since 1910) (Buchwald, 1984). The difference with the catchment area of the Drentsche A river system in 1810 (see Table III.2) is the smaller proportion of grassland which is due to the fact that less rivulets cross the Lüneburger Heide Plateau and hence the area of heathland was larger than in the Drentsche A system. After 1850 large coniferous plantations were set up resulting in 43% coverage of the whole area in 1900 and 65% in 1960, whereas in the same period the heathland area decreased to 48% and 22%, respectively. Little reclamation of the heathland took place since the area of arable land was 9% and that of grassland 3% in 1960 in contrast to the large reclamations in the Drentsche A area.

A general character of changes in the landscape, despite local differences, can be found in north-

Table III.3. Area (ha) and percentages of landscape types and hedge-banks (km) in the nature reserve Stroomdallandschap Drentsche A from 1650 onwards.

Year	1650		1810		1850		1900		1930		1970	
	ha	%	ha	%	ha	%	ha	%	ha	%	ha	%
Heathland/peatbog	1965	56	1800	51	1650	47	1395	40	1110	32	520	15
Woodland	1160	33	265	8	160	5	390	11	225	6	535	15
Arable field	115	3	95	3	95	3	120	3	280	8	200	6
Grassland	460	13	1290	37	1615	46	1615	46	1905	54	2265	64
Meadow	455		950		?		?		?		?	
Pasture	5		340		?		?		?		?	
Unknown	70	2	75	2	0		0		0		0	
Hedge-bank (km)	?		62		84		89		90		53	
Total	3520		3520		3520		3520		3520		3520	

Fig. III.8A–E. Changes in the location of hedge-banks in the Drentsche A catchment area in the period between 1810 and 1970.

western Europe which can be illustrated for the United Kingdom. Since c. 1750 lowland acidic heathland has declined 78% in area, lowland rich in fens have been reduced to scattered remnants, only 5% of permanent grasslands have remained agriculturally unimproved, 20% of the extent of pre-1940 chalk grassland is still present, the total length of hedges was reduced by 28% between 1946 and 1974 (Ratcliffe, 1984).

III.B.9. Historical geography of the 'esdorpen' landscape

Around 1650 a ground tax register was established. Unfortunately, no maps are available, since every cultivated allotment was registrated not by its area but by its toponyme. Toponymes indicate the character of the soil, previous and/or present land utilization and roughly the vegetation of the concerned area at the time of reclamation in the case of 'bos', 'holt' and 'heest' (Edelman-Vlam, 1969). This is shown in Table III.4.

It is known, that in the early middle ages there was a subdivision among private 'prata' (meadows; French: 'prés') for hay-making and grazing the aftermath, and common 'pascua' (pastures; French: 'pâturages', 'paquis') for rough grazing of waste areas (Slicher Van Bath, 1963; De Dainville, 1964). The Netherlands were mapped in detail (1811–1817) during Napoleontic times involving a subdivision of 'prés', 'pâturages' (cultivated) and 'paquis' (spontaneously developed under grazing) (Koeman, 1970). A comparison of the Napoleontic map and toponymes can give relevant information about the real land utilization indicated by the toponymes. Since the toponyme locations did not move an approximate estimation can be made of which parts of the valley were cultivated around 1650. Non-cultivated areas must have been *Alnus* forests or bogs as has been suggested in section III.B.2. These unreclaimed areas amounted to at least 50% of the area of the 'Stroomdallandschap Drentsche A' (Figure III.7A).

Table III.4. The most common toponymes in northern Drenthe and their interpretation according to assessments in the literature (Alstorphius Grevelink, 1844; Slicher Van Bath, 1945; Garming, 1955; Keuning & Naarding, 1963; De Vries, 1964; Wieringa, 1968; Edelman-Vlam, 1969; Smit et al., 1977; Booy, 1979; Wieringa, 1979).

Toponyme	Interpretation
made	hay-field with aftermath grazing
hem	hay-field in brook meander
stuk	irregular area, newly divided hay-field
broek	valley forest, swampy pasture
veen	swamp, bog, area for peat-cutting
vledder	swamp area, former peat in valley
land	forest, uncultivated area, pasture
bos	forest
heest	brush-wood, oak, beech
holt	forest at higher soils, grazed forest
lo	grazed forest
kamp	fenced allotment at village edge, allotment reclaimed from heathland
veld	heathland, sand, peat

It appeared from Fig. III.9 that the cultivated area in the valley in 1650 was nearly all hay-field as indicated by the toponymes 'made' and 'hem'. Moreover, the toponymes 'broek', 'stuk', 'bos', 'heest' and 'holt' should be associated with hay-making around 1650. Most areas with the toponymes 'bult', 'land', 'stuk', 'broek' and 'heest' were reclaimed after 1650. In that period 'heest' tyoponymes were mainly used as grazing areas, 'stuk' half and half for grazing and hay-making, and 'bult', 'land' and 'broek' predominantly for hay-making. The different interpretation of some toponymes in 1650 and 1800 is due to the fact that those toponymes are found in both 'prés' and in 'pâturages/paquis'. Figure III.10 summarizes the results and indicates that the cultivated area in the valley was almost only hay-field in 1650. In 1850 the number of toponymes indicating hay-fields equalled 50% and indicating pastures 25%. This change is in agreement with the findings of Bieleman (1987).

The characteristic lay-out of the esdorpen landscape featuring five village territories bordering the Drentsche A river on the Drenthe Plateau is depicted in Fig. III.11. The location of 'made', 'stuk' and 'broek' toponymes indicate the sequence of reclamation of the valley. This is illustrated by the older irregular 'ma(de)' allotments adjacent to the brook and the regular 'stuk' and 'broek' allotments further away from the brook (Photo III.5) (Fig. III.12). The location of 'woert', 'lo' and 'holt' topo-

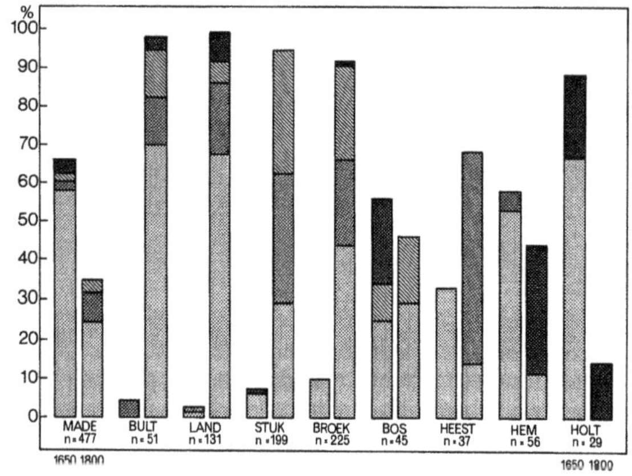

Fig. III.9. Percentages of land utilization in the area of the nature reserve Stroomdallandschap Drentsche A (meadow = light, pasture = dotted, woodland = hatched, heathland = darkest) for several toponymes in 1650 (left) and in 1810 (right). The location of toponymes as known in 1970 is assumed not to change from 1650 to 1810. The land utilization as known in 1810 is assumed to be the same in 1650 as far as cultivation had been taken place, which is indicated by the mentioning of a toponyme in that year.

nymes indicates the sequence of reclamation of the es (see section III.B.6).

It is apparent from the previous sections that the Drenthe Plateau had a long (pre)history of occupation. This occupation and its continuation are summarized in section III.B.10.

III.B.10. Continuity or discontinuity of settlements?

Since neolithic times the intersection of the Drenthe Plateau with numerous impassable valleys resulted in a natural pattern of comparable units suitable for settlement. During the late Bronze Age and early Iron Age these settlements exerted some territorial claims by their urn-fields and by their celtic fields. These territories partly coincided with the historic village territories. Only the Bronze and Iron Age territories were smaller (Fig. III.13). In the case of funeral fields during the late Iron Age, two or three small territories fused and corresponded completely with the village territories of the 'esdorpen' which developed from the ninth century onwards and remained to the present day (Fig. III.14) (Waterbolk, 1979, 1982; Harsema, 1984).

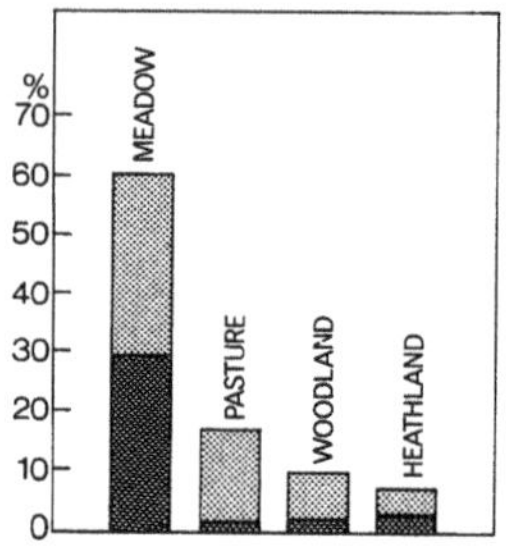

Fig. III.10. Total percentage of toponymes in the area of the nature reserve Stroomdallandschap Drentsche A (n = 1250) used as meadow, pasture, woodland and heathland. Percentages are subdivided for 1650 (dark) and 1810 (light).

This continuous settlement period did not occur on the whole Plateau. Van Gijn & Waterbolk (1984) hypothesized that due to sand-drifts and peat-formation some competition between the agricultural and pastoral component might have been arosen which forced the herdsman to move farther away from the village. The suitability of the salt marshes for grazing at that time might have resulted in a 'transhumance' i.e. the movement of herds of domestic livestock to complementary seasonal pastures. This system must gradually have changed into year-round occupation of the salt marshes. This particularly involved people living in territories adjacent to navigable parts of the little rivers Peizer Diep, Eelder Diep and Drentsche A which flowed into the salt marsh areas (Fig. III.14). The fringes of the Drenthe Plateau would have become occupied again in the second and third centuries when the marshes became less habitable during a transgression period (Waterbolk, 1979, 1982).

It seemed that the villages in the central part of the Drenthe Plateau had a long history of uninterrupted occupation. These villages revealed a continuous typological series of house plans during the first millenium, which suggests that a population change was very improbable. The development in a number of villages showed that their location was semi-permanent and that several moves had taken place withing their territories. Since these shifts were synchronous for most villages it seems likely that there was some general reason for it (Waterbolk, 1982). The move in the

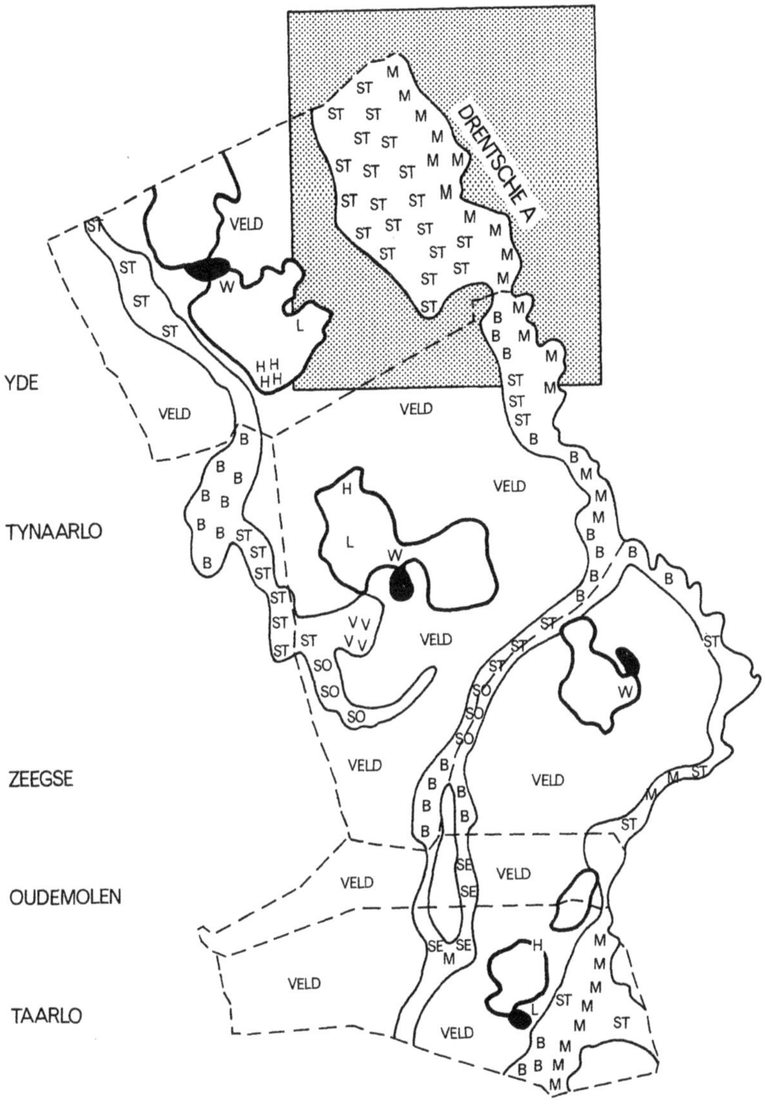

Fig. III.11. Lay-out of the esdorpen landscape concerning five village territories along the middle and lower course of the Drentsche A brook. The es (fat outline), river valley and heathland (veld) are indicated. The location of the toponymes woert (W), lo (L), holt (H) on the es and made (M), stuk (ST), broek (B), veen (V), stroet (SO, indicating scrub in small source areas), streng (SE, indicating very narrow valley) is depicted (after Wieringa, 1979).

2nd/3rd century A.D. might have been connected with the return of people from the marsh areas. The celtic fields had probable been abandoned because the fertility of the soil had become too low, and new places for the arable fields were chosen. The shift at the transition of the 5th and the 6th century coincided with an economic and population decline which might have been connected with the termination of the Roman Empire. The population increased during the 9th century and the 'esdorpen' agricultural system, which can be considered an agricultural revolution was adopted. New village locations were chosen closer to the valleys, and this probably marks the beginning of the valley exploitation. In the course of the 12th century the main villages had found their definite location, whereas their smaller hamlets were built near the streams indicating the definite involvement of the valleys in the agricultural system.

The idea of an uninterrupted occupation of vil-

Fig. III.12. Detail of Fig. III.11. (after data from J. Wieringa).

lage territories is not only based on the comparison of a large number of different villages on the Drenthe Plateau. It is also substantiated by the excavations in Peelo (Fig. III.15) which revealed that there had been continuous occupation from 400 B.C. until the present day (Bardet et al., 1983; Waterbolk, 1984).

The development of the Drenthe Plateau is not an isolated phenomenon. Behre (1976) and Waterbolk (1979) suggested that, with the exception of the fringes, there has been continuous occupation of all the uplands of the northern part of the Netherlands and the bordening north west part of Germany.

Recently a survey was published of the different cultural historical landscapes of the Province of Drenthe. Both the changes in the landscapes and the remnants which can nowadays be recognized had been mapped (Stichting voor Bodemkartering/ Provinciaal Planologische Dienst Drenthe, 1987).

III.B.11. Development of anthropogenic valley grasslands

Having summarized the agricultural practices and series of settlements, the origin and development of anthropogenic grasslands will be discussed below. Some kind of subdivision between arable fields and grazing areas is a prerequisite for anthropogenic grassland development. Arable fields be-

Photo III.5. Middle course area of the Drentsche A river valley featuring the Taarlo study area. The irregular lots ('made' toponyme) adjacent to the brook were already cultivated around 1600. The small regular lots with many parallel ditches ('stuk' and 'broek' toponymes) were reclaimed from heathland in the 19th century. The large lots in the background were reclaimed from heathland in the 20th century.

came semi-permanent with fallow periods during the celtic field agricultural system. Pasture-like plant communities with *Phleum* cf. *pratense, Poa* cf. *trivialis* and *Poa* cf. *pratensis* (Knörzer, 1971) could have originated from fallowing arable fields. Pennington (1969) suggested the occurrence of farmsteads with arable fields, fallow areas and pasture-like areas in the United Kingdom during the Iron Age.

Important changes in agricultural practices occurred at the transition of celtic field period and Roman period. The mouldboard plough facilitated arable field cultivation, but required the tractive power of oxen or horses. Good quality winter forage was necessary for these animals. Hence permanent arable fields with the need for tractive power enhanced the development of hay-meadows in areas with cold winters (Slicher Van Bath, 1963; Lange, 1976). Moore (1966) even said that meadows became the key to the farming economy. The number of livestock that could survive the winter period was determined by the amount of hay (Waterbolk, 1985).

The first hay-fields are known to date back to the Roman period thanks to descriptions by a Roman author as quoted by Lange (1976):

> Small animals are allowed on the field after hay-making from the second year of establishment onwards. Larger animals are allowed for grazing from the third year onwards when the hay-field has become firmer. Shrubs and fast growing forbs should, in particular, be prevented becoming established. Blackberries and rushes should be pulled up by the roots in the autumn. Tall forbs should be eliminated in the spring. No pigs are allowed in the hay-field because they destroy the turf. Large animals are not allowed for grazing during wet periods because they trample the grass roots. Hay-fields overgrown with mosses are best cultivated by the removal of the mosses and the sowing of hay-field seeds or dunging. The best time for hay-making, from both the quantity and palatability view point, is before the grass is too dry.

This type of hay-field seems rather artificial and is apparently combined with grazing. Further evidence of hay-making comes from finds of scythe blades from a number of Roman sites in the United Kingdom (Greig, 1984). Lange (1976) suggested that the very wet *Carex*-communities, which are not suited for grazing, were used for hay-making,

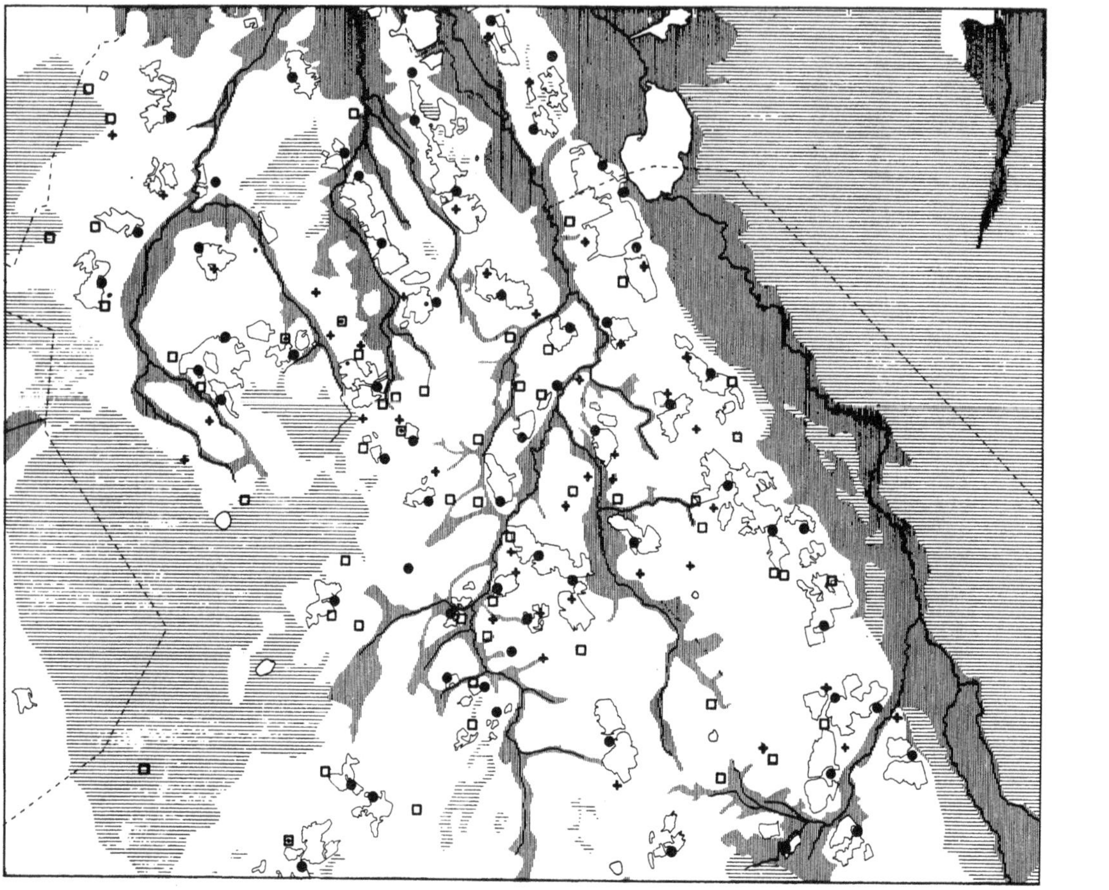

Fig. III.13. The northern part of the Drenthe Plateau with the location of recent villages (circles) and arable fields, celtic fields (squares), and ceramics (cross) from the early Iron Age (c. 700 B.C.). The sandy plateau is white, the river valleys are hatched vertically, the moors horizontally (after Waterbolk, 1982).

until the beginning of the middle ages, but she provided no evidence. Only after 1000 the drier *Arrhenatheretum*-like meadows had developed in Central Europe (Lange, 1976).

Macro-fossil finds, from settlements of various periods are summarized in Table III.5. The species concerned are arranged according to the recent ecology of these species. The *Molinietalia* species are thought to be characteristic for hay-fields (*Filipendulion* and *Calthion palustris*) and the *Arrhenatheretalia* species for pastures (*Arrhenatherion elatioris* and in particular *Cynosurion cristati*). This subdivision, however, is not very sharp. It is obvious that the proportions of *Molinietalia* and *Arrhenatheretalia* species are equal in all periods. Little indications for anthropogeneous grasslands can be traced in pre-Roman periods, but the more from the Roman period onwards at least in Germany (Knörzer, 1975) and the United Kingdom (Greig, 1984). Some evidence for the development of pastures and meadows is found from the middle ages onwards in the Netherlands. Van Zeist & Palfenier-Vegter (1979) mentioned macro-fossils (9th–12th century) from Gasselte and from Odoorn (7th century) and Pesse (7th–13th century) (Van Zeist et al., 1986). They think that those species originated from grasslands which were not natural, but must have resulted from the clearing of the forest near the settlements, probably in small valleys, especially alder and willow carr in the cases of *Molinietalia* grasslands.

Jansen (1972; 1979) published pollen diagrams of

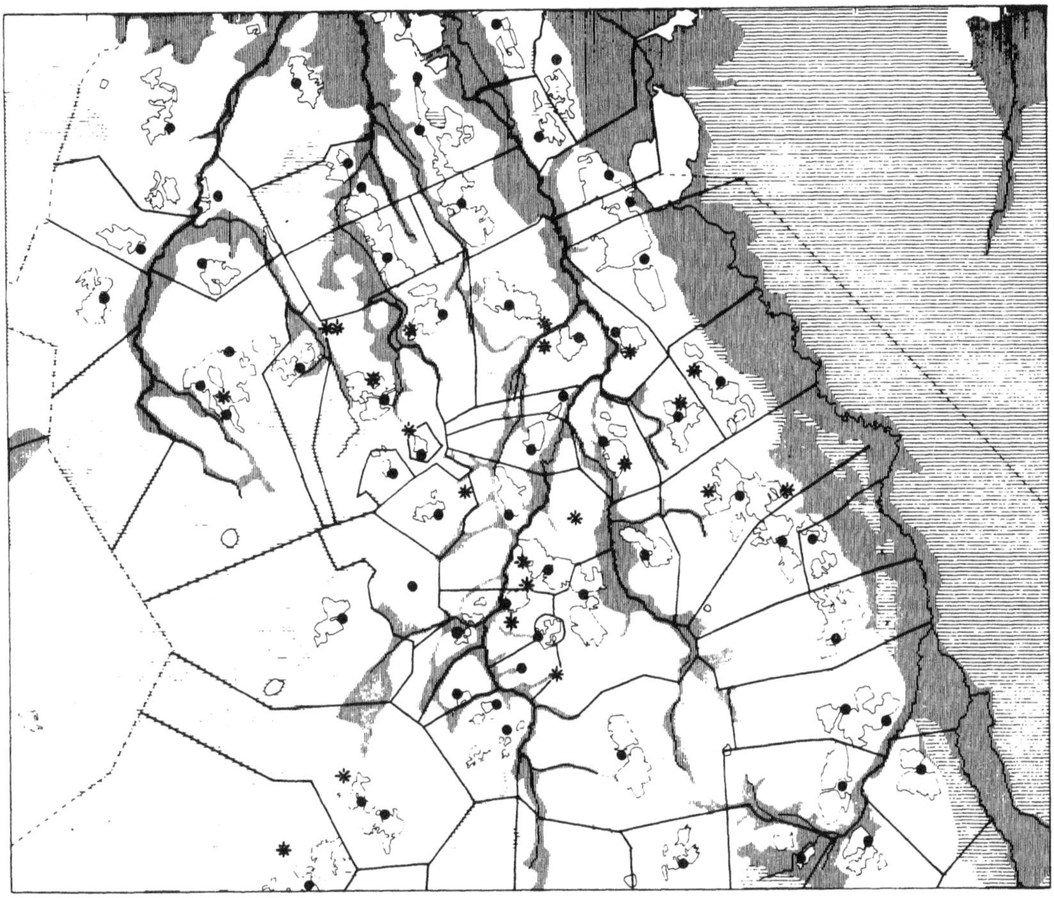

Fig. III.14. The northern part of the Drenthe Plateau with the location of recent villages (circles) and arable fields, the historic village territories and funeral fields (asterisk) in the late Iron Age (c. 200 B.C.). Note the absence of funeral fields along the northern fringe of the plateau. The sandy plateau is white, the river valleys are hatched vertically, the moors horizontally (after Waterbolk, 1982).

a brook valley in the south of the Netherlands. The hay-field communities *Calthion palustris* and *Cirsio-Molinietum* may have been present in the lowlands in the early middle ages, but later on the lowlands were covered by pastures (Fig. III.16).

In northwest Germany Behre (1976) demonstrated an *Alnus* pollen decrease from 900 onwards, which probably indicates clearances in the valleys. At the edge of the Dinkel valley, about 70 km south of the Drenthe Plateau a medieval clearing around 1200 was reconstructed from pollen analytical data by Van der Woude (1983). Wet hay-fields with *Cyperaceae, Caltha palustris* and *Polygonum* cf. *bistorta* replaced *Alnus* swamps, whereas *Alno-Padion* like forests developed into pastures in the somewhat higher areas. The later marked decrease of wet hay-field species and increase of *Gramineae, Plantago lanceolata* and *Rumex acetosa* could be attributed to artificial drainage which enabled more grazing to take place.

Many species are known from the Roman period which can be assigned to grassland communities. Knörzer (1975) suggested that their diverging ecology points to a characteristic cultivation of different soil types. The grassland vegetation was probably as diverse as it is now. Ellenberg (1978) drew a number of schemes featuring the ecology of present central European forests and their substitute communities with respect to the moisture level and acidity of the soil. Fig. III.17 shows the ecology of forest-building tree species in central Europe. It should be realized that *Fagus sylvatica* only appeared from 2000 B.C. onwards in pollen diagrams in the Drenthe Plateau regions. It is assumed that this species was able to become established mainly in forest regenerating from agricultural practices (Casparie & Groenman-Van Waateringe, 1980). The synecology of forest commu-

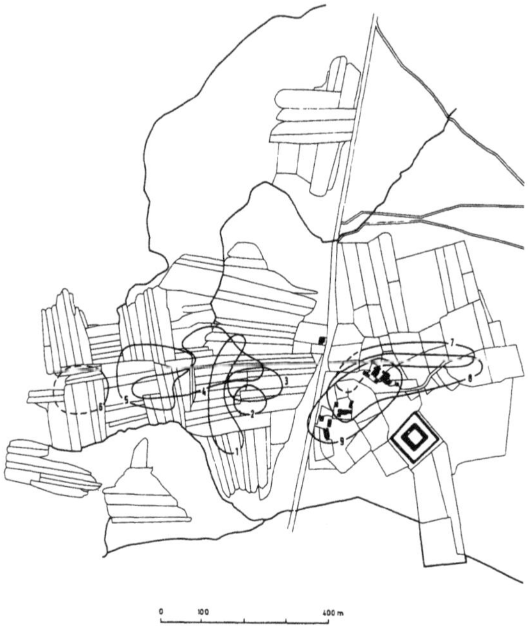

Fig. III.15. The settlement drift of Peeloo during the period from c. 400 B.C. onwards, projected against the cadastral registration map. 1: c. 400–100 B.C., 2: 100 B.C.–100, 3: 100–200, 4: c. 200–400, 5: c. 400–500, 6: c. 500–800, 7: c. 800–1300, 8: c. 1300–1600, 9: c. 1600–1800. (after Waterbolk, 1984).

nities at the alliance level is shown in Fig. III.18. The synecology of forest undergrowth communities that can be regarded as the intitial substitute after reclamation of the primeval forest is shown in Fig. III.19. It can be concluded that most recent groups of grassland species were already present in the undergrowth of the primeval forest or along fringes with open landscapes although they were scarce (Bottema, 1987). After centuries of continual human interference viz. after the middle ages, the semi-natural substitute communities could have developed into the well-defined hay-field syntaxa graded to the main ecological parameters (Fig. III.20). Some manuring of the most intensively cultivated hayfields could have led to the arrangement of grassland communities around 1900 as indicated in Fig. III.21 by Ellenberg (1978). Areas enabling grazing have been used quite intensively over the centuries. Therefore, various forms of the *Lolio-Cynosuretum* developed with species indicating dry, wet or acidic circumstances depending on local edaphic conditions (Ellenberg, 1978). It is thought by Van Zeist et al. (1986) that medieval *Molinietalia* meadows may not have differed very much from similar grasslands which up to 40 years ago were still found quite extensively in some parts of the Netherlands. Greig (1984) thinks that hay meadows of a similar type as today are found in some reserves already existed in Roman times. The *Arrhenatheretalia* grasslands in the vicinity of the settlements may have shown little resemblance to the more recent *Cynosurion cristati* partures (Van Zeist et al., 1986). In recent years both hay-fields and pastures on widely diverging substrates rapidly changed into a uniform *Poo-Lolietum* community owing to artificial drainage, ploughing, sowing and the application of artificial fertilizers.

After the generalizations about anthropogenic grasslands attention will be focused on the valley grasslands of the Drenthe Plateau. Grootjans (1980) described the character of the original semi-natural plant communities in detail along with their distribution in relation to the hydrological conditions of the Drenthe Plateau brook systems. The hydrological conditions of these brook systems and their background are summarized in section III.B.1. Figures III.22A–D show vegetation gradients and hydrological situations of cross-sections of the source area, upper course, middle course and lower course, respectively, of these brook systems. Only semi-natural plant communities without excessive application of manure or artificial fertilizers have been considered. These figures, therefore, give an impression of the natural semi-landscape around 1900, if one omits the more natural brook- and fen communities.

Grootjans (1980) emphasized hay-making as being the controlling anthropogenic factor, but grazing was also important. Comparing the detailed maps around 1810 with grassland subdivision in hay-fields and pastures (Fig. III.23) with soil maps one finds that pastures were mostly located on soils with a peat layer less than 1 m thick, wereas hay-fields also occurred on wet soils often with a continuous seepage flow and with a peat layer over 1 m thick (Fig. III.24). The first cut was done at the end of June or the beginning of July and the second cut in August. The aftermath was grazed from September onwards. Hay-making by scythe was still prac-

Table III.5. Presence of macro-fossils from Bronze Age, Iron Age, Roman period and Middle ages settlements indicating the occurrence of anthropogeneous grassland communities according to the recent syntaxonomy (character species of the class *Molinio-Arrhenatheretea,* the order *Molinietalia* and alliances *Filipendulion* (F) and *Calthion palustris* (Ca) and the order *Arrhenatheretalia* and alliances *Arrhenatherion elatioris* (A) and *Cynosurion cristati* (Cy). Assessments 1: United Kindgom (Greig, 1984), 2: Rheinland, Germany (Knörzer, 1975), 3: Gasselte, The Netherlands (Van Zeist & Palfenier-Vegter, 1979), 4: Odoorn en Pesse, respectively, The Netherlands (Van Zeist, et al. 1986).

	Period	Bronze Age	Iron Age		Roman period		Middle Ages				
	Assessment	1	1	2	1	2	1	2	3	4	4
	Century								9-12	7	7-13
	Molinio-Arrhenatheretea										
	Holcus cf. lanatus	-	-	-	-	+	-	+	-	-	-
	Rumex acetosa	-	-	-	+	+	-	-	-	-	-
	Ranunculus cf. acris	+	+	-	+	+	+	+	+	-	+
	Cerastium fontanum	-	-	-	-	+	-	+	-	-	+
	Festuca cf. pratensis	-	-	-	-	+	-	+	-	-	-
	Festuca rubra	-	-	+	-	+	-	-	-	-	-
	Plantago lanceolata	-	-	+	+	+	-	+	+	+	+
	Trifolium pratense	-	-	-	-	+	-	-	-	+	-
	Prunella vulgaris	-	+	-	+	-	+	-	-	-	-
	Centaurea jacea	-	+	-	+	+	+	+	-	-	-
	Vicia cracca	-	-	-	-	-	-	-	+	+	+
	Molinietalia										
	Lysimachia vulgaris	-	-	-	-	+	-	+	-	-	-
	Angelica sylvestris	-	-	-	-	-	-	+	-	-	-
	Juncus effusus	-	-	-	-	+	-	+	-	-	-
	Valeriana dioica	-	-	-	-	+	-	-	-	-	-
	Rhinanthus cf. minor	-	+	-	+	+	+	-	+	+	+
F	Filipendula ulmaria	-	+	-	-	+	+	+	-	-	+
F	Hypericum tetrapterum	-	-	-	-	+	-	+	-	-	-
F	Lythrum salicaria	-	-	-	-	+	-	-	-	-	-
F	Valeriana officinalis	-	-	-	-	+	-	+	-	-	-
Ca	Lychnis flos-cuculi	-	-	-	-	+	-	+	-	-	+
Ca	Caltha palustris	-	+	-	-	+	+	+	-	-	-
Ca	Senecio aquaticus	-	-	-	-	-	-	+	-	-	-
Ca	Scirpus sylvaticus	-	-	-	-	+	-	+	-	-	-
Ca	Bromus cf. racemosus	-	-	+	-	+	-	-	-	-	-
	Arrhenatheretalia										
	Achillea millefolium	-	-	-	-	+	+	+	-	-	+
	Trifolium dubium	-	-	-	-	+	-	-	-	-	-
	Carum carvi	-	-	-	-	+	-	+	-	-	-
	Leucanthemum vulgare	-	+	-	-	+	-	+	-	+	+
	Stellaria graminea	-	-	-	-	-	-	-	+	+	+
	Dactylis glomerata	-	-	-	-	+	+	+	-	-	-
	Taraxacum cf. vulgare	+	+	-	+	+	-	-	-	-	-
	Phleum cf. pratense	-	-	+	-	+	-	-	-	-	-
A	Anthriscus sylvestris	-	-	-	-	+	-	-	-	-	-
A	Heracleum sphondylium	-	-	-	+	+	-	-	-	-	-
A	Galium mollugo	-	-	-	-	+	-	+	-	-	-
A	Knautia arvensis	-	-	-	-	-	-	+	+	+	+
Cy	Bellis perennis	-	-	-	+	-	+	+	-	-	-
Cy	Cynosurus cristatus	-	-	-	-	+	-	-	-	-	-
Cy	Leontodon cf. autumnalis	-	+	-	+	+	+	+	-	-	+
Cy	Hypochaeris cf. radicata	-	+	-	+	-	-	-	-	-	+

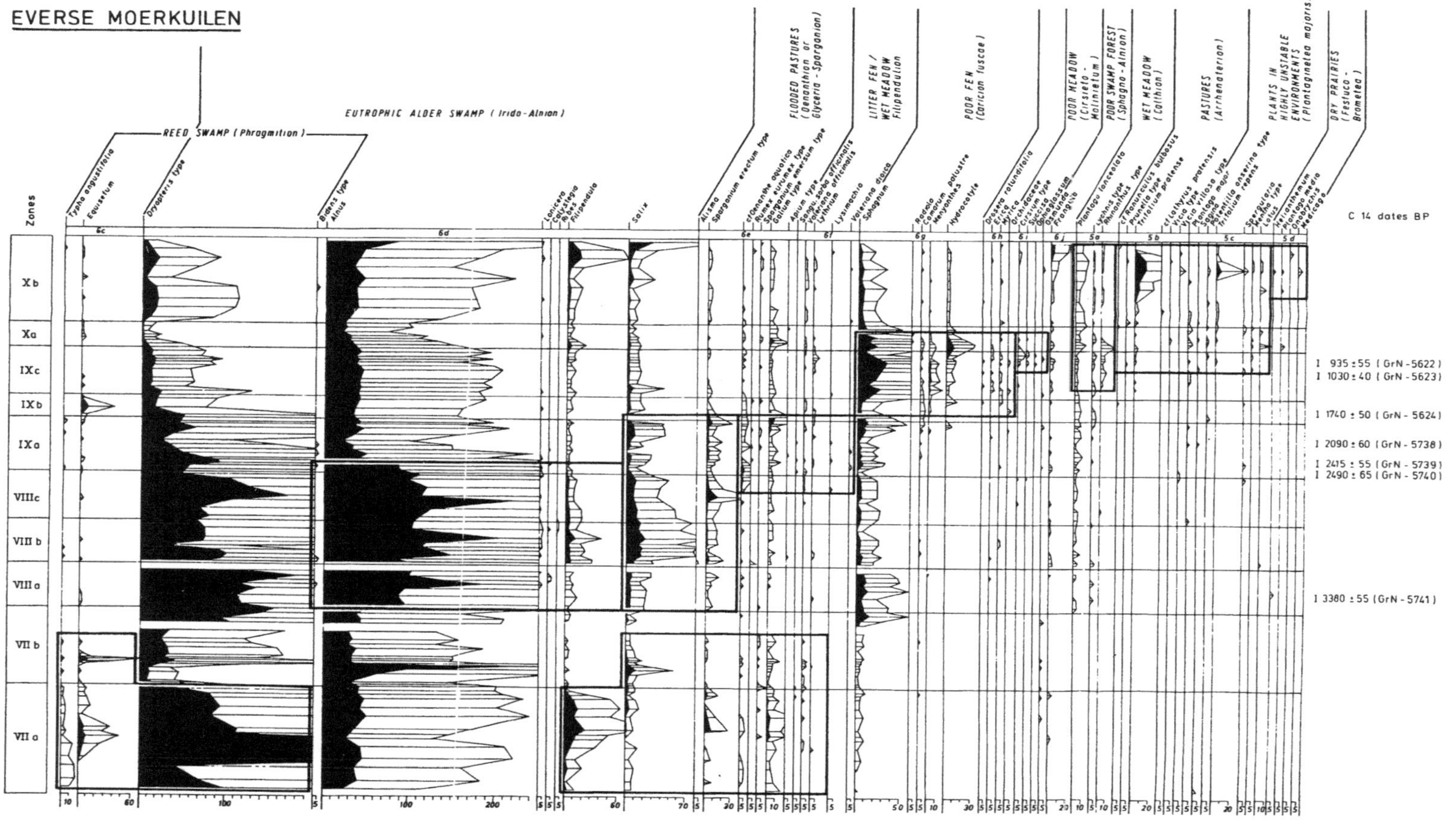

Fig. III.16. Pollen diagram of a brook valley in the southern part of the Netherlands, showing local pollengroups in different periods (after Janssen, 1979).

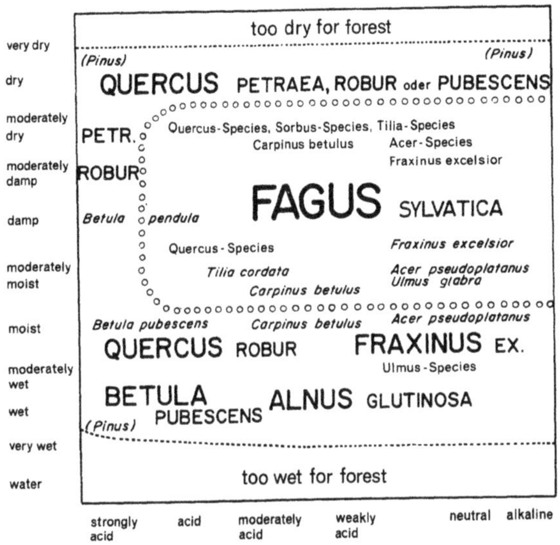

Fig. III.17. Diagram representing the ecology of forest-building three species in Central Europe with respect to groundwater table and soil acidity. The importance of species is indicated by the size of the characters. The position of *Fagus sylvatica* is indicated by the line consisting of circles. (after Ellenberg, 1978).

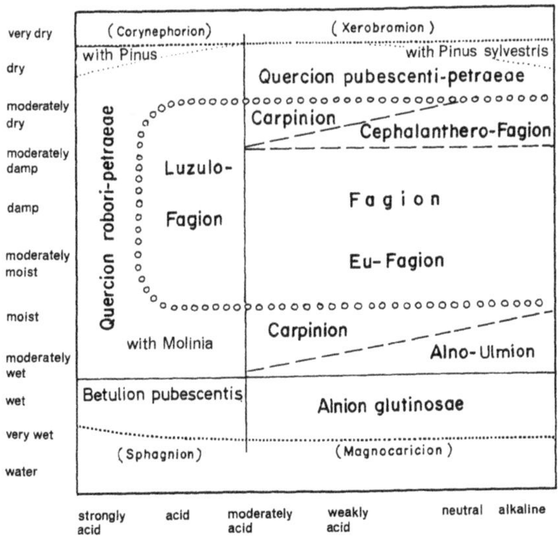

Fig. III.18. Diagram representing the synecology of forest communities (alliances) in Central Europe. Compare Fig. III.17 (after Ellenberg, 1978).

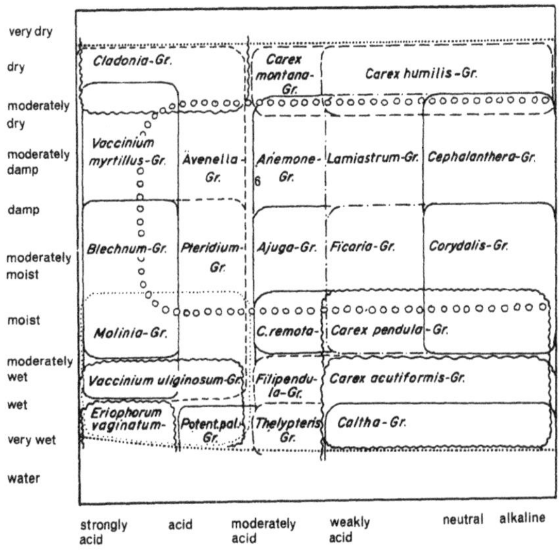

Fig. III.19. Diagram representing the synecology of forest understory communities in Central Europe. Compare Fig. III.17 (after Ellenberg, 1978).

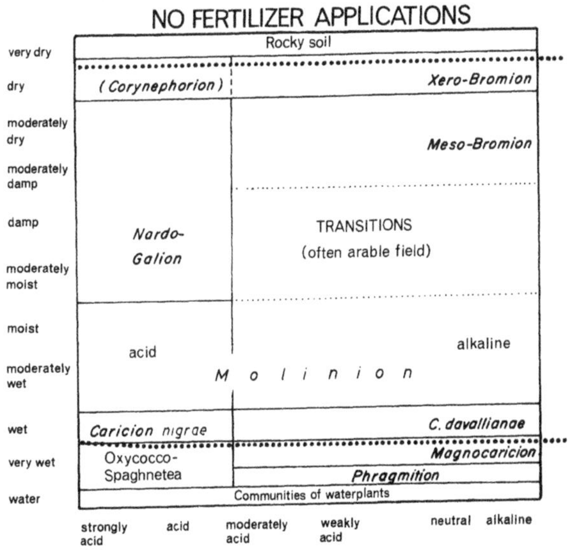

Fig. III.20. Diagram representing the synecology of meadow communities (alliances) which receive no fertilizer applications. Compare Fig. III.17 (after Ellenberg, 1978).

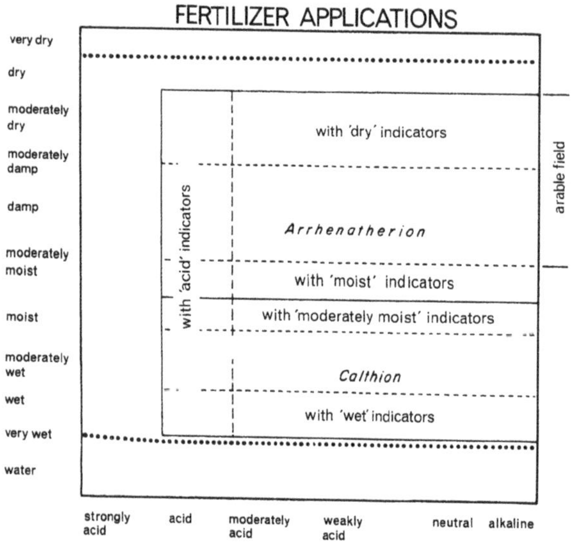

Fig. III.21. Diagram representing the synecology of meadow communities (alliances) with fertilizer applications. Compare Fig. III.17 (after Ellenberg, 1978).

tised by small farmers around 1925 but farmers having more land with often better drained meadows used hay-making machinery (Edelman, 1974). In the river valley of the reserve 'Zwarte Beek' in northern Belgium, which is comparable to the Drentsche A reserve, all hay-making was done by scythe until c. 1960 when machines were introduced (Burny, 1986).

The meadows in the river valleys consisted of 'stroomland' (20%) which was often inundated (Photo III.6) and of 'bovenland' (80%) which was drier but received no nutrients from the inundation water (Bieleman, 1987).

Probably from the late middle ages until the last century the farmers used to inundate the hay-meadows through a system of small ditches using barrages in order to catch the brook-mud and relatively fertile water for natural manure. The barrages were opened at the latest on the first of June,

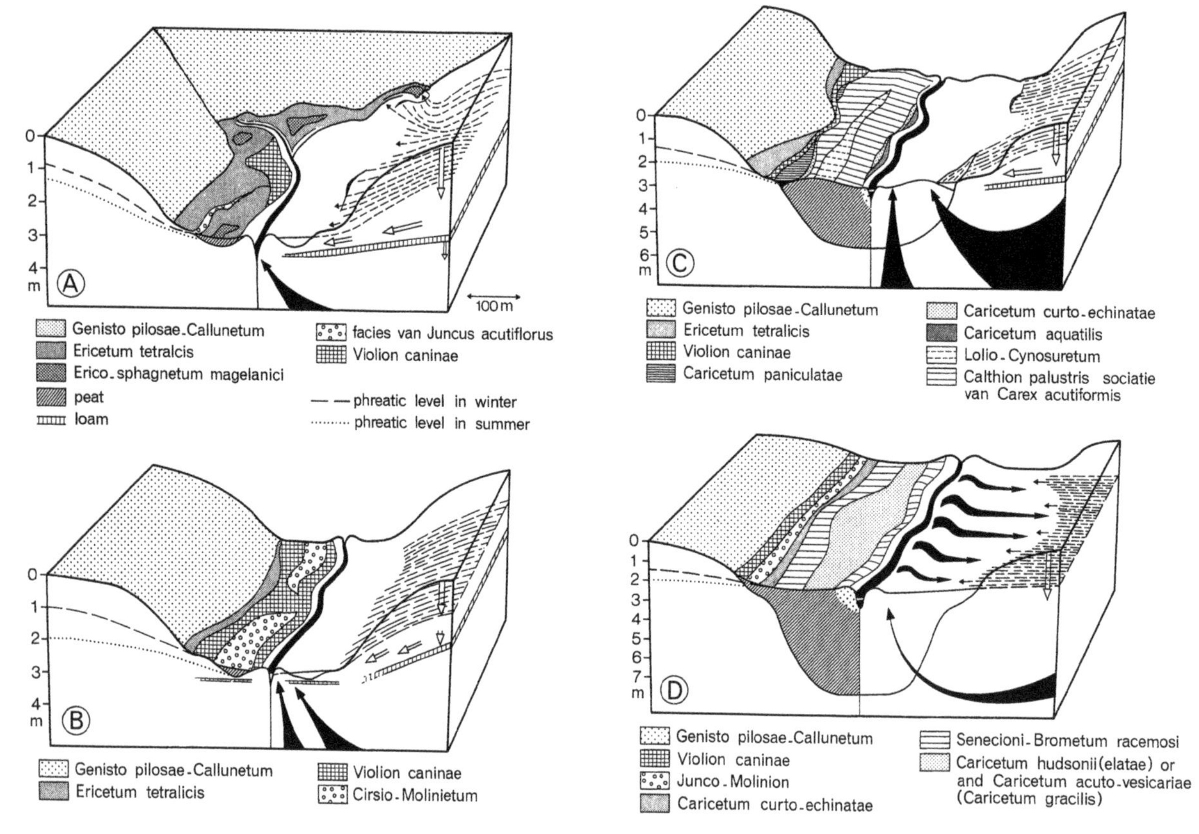

Fig. III.22A–D. Cross section of a river valley on the Drenthe Plateau showing gradients of plant communities (associations or alliances) on the left and hydrological situations on the right in A: source area, B: upper course, C: middle course and D: lower course. White arrows indicate the direction of atmocline groundwater. Black arrows indicate the direction of lithocline groundwater or flood water from the stream (after Grootjans, 1980).

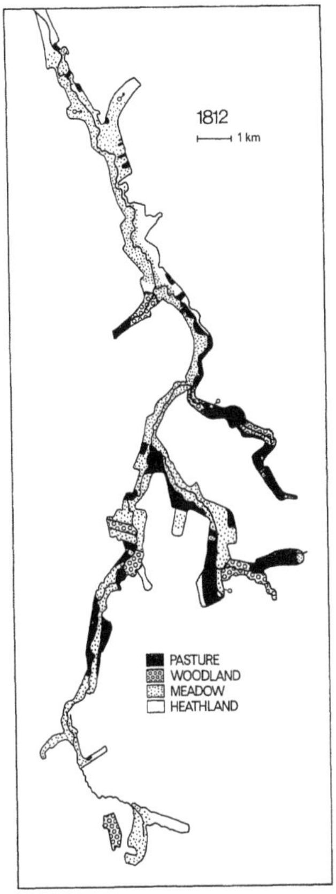

Fig. III.23. Location of pasture (paquis), meadow (pré), woodland and heathland in 1812, projected into the recent grassland area of the nature reserve Stroomdallandschap Drentsche A.

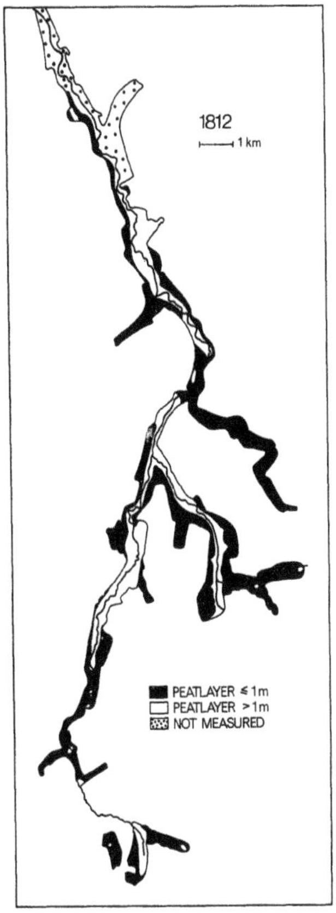

Fig. III.24. Location of soils with a peat layer less than 1 m thick and soils with continuous seepage flow and a peat layer over 1 m thick, projected in the grassland area of the nature reserve Stroomdallandschap Drentsche A (after Cultuurtechnische Dienst, 1958).

otherwise no hay-making was possible in mid-July (Alstorphius Grevelink, 1844; Cancrinus, 1950; Edelman, 1974). This system was even practised until the 1930's in the river valleys of the 'esdorpen' landscape in northern Belgium (Burny, 1986). Manure was applied from the middle of the 19th century onwards and at the end of the past century already some application of artificial fertilizer took place (Edelman, 1974).

By inclusing grazing the schemes published by Grootjans (1980) should be extended by incorporating *Arrhenatherion* and *Lolio-Potentillion* communities at the transition of sand and peat layer. If permanent grazing was involved scrub communities of the *Salicion cinereae* also occurred and could cover c. 50% of the pasture. In the 19th century the animals, therefore, wore bells so that the farmers could find them (Garming, 1955; Edelman, 1974).

Some type of permanent grazing with litter accumulation remained until the 1930's on an area of about 50% of Dutch pastures ('t Hart, 1943). Rotational grazing from that time onwards became adopted gradually resulting in intensively utilized pastures.

III.B.12. Explanation of species diversity in the Drentsche A valley

In this section the high species diversity on the Drenthe Plateau and the intersecting valleys will be discussed with references to some theoretical considerations mentioned in Chapter I. Viz (i) spatial heterogeneity, (ii) gradients concerning both abiotic factors and human impact, and (iii) continuity of human impact.

Photo III.6. Lower course area of the Drentsche A river valley featuring the Lammertshem study area. The irregular lots adjacent to the brook are inundated in the winter period. A dike protects, therefore, the large lots (in the lower right of the photo) resulting from a reallotment program in the 1970's, whereas the ditch drains them.

(i) The species diversity in the Drentsche A valley is very high by Dutch standards: 608 species of vascular plants occur in an area of 87 km^2, 3 areas of 1 km^2 containing >300 and 15 areas of 1 km^2 >250 species (Grootjans, 1985). This implies that the number of species found is more than three times that which would be expected on the average from such an area in the Dutch landscape (Figure III.25). Species diversity is low in the infiltration areas. Despite the fact that the lower course reveals a high diversity, the middle course, with unaffected seepage areas, harbours by far the most species (and communities). This must be attributed to large amounts of seepage water and a high Ca^{2+} level in the discharging groundwater, because areas severely affected by deep drainage and other amelioration works are much poorer in species (and communities) and these grid cells of 1 km^2 contain <200 species (Grootjans, 1985).

A number of rare phreatophytic species is found in the Drentsche A valley. These species are associated with high groundwater levels (phreatic levels) (Londo, 1988). Very rare species were defined as those occurring in 1 to 29 areas of the 5 × 5 km grid involving in total 1677 grid cells for the whole of the Netherlands, rare species in 30 to 79 areas in the Netherlands. In the Drentsche A valley 33 very rare and 33 rare species are found. They are not distributed randomly in the valley area, but are found in particular in areas with pronounced seepage, which are concentrated in the middle course (Grootjans, 1985).

It was argued in section III.B.1 that the complicated and interrelated geomorphological conditions, soils and hydrological conditions represent an enormous spatial heterogeneity: source, upper course, middle course and lower course areas. Moreover, gradients of abiotic conditions are found everywhere perpendicular to the brook.

(ii) Abiotic gradients are abundant in north Drenthe. It is one of the areas in the Netherlands with great botanical variety. According to Van Leeuwen in Westhoff et al. (1970) this is due to (i) bordering of the oligotrophic Pleistocene hinterland on the eutrophic Holocene coastal area, together forming a large-scale gradient, and (ii) the

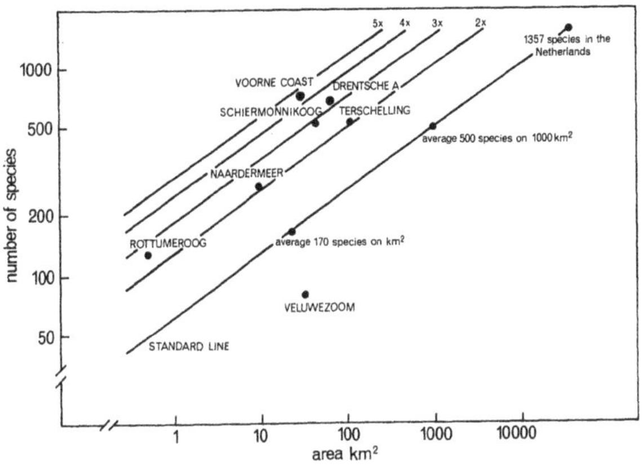

Fig. III.25. The number of species in the nature reserve Stroomdallandschap Drentsche A and other nature reserves as compared to the standard line for the species-area relationship in the Dutch flora (after Grootjans, 1985).

intersection of eutrophic brooks in the oligotrophic Pleistocene Plateau featuring small-scale gradients (Figure III.26). The large-scale gradient is of course very stable, but the small-scale gradients suffer from destruction. It can be deduced from the maps in Figure III.7 that the brook valley in the Drentsche A catchment area was bordered by the oligotrophic hinterland for c. 85% of its length in 1850, for c. 25% in 1930 and for c. 5% in 1970. The reclamation of the heathland areas influences the quantity and the quality of the infiltrated water owing to agricultural activities and, therefore, disturbs the hydrological balance and hence the gradient (Grootjans, 1985). Nevertheless, north Drenthe is one of the highlights (still) with respect to the *Carex* flora in the Netherlands. This turned out to be very representative for the whole of the Dutch flora (De Bruijn, 1980).

Human impact gradients around settlements can be expected to occur as a result of economical management especially with respect to the costs of carriage of agricultural products. Von Thünen (1842) described a theoretical agricultural system on boulder clay soils in the present DDR, where different types of agricultural practices occurred in concentric circles (German: 'Thünische Ringe') with the village as the centre. The circles from the centre outwards included: horticulture and cattle on the farmstead allotment, forestry for firewood, use on the farm and for building, arable fields, cattle or sheep-grazing for manure production, woodland for hunting (see also Bouwer, 1985; Bieleman, 1987).

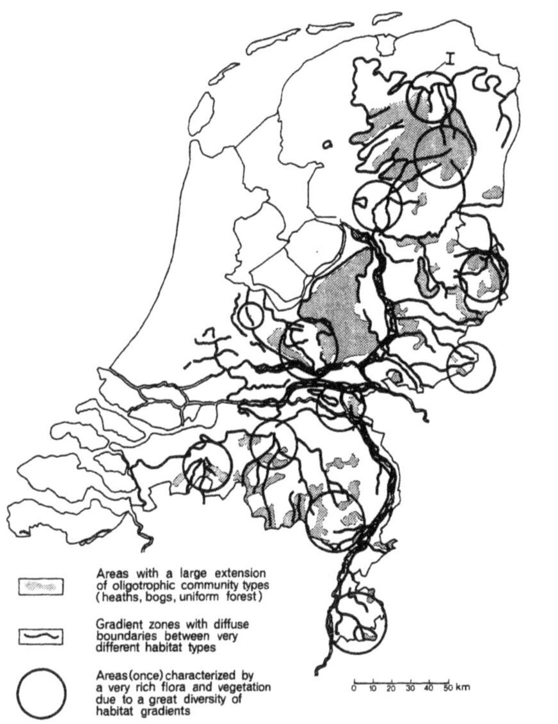

Fig. III.26. The ecological basis of botanical variety in the Netherlands (outside the coastal region) and the position of north Drenthe (I) (after Westhoff et al., 1970).

Some indications of human impact gradients can be found e.g. as a result of a grassland study in the southwestern part of the Netherlands in 1953, which probably reveals a general trend. Soil chemical analyses showed lower phosphate- and potassium-conditions when the allotments were further away from the farmstead (Vermeulen, 1954) (Fig. III.27). Similar gradients of human impact were found elsewhere from grassland (Koopmans, 1960; Pieters, 1971) and arable fields (Draaisma, 1958). These gradients disappeared after re-allotment programmes because they involved the removal of farmsteads into the former open rural landscape. The village territories in the Drentsche A valley always had the 'es' adjacent to the village with some woodland fringing the 'es' which was in turn surrounded by heathland. The grassland was always very close to the villages and often adjacent to the 'es', but the position of meadow and pasture depended more on the type of soil in the valley than on the distance to the village.

(iii) The continuity of human impact cannot al-

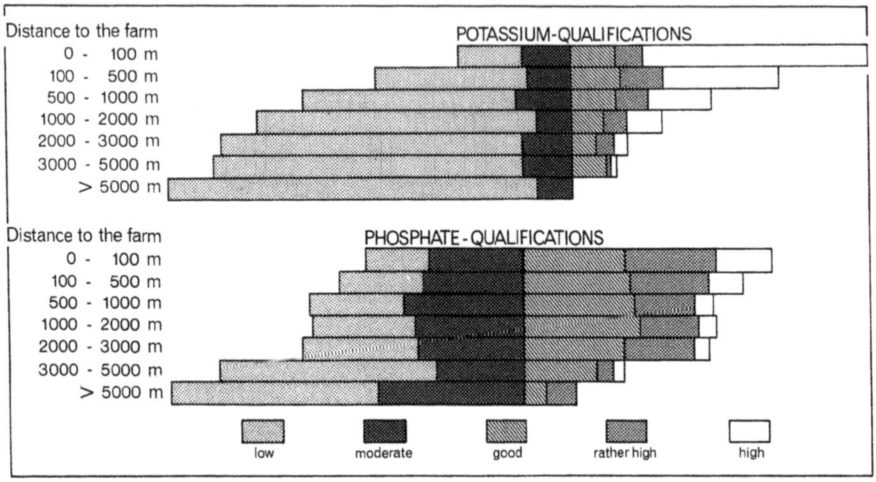

Fig. III.27. A subdivision of the phosphate- and potassium qualification into five groups as related to the distance of 2671 allotments from the farms (from <100 m to >5000 m) (after Vermeulen, 1954).

ways be traced on the individual allotment level. Around 1650 80% of the valley was reclaimed to meadows and hay-making had probably been practiced for 300 years. The valley fringes were reclaimed later and often became hay-pasture or pasture. Until 1700 sods were cut from the valley grasslands (Bieleman, 1987). In 1810 several tens of ha of the valley were still unreclaimed. Bog-ore had accumulated due to the continuous seepage flow and it was mined in the Drenthian valleys from 1870 to 1960 (Edelman, 1974; Booy, 1986) and in similar Belgian valleys (Burny, 1986). Surface irregularities were leveled and many lots were covered with sand (Edelman, 1974) in the 20th century. Agricultural grassland surveys made in the early 1960's revealed poor agricultural values for the peaty parts of the lots and a moderate value for the drier sandy parts (Proefstation voor Akker- en Weidebouw, 1965, 1966). These surveys showed that all former meadows had changed into hay-pastures, probably thanks to superficial drainage. Schimmel (1955) described parts of the valley as being strongly influenced by agricultural practices.

Generally speaking it can be concluded that human impact lasted for over 300 years, but that this impact was far from constant on the individual allotment level. Wells et al. (1976) reached similar conclusions for a chalk grassland complex in the United Kingdom. The traditional view that most areas of present-day chalk grasslands were used as sheepwalks for many centuries was not supported by studies of land-use history, vegetation and soils on the Porton Ranges. Written evidence and the plough marks visible on air photographs indicate that over 75% and probably more of the Ranges were cultivated. Only few examples of longlasting continuous land use are known such as the reserves 'Haselüner Kuhweide' (Tüxen, 1974) and 'Borkener Pardies' (Burrichter et al., 1980) in northwest Germany which both happened to be areas used for common grazing since medieval times.

Despite the often irregular and short lived human impact a high species diversity could be reached in the Drentsche A valley which can to a great deal be attributed to the afore-mentioned complex hydrological conditions. In 1850 one of the present highlight meadows of the nature reserve was partly woodland and the soil was ploughed at the beginning of the 20th century, but it features 40 species · m^{-2} nowadays. The species diversity should, therefore, be primarily attributed to the complex hydrological conditions and secondary to the human impact, because the hydrological conditions set limits to the intensity of agricultural practices.

III.B.13. Possibilities for restoration

Since the nature reserve Stroomdallandschap

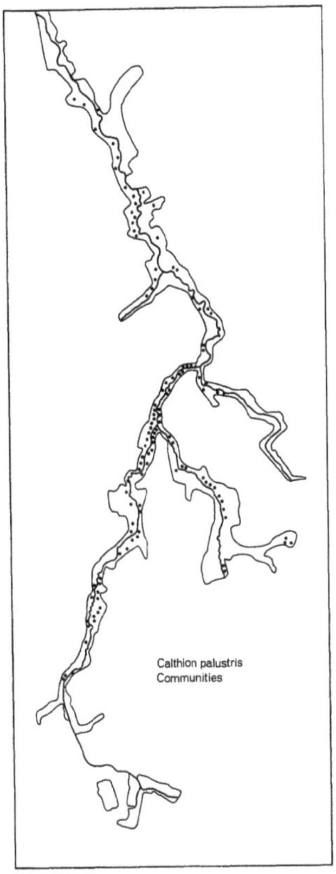

Fig. III.28. Location of *Calthion palustris* communities in the nature reserve Stroomdallandschap Drentsche A (after Everts et al., 1984).

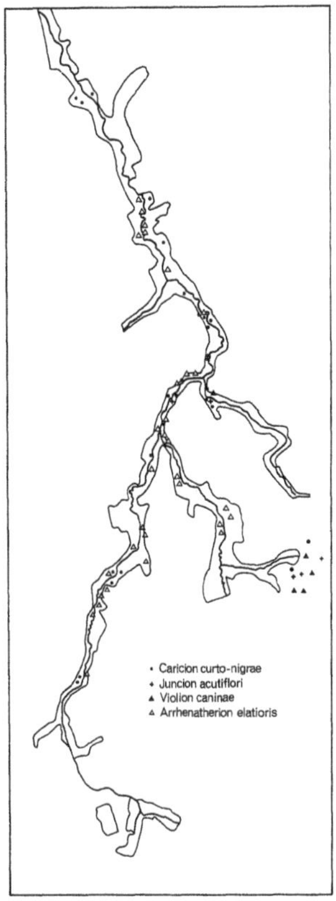

Fig. III.29. Location of *Caricion curto-nigrae, Juncion acutiflori, Violion caninae* and *Arrhenatherion elatioris* communities in the nature reserve Stroomdallandschap Drentsche A (after Everts et al., 1984).

Drentsche A (founded in 1965) is surrounded by re-allotment areas, the possibilities for restoration might be limited. This is particularly true in areas severely affected by deep drainage and other amelioration works which result in the disappearance of the hydrological and trophic gradients (Grootjans, 1985). Since the main hydrological character still persists the potential value of the nature reserve is considered high and could eventually result in the lay-out of plant communities shown in Fig. III.22A–D.

Although the nature reserve was mapped in the 1960's (Proefstation voor de Akker- en Weidebouw, 1965; 1966) it was done for agricultural purposes. Therefore, no adequate vegetation map existed when the reserve was established in 1965. In 1982 Everts et al. (1984) completed a vegetation survey at the allotment level showing the distribution of plant communities. Probably owing to the (previous) application of fertilizers the *Lolio-Potentillion* and to a somewhat lesser extent *Cynosurion* communities were found all over the reserve. *Calthion* communities (Fig. III.28) occurred in many lots, particularly in the middle course area, but many gaps were found in their distribution. The distribution of *Caricion curto-nigrae, Juncion acutiflori, Arrhenatherion elatioris* and *Violion caninae* (Fig. III.29) is very patchy. Although some character species of these syntaxa are widely distributed over the nature reserve, others reveal a very patchy pattern. The occurrence of several species is limited to the edges of ditches. The island character of the remaining heathland areas has already been shown (Fig. III.7).

The conclusion from the above must be that the success of restoration management in the study area not only depends on the abiotic environmental conditions but also on the dispersal capacity of those species still occurring.

III.C. The nature reserve Stroomdallandschap Drentsche A

III.C.1. Acquirement

The nature reserve Stroomdallandschap Drentsche A was established in 1965. It was the result of re-allotment programmes for northern Drenthe in which the State wanted to save one of the last more or less complete valley systems of the Drenthe Plateau. The nature reserve will eventually measure approximately 3500 ha, involving 1500 ha of grassland area. From 1965 onwards the State acquired lots (about 200 ha of grassland) despite the agricultural organizations advising their members not to sell until agreements were reached about reserve boundaries and prices. In 1969, however, an agreement was made based on ground valuations by both the State and the agricultural organizations as well as an extra Dfl. 4,000 per ha to make the farmers more interested. It meant prices varying from Dfl. 5,250 to Dfl. 12,000 per ha depending on the agricultural value of the allotments (Ernst, 1976). The agreement resulted in a rapid increase of acquired lots and in 1980 c. 840 ha of grassland were owned by the State (Fig. III.30).

Up until 1970 allotments were acquired particularly in the very wet lower course. Their agricultural value was very low, and they were partly abandoned. Many lots were acquired in the middle course between 1970 and 1976. From 1976 onwards scattered lots were acquired. It seems from Figure III.31 that management by the State Forestry Commission could be practiced in extended units, but in fact the management units are very dispersed due to long-term lease contracts.

III.C.2. Nature management practices

It is assumed that the grasslands adjacent to the brook revealed the maximal species and community diversity around 1900. The aim of the manage-

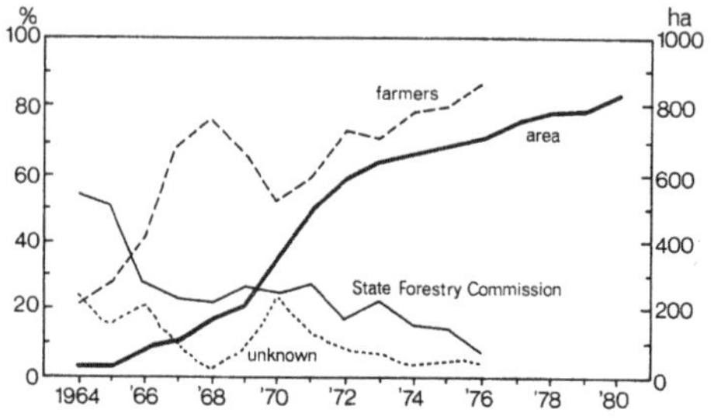

Fig. III.30. Total area of acquired valley grasslands in the nature reserve Stroomdallandschap Drentsche A until 1980 and the percentages managed by farmers and by the State Forestry Commission in each year.

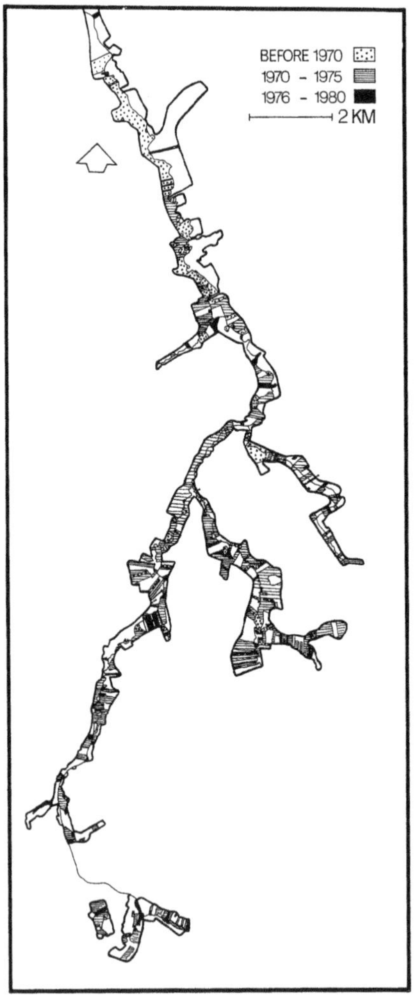

Fig. III.31. Acquirements of valley grasslands by the State Forestry Commission in the nature reserve Stroomdallandschap Drentsche A in the periods 1965 to 1970, 1970 to 1975 and 1976 to 1980.

ment is, therefore, to restore these qualities which meant no application of fertilizers of herbicides and no lowering of the groundwater table. Since fertilizers have often been applied for several decades the management's aim is to impoverish the soil mainly by hay-making, although sometimes by grazing: this is known as impoverishing management. Hay-making and grazing with the application of fertilizers is also done due to the long-term lease contracts: this type is called non-impoverishing management. Abandoning and mowing without the removal of the cuttings are also included in this latter management type because no nutrients are removed.

Impoverishing management implies hay-making once or twice a year, hay-making and grazing the aftermath or only grazing. During the first nine years of the reserve hay-making was the only important practice (Fig. III.32), but from 1973 onwards, the aftermath gradually became more grazed and only grazing became an important management practice. It appears from Fig. III.30 that it was only the first few years that the highest proportion of managed areas was that which was controlled by the State Forestry Commission. Within a few years the area to be managed became too large for the technical and personal equipment of the State Forestry Commission. This problem was solved by involving the farmers adjacent to the nature reserve who became more and more interested in a cheap supplement to the ordinary forage. These farmers both collect the cuttings left by the State Forestry Commission and buy hay on the stem. Since the farmers get their hay from other allotments year by year illegal application of fertilizers is effectively prevented, because they will not run the risk of applying fertilizers for an other farmer.

Non-impoverishing management mainly implied the abandoning of the reserve for the first years (Fig. III.33). From 1967 onwards, however, more and more allotments with long-term lease contracts were acquired implying the continuation of normal

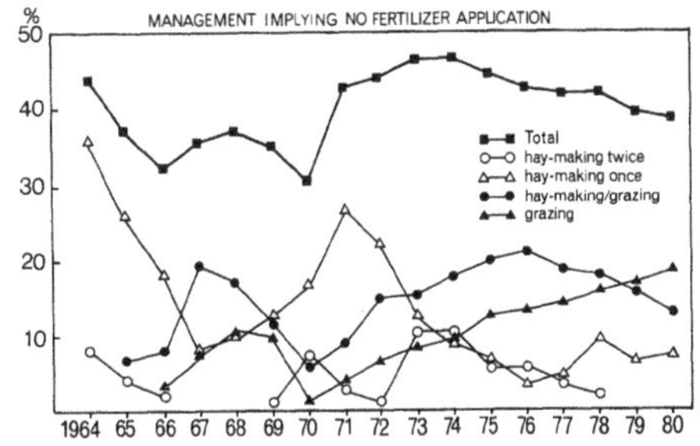

Fig. III.32. Percentages of the acquired valley grassland area in the nature reserve Stroomdallandschap Drentsche A with a management implying no fertilizer applications subdivided into hay-making twice a year (open circles), hay-making once a year (open triangles), hay-making and grazing the aftermath (closed circles) and grazing (closed triangles) until 1980.

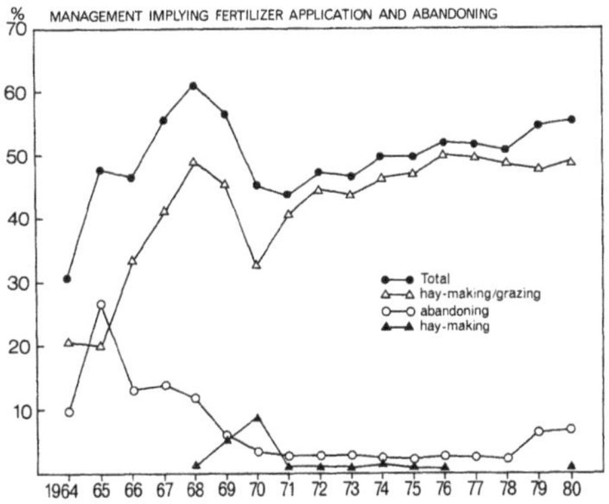

Fig. III.33. Percentages of the acquired valley grassland area in the nature reserve Stroomdallandschap Drentsche A with a management implying abandoning (open circles) and fertilizer applications subdivided into hay-pasture (open triangle) and hay-making (closed triangles).

agricultural practices for a maximum of six years. Due to the gradual termination of these contracts the non-impoverishing management proportion will decrease in the near future. Moreover, the possibilities of large management units will then increase.

An important conclusion of this section is the large involvement of local farmers in management practices, concordant with the agricultural history of the area.

CHAPTER IV

The Schiermonnikoog salt marsh

IV.A. Introduction

In this chapter the second area in which restoration experiments of former species-rich grassland took place is described. It is found in the coastal nature reserve 'Oosterkwelder' which is part of the salt marsh of Schiermonnikoog, one of the Dutch West Frisian Islands (53° 30′ N, 6° 10′ E). Its utilization history only goes back to the late middle ages and was deduced from historical geographical data, old archives and pollen analytical data.

Firstly the geomorphological and the hydrological conditions will be described and then the settlement on Schiermonnikoog from c. 1300 onwards and the land use of the salt marshes. Particular attention has been paid to the eastward shift of the island of Werveroog, which only later acquired the present name of Schiermonnikoog.

The vegetation development and species diversity will be discussed referring to spatial heterogeneity, gradients and human impact.

The acquirement of the nature reserve and finally the application of management practices will also be covered.

IV.B. Land use of Schiermonnikoog

IV.B.1. Geomorphological and hydrological conditions and vegetation of island salt marshes

The rapid rise in sea level, after the last glacial period, caused sand movement towards the coast resulting in bare beach barriers between 3000 and 1000 B.C. When the rise in sea level decreased the vegetated old dunes developed between 1000 B.C. and 0. The old dunes maintained themselves until c. 1200 (Fig. IV.1) (Veenstra, 1980). A climatic change and a transgression period caused much coastal erosion and the transported sand formed the high young dunes which often covered the older ones (Jelgersma et al., 1970; Zagwijn, 1974) (Fig. IV.2). The geological conditions of the island of Schiermonnikoog are shown in Fig. IV.3.

Vast brackish swamps and later on peat-bogs developed behind this coastal defense. This coast type is known as tidal flat and barrier-island coast (Bird, 1969). As a result of the continual rise in the sea level, the early beach barriers and later dune ridges were pushed back as far as their present location, which was reached over a period of a thousand years (Fig. IV.1) (Veenstra, 1980). Moreover, the peat-bogs became inundated and eroded during the late middle ages during the phases in development of the Wadden Sea and the present West Frisian Islands. Salt marshes could develop behind the dune ridges.

The tidal flat and barrier-island coasts are characterized by a waterbasin parallel to the coastline and connected with the open sea through inlets. The run-off from the mainland can be rather small or even non existent, but tidal action is quite large. The mud flats and marshes are situated parallel to the coastline (Veenstra, 1980).

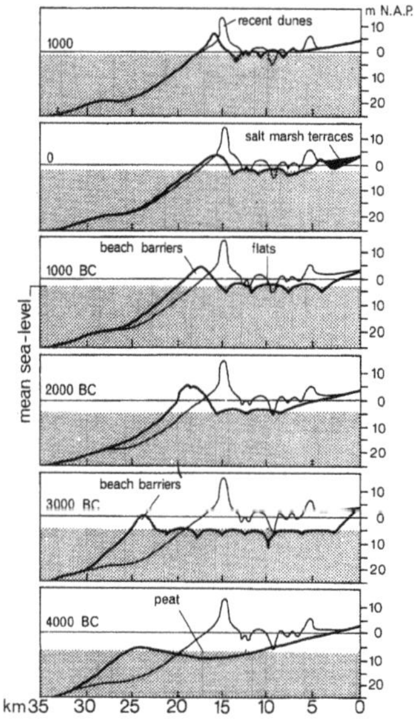

Fig. IV.1. Evolution of the Wadden Sea area between 400 B.C. and 1000. NAP = Dutch Ordnance Level (After Veenstra, 1980).

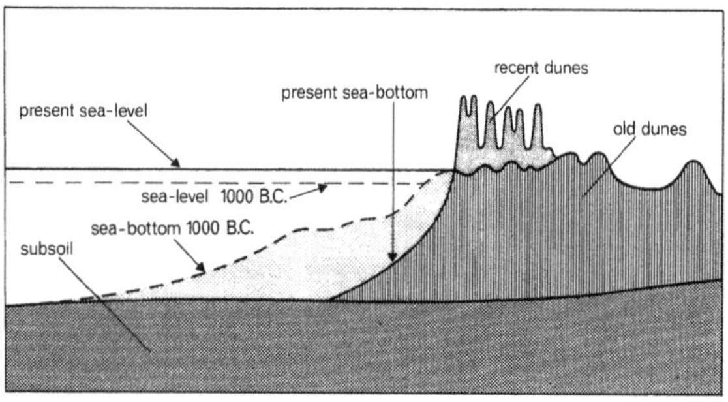

Fig. IV.2. Sea-level and sea-bottom along the Dutch coast between 1000 B.C. and present. After 1100 the sea-bottom eroded and young dunes were built from the released sand (after Zagwijn, 1974).

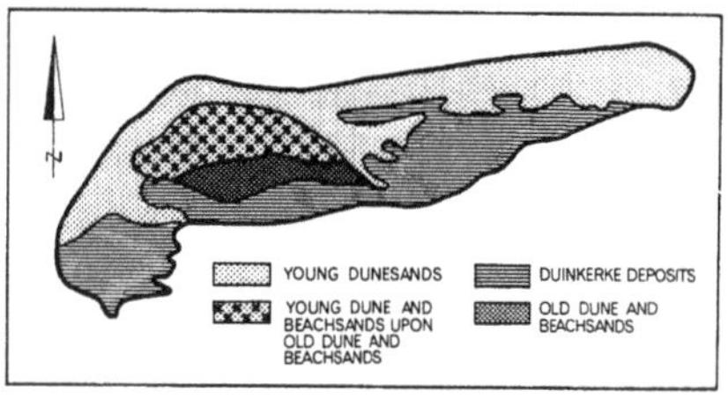

Fig. IV.3. Geological map of the Island of Schiermonnikoog (after Rijks Geologische Dienst, 1977 in Bakker et al., 1979).

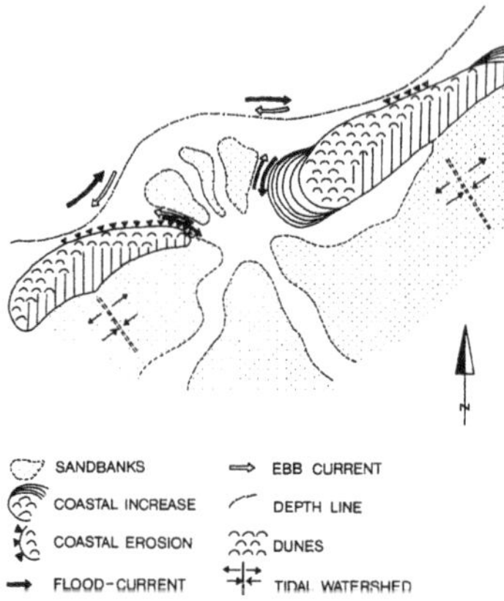

Fig. IV.4. Scheme of coastal forms and processes of a tidal flat and barrier-island coast (after Bakker et al., 1979).

An outer and inner delta of sand banks were formed in the tidal inlets between the islands. Due to the prevailing direction of water current these sand banks gradually moved eastwards, intermittently causing accretion and erosion of the island's western point (Fig. IV.4) (Bakker et al. 1979).

In c. 1300 the island of Schiermonnikoog was first mentioned in historical documents and was then called Warvteroe, Wervneraeghe or Werveroog (see section IV.B.2 and 3). It was then five or six km west of the present position, and it was possible to walk across the tidal watershed from Friesland to the island (Fig. IV.5) (De Haan et al., 1983). The island of Basseaeghe, at the present position of Schiermonnikoog, moved eastwards. It is very likely that this tidal watershed was broken during the first half of the fourteenth century, due to frequent winter gales. Moreover, a slight transgression occurred in the same period (Waterbolk, 1976). The burst of the tidal watershed of Werveroog considerably enlarged the catchment area of the Scudbalcke inlet, since the Lauwerszee and its inland tributaries were at that time not embanked. The search for a new balance resulted in an eastward movement of the island and a new tidal watershed connecting it with the Groningen mainland. Initially the average annual erosion of the western point of the island amounted to twelve to fifteen m between 1350 and 1550 (De Haan et al., 1983), whereas it was later reduced to 7 m between 1550 and 1920 (Isbary, 1936). The eastward movement of the island is reflected in the $CaCO_3$ content of the dune

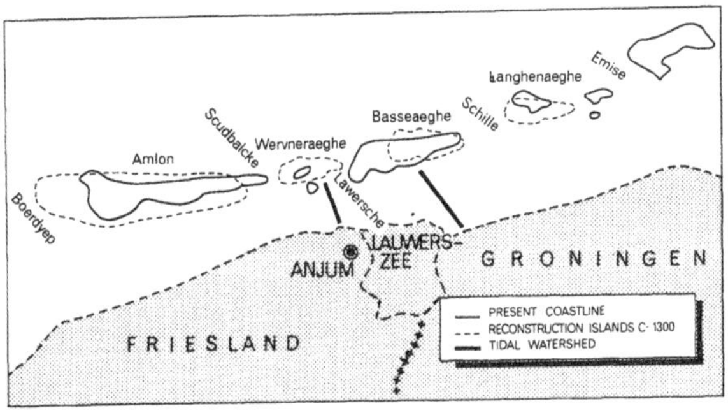

Fig. IV.5. Position of some Dutch Wadden Islands c. 1300 with the tidal watershed to Groningen (after Reitsma & Bakker, 1986). Indications for sailors in order to reach the Frisian mainland around 1300 only mentioned the Lawersche, apparently the Scudbalcke was not important for through traffic: it was a blind road.

sand, namely, old dunes (1600) in the western part contain 0.1–0.9% $CaCO_3$, dunes (1885) in the middle part 0.5–1.0% $CaCO_3$ and young dunes (1920) in the eastern part 1.3–1.8% $CaCO_3$ (Leertouwer, 1966).

The catchment area of the inlet west of Schiermonnikoog was strongly diminished, after the seventeenth century, by the embankment of the estuaries of the Lauwerszee. This had a multiple effect. The water movements of the tidal inlet decreased resulting in the deposition of coarse material closer to the axis of the inlet with a subsequent contraction of the outer and inner delta. The inner delta sandbanks, which were moving in an easterly direction, joined up with a large sandbank south of Schiermonnikoog. The inner delta shift towards the axis of the tidal inlet initially replaced the Noorder Noorman channel towards the southwestern point of the island, with subsequent erosion of dunes and salt marshes. Later, during the period 1825–1840, the sandbanks grew together with the island resulting in the present day Rif (de Haan et al., 1983). The development of dune slack communities after the attachment of a sandbank c. 1960 from the outer delta was described by Joenje & Thalen (1968). The different positions of the island are compared in Fig. IV.6 (Reitsma & Bakker, 1986).

The reduction of the catchment area of the tidal inlet caused an eastward shift of the tidal watershed with a subsequent large growth of the eastern part

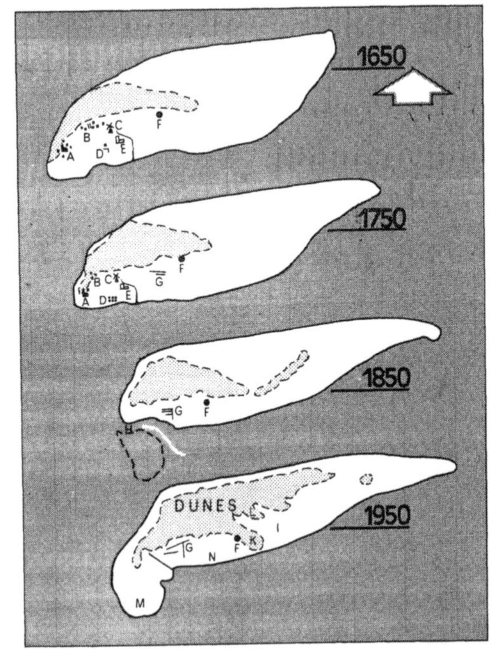

Fig. IV.6. Movement of the Island of Schiermonnikoog between 1650 and 1950. The maps are depicted in such a way that a ruler can indicate the position of each point in different periods. A = Westerburen, B = Oosterburen, C = Moolenburen, D = de Dompen, E = Binnendijken, F = Duck decoy, G = Hoochdorp (= later Schiermonnikoog), H = Noorder Noorman, I = Oosterkwelder, K = Kooiduinen, L = Kobbeduinen, M = Rif, N = Bancksp older (after Reitsma & Bakker, 1986).

of Schiermonnikoog. Until c. 1850, however, the Kooiduinen formed the coastal defense at the eastern part of the island enclosing the old salt marsh area south of it (Fig. IV.6). The whole island east of the Kooiduinen was a vast bare sand flat with a few low dunes. After c. 1870, a great deal of sand probably became available for the creation of the Steinsdúne (now Kobbeduinen) dune ridge, after the joining together of a sand bank from the outer delta and the island. The old salt marsh was embanked in 1860. From that period onwards our study area, the Oosterkwelder salt marsh, could develop and it is therefore a very young salt marsh.

Due to erosion by wave action resulting in a salt marsh cliff (Photo IV.1) no gradual transition is found between mud flats and marshes. During spring-tides and gales coarse sand is deposited on the cliff forming a levee, whereas clay is deposited on the lower depressions behind the levee (Fig.

Photo IV.1. The transition between the salt marsh and the tidal flats features a cliff (50 cm) in the Oosterkwelder salt marsh.

IV.7). The same process takes place, on a smaller scale, along the creeks where creekbank levees are created. The different soil profiles of creekbank levees and depressions in the Oosterkwelder salt marsh are shown in Fig. IV.8.

The present geomorphological conditions of the Oosterkwelder salt marsh is depicted in Fig. IV.9 featuring the daily inundated salt marsh in front of the cliff, the salt marsh levees, depressions and creeks. The rectangular low salt marsh areas adjacent to the dike indicate the clay pits excavated during the building of the dike in 1860. The higher dunes on the salt marsh are seldom flooded and contain a fresh water sack, which is not found in the lower dunes.

The height of the mud flats and lower salt marsh varies between 35 cm – MHW (Mean High Water) and over 200 cm + MHW for the low dunes which occur in the salt marsh. The frequency of flooding in the lower salt marsh is several hundreds of times a year, whereas the low dunes have a flooding chance of c. once a year (Fig. IV.10).

Salinity of the Wadden Sea water originating from the North Sea fluctuates during the year between 27 and 34‰ depending on the hinterland run-off. It is always lowest at the tidal watershed because of the distance from the tidal inlets (Postma, 1982).

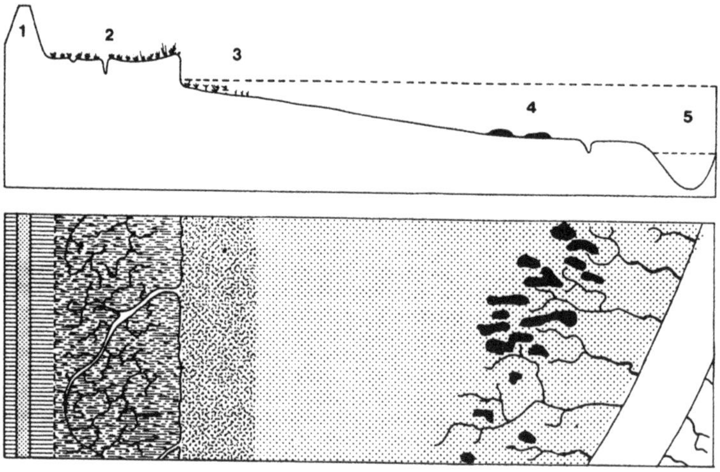

Fig. IV.7. Cross-section and map of Wadden Sea environment. 1 = sea dike or dunes, 2 = salt marsh with creeks and creek bank levees, 3 = high tidal flats with *Salicornia* and *Spartina,* 4 = mud flats with gullies and musselbeds, 5 = channel (after Veenstra, 1980).

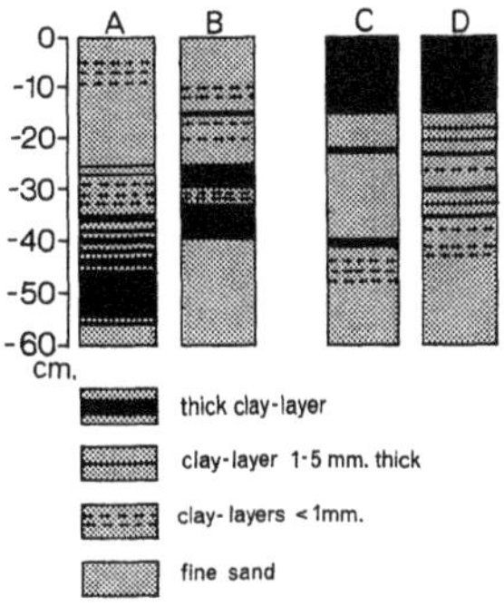

Fig. IV.8A–D. Soil profiles of creek bank levees (A, B) and the depression behind them (C, D) on the Oosterkwelder of Schiermonnikoog (after Bleuten, 1971).

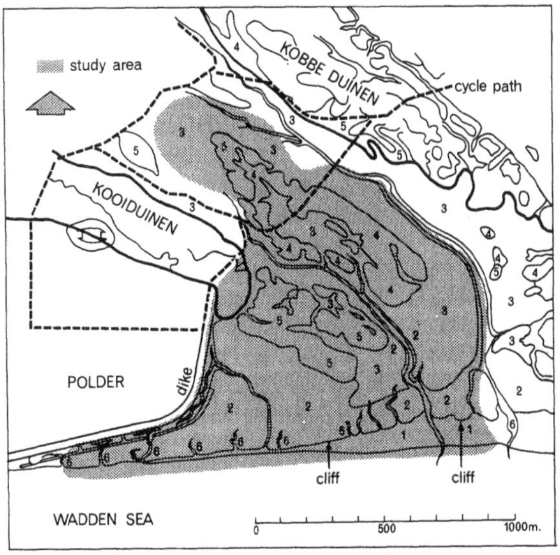

Fig. IV.9. Geomorphological map of part of the Oosterkwelder of Schiermonnikoog. 1 = daily inundated salt marsh, 2 = levees, 3 = depression, 4 = low dune, 5 = low dune containing fresh water, 6 = creek (after Bleuten, 1971).

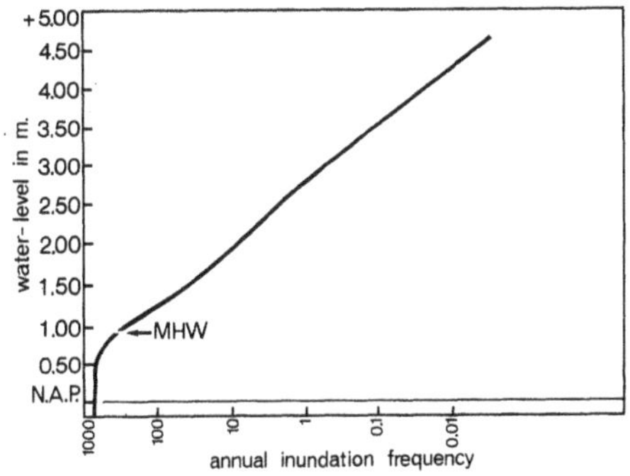

Fig. IV.10. Inundation frequency of various N.A.P. levels at the Oosterkwelder salt smarsh of Schiermonnikoog derived from the average frequencies during the period 1967 to 1974 (after Rijkswaterstaat Groningen).

Starting from the tidal flats, a characteristic zonation in the halosere of salt marsh communities on the sandy Frisian islands is found (Westhoff & den Held, 1969; Beeftink, 1977A; Dijkema, 1983). The *Salicornietum strictae* and *Spartinetum townsendii* occur on tidal flats with daily submergences. The lower salt marsh, between Mean High Water level and spring tide level is characterized by the *Puccinellion maritimae* with the *Puccinellietum maritimae* in depressions, the *Halimionetum portulacoidis* on creek bank levees and the *Plantagini-Limonietum* in the intermediate positions. The *Armerion maritimae* occurs on the mid- and upper salt marsh from just below the spring tide level upto the level reached by winter gales. The *Juncetum gerardii* is found on the clay soils, whereas the *Artemisietum maritimae* occurs on sand ridges and creek bank levees. The *Puccinellio-Spergularion salinae* community can develop in places with irregulary changing conditions such as occasionally stagnant water. The *Atriplici-Agropyretum pungentis* can develop higher on the salt marsh after the deposition of drift material.

Lolio-Potentillion communities can be found on the upper slightly desalinated salt marsh (community of *Agrostis stolonifera* and *Trifolium fragiferum*) and on adjacent low dunes (community of *Ononis spinosa* and *Carex distans*). Also a more ephemerous community can be found at the transition of wet/salt and dry/fresh viz. the *Saginion maritimae*. Both of the afore-mentioned communities form transitions to the low dune communities of the *Galio-Koelerion* and *Hippophaë* shrubs on slightly higher dunes. Details of the plant communities are given in chapter VII.

On the basis of the recent synecology of the communities (Westhoff & den Held, 1969) it can be assumed that the extent of *Puccinellietum maritimae, Juncetum gerardii, Puccinellio-Spergularion salinae, Lolio-Potentillion* communities, and *Saginion maritimae* was rather small in the original unexploited salt marsh vegetation because these communities mainly depend on grazing. Some large herbivores and particularly wild fowl (geese, ducks) must have occurred as the natural grazers. *Salicornietum strictae, Plantagini-Limonietum, Halimionetum portulacoidis, Artemisietum maritimae,* and *Atriplici-Agropyretum pungentis* which are characteristic for unexploited salt marshes must have been wide spread before the beginning of human exploitation.

IV.B.2. Human exploitation on mainland salt marshes (600 B.C.–1300)

Information concerning the first human exploitation of the salt marshes can be gained from the extensive mainland salt marshes that became inhabited by people from the Drenthe Plateau during the early Iron Age (cf. III.B.10). Initially (c. 600 B.C.)

farm-steads were built on the salt marsh (Dutch: vlak nederzetting; German: Flachsiedlung), but, because of periods of transgression from c. 500 B.C. onwards, the dwelling places were also built on a mound of salt marsh sods (Dutch: terp; German: Wurt). These mounds gradually grew higher through the accretion of waste and dung and therefore, particularly through macro-fossils, became excellent records of former settlement and land use. The mounds sometimes bore a single farm-stead (Dutch: huisterp), but also sometimes joined together and often had three, sometimes four farm-steads or an entire settlement (Dutch: dorpsterp) (Miedema, 1983).

The dwelling places were situated at the highest parts of the salt marsh viz. salt marsh ridges and creek bank levees. These were also the only places to grow arable crops and were preferred as cattle-grazing areas. The salt marsh depressions between the creek bank levees were mainly used for hay-making and sheep-grazing (Miedema, 1983). Despite the finding of unthreshed arable crop remnants, it was long considered improbable that they originated from locally grown crops. Recently it has been demonstrated that crops grown from species found in mounds can, indeed, be cultivated on the upper salt marsh in the Netherlands (Van Zeist et al., 1976; Van Zeist, 1980) and in Germany (Körner-Grohne, 1967). Moreover, ploughing marks have been discovered on former salt marshes (Haarnagel, 1979).

Miedema (1983) studied a salt marsh area of c. 150 km^2 northwest of the city of Groningen. Due to several transgression periods most mounds were only inhabited for a few centuries (Waterbolk, 1976) but the area, as a whole, was inhabited continuously from the early Iron Age onwards (c. 600 B.C.). This permanent occupation might have been preceded by a transhumant phase in which people from the Drenthe Plateau herded cattle and sheep in the summer period on the adjacent salt marshes (Van Gijn & Waterbolk, 1984). Each farm-stead needed food for seven to eight people, including children, realised on 5 ha of arable field including the dwelling place, and 30 ha of grassland (20 cattle, 5 horses, 20 sheep/goats) (Haarnagel, 1979); Miedema, 1983). During the first settlements the salt marsh was not completely exploited but it gradually became overcrowded. Table IV.1 shows that the whole area suitable for agricultural practices was not needed during the early Iron Age. Only a minor part (c. 600 ha) of the inferior lower salt marsh needed to be exploited. During the late Iron Age half of the area suited for arable field was needed and c. 7000 ha of the lower salt marsh was exploited as grassland. During the Roman Period all the suitable grassland was exploited and there was probably even a shortage of 700 ha which was possibly compensated for by stubble grazing of the arable fields (Miedema, 1983). It seems, therefore, not unlikely that people looked for new salt marsh areas on the Wadden islands during the second half of the Middle Ages.

Körner-Grohne (1967), Van Zeist (1974) and Behre (1979; 1985) reconstructed the salt marsh plant communities in the north of the Netherlands and the northwest of the German mainland from the time of the mound settlements based on present synecology (Westhoff & Den Held, 1969). The lower salt marsh contained the *Puccinellietum maritimae* and the mid salt marsh the *Juncetum gerardii*. It was assumed that the *Juncetum gerardii leontodontetosum autumnalis* with *Lolio-Potentillion* and *Saginion maritimae* elements occurred on the upper salt marsh. Since the *Plantagini-Limonietum* and *Halimionetum portulacoidis* and the *Artemisietum maritimae* were probably not found, it must be assumed that the salt marshes were heavily exploited (Waterbolk, 1976) like Miedema (1983) already suggested. *Halimione portulacoides, Artemisia maritima* and *Elymus pycnanthus,* dominant species on ungrazed salt marshes, were rarely or never found as macro-fossils.

For centuries mounds were the only defence against inundations and the settlements did not cooperate in the building of dikes as protection against the sea. A first start was, possibly, made c. 1000 (Waterbolk, 1976). It started in the adjacent northwest Germany in the 11th century (Behre, 1985). The first low dikes enclosed cultivated salt marsh areas around one or two mounds. C. 1100, a closed sea-defence probably existed in Groningen and Friesland, featuring dikes with a maximum height of 2.40 m. Vast salt marsh areas, however, remained unprotected by dikes and had therefore no owners.

During the twelfth century Cistercian monks founded the Klaarkamp Abbey near Rinsumageest in the neighbourhood of Dokkum in Friesland. They had just returned from the Crusades and introduced a locally unknown skill which enabled them to build dikes extensively and reclaim the unprotected salt marsh. The Cistercian Order, therefore, became owner of large areas of northeast Friesland and northwest Groningen and founded more abbeys of which the one in Aduard was the most famous (Reitsma & Bakker, 1986). It is not surprising that these monks discovered the unprotected salt marsh of the island of Werveroog after crossing the tidal watershed from Friesland.

Table IV.1. Area suited and needed for arable field and grassland on the salt marsh northwest of Groningen during three settlement periods (after Miedema, 1983).

	Area suited		Area needed	
	Arable field	Grassland	Arable field	Grassland
Early Iron Age	730	5400	160	1200
Late Iron Age	1940	10950	1000	6630
Roman Period	1940	10950	1980	11660

IV.B.3. Settlement on Schiermonnikoog (1300–present)

The activities of the Cistercian monks resulted in a contract between the abbot of the Klaarkamp Abbey and the Tannerguild in Bremen c. 1300 concerning the transport of dried fells and leather from the island Wervteroe to the Frisian mainland (De Haan et al. 1983). Since the island had been given a name it is possible that it had been inhabited but nothing is known about these people. The present name of Schiermonnikoog was derived from the Cistercian monks. The name 'Scirca monch each' was first mentioned in 1440 ('Schier' = grey (the colour of the monk's habit), 'monnik' = monk, 'oog' = island).

A map made by Cristiaan Sgrooten, in 1570, as a part of Philips II's Empire revealed the first picture of Schiermonnikoog's settlement (Fig. IV.11). It is apparent that the salt marsh was inhabited and a mill existed indicating agricultural activities (Reitsma & Bakker, 1986). Although private properties always remained, the Cistercian Order owned most of the island.

The religious revolution during the Eighty Years' War caused great changes with respect to properties c. 1580. The island fell under the jurisdiction of the States of Friesland who sold it, including tax on its products, to the Stachouwer family in 1640. The family lived on the estate of 'Binnendijken'.

This estate 'Binnendijken' was enclosed by low dikes but the 1702 map made by Hindrik Lammertz (Fig. IV.12) also showed the hamlet 'Dompen' in the unprotected salt marsh. Other hamlets were found in the dunes or at their feet viz. 'Oosterburen' and 'Westerburen' (see also Fig. IV.6). The hamlet 'Moolenburen' was probably situated in the upper salt marsh and included a mound with the mill on top of it. The eastern salt marsh was accessible by a bridge crossing the creek. The eastward shift of the island was, again, illustrated by the location of the 'Monckendyke' at the southwest point in a salt marsh area which was nearly completely eroded (Reitsma & Bakker, 1986).

Westerburen, in particular, was buried by sand and flooded during the end of the seventeenth and the beginning of the eighteenth century. Initially houses were rebuilt in Oosterburen and in the Dompen, but finally a new settlement was chosen more to the east and higher in the salt marsh. This 'Hoochdorp' was situated where the present village is found (see also Fig. IV.6). The first houses were built c. 1720, whereas the last house in Oosterburen was pulled down in 1782. From that time onwards the present village was the only one on the island of schiermonnikoog. From a map from 1769 (Fig. IV.13) one can see that the salt marsh south of the village was protected by a system of low dikes enabling agricultural practices as appeared (Reitsma & Bakker, 1986).

During the nineteenth century the system of low dikes was extended eastwards in the salt marsh. Only a small part of the former dike protecting the estate of 'Binnendijken' was left in c. 1850, as the major part had vanished into the sea (Fig. IV.14).

Mr. Banck bought the island in 1859. He embanked the major part of the existing salt marsh (Banckspolder) since the low dikes could not protect the arable fields from winter gales. The German Count von Bernstorff became the final private owner of the island in c. 1890.

Perhaps because of the afore-mentioned developments the people of Schiermonnikoog seemed to subsist mainly on agriculture. The number of inhabitants, however, depended mainly on the products of the sea. The population increased markedly during the period 1643–1680, probably due to the economic climate created by the Stachouwer family facilitating the establishment of skippers from Emden, Dokkum, Juist, Norderney, and Groningen (Reitsma & Bakker, 1986). In 1712 Schiermonnikoog had c. 1100 inhabitants and 110 ships. The ships were not specialized fishing-boats. They were also adapted as cargoboats for food transports from elsewhere, since the island was not selfsupporting for 1100 people. The population decreased after 1850 due to the moving of skippers to the southwestern part of the Netherlands.

Apart from the few full-time farmers many people who mainly depended on the products from the sea also had one or two cows and an arable field. Since the men worked on the ships, agriculture was predominantly carried out by women.

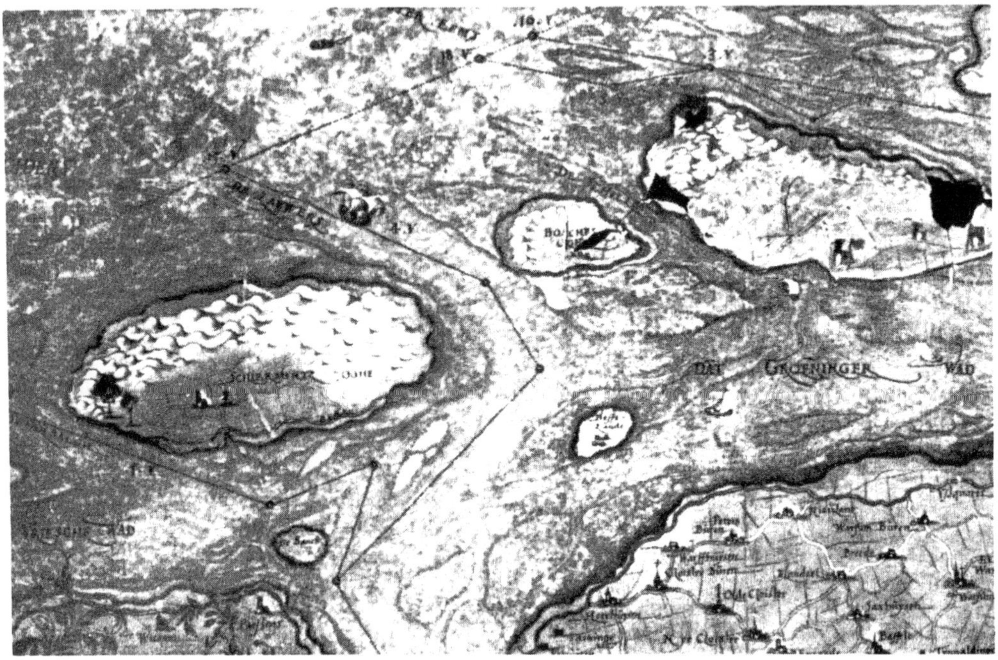

Fig. IV.11. Map of the Island of Schiermonnikoog by Christiaan Sgrooten made in the period between 1567 and 1570 by order of the Duke of Alva as a part of the map of the empire of Philips II. The church and the mill lie centrally. The building upon the edge of the salt marsh is remarkable, since the existence of the hamlet 'de Dompen' in the 16th century is not yet known from archives. (after Reitsma & Bakker, 1986).

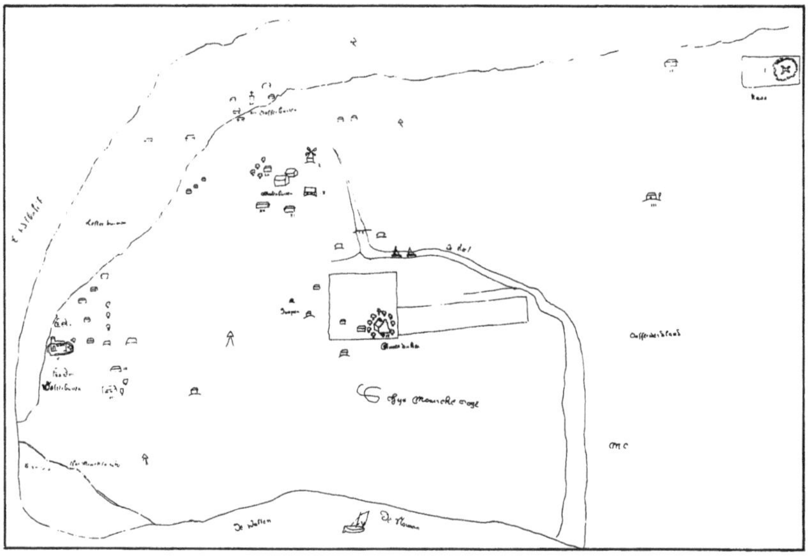

Fig. IV.12. Map of the western part of the Island of Schiermonnikoog by Hindrik Lammertz from 1702. Remnants of the old dike built by monks indicate the erosion of the south western part of the island. The hamlets Westerburen, Oosterburen, Molenbuurt and de Dompen are indicated besides the estate Binnendijken. (after Reitsma & Bakker, 1986).

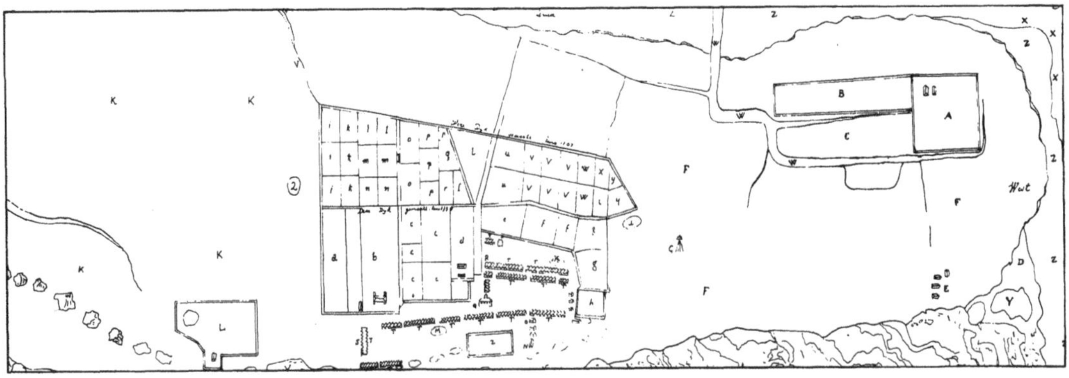

Fig. IV.13. Map of the Western part of the Island of Schiermonnikoog by Coppen Lambert Stachouwer from 1769. The last houses of the hamlets Oosterburen and de Dompen and some sheds of the former estate Binnendijken are still present. The first streets of the new village Hoochdorp and low dikes south of it are indicated. F means hayfield, K means pasture. Note the south-north orientation of the map! (after Reitsma & Bakker, 1986).

Fig. IV.14. Map of the Island of Schiermonnikoog from 1853, the first official Topographic and Military Map of the Netherlands. The study area Oosterkwelder will develop eastwards of the Kooiduinen dune ridge after 1870. A dike system protects the village and agriculturally used parts of the salt marsh. (after Reitsma & Bakker, 1986).

IV.B.4. Natural vegetation development

The natural vegetation development during the eastward shift of the island can best be illustrated by the only known pollen diagram. This diagram is derived from the present village and was covered by 2 m drift sand during the seventeenth century (De Jong, 1975). The oldest pollen originated from a *Caricion curto-nigrae* community which must have been deposited into a wet dune slack influenced by seepage water from the surrounding dunes (Fig. IV.15). Radio carbon dating revealed the start of the pollen deposition as being c. 1400, therefore, taking the shift of the island into consideration, the dune slack must have been situated at the far eastern point of the then dunes (see also Fig. IV.6).

The decrease of *Hippophaë rhamnoides* and the increase of *Salix,* presumably *Salix arenaria* probably indicates the acidification of the initially calcareous conditions. The more or less simultaneous

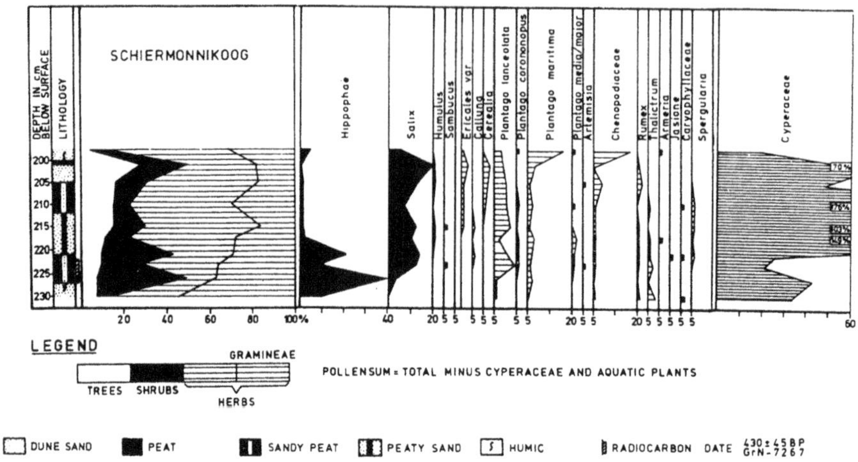

Fig. IV.15. Pollen diagram from Schiermonnikoog located at the present village (after De Jong, 1975).

appearance of *Calluna vulgaris* and *Erica tetralix* is consistent with this idea. The salt marsh at the eastern point of the dunes must have been very young at that time since only small amounts of clay were deposited. It must have been comparable with the present salt marsh in the neighbourhood of the Willemsduin, namely, some scarce salt marsh plants occurring at the sand plate.

The pollen of *Plantago maritima* and *Plantago coronopus,* which were found in relatively low values, probably came from salt marsh communities in the surrounding area. Although *Chenopodiaceae* could be connected with agriculture and treading by man and cattle, the contemporaneous expansion of *Plantago maritima* and *Chenopodiaceae* with many representatives in salt marshes, might give an indication of an increase in salt marsh communities (De Jong, 1975).

IV.B.5. Land use of the Schiermonnikoog salt marshes

The contract between the abbot of the Klaarkamp Abbey and the Tannerguild, mentioned in section IV.B.3, establishes that the island was grazed in c. 1300. Grazing was possible in both the salt marsh and the surrounding dunes. The occurrence of *Plantago lanceolata* in the pollen diagram (Fig. IV.15) indicated grazing from at least c. 1400 onwards. It is not known from historical documents, when agriculture started, but the Cistercian monks had storehouses at the beginning of the fourteenth century (Reitsma & Bakker, 1986).

In the pollen diagram the first *Cerealia* pollen came from c. 1500 indicating the beginning of local agriculture or for the whole island. The pollen diagram revealed that the fertile clay deposits enabled agriculture in the salt marsh (see section IV.B.2), which was formerly impossible in the sandy salt marsh in the area of the peat deposits. The winter gales transgressing the low dikes provided an annual new layer of fertile deposits and were, therefore, considered an advantage rather than a disadvantage. Only gales in the spring or in the summer period were disastrous for the grown crops (Van Zeist, 1974).

There are no known documents about land use of the salt marsh until the beginning of the seventeenth century. In 1608 there is mention of a common grazing east of the village. It is known that guide-lines for grazing and hay-making in the salt marsh were issued in 1643. The 1769 map (Fig. IV.13) revealed grazing east of the village and hay-making south and west of the village. Hay-making was, moreover, practiced in dune slacks north east of the village. The first indication that German grassmowers hired themselves out dates from 1724. More or less regular hay-making also took place on the present Oosterkwelder salt marsh. Hay-making always produced low yields, since most people only owned a few cows and for these one could scrape together just enough winter hay in the dunes and the salt marsh.

The major part of the then existing salt marsh was embanked in 1860. Farms were built in the new polder and from that period onwards a few full-time farmers owned most of the live-stock. The soil fertility of the new polder decreased since clay was no longer deposited during the winter gales. It, therefore, became necessary for manure to be applied not only to the potato crops in the dunes but also to oats, flax, rye, and rape-seed grown near the village. These arable fields became worthless after a dike burst in 1962 and from that time onwards the whole polder became pasture (Reitsma & Bakker, 1986).

After the creation of the Steinsdúne (Kobbeduinen) the salt marsh rapidly extented eastwards (see also Fig. IV.6). Cattle from the mainland, therefore, also grazed the salt marsh during the summer period. The first mention of the transport of the cattle was in 1872. A systematic picture of the numbers of live-stock cannot be given, therefore the stocking rate in the salt marsh is not known. The extension of the grazed salt marsh from c. 1600 onwards is reconstructed and depicted in Fig. IV.16 (Reitsma & Bakker, 1986).

IV.B.6. Explanation of species diversity on Schiermonnikoog

At present the species diversity on the island of Schiermonnikoog is very high, namely 560 species of vascular plants occurring in an area of 35 km^2. This implies that the number of species found is more than treble that which might be expected from that area for the Dutch situation (Figure III.25). A number of rare species is found on Schiermonnikoog, 29 very rare species, i.e. occurring in 1 to 29 grid cells of the 5 × 5 km grid (involving 1677 cells in the Netherlands) and 50 rare, i.e. species occur in 30 to 79 cells. The island of Schiermonnikoog revealed four species which occur on one or two out of the 40 islands or coastal areas in the whole Wadden Sea area (the Netherlands, Germany, Denmark) and 32 species occurring on three to six islands (Mennema & Weeda, 1983). This illustrates the local differences between the islands in the Wadden Sea area. Schiermonnikoog is, therefore, one of the most valuable islands with respect of the floristic diversity and the number of rare species both on a national and the local level of the Wadden Sea area.

The island of Schiermonnikoog revealed twenty plant species characteristic for beaches, see dunes and sandy flood-marks of which *Beta maritima, Euphorbia paralias* and *Glaucium flavum* were rare i.e. occurred on less than 10 out of 40 islands or coastal areas in the Wadden Sea area. Schiermonnikoog contains eighteen species characteristic for salt and strongly brackish waters, mud flats and low salt marshes most of these occurring in nearly the whole Wadden Sea area. Finally, twenty species characteristic for high salt marsh and contact situations between saline and desalinated environments were found on Schiermonnikoog, *Alopecurus bulbosus, Carex punctata* and *Hordeum marinum* being rare species (Mennema & Weeda, 1983). Although *Juncus maritimus* is not a rare species, it should be mentioned that the stands found on Schiermonnikoog are among the most extensive in northwest Europe (Bakker & Fresco, 1969). The subdivision in ecological groups is derived from Van der Meijden et al., (1983B). The ecological groups of beaches/sand dunes and high salt marshes/contact situations revealed the number of rare species and the total number of species per ecological group to be positively correlated with the size of the island or coastal area in the Wadden Sea area. The lower salt marsh group, however, had about the same figure for all the islands and coastal areas, indicating that the complete set of low salt marsh species is found in each of the salt marshes (Weeda & Mennema, 1983). Schiermonnikoog represents a high degree of spatial heterogeneity. The eastward shift of the island which has already been mentioned, attributed to dunes, dune valleys and salt marsh with different age and subsequent development. These phenomena resulted in various groundwater salinities and percentages of $CaCO_3$ in the soil of infiltration areas. As a result, the ionic composition of seepage water also varies considerably, depending on the route and the duration of the infiltrated water

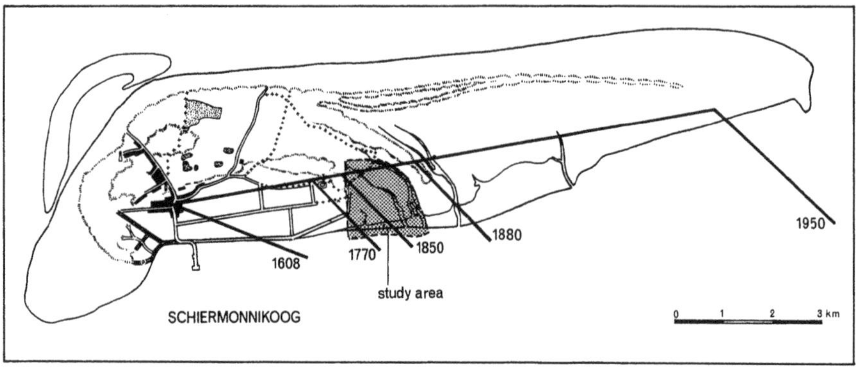

Fig. IV.16. Movement of the eastern boundary of the grazed part of the salt marsh during the period between 1600 and 1950. (after Reitsma & Bakker, 1986).

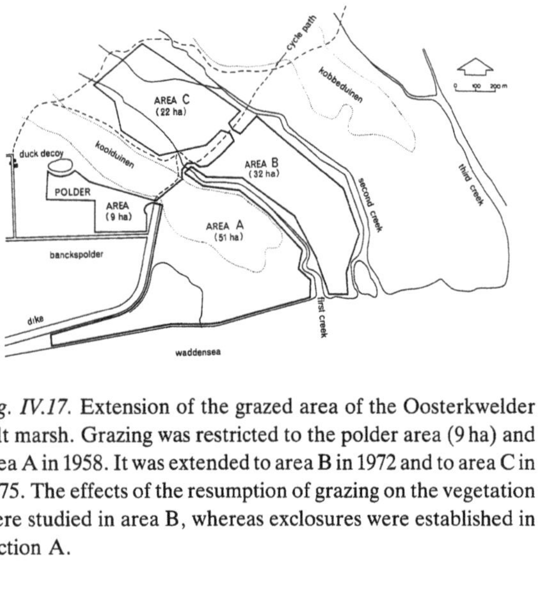

Fig. IV.17. Extension of the grazed area of the Oosterkwelder salt marsh. Grazing was restricted to the polder area (9 ha) and area A in 1958. It was extended to area B in 1972 and to area C in 1975. The effects of the resumption of grazing on the vegetation were studied in area B, whereas exclosures were established in section A.

(Wams & Wemelsfelder, 1984). Since this holds both for the macro-scale of the whole island and the locally superimposed micro-scale (e.g. Thalen, 1971), the spatial heterogeneity and its gradients will contribute to a great extent to the species diversity of Schiermonnikoog.

Due to the gradual eastward movement of the island no continuation with respect to human impact can be found at any place. Moreover, some ancient activities ceased in 1860 when the salt marsh area became largely a polder. Agricultural activities in the dunes ceased around the village because the manure was needed in the polder. In the polder itself drainage, and later on the application of fertilizers, changed the agricultural practices. Grazing in the dunes and on the greater part of the present salt marsh ceased. Some older dune valleys may have been cut for over a 100 years. Hay-making in the present salt marsh is recent because of the age of the particular salt marsh. A good human exploitation gradient could develop in the future since the Oosterkwelder salt marsh has been grazed and applied with artificial fertilizers upto the first creek, and just grazed between the first and second creeks. It is, however, cut locally between the second and fourth creek and left untouched eastwards from the fourth creek. Our conclusion is that the continuity and/or gradients in human impact play as yet a minor role in the species diversity of Schiermonnikoog.

IV.C. Acquirement and management of the nature reserve 'Oosterkwelder'

From c. 1890 onwards the whole island of Schiermonnikoog, was private property of the German Count Bernstorff, it was confiscated by the Dutch Government after World War II. The 'Dienst der Domeinen', a Service of the Finance Department, was charged with the nature management of the island. Recently the State Forestry Commission became in charge. The Oosterkwelder which remained unembanked in 1860 (see section IV.B.5) has been a nature reserve since 1954. It includes 2400 ha.

Photo IV.2. Cattle cross the creek which separates different parts of the grazed salt marsh.

Until 1958 the whole salt marsh was grazed by yearlings which were looked after by a herdsman. The yearlings came from the mainland in the spring and remained until the autumn. The number of yearlings varied year by year. They numbered 400 in 1952 (Den Hartog, 1952). Where the high salt marsh is adjacent to the dunes changes between dry/fresh and wet/brackish conditions are found favouring the occurrence of the liver-fluke. The yearlings, therefore, often returned infected to the mainland. This veterinary problem was the main reason for ceasing overall grazing in 1958. Only the most western salt marsh area A connected with an innerdike area with fresh water supply (Fig. IV.17) was fenced in and grazed by island yearlings.

From 1962 the farmers applied artificial fertilizers to 20 ha of the 50 ha fenced-in salt marsh area. The natural salt marsh production apparently became too small for the stocking rate at the time because the island farmers became more and more interested in salt marsh grazing. They therefore asked for an extension of the grazed area and the authority in charge of the management readily agreed because the salt marsh became rugged and diversity decreased after the cessation of grazing in 1958. In 1972 area B and in 1975 area C were added (Photo IV.2) resulting in an area of about 100 ha grazed by 150 yearlings from the end of May until October (Fig. IV.17). The farmers rent the salt marsh from the authorities in charge of the management.

Extensive hay-making is practiced on the salt marsh east of the grazed area. The managers apply the hay in order to prevent erosion of the dune paths through recreation pressure and the farmers use it as a coverage for the hay-stacks. In the dry summer of 1976 little hay could be harvested in the polder and the salt marsh hay was taken as forage as in former days.

PART THREE

Vegetation dynamics

CHAPTER V

The heathland and reclaimed grassland vegetation

V.A. Introduction

The study area Westerholt (53° 01′ N, 6° 11′ E) comprises 3 ha of *Calluna vulgaris*/*Erica tetralix* heathland, 2 ha of deciduous woodland (*Betula* spp. and *Salix* spp.) and 6 ha of pasture land which was originally heathland but had been ploughed to a depth of 40 cm and sown with grass in the 1950's (Photo V.1 and 2). Until 1972 the heathland area had been abandoned for several decades, whereas the grassland had been subjected to the application of fertilizers, and had been used for hay production. The aftermath had been grazed by cattle. Fertilizer application had been stopped in 1971. In October 1972 the area was fenced-in and sheep of the Schoonebeeker breed (Van Bemmel 1963; Clason 1980B) were introduced to graze the whole area at an average stocking rate of 3 sheep · ha^{-1}. This stocking rate resulted in local differences in grazing intensity which will be referred to several times in this chapter but will be discussed in detail in chapter VIII. Twenty-five sheep grazed during the period July to December and 40 between January and July. In the spring of 1980 the sheepflock was replaced by a new one. Since 1980 the area has received a winter rest period of 2 months from grazing.

In this chapter vegetation dynamics will be analyzed and discussed, in heathland and adjacent grassland which was reclaimed from heathland and then subjected to sheep-grazing after the cessation of fertilizer application. The heathland area remained abandoned locally as a comparison. The questions to be answered are (i) can grazing prevent succession to woodland?, and (ii) to what extent do detrimental effects from the adjacent grassland occur on the heathland communities? Grazing was compared to hay-making in the grassland area. The questions to be answered in this area are (iii) to what extent can the grassland be reverted to heathland-like communities? and (iv) how does species diversity change during this process?

It was expected that the restoration of grassland to heathland would depend on the removal of nutrients. Several experiments were, therefore, carried out. Part of the area under a hay-making regime was fertilized annually with nitrogen (50 kg N · ha^{-1}) since 1973 in order to grow a high crop and thus remove more phosphorus and potassium than would be the case without nitrogen application. In May 1973 the uppermost 5 cm of the grassland area was removed in another place. Finally, in September 1978 two small grassland areas were ploughed, one area to a depth of 20 cm and the other 60 cm in order to get the unchanged C-horizon to the surface. The whole ungrazed grassland area was mown once a year in the period from July to September. The lay-out of the management regimes is shown in Fig. V.1.

The abiotic conditions of the study area are described in section V.B. The effects of the different management regimes in both the heathland and grassland area were compared by sequential mapping in 1972 (before the start of the experiment), in 1977 and in 1982 (section V.E) and by year-to-year recording of permanent plots (Section V.D).

Classification of the plant communities was done by comparing c. 480 relevés. After classification, the syntaxonomical position of the coena and their synecology is discussed (Section V.C).

Vegetational trends were derived from permanent plot data by the dissimilarities between each year and with the initial year (section V.D). Species diversity and cover percentage increase or decrease under the management regimes in different plant communities were compared to findings from elsewhere (section V.F).

V.B. Site description

V.B.1. Geomorphological conditions

Details about the geomorphological conditions of the Drenthe Plateau were given by De Gans (1980), as has already been mentioned in section III.B.1.

The deposits of the geological topsoil of the study area originated from the last three Ice Ages. Loam and fine-grained sandy deposits, probably originating from melting water, remained from the

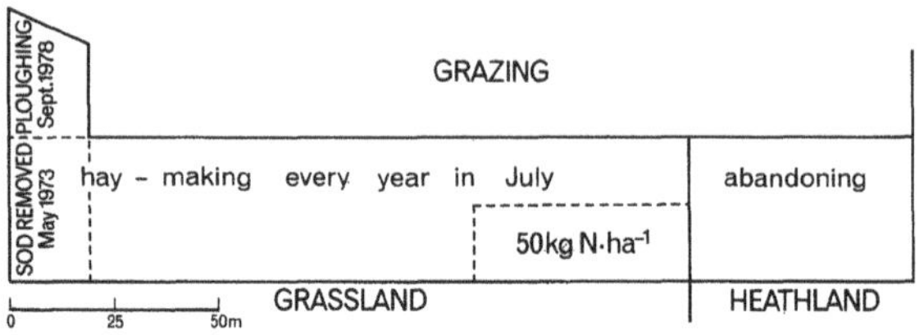

Fig. V.1. Lay-out of the different experiments in heathland and grassland.

Elsterian (600,000 B.P.). The loam is too light to be classified 'potklei', but heavy enough to be water impermeable.

Boulder-clay sediments from the Saalien Ice Age (200,000 B.P.) were not found in the study area. Only gravel layers on top of the Elsterian sediments were found, this, indicating erosion of the Saalien deposits.

Coarse-grained aeolean cover sand deposits, resulting from the Weichselian (70,000 B.P.), were found locally on top of the Elsterian deposits. Finally peat was formed in lower parts during the Holocene (10,000 B.P. until present).

All processes which have been mentioned resulted in virtually flat land with undulating areas, but the elevation differences did not exceed c. 3 m. The geomorphological characteristics of the study area are illustrated in the cross-section in Fig. V.2 (Van Wieren, 1977).

V.B.2. Soil

A rough soil map of the study area is shown in Fig. V.3 (Van Wieren, 1977). The units were derived from the higher levels of soil classification according to De Bakker & Schelling (1966) and refined by Van Heuveln (1980).

The 'Veld' podsol soils were found on the higher cover sands (Photo V.1 and 2). The 'Broek' earth soil developed in the peat deposit sediment during the Holocene period. The 'Goor' earth soil consists of fine-grained loamy sand. Both the latter soil types contain loam within a depth of 120 cm, representing the Elsterian deposits mentioned in section V.B.1. The old clay soil even contains old loam or clay from the Elsterian deposits within a depth of 40 cm below the soil surface.

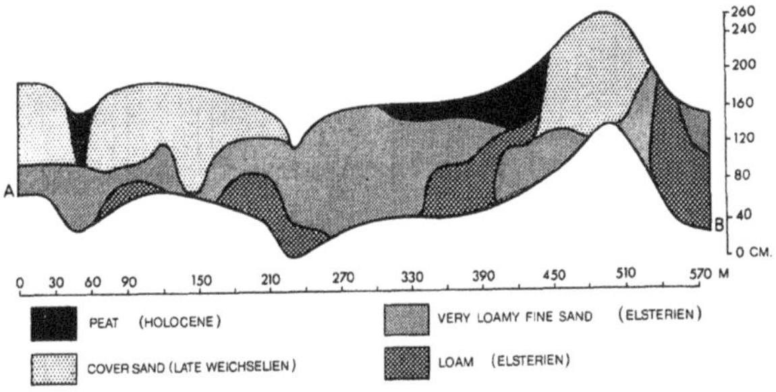

Fig. V.2. Cross-section showing the main soil characteristics at a depth of 0–120 cm, and the relief of the study area. The position of the transect is indicated in Fig. V.3 (after Van Wieren, 1977).

V.B.3. Hydrological conditions

All soil types in section V.B.2 showed a mean highest groundwater table of <40 cm and a mean lowest groundwater table of >120 cm below soil surface which could be seen from rust and gley features. The area is, therefore, characterized by low summer groundwater tables and very high winter groundwater tables. Locally the area even becomes inundated annually for several months due to the water impermeable loam layer in the subsoil mentioned in section V.B.1. The groundwater tables in

Photo V.1. Undisturbed 'Veld' podzol soil under heathland in the Westerholt study area.

Photo V.2. Ploughed 'Veld' podzol soil under the grassland area in the Westerholt study area. The former sods are recognizable in the sloping dark lines.

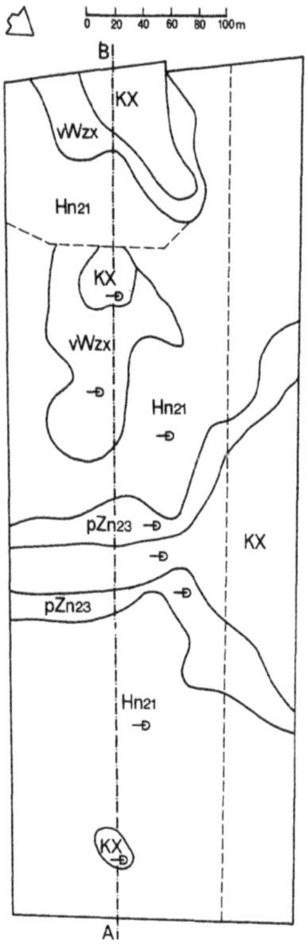

Fig. V.3. Soil map of the Westerholt study area with an indication of the cross-section (A–B) shown in Fig. V.2 (vWzx: 'Broek' earth soil'; Hn21: 'Veld' podzol soil; pZn23: 'Goor' earth soil'; KX: old clay soil; –Ð' reclaimed) (after Van Wieren, 1977).

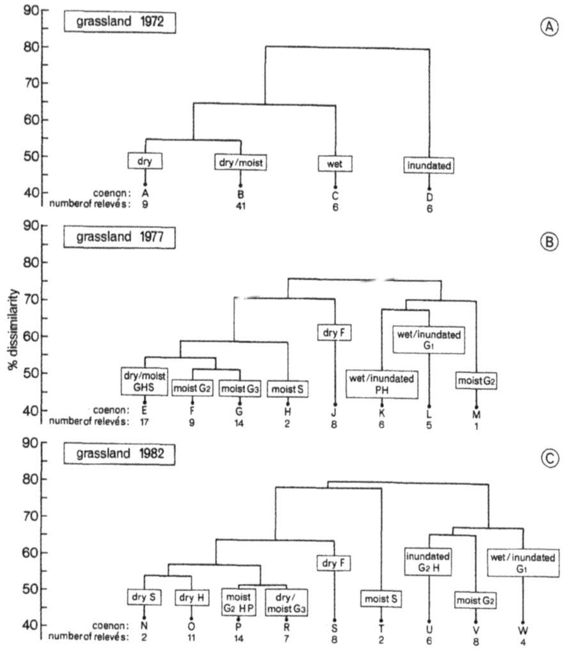

Fig. V.4. Simplified (all dissimilarities below 50% are not distinguished) dendrograms of all permanent plots in the grassland area in 1972, 1977 and 1982, respectively, with an indication of moisture and management regime for each coenon (H: haymaking, F: fertilizer applications, P: ploughed, S: sod removal, G: grazing (1 = hardly, 2 = intermediately, 3 = heavily)).

the area will be discussed in some more detail in relation to the distribution of plant communities in section V.C.2 and section V.C.3.

V.C. Vegetation classification, syntaxonomy and synecology

V.C.1. Vegetation classification

The vegetation description started with the recording of c. 90 relevés in 1972 (c. 60 in the grassland and c. 30 in the heathland area). The positions of these relevés were fixed and they were, thereafter, used as permanent plots. Permanent plot data from 1972, 1977 and 1982 were clustered as described in section II.C.3. Synoptic tables were made, starting from the results of these clusterings. The detailed relevé data were published in an internal report (Dallinga & Tjaden, 1981).

V.C.2. Grassland communities (Table V.1)

The clustering results of the permanent plot data in 1972, 1977 and 1982 are shown in the dendrograms in Fig. V.4A–C. Four clusters can be recognized in the 1972 data, eight in the 1977 data, and nine in the 1982 data at a dissimilarity level of 50%.

In the 1972 clustering (Fig. V.4A) coenon A is characterized by high cover percentages of *Poa pratensis* and *Lolium perenne*, coenon B reveals a high *Holcus lanatus* coverage and locally a high *Rumex acetosa* or *Elymus repens* coverage, coenon C is characterized by the high cover percentages of *Agrostis capillaris, Holcus lanatus* and *Poa pratensis*, and coenon D particularly by the abundance of *Agrostis stolonifera, Carex nigra* and *Juncus effusus*. The coena resulting from the clustering can be related to four groundwater table regimes (Fig. V.5A–D). Coenon A was considered to be dry, coenon B to be moist, coenon C wet and coenon D to be inundated. At the start of the experi-

Table V.1. Synoptic table of the coena (number 1–9) occurring in the grassland area during the period 1972 to 1982. The coena found in the years 1972, 1977 and 1982 are referred to by the letters A–W. Character and differential species of the Festuco-Cynosuretum (FC), the Lolio-Cynosuretum (LC), the Lolio-Potentillion (LP), the Poo-Lolietum (PL) and the Caricion curto-nigrae (Ccn) are indicated. The occurrence of the species is given in percentages of frequency and in mean cover percentages (+: 1–5; a: 6–10; 1: 11–20; 2: 21–30 etc.).

	Coenon	1	2	3	5	6	7	4	8	9
	1972 data	A					C	B		D
	1977 data	E		H	G	F	M	J	K	L
	1982 data	O	N	T	P	R	V	S	U	W
	Groundwater table	Dry	Dry	Dry/ moist	Moist	Moist	Moist	Moist	Wet/ inundated	Wet/ inundated
	Management	Hay-making/ heavily grazed	Sod removal	Sod removal	Heavily grazed	Hay-making/ heavily grazed	Inter- mediately grazed	Hay-making/ heavily grazed	Hay-making/ lightly grazed	Hardly grazed
	Number of relevés	37	2	4	28	16	15	57	12	15
	Poa pratensis	100/2	100/+	25/+	100/+	94/+	93/+	100/1	20/+	24/+
FC	Trifolium dubium	68/a	100/+	25/+	10/+	19/+	71/+	23/+	10/+	.
FC	Rumex acetosella	29/+	100/+	50/+	21/+	25/+	.	3/+	.	.
FC	Festuca rubra	10/+	100/+	.	8/+	.	.	.	.	.
FC	Festuca ovina	21/+	100/+	.	5/+	6/+	.	.	.	.
	Betula pubescens	.	100/+	25/+	.	6/+	.	.	.	.
	Erica tetralix	.	.	100/+	8/+	6/+	.	.	.	.
	Calluna vulgaris	5/+	.	100/a	18/+	6/+	7/+	.	.	.
	Juncus squarrosus	.	.	75/+	3/+	.	.	.	.	.
	Juncus conglomeratus	.	.	75/+	29/+	18/+	20/+	.	30/+	12/+
	Leontodon autumnalis	71/+	50/+	100/+	84/1	87/a	27/+	26/+	50/+	41/+
FC	Hypochearis radicata	79/+	100/a	100/a	84/1	81/+	.	7/+	10/+	6/+
LC	Bellis perennis	76/+	100/+	50/+	79/a	93/+	7/+	39/+	20/+	.
	Trifolium repens	60/+	100/+	.	89/1	81/a	47/+	37/+	30/+	12/+
FC	Anthoxanthum odoratum	63/a	100/+	50/+	79/+	56/a	47/+	19/+	20/+	12/+
FC	Agrostis capillaris	74/a	100/1	100/+	89/4	100/3	47/5	40/a	40/+	18/+
	Holcus lanatus	97/1	100/+	100/+	100/1	100/1	92/a	100/4	60/+	65/+
	Juncus effusus	26/+	.	100/+	66/+	69/+	100/1	23/+	100/2	100/3
Ccn	Galium palustre	.	.	.	13/+	18/+	33/+	2/+	90/+	94/+
LP	Alopecurus geniculatus	.	.	.	29/+	6/+	33/+	42/+	60/+	59/+
Ccn	Ranunculus flammula	.	.	.	3/+	12/+	27/+	.	40/+	59/+
	Potentilla anserina	.	.	.	.	.	.	.	40/+	29/+
LP	Agrostis stolonifera	5/+	.	.	50/+	37/+	53/+	65/+	70/1	100/2
Ccn	Carex nigra	.	.	50/+	8/+	44/a	93/a	9/+	80/a	100/2
	Bromus hordeaceus	56/+	.	.	10/+	9/+	.	46/+	.	.
LC	Festuca pratensis	10/+	.	.	13/+	.	13/+	46/+	.	.
	Phleum pratense	74/+	.	.	34/+	50/+	33/+	77/+	10/+	18/+
	Elymus repens	24/+	.	.	13/+	13/+	6/+	44/1	10/+	.
LC	Dactylis glomerata	34/+	.	.	5/+	.	.	24/1	.	.
LC	Cynosurus cristatus	.	.	.	.	26/+	.	.	.	.
PL	Lolium perenne	7/4	.	25/+	79/+	87/+	40/+	77/a	30/+	6/+
	Taraxacum spec.	100/+	100/+	75/+	81/+	87/+	47/+	95/+	40/+	23/+
PL	Plantago major	37/+	.	.	55/+	44/+	20/+	33/+	.	6/+
LC	Ranunculus repens	26/+	.	.	87/a	56/+	33/+	35/a	70/1	76/+
LC	Ranunculus acris	42/+	.	25/+	74/+	87/+	47/+	35/+	.	18/+
	Cerastium fontanum	100/+	100/+	.	89/+	100/+	40/+	95/+	30/+	.
LC	Poa trivialis	39/+	.	.	55/+	62/+	60/+	88/+	60/+	47/+
	Rumex acetosa	66/+	100/+	.	87/+	75/+	40/+	54/1	20/+	.
	Cirsium vulgare	.	.	.	8/+	25/+	.	10/+	.	2/+
	Cirsium palustre	21/+	.	50/+	60/a	37/+	20/+	40/+	10/+	.
	Juncus bufonius	.	.	.	5/+	6/+	13/+	2/+	10/+	.
	Trifolium pratense	3/+	.	.	3/+	2/+	.	3/+	.	.
	Cirsium arvense	3/+	.	.	13/+	12/+	13/+	5/+	20/+	6/+
	Cardamine pratensis	8/+	.	.	13/+	20/+	47/+	5/+	60/+	41/+
	Sagina procumbens	3/+	.	50/+	31/+	25/+	13/+	.	10/+	.
	Achillea millefolium	3/+	100/+	50/+	5/+	.	.	.	.	.
	Prunella vulgaris	26/+	.	.	8/+	6/+	.	.	10/+	.
FC	Luzula campestris	3/+	.	50/+	3/+	6/+	.	.	.	6/+
	Lotus uliginosus	.	.	.	16/+	31/+	27/+	.	30/+	12/+
	Carex ovalis	.	.	25/+	.	6/+	7/+	.	.	6/+
Ccn	Lycopus europaeus	.	.	.	.	6/+	7/+	.	.	12/+
	Rhinanthus angustifolius	18/a	100/+	59/+	.	.	.	.	20/+	.
PL	Poa annua	3/+	.	.	.	.	.	.	.	.
	Hieracium laevigatum	.	.	75/+	.	.	.	.	.	.
	Chamerion angustifolium	3/+	50/+	.	.	.	.	.	.	.
	Plantago lanceolata	5/+	.	25/+	3/+	19/+	.	.	.	.
	Stellaria media	3/+	.	.	.	.	.	16/+	.	.
	Urtica dioica	3/+	.	.	.	6/+	.	.	10/+	6/+
	Juncus acutiflorus	.	.	.	.	6/+	.	.	10/+	.
Ccn	Hydrocotyle vulgaris	.	.	.	.	.	13/+	.	.	.
Ccn	Viola palustris	.	.	.	3/+	6/+	7/+	.	.	.
	Achillea ptarmica	.	.	.	.	12/+	.	2/+	10/+	.
	Equisetum fluviatile	.	.	.	.	6/+	7/+	.	10/+	6/+
	Epilobium obscurum	.	.	.	.	.	.	.	10/+	12/+
	Angelica sylvestris	.	.	.	.	.	.	5/+	.	6/+
	Juncus articulatus	.	.	.	.	12/+	20/+	2/+	40/+	29/+
	Polygonum hydropiper	.	.	.	.	.	7/+	2/+	.	18/+
Ccn	Eleochaeris palustris	.	.	.	.	.	.	.	20/+	41/+
	Glyceria fluitans	.	.	.	.	.	.	3/+	20/+	41/+
Ccn	Myosotis caespitosa	.	.	.	.	.	7/+	.	10/+	23/+
Ccn	Agrostis canina	3/+	.	.	8/+	.	.	.	.	18/a

Addenda: species occurring in one or two relevés: Mentha aquatica, Carex rostrata, Molinia caerulea, Quercus robur, Arrhenatherum elatior, Potentilla erecta, Holcus mollis, Gnaphalium uliginosum, Salix aurita, Hieracium umbellatum, Glechoma hederacea, Trifolium arvense, Polygonum persicaria, Galium mollugo, Polygonum aviculare, Rumex crispus, Veronica scutellata, Potentilla palustris, Eriophorum angustifolium, Epilobium palustre.

Photo V.3. Stand with *Anthoxanthum odoratum* and *Rhinanthus angustifolius* after 8 years of cutting without fertilizer application.

Photo V.4. Stand with *Agrostis capillaris* and *Leontodon autumnalis* after 8 years of grazing without fertilizer application.

ment in 1972 the subdivision into different coena, therefore, appeared to be primarily related to hydrological conditions.

The 1977 clustering (Fig. V.4B) showed coenon E, characterized by *Holcus lanatus, Hypochaeris radicata* and *Trifolium dubium.* This coenon was found in dry and moist sites and included permanent plots in the pasture, the hayfield and sites where the sod had been removed. Coenon F was characterized by *Agrostis capillaris, Holcus lanatus, Juncus effusus* and *Carex nigra.* All the permanent plots were situated in moderately grazed, moist sites. Coenon G contained permanent plots in

Photo V.5. Stand dominated by *Holcus lanatus* after 8 years of cutting with application of nitrogen.

Photo V.6. Stand with *Calluna vulgaris* and *Erica tetralix* after 8 years of sod removal and cutting without fertilizer application.

heavily grazed, moist sites and showed high cover percentages of the rosette plant species *Hypochaeris radicata, Bellis perennis* and *Plantago major*. Coenon H was characterized by *Calluna vulgaris* and *Erica tetralix* and was found in moist sites were the sods had been removed. The permanent plots in dry sites which were subjected to hay-making with fertilizer application were included in coenon J. The deviating position of the coena K, L and M in the dendrogram is due to the wet or inundated sites in which they were found. Coenon K was characterized by *Ranunculus flammula, Alopecurus geniculatus, Juncus articulatus* and the high coverage of *Ranunculus repens* and was found in both the pasture and hayfield. Coenon L occurred only in the hardly grazed pasture area and was characterized by a high coverage of *Juncus effusus* and *Carex nigra*. In the moderately grazed area *Agrostis capillaris* reached dominance in coenon M.

The 1982 clustering (Fig. V.4C) showed coenon N, characterized by *Agrostis capillaris, Hypochaeris radicata, Festuca rubra* and *F. ovina*. This coenon was found in dry sites where the sods had been removed. Coenon 0 was characterized by *Prunella vulgaris, Holcus lanatus, Anthoxanthum odoratum* and *Rhinanthus angustifolius* and all the permanent plots occurred in dry sites with hay-making involved (Photo V.3). The permanent plots of coenon P were often found in moderately grazed, sometimes in hayed, moist sites and showed a coverage of more than 50% for *Agrostis capillaris* accompanied by *Holcus lanatus* and many rosette plant species (Photo V.4). The permanent plots in the ploughed area were also part of coenon P. Coenon R contained grazed permanent plots with a high *Holcus lanatus* and *Lolium perenne* coverage in dry and moist conditions. Coenon S was found in dry sites with fertilizer application and was characterized by *Dactylis glomerata, Holcus lanatus, Rumex acetosa* and *Poa pratensis* (Photo V.5). The two permanent plots of coenon T were found in moist sites where the sods had been removed and were characterized by *Agrostis capillaris, Erica tetralix* and *Calluna vulgaris* (Photo V.6). Coenon U featured *Carex nigra, Agrostis stolonifera* and *Ranunculus flammula* in inundated sites under a hay-making regime and in moderately grazed areas. The permanent plots of coenon V were found in moist, moderately grazed sites and revealed cover percentages between 50 and 90 for *Agrostis capillaris* and were species-poor. Coenon W was characterized by high coverages of *Juncus effusus* and *Agrostis capillaris* and was found in wet and inundated sites in the hardly grazed area.

Although the hydrological conditions can still be recognized in the subdivision of the coena, these seem to be further differentiated by the impact of the different management practices viz. hay-making, sod removal, ploughing and different grazing intensities. The effects of these management practices will be discussed in more detail in later sections.

The 1972, 1977 and 1982 data revealed widely diverging coena and it was difficult to connect and interrelate them over the whole study period. In order to solve this problem, a qualitative comparison was made of three synoptic tables. The resulting synoptic table is shown in Table V.1 with the corresponding coena derived from the 1972, 1977 and 1982 permanent plot data at the top of the table.

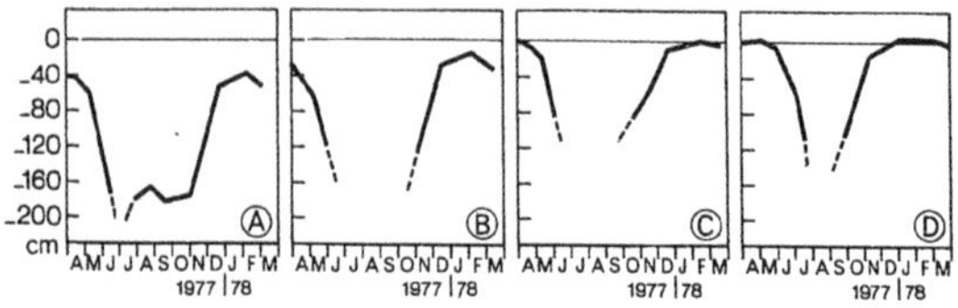

Fig. V.5A–D. Groundwater table in 1977/78 in the main grassland vegetation types. A: *Poa pratensis* community, B: *Holcus lanatus* community, C: *Agrostis capillaris* community, D: *Juncus effusus/Carex nigra* community (after Dallinga et al., 1982A).

It is not easy to classify the coena of the grassland area of the Westerholt because of their fragmented development. Most coena can be considered as basal communities of the *Molinio-Arrhenatheretea* (Basal Gesellschaft sensu Kopecký & Hejný (1974, 1975)). A subdivision is possible by taking character and differential species of the *Lolio-Potentillion anserinae* and the *Caricion curto-nigrae* (both including character species of the order and class) into account. Moreover, differential species of the *Festuco-Cynosuretum* Tx. apud Büker 1942 and the *Lolio-Cynosuretum* according to Oberdorfer (1983) enabled a further subdivision (Table V.1) by considering the total presence value of the species indicated. Thus, the coena 1, 2, and 3 could be assigned to the *Festuco-Cynosuretum* (with *Nardo-Callunetea* and *Ericion tetralicis* elements in coenon 3), coenon 4 to *Lolio-Cynosuretum* and the coena 5, 6 and 7 as transitions between both associations. Coenon 8 had more *Lolio-Cynosuretum* and *Festuco-Cynosuretum* characteristics, but included elements of *Lolio-Potentillion anserinae* and *Caricion curto-nigrae*, whereas the *Caricion curto-nigrae* elements dominated in coenon 9.

Coenon 1 was characterized by the high cover percentages of *Poa pratensis* and *Trifolium dubium*, coenon 2 by the occurrence of *Festuca rubra, Rumex acetosella, Festuca ovina* in combination with *Agrostis capillaris*. Both coena occurred in the dry areas (Fig. V.5A). Coenon 3 was characterized by the occurrence of *Erica tetralix, Calluna vulgaris, Juncus squarrosus* and *Hypochaeris radicata*, and coenon 4 by the local abundance of *Elymus repens, Festuca pratensis* and the high coverage of *Holcus lanatus*. These coena were found is somewhat moister areas. Coenon 5 was characterized by high cover percentages of rosette plant species (e.g. *Leontodon autumnalis* and *Hypochaeris radicata*) and coenon 6 by high cover percentages of *Agrostis capillaris* and *Holcus lanatus*. The very high cover percentage of *Agrostis capillaris* was characteristic for coenon 7. These three coena were found on moist soils whereas coenon 8 was found in wetter areas and was characterized by *Juncus effusus* and *Agrostis stolonifera*. Coenon 9 was found in both wet and inundated areas and was characterized by *Juncus effusus* and *Carex nigra*.

V.C.3. Heathland communities (Table V.2)

The results of the clustering of the permanent plot

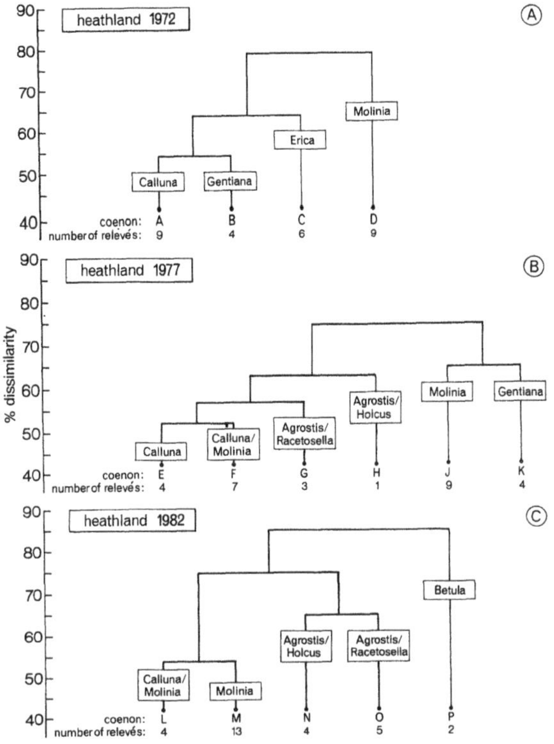

Fig. V.6. Simplified (all dissimilarities below 50% are not distinguished) dendrograms of all permanent plots in the heathland area in 1972, 1977 and 1982, respectively, with an indication of the dominant species.

data in 1972, 1977 and 1982 are shown in the dendrograms in Fig. V.6A–C. Four clusters can be recognized in the 1972 data, six in 1977 and five in the 1982 data at a dissimilarity level of 50%.

In the 1972 clustering (Fig. V.6A) coenon A was characterized by a high cover percentage of *Calluna vulgaris,* coenon B by the occurrence of *Potentilla erecta* and *Gentiana pneumonanthe*, coenon C by the high cover percentage of *Erica tetralix* and coenon D by the dominance of *Molinia caerulea.* The coena A and C were species-poor and occurred on the drier soils, whereas the species-richer coena B and D were found on moister soils.

The 1977 clustering (Fig. V.6B) showed coenon E, characterized by *Calluna vulgaris,* coenon F by the co-dominance of *Calluna vulgaris* and *Molinia caerulea,* coenon G by *Agrostis capillaris* and *Rumex acetosella,* coenon H by *Agrostis capillaris* and *Holcus lanatus,* coenon J by *Molinia caerulea,* and coenon K by *Potentilla erecta* and *Gentiana pneumonanthe.* The coena E, F, G and H were found on drier soils and were species-poor, whereas the coena J and K were richer in species and occurred on moister soils (Fig. V.7).

In the 1982 clustering (Fig. V.6C) coenon L was characterized

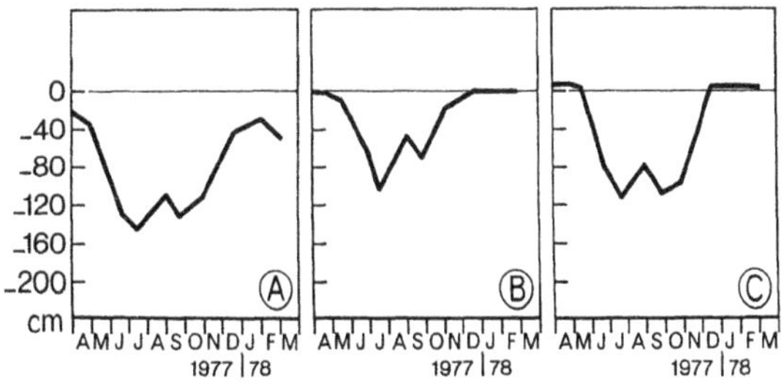

Fig. V.7A–C. Groundwater table in 1977/78 in the main heathland vegetation types. A: *Calluna vulgaris* community, B: *Molinia/Gentiana* community, C: *Molinia caerulea* community (after Dallinga & Tjaden, 1981).

by the co-dominance of *Calluna vulgaris* and *Molinia caerulea,* coenon M by the dominance of *Molinia caerulea,* coenon N by *Agrostis capillaris* and *Holcus lanatus,* coenon O by *Agrostis capillaris* and *Rumex acetosella* and coenon P by *Betula pubescens.* Coenon M was species-rich and occurred on moister soils, whereas all the other coena were species-poor and occurred on both drier and moister soils.

Although the hydrological conditions could still be recognized in the subdivision of the coena, these seemed to be further differentiated by the impact of different grazing intensities (increase of *Holcus lanatus* and *Agrostis capillaris* with heavy grazing and of *Betula pubescens* with laissez-faire).

As the grassland communities a qualitative comparison was made for the heathland area. The resulting synoptic table is shown in Table V.2 with the corresponding coena derived from the 1972, 1977 and 1982 permanent plot data at the top of the table.

All the coena include *Genisto-Callunetum* and *Ericetum tetralicis* elements and cannot easily be subdivided because of the often fragmented development and only the vascular plants are known. A subdivision was made taking character and differential species of the *Genisto-Callunetum* and *Ericetum tetralicis* into account including alliance character species. Thus, the coena 10 and 11 can be classified as *Ericetum tetralicis,* coenon 15 as a *Molinia caerulea* facies of the *Ericetum tetralicis* and coenon 16 as *Ericetum tetralicis* with *Genisto-Callunetum* elements. Coenon 12 can be considered as a fragmented development of coenon 16. The coena 13 and 14 feature a fragmented developed *Ericetum tetralicis* with *Festuco-Cynosuretum* elements indicating successional stages from heathland to grassland. Coenon 17 is regarded as a *Molinia caerulea/Betula pubescens* sociation representing a successional stage from heathland to woodland.

Coenon 10 was characterized by *Calluna vulgaris,* coenon 11 by *Erica tetralix,* coenon 12 by *Calluna vulgaris* and *Molinia caerulea,* coenon 13 by *Agrostis capillaris, Rumex acetosella,*

Table V.2. Synoptic table of the coena (numbers 10–17) occurring in the Westerholt heathland area during the period 1972 to 1982. The coena found in the years 1972, 1977 and 1982 are referred to by the letters A–P. Character and differential species of the Ericion tetralicis (E) and the Genisto-Callunetum (GC) are indicated. The occurrence of the species is given in percentages of frequency and in mean cover percentages (+: 1–5; a: 6–10; 1: 11–20; 2: 21–30 etc.).

	Coenon	16	15	11	10	12	13	14	17
	1972 data	B	D	C	A				
	1977 data	K	J		E	F	G	H	
	1982 data		M			L	O	N	P
	Number of relevés	8	31	6	13	11	8	5	2
E	Erica tetralix	100/+	81/a	100/2	93/a	100/+	87/+	40/+	.
GC	Calluna vulgaris	100/+	59/a	100/3	93/4	100/2	87/+	20/+	.
E	Molinia caerulea	100/2	100/5	83/a	93/+	100/2	75/1	80/1	100/+
	Potentilla erecta	87/+	18/+	.	.	.	.	40/+	.
E	Gentiana pneumonanthe	75/+	15/+	.	.	.	.	.	.
GC	Festuca ovina	87/+	11/+	.	7/+	27/+	.	.	.
E	Luzula multiflora	25/+	.	.	7/+	.	.	.	.
E	Salix repens	62/+	11/+	.	.	.	.	.	.
E	Dactylorhiza maculata	62/+	15/+	.	.	.	.	.	.
	Arnica montana	12/+	4/+	.	.	.	.	.	.
	Genista anglica	12/+	4/+	.	.	.	.	.	.
E	Pedicularis sylvatica	12/+	4/+	.	.	.	.	.	.
	Succisa pratensis	12/+	11/+	.	.	.	.	.	.
E	Carex panicea	12/+	7/+	.	.	9/+	.	.	.
	Agrostis capillaris	12/+	44/+	.	21/+	18/+	100/2	100/1	.
	Poa pratensis		18/+	.	21/+	9/+	50/+	60/+	
	Holcus lanatus	50/+	67/+	17/+	28/+	9/+	87/+	100/1	.
	Anthoxanthum odoratum	25/+	30/+	.	.	42/+	62/+	20/+	.
GC	Rumex acetosella	12/+	11/+	.	7/+	36/+	100/+	20/+	.
	Betula pubescens	75/+	78/+	50/+	35/+	82/+	50/+	80/+	100/8
E	Juncus squarrosus	37/+	41/+	33/+	7/+	36/+	.	.	.
E	Scirpus cespitosus	12/+	26/+	33/+	42/+	45/+	25/+	.	.
GC	Carex pilulifera	62/+	11/+	17/+	13/+	27/+	50/+	40/+	.
	Quercus robur	12/+	15/+	17/+	.	9/+	25/+	20/+	.
	Agrostis canina	87/+	52/+	.	13/+	64/+	87/+	60/+	.
E	Rhynchospora alba	.	4/+	.	13/+	.	.	.	.
E	Drosera rotundifolia	12/+	7/+	17/+	13/+	.	.	.	.
E	Eriophorum angustifolium	12/+	15/+	33/+	7/+	18/+	.	.	.
	Salix aurita	12/+	7/+	.	.	.	.	.	.
E	Narthecium ossifragum	.	7/+	.	.	.	.	20/+	.
	Danthonia decumbens	12/+	4/+	.	.	.	.	.	.
	Hydrocotyle vulgaris	12/+	11/+	.	.	.	.	20/+	.
GC	Deschampsia flexuosa	37/+	33/+	.	21/+	45/+	37/+	20/2	.
E	Carex nigra	62/+	55/+	33/+	42/+	45/+	37/+	40/+	.
	Juncus effusus	37/+	11/+	.	.	9/+	.	40/+	.
	Cirsium palustre	.	7/+	.	.	.	.	20/+	.
	Cerastium fontanum	12/+	18/+	.	7/+	18/+	37/+	60/+	.
GC	Luzula campestris	25/+	4/+	.	.	27/+	.	.	.
	Rubus spec.	.	.	.	.	.	12/+	.	50/+
	Plantago major	.	.	.	7/+	9/+	.	.	.
	Rumex acetosa	.	4/+	.	.	.	.	20/+	.
	Galium palustre	.	4/+	.	.	.	.	20/+	.
	Leontodon autumnalis	.	.	.	.	9/+	.	20/+	.
	Equisetum fluviatile	25/+	.	.	.	.	.	.	.
	Trifolium repens	.	7/+	.	.	.	.	.	.
	Frangula alnus	.	4/+	.	.	18/+	.	.	.
	Ranunculus acris	.	4/+	.	.	.	.	20/+	.
	Galeopsis bifida	.	.	.	.	.	.	80/+	.
	Poa trivialis	12/+	7/+	.	21/+	.	.	.	.

Addenda: Species occurring in one relevé: Poa annua, Galium saxatile, Peucedanum palustre, Hieracium umbellatum, Urtica dioica, Lolium perenne, Festuca rubra, Galeopsis spec., Galeopsis tetrahit, Sorbus aucuparia, Drosera intermedia, Epilobium spec., Taraxacum spec., Angelica sylvestris, Agrostis stolonifera, Lamium album, Stellaria media, Senecio vulgaris, Cardamine pratensis, Hypochaeris radicata, Thelypteris palustris, Dryopteris carthusiana, Juncus conglomeratus.

Erica tetralix and *Calluna vulgaris*, coenon 14 by *Molinia caerulea* and *Holcus lanatus*, coenon 15 by *Molinia caerulea*, coenon 16 by *Molinia caerulea* and *Gentiana pneumonanthe* and coenon 17 by *Betula pubescens* and other trees.

V.D. Permanent plots

V.D.1. Subsets of permanent plots

At the start of the experiment in 1972 the grassland area was subdivided into dry, moist, wet and inundated plant communities (Fig. V.4A) and the heathland into dry and moister communities (Fig. V.6A). Various numbers of permanent plots were situated in these different plant communities. The imposition of different management regimes resulted in the subsets of permanent plots shown in Table V.3. The lay-out of management regimes was shown in Fig. V.1.

V.D.2. Cover changes of abundant species

In the dry *Poa pratensis* community (Fig. V.8A) *Agrostis capillaris* increased strongly in the grazed area, whereas *Elymus repens* disappeared and *Taraxacum spec.* and *Hypochaeris radicata* temporarily reached coverages of 10%. *Lolium perenne* and *Poa pratensis* disappeared under sod removal and hay-making regimes, whereas different species temporarily appeared in both regimes. The application of fertilizer followed by hay-making resulted in an enormous increase of *Holcus lanatus* followed by the appearance of *Rumex acetosa* and *Dactylis glomerata*.

The moist *Holcus lanatus* community (Fig. V.8B) featured a large increase of *Agrostis capillaris* and temporarily increases of *Trifolium repens* and *Leontodon autumnalis* in the grazed plots. The increase of *Agrostis capillaris* was less in the cut plots, whereas *Anthoxanthum odoratum* appeared for the first time. *Hypochaeris radicata* reached high abundances after sod removal, whereas the *Agrostis capillaris* cover remained low. *Agrostis capillaris,* however, recovered rapidly after ploughing.

In the wet, grazed, *Agrostis capillaris* community (Fig. V.8C) *Agrostis capillaris* and *Juncus effusus* became dominant.

Juncus effusus became the only dominant species in the grazed, inundated *Juncus effusus/Carex nigra* community (Fig. V.8D), whereas *Agrostis stolonifera* rapidly disappeared. *Agrostis stolonifera* recovered itself and *Carex nigra* persisted better in the cut plots, while *Ranunculus repens* also became an important species. *Juncus effusus* reached only small cover percentages in the hay-making regime.

In the *Calluna vulgaris* community (Fig. V.8E) both *Calluna vulgaris* and *Erica tetralix* disappeared in the grazed plots, whereas *Agrostis capillaris* and *Molinia caerulea* took over. *Molinia caerulea* became the only dominant species in the abandoned plots after an increase and subsequent set back of *Calluna vulgaris.*

The *Molinia Gentiana* community (Fig. V.8F)

Table V.3. Numbers of permanent plots in different regimes in various plant communities.

Coenon	Regime					
	Grazing	Hay-making	Sod removal/ hay-making	Ploughing/ hay-making	Fertilizer applications/ hay-making	Abandoning
Poa pratensis community	4	12	2	–	8	–
Holcus lanatus community	28	4	2	4	–	–
Agrostis capillaris community	4	–	–	–	–	–
Juncus effusus/Carex nigra community	4	4	–	–	–	–
Calluna vulgaris community	11	–	–	–	–	3
Molinia/Gentiana community	12	–	–	–	–	2

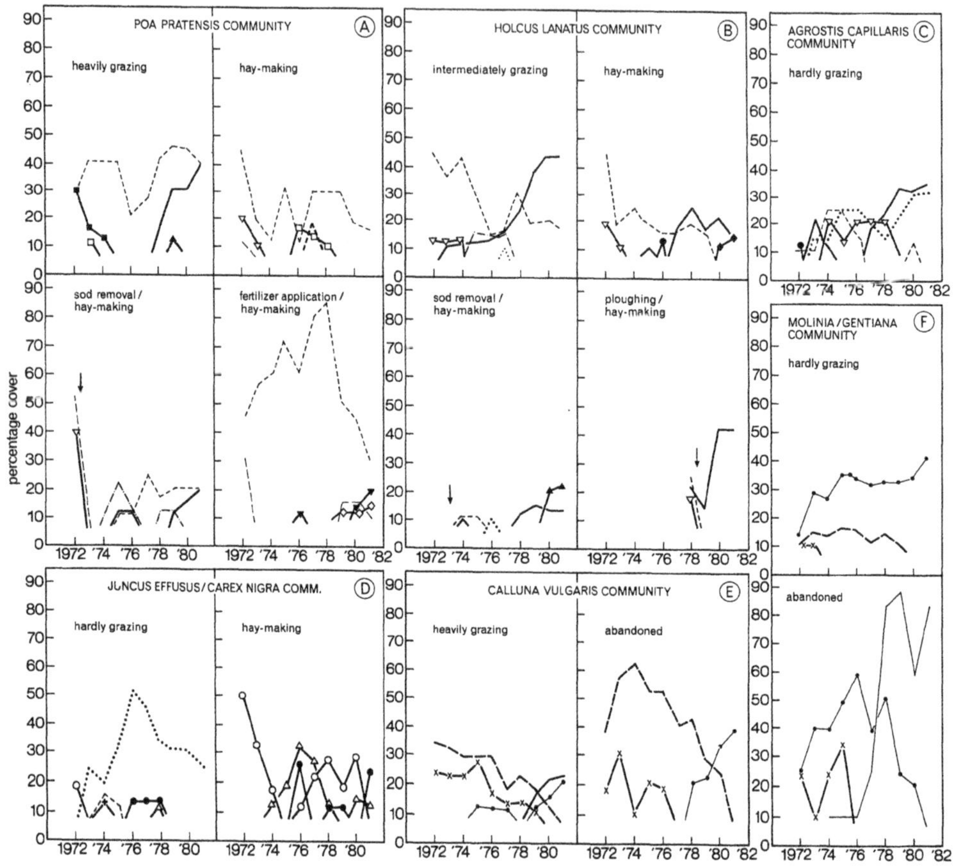

Fig. V.8A–F. Changes of cover percentages of abundant (coverage ⩾10%) species under different management regimes in various plant communities (after Veldman, 1983). Juncus effusus ●●●●, Agrostis stolonifera ○——○, Galium palustre +——+, Carex nigra ●——●, Holcus lanatus — — —, Ranunculus repens △——△, Agrostis capillaris ———, Hypochaeris radicata ▲——▲, Taraxacum spec. □——□, Elymus repens ■——■, Lolium perenne ▽——▽, Poa pratensis ----, Trifolium dubium — — —, Rumex acetosa ▼——▼, Dactylis glomerata ◇——◇, Anthoxanthum odoratum ◆——◆, Trifolium repens ·—·—, Leontodon autumnalis ...., Calluna vulgaris — —, Betula pubescens ———, Molinia caerulea •——•, Erica tetralix ×——×.

featured *Molinia caerulea* rapidly taking over in the grazed plots. The same happened in the abandoned plots during the first years, but *Betula pubescens* later became the dominant species.

In all plant communities most management practices have a sometimes large year-to-year fluctuation in cover percentages of the abundant species in common. The next section deals with the question as to whether the vegetational changes found in the permanent plots reveal trends, taking the full species composition in the permanent plots into account.

V.D.3. Vegetational trends

Vegetational changes in the permanent plots were quantified to detect trends or gradual deviations from the initial stand as indicated in section II.C.3. The vegetational trends are shown in Fig. V.9A–F.

The vegetation in the dry *Poa pratensis* community (Fig. V.9A) did not change markedly in the plots where the sods had been removed, but changed significantly in the grazed, cut and fertilized permanent plots.

With the exception of the ploughed plots, the vegetation changed significantly in the moist *Hol-*

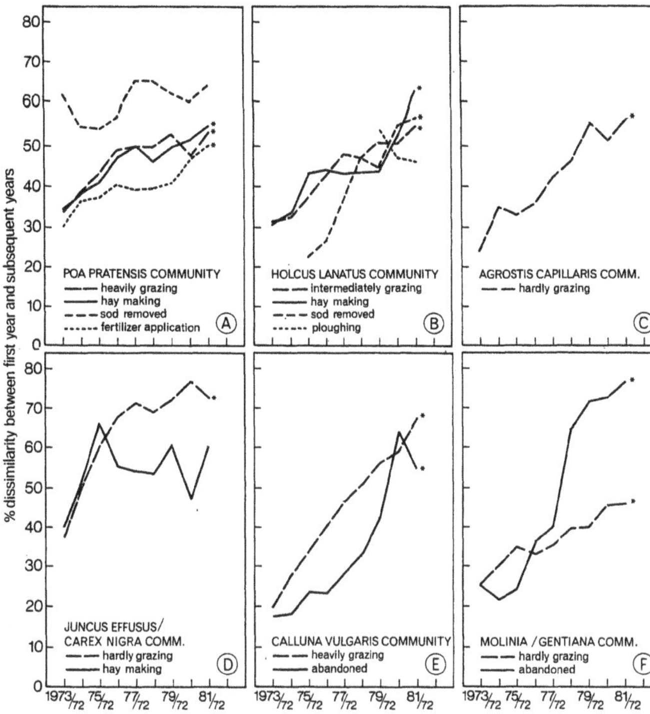

Fig. V.9A–F. Changes of percentages dissimilarity between the first year and subsequent years under different management practices in various plant communities. * Significant change according to Spearman's rank correlation test ($P<0.05$) (after Veldman, 1983).

cus lanatus community (Fig. V.9B) under all management practices.

Grazing caused a significant change in the wet *Agrostis capillaris* community (Fig. V.9C).

The degree of dissimilarity changed significantly in the inundated *Juncus effusus*/*Carex nigra* community (Fig. V.9D) in the grazed permanent plots. Hay-making only revealed a sudden large deviation from the initial stand with subsequent large fluctuations.

It should be kept in mind that, although the former agricultural practices included fertilizer applications, hay-making and grazing the aftermath, the newly imposed practices deviate strongly, not only because of the cessation of fertilizer applications. Hay-making is carried out later than in former times and grazing not only takes place for a much longer period, but implies, in fact, abandoning at the places with moderate and in particular hardly any grazing. This is also the case in the comparison of species diversity and equitability which will be discussed in the next two sections. The differences in grazing intensity are considered in detail in chapter VIII.

The vegetation changed significantly in the *Calluna vulgaris* community (Fig. V.9E) in both the grazed and abandoned plots. Eventually the deviation from the initial stand was similar, but initially the changes under grazing were higher than in the abandoned plots, in which no new regime was imposed.

The abandoned wet/inundated *Molinia*/*Gentiana* community (Fig. V.9F) deviated much more from the initial stand than the grazed plots. Changes in both regimes were, however, significant.

The pattern of vegetational change after the imposition of new management practices can also be demonstrated by the dissimilarity percentages between consecutive years for various parts of the study area. The *Poa pratensis* community (Fig. V.10A) showed a significantly decreasing dissimilarity after sod removal and hay-making and a pattern of small changes with grazing, hay-making and fertilizer applications and purely hay-making. This

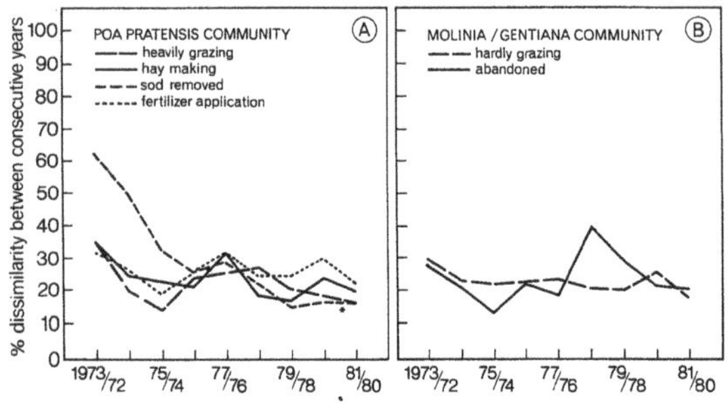

Fig. V.10A–B. Changes of percentages dissimilarity between consecutive years under different management practices in various plant communities. * Significant change according to Spearman's rank correlation test (P<0.05) (after Veldman, 1983).

pattern of small year-to-year changes is representative for all other communities of which the *Molinia/Gentiana* community is given as another example (Fig. V.10B).

V.D.4. Species diversity

This section covers the changes in species diversity in the plant communities which were studied. The development under the management regimes mentioned in section V.D.1 will be discussed (Fig. V.11A–F). Changes in species diversity were considered to be trends as indicated in section II.C.3. Statistical significance of differences among the treatments in the communities were assessed by applying the sign test (P<0.05).

In the dry *Poa pratensis* community (Fig. V.11A) species diversity increased significantly in the cut plots and in the plots with sod removal. Species diversity seemed to decrease after an initial increase in the grazed plots. The application of fertilizers resulted in a significantly lower species diversity than in all the other experiments. Grazing resulted in a significantly higher species diversity than sod removal.

The moist *Holcus lanatus* community (Fig. V.11B) showed no trends in species diversity. Grazing seemed to show an increase and subsequent decrease as in the dry *Poa pratensis* community. Species diversity differed significantly between the management practices namely, sod removal >grazing >hay-making.

Species diversity decreased significantly in the wet *Agrostis capillaris* community (Fig. V.11C) under grazing. Although it was significantly higher than in the cut permanent plots, species diversity decreased significantly in the inundated *Juncus effusus/Carex nigra* community (Fig. V.11D) under grazing, becoming equal to the mown plots in the end.

Species diversity increased significantly in both the grazed and abandoned plots in the *Calluna vulgaris* community (Fig. V.11E). The number of species was significantly highest in the grazed plots. In the *Molinia/Gentiana* community (Fig. V.11F) no trend was found with respect to changes in species diversity. Species diversity was significantly higher in the grazed plots than in those which were abandoned.

It may again be emphasized that all considerations on species diversity only include vascular plants.

V.D.5. Equitability

This section deals with the changes in relative equitability i.e. a measure of the concentration of dominance in the group of species in a relevé. Changes in relative equitability were called trends as indicated in section II.C.3. Statistical significance of differences among the treatments in the communities were assessed by applying the sign test (P<0.05) (Fig. V.12A–F).

The relative equitability was significantly lower in the fertilized site than in all other regimes in the *Poa pratensis* community. In the *Holcus lanatus* community it was higher with sod removal than with grazing, but it showed a decreasing trend with both regimes. The relative equitability decreased significantly in the hardly grazed *Agrostis capillaris* and the *Juncus effusus/Carex nigra* communities. It was significantly lower in the grazed than in the mown sites in the *Juncus effusus/Carex nigra* community.

The relative equitability fluctuated in the grazed and abandoned *Calluna vulgaris* community, but showed a decreasing trend in the *Molinia/Gentiana* community.

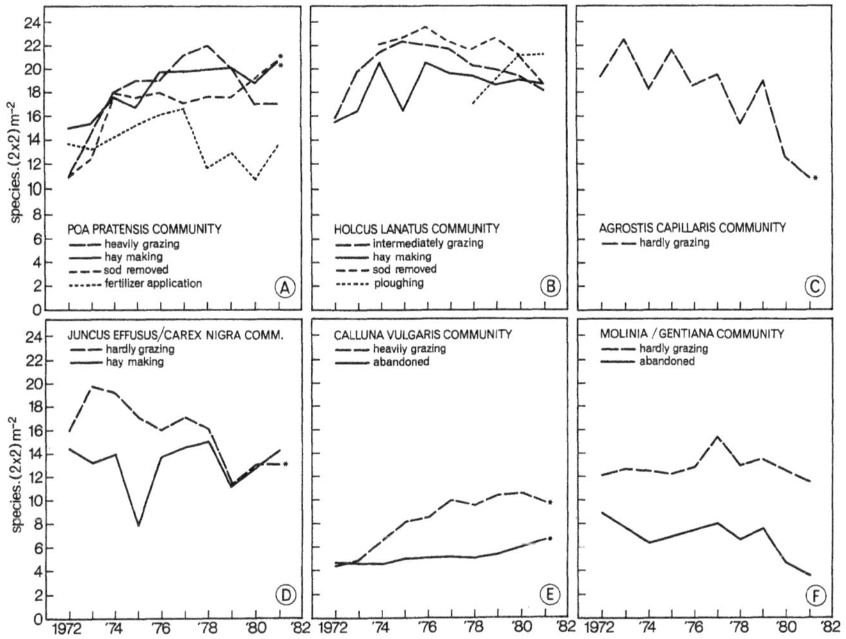

Fig. V.11A–F. Changes of numbers of species · 4 m^{-2} under different management practices in various plant communities. * Significant change according to Spearman's rank correlation test ($P<0.05$).

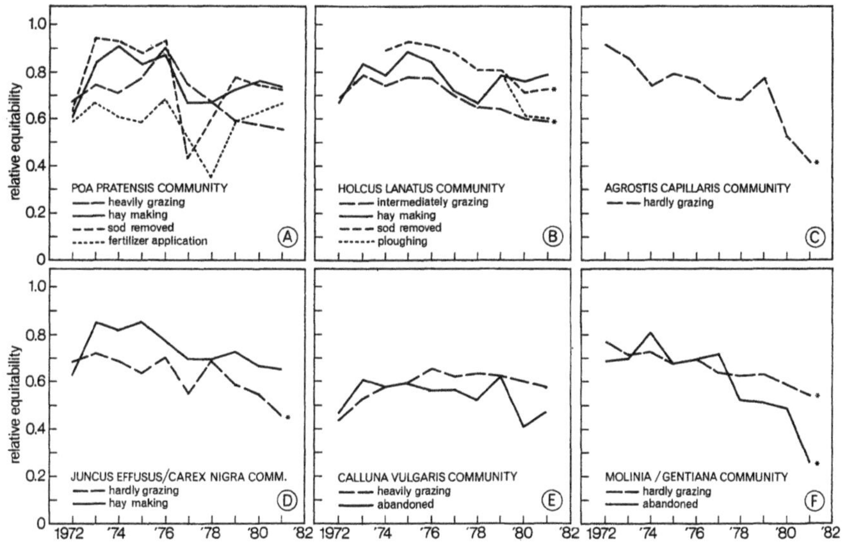

Fig. V.12A–F. Changes of the relative equitability under different management practices in various plant communities. * Significant change according to Spearman's rank correlation test ($P<0.05$) (after Veldman, 1983).

Table V.4A. Cover changes of all species occurring in ⩾33% of the permanent plots in the Poa pratensis community. P: species indicating nutrient-poor soil conditions; R: nutrient-rich conditions; I: intermediate. +: increase 1–3%, ++: increase 4–10%, +++: increase >10%, −: decrease 1–3%, −−: decrease 4–10%, −−−: decrease >10%, Λ: temporarily increase with (−) eventually smaller cover percentage or (+) eventually higher cover percentage, C: constant, .: absent.

Poa pratensis community		Grazing	Hay-making		
				Fertilizer applications 50 kg $N.ha^{-1}$	Sod removal
R	Poa pratensis	C	C	C	C
I	Festuca rubra	.	C	C	+
I	Holcus lanatus	C	C	Λ	+ + +
R	Bromus hordeaceus	Λ	Λ	Λ	Λ (+)
I	Bellis perennis	Λ	Λ	Λ	Λ (+)
R	Cirsium vulgare	C	Λ	Λ	.
R	Lolium perenne	-	- - -	- -	C
R	Phleum pratense	-	-	-	.
R	Elymus repens	- - -	- -	C	.
R	Alopecurus geniculatus	.	-	-	.
I	Cerastium fontanum	-	-	Λ (-)	C
I	Agrostis capillaris	+ + +	+ +	Λ	+ + +
R	Trifolium repens	Λ	-	-	C
P	Calluna vulgaris	Λ	.	.	.
R	Plantago major	Λ	-	-	- -
R	Poa trivialis	Λ	-	C	C
R	Cirsium arvense	Λ	.	.	.
I	Sagina procumbens	Λ	.	.	+
P	Hypochearis radicata	+ +	+ +	Λ (+)	+
I	Leontodon autumnalis	Λ	+	Λ (-)	Λ
I	Anthoxanthum odoratum	Λ	+ +	C	+
P	Rumex acetosella	Λ	+ +	.	Λ (+)
I	Trifolium dubium	.	Λ	Λ (-)	C
R	Poa annua	.	Λ	.	-
	Taraxacum spec.	-	Λ (+ +)	Λ (-)	Λ
R	Ranunculus repens	Λ	+ +	C	-
I	Ranunculus acris	Λ	+	Λ (+)	.
P	Rhinanthus angustifolius	.	+	.	+
I	Plantago lanceolata	.	+	.	.
I	Rumex acetosa	Λ	+	+ + +	+
R	Festuca pratensis	-	C	Λ	.
I	Dactylis glomerata	.	+	+ + +	.
P	Cirsium palustre	C	Λ	+ +	.
R	Stellaria media	-	Λ (-)	Λ	-
I	Achillea millefolium	.	.	.	+
P	Betula pubescens	.	.	.	+

Table V.4B. Cover changes of all species occurring in ≥33% of the permanent plots in the Holcus lanatus, Agrostis capillaris and Juncus effusus/Carex nigra communities. P: species indicating nutrient-poor soil conditions; R: nutrient-rich conditions; I: intermediate. +: increase 1–3%, ++: increase 4–10%, +++: increase >10%, −: decrease 1–3%, −−: decrease 4–10%, −−−: decrease >10%, Λ: temporarily increase with (−) eventually smaller cover percentage or (+) eventually higher cover percentage, C: constant, .: absent.

		Holcus lanatus community				Agrostis capillaris community	Juncus effusus/Carex nigra community	
		Grazing	Hay-making	Sod removal and hay-making	Ploughing and hay-making	Grazing	Grazing	Hay-making
I	Holcus lanatus	- - -	- - -	- -	Λ	Λ (- -)	C	C
I	Cerastium fontanum	-	-	-	C	Λ	.	.
R	Lolium perenne	- -	- - -	Λ (-)	-	Λ (- -)	.	C
R	Poa pratensis	- -	C	-	-	- -	.	C
R	Bromus hordeaceus	Λ (-)	.	.	.	.	.	.
R	Festuca pratensis	-	.	.	.	-	.	.
R	Phleum pratense	-	C	.	.	-	.	C
R	Poa annua	-	.	- -	-	-	.	.
R	Plantago major	Λ (-)	-	- -	.	-	.	.
I	Agrostis capillaris	+ + +	+ + +	+	+ + +	+ + +	.	.
I	Leontodon autumnalis	Λ (+ +)	+ +	Λ	Λ (+)	-	Λ	C
P	Hypochaeris radicata	Λ (+)	+	+ + +	+ +	.	.	.
I	Bellis perennis	Λ	Λ (-)	- -	C	-	.	.
R	Trifolium repens	Λ (+)	-	- -	-	-	-	-
I	Cynosurus cristatus	Λ (+)	.	+	.	.	.	.
I	Galium palustre	Λ	C	.	.	-	-	-
	Taraxacum spec.	Λ	C	-	C	-	-	-
R	Ranunculus repens	+	C	.	C	Λ (-)	C	Λ (+ +)
I	Ranunculus acris	+	C	.	C	C	C	C
I	Juncus bufonius	C	.	-	-	Λ (-)	.	.
I	Sagina procumbens	Λ	.	-	.	-	.	.
R	Cirsium vulgare	Λ	.	.	.	.	.	.
I	Trifolium dubium	Λ (-)	Λ (+)	Λ (+)	.	.	.	.
I	Anthoxanthum odoratum	C	+ + +	+	+	-	.	+
I	Rumex acetosa	-	+ +	.	C	Λ	-	C
I	Cardamine pratensis	C	Λ (+)	.	.	C	-	C
P	Prunella vulgaris	.	Λ (+)	.	.	.	.	.
P	Rhinanthus angustifolius	.	+ +	+	.	.	.	.
P	Erica tetralix	.	.	+	+	.	.	.
P	Calluna vulgaris	.	.	+ +	+	.	.	.
P	Juncus squarrosus	.	.	+	+	.	.	.
I	Carex ovalis	.	.	Λ	.	.	.	C
P	Cirsium palustre	C	.	Λ	.	Λ	.	.
R	Agrostis stolonifera	.	.	.	.	.	- - -	C
P	Juncus effusus	Λ	C	C	+	+ + +	Λ (+ + +)	C
P	Juncus articulatus	.	.	.	.	-	C	-
R	Poa trivialis	Λ	-	C	.	Λ (+)	C	C
P	Juncus conglomeratus	+	.	+	+	+	.	.
P	Carex nigra	Λ	+ +	Λ (+)	.	C	Λ (-)	+ +
R	Alopecurus geniculatus	Λ (-)	-	.	.	- -	- - -	Λ (+)
P	Ranunculus flammula	.	.	.	.	-	-	C

Table V.4C. Cover changes of all species occurring in ⩾33% of the permanent plots in the Calluna vulgaris and Molinia caerulea/Gentiana pneumonanthe communities. P: species indicating nutrient-poor soil conditions; R: Nutrient-rich conditions; I: intermediate. +: increase 1–3%, ++: increase 4–10%, +++: increase >10%, −: decrease 1–3%, −−: decrease 4–10%, −−−: decrease >10%, Λ: temporarily increase with (−) eventually smaller cover percentage or (+) eventually higher cover percentage, C: constant, .: absent.

		Grazed		Abandoned	
		Calluna vulgaris community	Molinia/ Gentiana community	Calluna vulgaris community	Molinia/ Gentiana community
P	Calluna vulgaris	- - -	Λ (- -)	- - -	- -
P	Erica tetralix	- - -	C	- - -	- - -
P	Eriophorum angustifolium	.	-	.	-
P	Potentilla erecta	.	-	.	-
P	Gentiana pneumonanthe	.	-	.	-
P	Dactylorhiza maculata	.	-	.	-
P	Salix repens	.	-	.	-
P	Salix aurita	.	-	.	-
P	Juncus conglomeratus	.	-	.	.
P	Narthecium ossifragum	.	-	.	.
P	Drosera rotundifolia	.	-	.	.
P	Rhynchospora alba	.	-	.	.
P	Carex panicea	.	-	.	.
P	Succisa pratensis	.	-	.	.
P	Pedicularis sylvatica	.	-	.	.
P	Genista anglica	.	-	.	.
P	Molinia caerulea	+ + +	+ + +	+ + +	Λ (- - -)
P	Betula pubescens	+ +	+	+ +	+ + +
P	Deschampsia flexuosa	+ +	+	.	.
I	Holcus lanatus	+ +	+	.	+
I	Agrostis capillaris	+ + +	C	.	.
P	Agrostis canina	+	C	.	- -
R	Poa trivialis	Λ	C	.	.
	Quercus robur	Λ	C	.	.
P	Carex nigra	+ +	C	C	.
I	Cerastium fontanum	Λ	-	.	.
R	Poa pratensis	+	-	.	.
P	Rumex acetosella	+	Λ	.	.
I	Anthoxanthum odoratum	+	Λ	.	.
P	Carex pilulifera	Λ	Λ	.	.
P	Scirpus cespitosus	C	Λ	Λ	.
P	Juncus squarrosus	.	Λ	.	.
P	Luzula campestris	.	Λ	.	.
P	Danthonia decumbens	.	Λ	.	.
	Sorbus aucuparia	.	Λ	.	+
P	Festuca ovina	.	C	.	.
P	Juncus effusus	.	C	.	.
P	Luzula multiflora	.	C	.	.
R	Ranunculus repens	.	+	.	.

V.D.6. Survey of species changes

The cover changes of abundant species (cover percentages ≥10% during part of the study period) were discussed in section V.D.2. This section gives a survey of groups of species with a similar successional response under different management practices in some plant communities.

The successional response of each species is shown in Table V.4A–C indicating the period in which the species reached maximal cover percentage. If maximal cover percentage was reached in one of the first three years the species was considered as a decreaser whereas if maximal cover percentage was reached in one of the last three years the species was considered to be an increaser. If, however, maximal coverage was reached in the intermediate years the species was considered to be fluctuating and if no maximal cover percentage was found the species was regarded as being constant. Some additional quantification was added by indicating the initial coverage as well as small or large increases or decreases of cover percentages. The large changes are in agreement with the information provided in section V.D.2. In the dry *Poa pratensis* community (Table V.4A) several groups of plant species can be discerned, each behaving characteristically in the different management regimes. Only the *Poa pratensis*-group, including both 'rich' and 'intermediate' species reacted in a similar way in all management practices, remaining constant or showing temporary cover increase. The *Lolium perenne*-group with 'rich' species showed a cover decrease in all regimes, except for sod removal where these species were often not present at all. The *Agrostis capillaris*-group showed a cover increase particularly in the grazed plots, whereas the species of the *Leontodon autumnalis*-group revealed increasing cover percentages in the haymaking regime. Species of the *Rumex acetosa*-group increased particularly in the plots with fertilizer application. *Achillea millefolium* and *Betula pubescens* only increased in the plots with sod removal. The four last groups in the table included 'poor', 'intermediate' and 'rich' species. This discrepancy also occurred in the groups of the communities mentioned in tables V.4B–C).

The moist *Holcus lanatus* community, the wet *Agrostis capillaris* and the inundated *Juncus effusus*/*Carex nigra* communities (Table V.4B) also revealed groups of species with similar responses. Species of the *Holcus lanatus-group* showed decreasing cover percentages in most moist plots, whereas species of the *Agrostis capillaris*-group revealed the opposite trend. The *Bellis perennis*-group showed a cover increase in the grazed permanent plots in the moist community, a decrease under all other treatments. The *Trifolium dubium*-group and the *Erica tetralix*-group revealed a cover increase in the plots with hay-making and sod removal, or ploughing and hay-making, respectively.

Agrostis stolonifera remained constant in the inundated community. The *Juncus effusus*-group showed a cover increase in the grazed wet/inundated plots, whereas the species of the *Carex nigra*-group revealed a cover percentage increase in the inundated plots with hay-making.

In the heathland communities (Table V.4C) *Calluna vulgaris* and *Erica tetralix* both decreased in the grazed and the abandoned plots and *Erica tetralix* only maintained itself in the grazed *Ericetum tetralicis*. The *Eriophorum angustifolium*-group, with characteristic heathland species, also decreased in the grazed plots. *Molinia caerulea* and *Betula pubescens* increased in all the management regimes. The decrease of *Molinia caerulea* in the abandoned *Ericetum tetralicis* was caused by the large increase of *Betula pubescens*. *Agrostis capillaris* and *A. canina* increased particularly in the grazed *Genisto-Callunetum* as is the case of the species of the *Poa trivialis*-group. It can be concluded that 'intermediate' and 'rich' species appear to increase more in the grazed dry heathland than in the wet heathland community. This phenomenon is descussed later in chapter VIII.

V.E. Vegetation mapping

V.E.1. Introduction

The permanent plot data discussed in the preceding section gave detailed information for limited areas which are not representative in all the details for

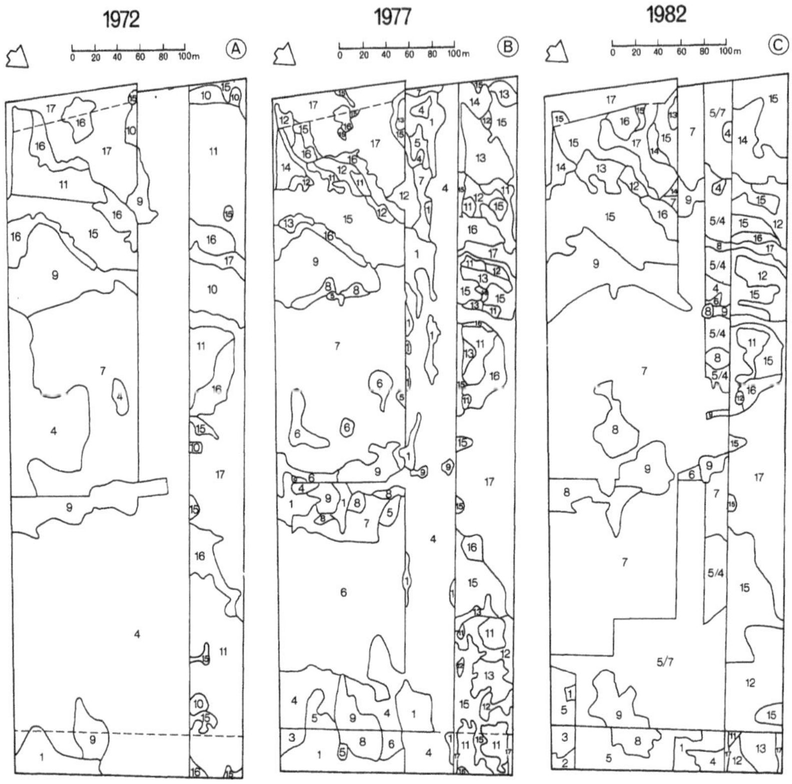

Fig. V.13A–C. Simplified vegetation map of the Westerholt study area. A: 1972, B: 1977, C: 1982 (after Dallinga et al., 1982A; Ter Heerdt et al., 1986). Coenon 1 *Poa pratensis* community, 2 *Agrostis capillaris/Rumex acetosella* community, 3 *Hypochaeris/Calluna* community, 4 *Holcus lanatus* community, 5 *Leontodon/Hypochaeris* community, 6 *Agrostis/Holcus* community, 7 *Agrostis capillaris* community, 8 *Juncus/Agrostis stolonifera* community, 9 *Juncus effusus/Carex nigra* community, 10 *Calluna vulgaris* community, 11 *Erica tetralix* community, 12 *Calluna/Molinia* community, 13 *Agrostis capillaris/Rumex acetosella/Calluna vulgaris* community, 14 *Molinia/Holcus* community, 15 *Molinia caerulea* community, 16 *Molinia/Gentiana* community, 17 *Betula*/woodland community.

the whole area. Information about changes in the whole study area can be gained from vegetation mapping. This is considered in this section.

In two cases, a complex was mapped because the coena 5 and 7 and 5 and 4 occurred in a mosaic of which the components could not be mapped at the 1 : 1000 scale which had been adopted. The coena described in section V.C. could easily be recognized in the grassland area during mapping taking the different cover percentages of *Agrostis capillaris, Holcus lanatus,* rosette plants, *Poa pratensis, Carex nigra* and *Juncus effusus* into account. The cover percentages of *Erica tetralix, Calluna vulgaris, Agrostis capillaris, Holcus lanatus, Molinia caerulea, Betula pubescens* and the occurrence of species like *Gentiana pneumonanthe* and *Potentilla erecta* proved to be useful for differentiating the heathland area into mappable units throughout the study.

V.E.2. Changes in the grassland area

A comparison of the vegetation maps made in 1972, 1977 and 1982 (Fig. V.13A–C) clearly shows that in the hay-making area communities with *Anthoxanthum odoratum* and rosette plant species (coenon 5) were established mainly at the cost of communities with abundant *Holcus lanatus* (coenon 4). In the wet area the *Juncus effusus* dominance (coenon 9) was replaced by a community with *Juncus effusus* and *Agrostis stolonifera* (coenon 8). The rosette plant community (coenon 5)

Photo V.7. Grassland after sod removal. Cutting without fertilizer application takes place at the right side of the fence, sheep-grazing at the left side.

Photo V.8. The same situation as Photo V.7 after 4 years. Heathland species appeared in the area in which the sods were removed. *Juncus effusus* spread in the grazed area.

also appeared in the ploughed area. Coenon 3 with *Hypochaeris radicata* and *Calluna vulgaris* was found immediately after sod removal in the moist area (Photo V.7 and 8), whereas coenon 2 with *Agrostis capillaris, Festuca rubra* and *F. ovina* only appeared later in the dry area. Only in the area with fertilizer application did coenon 4 persist.

Table V.5 shows the cover of the different communities under the grazing regime. It is clear that the community dominated by *Holcus lanatus* (coenon 4) decreased and the community dominated by *Agrostis capillaris* (coenon 7) and co-dominated by *Juncus effusus* and *Agrostis stolonifera* (coenon 8) increased. The community with abundant *Poa pratensis* (coenon 1) spread in 1977 but had disappeared in 1982. Since the mosaic of the coena 5 and 7 and 5 and 4 contained an equal area of both, it can be concluded that the community with rosette plant species (coenon 5) spread considerably in 1982.

A few ideas about grazing intensity should be mentioned before discussing the transitions between the coena. Hutchinson & King (1980) regarded the proportion of closely grazed areas to be a measure of grazing intensity. Hence the grazing intensity shows a negative relationship with the canopy height. The height of the canopy was estimated during mapping of the coena (in case of a mosaic by stratified sampling) and a mean height of less than 15 cm was considered as heavily grazed, areas with a mean canopy height between 16 and 35 cm as intermediately and areas with a mean height over 36 cm as hardly grazed.

Sheep do not graze the entire stand to a uniform height but tend to form mosaics with closely grazed areas (canopy height less than 5 cm) alternating with taller parts. It is, therefore, possible to measure the grazing intensity during mapping by estimating the cover percentage of closely grazed areas in the vegetation types. Mapping units containing 6–60% closely grazed areas were regarded being intensively grazed, units containing 1–5% closely grazed areas as intermediately and units with no closely grazed patches as hardly grazed. According to the afore-mentioned criteria the drier parts of the grassland were both intensively and intermediately grazed for c. 50%, whereas the wetter parts were hardly grazed over their whole area (Ter Heerdt et al., 1986). Fourty percent of the heathland area was intensively grazed (mainly the drier parts), 10% intermediately and 50% was hardly grazed (mainly the wetter parts (Schutter et al., 1987).

The transitions found in at least 5% of the grassland area are shown in Fig. V.14 for the periods 1972 to 1977 and 1977 to 1982. The composition of the coena changed during the whole study period. The main changes took place in the dry *Poa pratensis* and the moist *Holcus lanatus* communities. These changes can be summarized as being within communities and transitions from communities dominated by *Holcus lanatus* towards communities dominated by *Agrostis capillaris* with a temporary

Table V.5. Area (m²) of the plant communities in the grazed grassland in 1972, 1977 and 1982 (the coena 2 and 3 are not found in the grazed area). * Area including the proportion of the respective coena in the mosaic of the coena 5 and 7.

Coenon		1972	1977	1982
1	Poa pratensis community	100	6300	0
4	Holcus lanatus community	54700	22900	1300 (3500)*
5	Leontodon/Hypochaeris community	0	1200	0 (9750)*
6	Agrostis capillaris/Holcus community	0	14900	200
7	Agrostis capillaris community	6100	16300	37100 (44650)*
8	Juncus effusus/Agrostis stolonifera community	0	500	3100
9	Juncus effusus/Carex nigra community	8400	7200	7500
	Mosaic 5/7	0	0	15100
	Mosaic 5/4	0	0	4400
	Total (m²)	69300	69300	68700

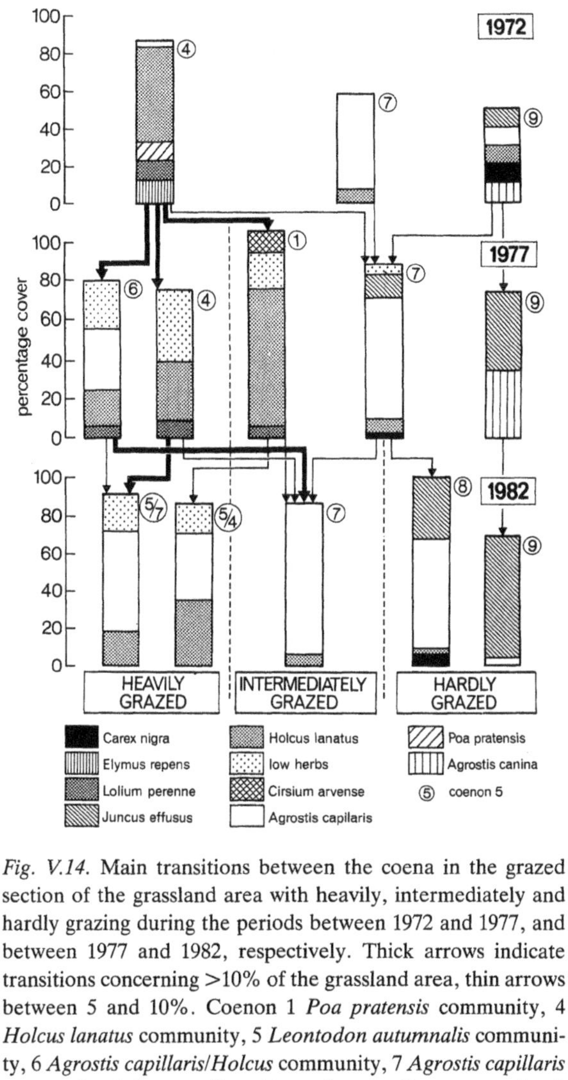

Fig. V.14. Main transitions between the coena in the grazed section of the grassland area with heavily, intermediately and hardly grazing during the periods between 1972 and 1977, and between 1977 and 1982, respectively. Thick arrows indicate transitions concerning >10% of the grassland area, thin arrows between 5 and 10%. Coenon 1 *Poa pratensis* community, 4 *Holcus lanatus* community, 5 *Leontodon autumnalis* community, 6 *Agrostis capillaris/Holcus* community, 7 *Agrostis capillaris* community, 8 *Juncus effusus/Agrostis stolonifera* community, 9 *Juncus effusus* community. Cover percentages of the main species are indicated (after ter Heerdt et al., 1986).

and local increase in rosette plant species. These phenomena will be discussed later, in chapter VIII.

V.E.3. Changes in the heathland area

The comparison of the vegetation maps made in 1972, 1977 and 1982 shows that communities with a high *Molinia caerulea* cover percentage, coena 12 and 15 took over in the abandoned area in the *Erica* and *Calluna* communities. In the *Molinia/Gentiana* community (coenon 16) the *Betula*/woodland community took over (Photo V.9 and 10).

Table V.6 shows the cover of the different communities under the grazing regime. The area of the species-poor communities with little *Molinia caerulea* and grassland species (coena 10 and 11) decreased strongly. The same is true, to a lesser degree, for the species-rich community with few grassland species (coenon 16). The area of the species-poor community with intermediate *Molinia caerulea* cover (coenon 12) increased and the same was the case for the communities with high grassland species and *Molinia* cover (coena 13, 14 and 15). However, the area of coenon 13 decreased again.

The transitions found in at least 5% of the heathland area are shown in Fig. V.15A–B for the periods 1972 and 1977 and 1977 to 1982, respectively. During the first period coena 10 and 11 lost area to the coena 15 and 12, 13, 15, respectively. This particularly indicates an increase in grassland species and *Molinia caerulea* in the species-poor heathland.

Table V.6. Area (m^2) of the plant communities in the grazed heathland in 1972, 1977 and 1982 (the woodland community (coenon 17) only developed in the abandoned area).

Coenon	1972	1977	1982
10 Calluna vulgaris community	7600	0	0
11 Erica tetralix community	13050	2900	450
12 Calluna/Molinia community	0	3750	5950
13 Agrostis capillaris/Rumex acetosella/Calluna community	0	5650	1150
14 Molinia/Holcus community	0	1500	3500
15 Molinia caerulea community	3500	11350	19650
16 Molinia/Gentiana community	5750	4750	2300
Total (m^2)	29900	29900	33000

Photo V.9. Heathland with young saplings of *Betula pubescens* after six months of sheep-grazing (right) and abandoning (left).

Photo V.10. The same situation as Photo V.9 after 4 years of grazing (right) and abandoning (left).

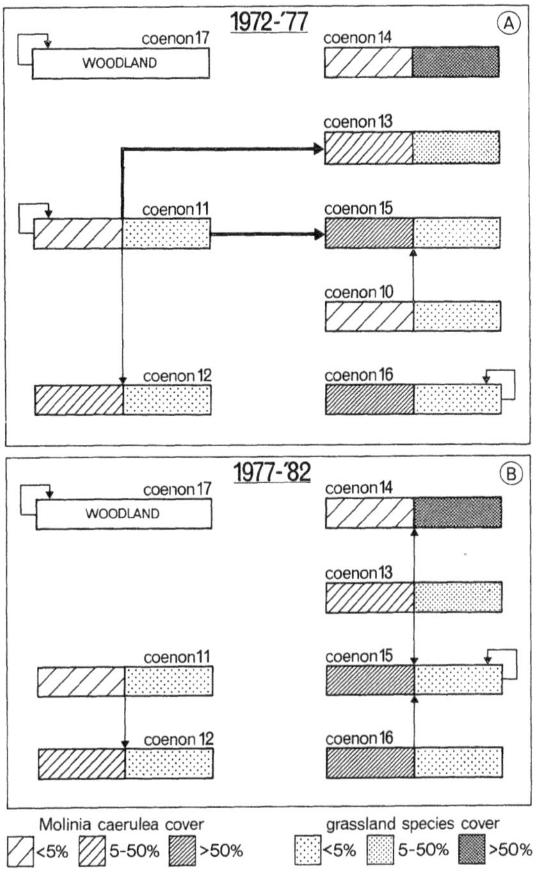

Fig. V.15. Main transitions between the coena in the grazed section of the heathland area, A: during the periods between 1972 and 1977, B: between 1977 and 1982, respectively. Thick arrows indicate transitions concerning >10% of the heathland area, thin arrows between 5 and 10%. Coenon 10 *Calluna vulgaris* community 11 *Erica tetralix* community, 12 *Calluna/Molinia* community, 13 *Agrostis capillaris/Rumex acetosella/Calluna* community, 14 *Molinia/Holcus* community, 15 *Molinia caerulea* community, 16 *Molinia/Gentiana* community, 17 *Betula*/woodland community. Relative cover percentages of some species are indicated (after Ter Heerdt & Schutter, 1984).

The decrease of coenon 11 to the benefit of coenon 12 continued from 1977 to 1982. All other transitions indicate an increasing coverage of grassland species and *Molinia* in both the species-poor and the species-rich heathland communities.

V.F. Comparison of grazing and hay-making and discussion

V.F.1. Regeneration of heathland

Species indicating nutrient-rich soil conditions decreased in the grassland area, whereas species indicating nutrient poor conditions increased in the grassland area (Table V.4A–C). This phenomenon seems to be correlated to the cessation of the application of artificial fertilizers. The vegetational succession has not, however, as yet lead towards the appearance and increase of heathland species with the exception of the treatments of sod removal or ploughing with subsequent hay-making.

In the Netherlands Heijink (1974) surveyed nearly 90 abandoned areas which had originally been reclaimed from heathland. The length of time when no agricultural practices had been carried out varied from a few years to over 25 years. The management which was practiced was aimed at the restoration of heathland communities. Former arable fields and pastures were both involved. No heathland regeneration was found in the abandoned pastures, but a return of *Erica tetralix* and *Calluna vulgaris* occurred in the abandoned arable fields. Grazing and hay-making barely resulted in the return of heathland species in former pastures. In the previously arable fields, grazing, hay-making and ploughing with subsequent hay-making resulted in the occurrence of *Erica tetralix, Calluna vulgaris* and *Juncus squarrosus.* These species also rapidly emerged after deep ploughing, but were overgrown by grasses. It must be emphasized that heathland regeneration varied from the occurrence of only seedlings to cover percentages of maximal 40% of *Calluna vulgaris* and/or *Erica tetralix.*

Heijink's (1974) comparative study in heathland regeneration resulted in certain conclusions. Regeneration from pasture, if it occurs at all, takes more time than from arable fields. It occurred better in dry areas than in wet ones. Finally, the longer the period that the reclaimed heathland had been under agricultural practices, the longer heathland regeneration took. When applying these conclusions to the soil types, groundwater table and agricultural history of the Westerholt, it seems that heathland regeneration is possible in most of the study area (Fig. V.16).

The finding of scarce individuals of *Erica tetralix* and *Calluna vulgaris* in the grazed and cut areas is in agreement with Heijink (1974). The same is the case for the area with ploughing and subsequent hay-making which results in some regeneration of heathland species. The deep ploughing mentioned by Heijink, involved the return of the former

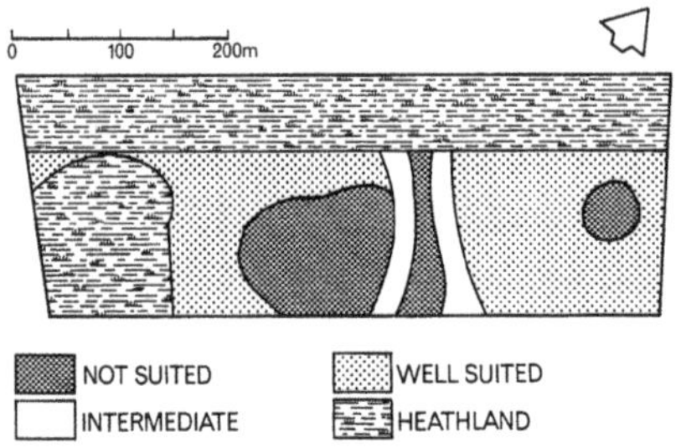

Fig. V.16. Suitability for heathland restoration in the study area on the basis of occurring soil types according to criteria of Heijink (1974) (after Van Wieren, 1977).

heather sod to the soil surface, thus, resulting in the mineralization of that former sod and the subsequent spreading of grasses overgrowing the emerged *Erica tetralix* and *Calluna vulgaris* seedlings. In this study we ploughed deeper than the former sod and, therefore, brought up the nutrient-poor C-horizon resulting in a spreading of heathland species. Since Heijink did not mention sod removal experiments, our results can not be compared.

V.F.2. Comparison of management practices in the grassland area

A comparison between the experiments showed that a number of species indicating nutrient-rich soil conditions (cf. Ellenberg 1979, Klapp 1965, Kruijne et al. 1967) decreased in most management experiments, probably due to the cessation of fertilizer application: viz. *Lolium perenne, Phleum pratense, Poa pratensis, Elymus repens, Alopecurus geniculatus*. The application of nitrogen fertilizers did, however, only reverse this phenomenon in the case of *Elymus repens*. The other species decreased in this experiment as well and others viz. *Rumex acetosa* and *Dactylis glomerata* increased. Species indicating intermediate and nutrient-poor soil conditions increased in both grazed and cut plots, viz. *Agrostis capillaris*, but sometimes the increase was earlier and greater in grazed than in cut plots viz. *Hypochaeris radicata*. Some species react positively in grazed plots viz. *Trifolium repens, Cynosurus cristatus, Sagina procumbens*, others in cut plots viz. *Anthoxanthum odoratum, Rumex acetosa, Carex nigra*. Heathland species, viz. *Erica tetralix, Juncus squarrosus, Calluna vulgaris* only appeared in moist areas with sod removal, with the exception of *Calluna vulgaris* which also appeared in dry, grazed plots.

The species diversity and equitability in the plots with fertilizer application was lower than in the other management experiments. Hay-making resulted in an increase or maintenance of species diversity, whereas in the grazed plots species diversity and equitability decreased, particularly in the wet and inundated communities. Hay-making and grazing apparently affect the plant communities in different ways and, moreover, reveal no similar effects in the whole study area. These phenomena will be discussed in detail in chapter VIII.

The comparisons between grazing and hay-making in all the figures and tables in this chapter do not take different grazing intensities into account. The only exception is the transition scheme of the grassland area (Fig. V.14). Canopy height was taken as a measure of grazing intensity according to Hutchinson & King (1980). Mean canopy height can easily be estimated and can, therefore, be used in indicating grazing intensity in the whole grassland area. Another parameter for grazing intensity is litter accumulation, as pointed out by Job & Taylor (1978). Since litter thickness and cover in permanent plots were positively correlated ($r = 0.909$, $P<0.01$), the latter could easily be used as a measure for grazing intensity at the level of permanent plots. This estimation of grazing intensities is in accordance with that mentioned in section V.E.2. involving the canopy height of the coena. The method of estimating litter coverage could not be applied for large areas.

Species diversity (Fig. V.17A) and equitability (Fig. V.17B) both show a positive relationship with grazing intensity in the moist grassland area. It depends on the intensity whether grazing results in a higher species diversity or equitability than hay-making.

Species composition and abundancy is shown in detail in Table V.7. The *Lolium perenne*-group characterizes the heavily grazed areas, the *Ranunculus repens*-group the intermediate grazing intensity, the *Agrostis capillaris*-group the light grazing

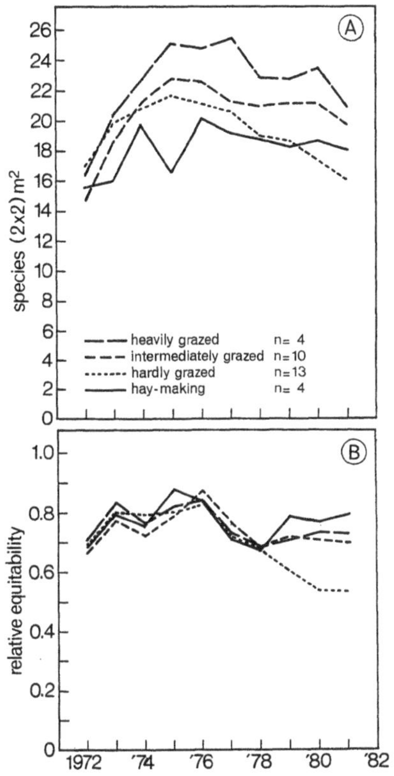

Fig. V.17A–B. A: number of species · 4 m^{-2} and B: relative equitability in the *Holcus lanatus* community under hay-making and grazing with different intensities (see also Table V.7) (after Veldman, 1983).

intensity and the *Rumex acetosa*-group the hay-making areas.

Our labeling as grazing or hay-making species agrees with the assessments in the literature for most species. *Hypochaeris radicata, Plantago lanceolata* and *Cirsium palustre,* however, performed best in the grazed area, but according to Kruijne et al. (1967) they are hayfield species. These authors labelled *Phleum pratense* and *Trifolium dubium* as pasture species, whereas, in this study, they performed best in the hay-making regime. The agricultural practices are probably less important for some species than other environmental factors, but which particular factor this is, is difficult to indicate. It should be mentioned here that discrepancies between our data and those of Kruijne et al. (1967) could be due to different approaches, as mentioned by Elberse et al. (1983). Kruijne et al. used the correlative approach, in measuring the greatest possible number of habitat factors and correlating these to the botanical composition of a great number of grassland plots. It was not easy to indicate the causal factor for the presence or absence of a particular plant species because of the interrelationship of the different habitat factors. In the Netherlands for instance until recently the soil fertility of permanent pastures near the farm buildings was much higher than that of hay-fields which were further away (cf. section III.B.12). The exploitation and mineral supply of grassland are, therefore, distinctly related. Elberse et al. (1983), when comparing hay-fields and pastures with and without fertilizer dressing, therefore, decided that *Holcus lanatus* and *Plantago lanceolata* primarily indicate moderate soil poverty and not hay-field use as Kruijne et al. (1967) suggested. Large year-to-year fluctuations in cover percentages of the dominant species were also reported by Van Den Bergh (1979).

Cover percentages of *Lolium perenne, Alopecurus geniculatus, Festuca pratensis, Elymus repens, Poa pratensis* and *Phleum pratense* decreased with both grazing and hay-making. Since these species indicate nutrient-rich soil conditions their decrease could be attributed to the cessation of fertilizer application. The increase, with both grazing and hay-making, of *Cynosurus cristatus, Cardamine pratensis, Leontodon autumnalis, Ranunculus acris, Agrostis capillaris, Rumex acetosa, Galium palustre,* and *Anthoxanthum odoratum* indicating intermediate, and *Hypochaeris radicata* and *Cirsium palustre* indicating nutrient-poor soil conditions seems to fit into this consideration. However, the increase with both grazing and hay-making of nutrient-rich *Ranunculus repens* and the decrease of intermediate *Holcus lanatus,* and *Cerastium fontanum,* are contradictory. It can be assumed that hay-making removes more nutrients than grazing, including the return of dung. It therefore, seems remarkable that the intermediate *Holcus lanatus, Sagina procumbens, Bellis perennis, Agrostis capillaris,* and even *Hypochaeris radicata* (indicating nutrient-poor soil conditions) show higher cover percentages in the grazed than in the cut plots. It is obvious from the results of Table V.7 that it is very

Table V.7. Species composition and mean cover percentage, originating from an initially uniform sward in 1972 at hay-making and different grazing intensities: (A) heavily grazed (area covered by litter ≤30%); (B) intermediately grazed (<70% area covered by litter >30%); (C) lightly grazed (area covered by litter ≥70%; and the management practices under which the species grew best in comparative experiments according to references: 1, Klapp (1965); 2, Schulz (1974); 3, Elberse et al. (1983); 4, Wind (1980) and in comparing lots with known agricultural practices: 5, Kruijne et al. (1967).

Year		1972		1981					
Management practice		Initial sward	Grazing			Hay-making	Assessment in the literature		
			A	> B	> C				
Number of permanent plots		(n=32)	(n=4)	(n=10)	(n=14)	(n=4)	Pasture	Hay-field	Indifferent
R	Lolium perenne	13	6	2.9	0.7	.	1,2,3,4,5		
I	Holcus lanatus	44	22	24	8	1.5	3	1,4	5
R	Poa trivialis	1	2	1.8	0.9	0.3	3,5		
I	Cynosurus cristatus	.	0.5	0.2	0.3	0.3	1,3,5		
R	Alopecurus geniculatus	1	0.5	0.3	.	.	5		
R	Festuca pratensis	1	0.3	0.2	0.1	.	3		
R	Trifolium repens	2	8	5	1.2	0.5	1,3,4,5		
I	Sagina procumbens	0.1	0.8	0.4	0.1	.			5
I	Cardamine pratensis	0.1	0.8	0.7	0.3	0.8		5	3
I	Leontodon autumnalis	1	13	6	3	8	3		5
I	Bellis perennis	2	3.5	1.1	0.6	0.3	1,3,5		
R	Plantago major	1	3	0.6	0.1	.	1,5		
R	Ranunculus repens	1	5	7	2.6	7	1,3,4,5		
P	Hypochaeris radicata	0.1	3	3.3	0.8	1.8		5	3
I	Ranunculus acris	1	1.5	9	1.8	5	3		5
R	Elymus repens	2	.	0.1	.	.		5	3
I	Cerastium fontanum	2	1.0	1.3	0.8	0.8	5		3
I	Plantago lanceolata	.	.	0.5	0.1	0.3	3	5	
R	Cirsium arvense	0.1	.	0.5	0.2	.	5		
I	Agrostis capillaris	1	27	25	60	15	5		
R	Poa pratensis	7	1.3	1.4	2.5	2	5	3	
P	Cirsium palustre	0.1	2	3.2	3.5	0.3		5	
I	Lotus uliginosus	0.1	.	0.6	1.5	.		3,5	
I	Rumex acetosa	1	1.3	1.6	1.4	5		3,5	
R	Phleum pratense	3	.	0.4	0.4	2	5		3
I	Galium palustre	0.1	0.2	.	0.2	0.5		5	
I	Anthoxanthum odoratum	.	1	1.5	1.2	15		1,4,5	3
P	Juncus effusus	1	1.3	0.5	1.9	3.3		5	
	Taraxacum spec.	3	1.8	2.5	1.3	4.5	1,3,5		
P	Rhinanthus angustifolius	.	.	.	.	7		5	
I	Trifolium dubium	1	.	.	.	0.3	5		
I	Trifolium pratense	0.1	.	.	.	0.3		5	3

Photo V.11. Grass species have established in the heavily grazed heathland area.

important that some measure of grazing intensity is indicated in comparisons of hay-making and grazing.

The application of Ammonium Nitrogen Limestone (50 kg $N \cdot ha^{-1}$) initially resulted in an increase of intermediate species viz. *Holcus lanatus, Agrostis capillaris* and nutrient-rich species viz. *Bromus hordeaceus, Cirsium vulgare, Festuca pratensis* but eventually *Rumex acetosa* and *Dactylis glomerata* took over. This is in agreement with the *Poa pratensis* and *Dactylis glomerata* dominance after c. 10 years mentioned by Elberse et al. (1983) applying 160 kg $N \cdot ha^{-1}$ (in NPK) and it is eventually followed by the dominance of the very tall species *Arrhenatherum elatior* and *Anthrisus sylvestris*. In the grassland experiment in Rothamsted (U.K.) which lasted over 100 years, *Poa trivialis* initially dominated with NPK (96 kg N, 35 kg P_2O_5 and 225 kg $K_2O \cdot ha^{-1}$) fertilizers. This was followed by *Bromus hordeaceus* and *Arrhenatherum elatior* which eventually became dominant after 60 years and maintained themselves until the present day (Van Den Bergh, 1979).

Sod removal, particularly in the moist communities, is the only management practice where considerable numbers of species, indicating nutrient-poor soil conditions, appeared. The same holds, to a lesser degree, for the ploughed area where nutrient-poor sand from the C-horizon was brought to the surface.

There is little documentation concerning the succession from grassland which is comparable with our study area. Some details, found on similar soil types are given in the survey carried out by Heijink (1974). A pasture which had been farmed for five years showed a vegetation with *Juncus effusus, Molinia caerulea, Holcus lanatus* and *Calluna vulgaris* covering 30% locally 15 years after being abandoned. If the farming period was 35 years, a vegetation developed which was dominated by *Festuca rubra* and *Juncus effusus* and no heather had yet emerged 15 years after farming had been abandoned. Hay-making after the cessation of fertilizer dressing resulted in communities with *Anthoxanthum odoratum, Holcus lanatus, Agrostis canina,*

Plantago lanceolata, Luzula campestris in dry areas and *Juncus effusus, Agrostis stolonifera, Anthoxanthum odoratum, Ranunculus acris, Lotus uliginosus* in moist areas. *Erica tetralix* was only found in one lot. No heather was found at all in grazed pasture. It can be concluded that the grassland succession in the study area includes the findings mentioned in Heijink's (1974) survey.

The comparison of the information about succession gained from sequential mapping and from permanent plots in the grazed area (see sections II.C and VII.E.1) was studied by a modified transition model in which a plot was labeled with the name of one single unit of the vegetation types. The comparison was made between data from 80 permanent plots and the repeated mappings from 1972, 1977 and 1982. As a result the conclusion could be drawn that although a limited number of permanent plots can give a relatively large amount of information, the minimum number of permanent plots needed for a reliable estimate of the probability of the most frequently occurring transitions would have been 248. Hence for the purpose of quantitative estimation of vegetational changes in a dynamic habitat a yearly mapping (legend units including dominant species) is to be preferred (Van Den Berg et al., 1985).

V.F.3. The heathland area

The abandoned heathland changed into woodland, first in the moist areas and later in the dry ones. This process started in the lower parts dominated by *Molinia caerulea* and continued. The area with *Betula pubescens* saplings increased markedly to covering nearly the whole heathland area. Grazing prevented these saplings from growing upwards.

If the canopy height is taken as a relative measure for grazing intensity it can be concluded that a large part of the area with grassland species was grazed intensively. These species indicate intermediate soil conditions with respect to nutrients or even nutrient-rich soil conditions viz. *Poa trivialis* and *P. pratensis* and were facilitated by the disappearance of heathland species (Photo V.11). The latter maintain themselves better in lighter grazed areas, unless these become overgrown by *Molinia caerulea.*

Although *Molinia caerulea* spread particularly in the grazed area, this phenomenon does not seem to be only correlated with grazing, since spreading also took place in the abandoned area. There is a drop in the groundwater table in the adjacent agricultural area as a result of land consolidation programmes and this seems to favour the *Molinia caerulea* growth (Westhoff et al. 1973). The nutrients deposited by the'acid rain' may also play a role.

The heathland area was abandoned for several decades, and the remaining heather plants had therefore become old and tall (up to 50 cm) and litter had accumulated. Grazing and accompanying trampling resulted in the die back of old heather plants and destruction of the litter layer. This process resulted locally in the emergence of seedlings and the tillering of old plants with a subsequent higher coverage (see also Gimingham, 1972). Since grazing involves the local voiding of dung, some nutrient enrichment will take place by the more rapid turnover of nutrients (Perkins, 1978). Miles (1974) found that the application of fertilizers also enhanced the establishment of species which had been introduced in the heathland communities. The connection of pasture and heathland in the fenced-in area can promote the transport of grassland species via dung which is voided (see also Welch, 1985).

Although some positive effects of grazing on the heathland area were found, the negative effects predominate, namely the increase of saplings of *Betula pubescens* all over the area and the appearance of grassland species on large areas. Cessation of grazing is no solution since the *Betula* saplings rapidly turned into woodland in a exclosure in the study area (Dallinga et al., 1982B). A better solution, but more difficult to carry out, should be sod removal in the heathland area, since a good recovery of heathland species was found in the study area (Dallinga et al., 1982B).

CHAPTER VI

The valley grassland vegetation

VI.A. Hay-making and other regimes

In this chapter the succession in meadows subjected to hay-making after the cessation of fertilizer application is discussed. The results of regular management practices carried out by farmers or by the State Forestry Commission will be dealt with. These practices often imply hay-making in July without fertilizer application. A comparison was made in lots where hay-making was started in different years and was carried out with different regimes. The questions to be answered are, (i) to what extent can meadows which were formerly applied with fertilizers be restored to species-rich grasslands?, (ii) how long does this take?, (iii) how do different substrates influence successional lines? Another question will also arise (iv) what are the successional lines with cutting regimes other than the July one e.g. hay-making twice, hay-making in September, cutting and mulching or even abandoning? Comparisons between single lots will always imply differences of substrate or of former agricultural activities. Different management experiments within single lots were, therefore, started in the upper course and the lower course area in order to answer this question.

The study areas are situated in different parts of the 'Stroomdallandschap Drentsche A' along various tributaries of the brook system. The assignment to upper, middle and lower course of the valley system was derived from Everts et al. (1984). The upper course lots managed by the State Forestry Commission along the Anlooër Diepje (53° 05′ N, 6° 40′ E) are called 'Anlo' for convenience. The middle course lots along the Taarlosche Diep (53° 02′ N, 6° 38′ E) are referred to as 'Taarlo'. The upper course lots with different management regimes are known under the toponym 'Loefvledder' and are located (53° 05′ N, 6° 40′ E) along a tributary of the Anlooër Diepje. The lower course lots with different management regimes are known under the toponym 'Lammertshem' and are located (53° 08′ N, 6° 39′ E) along the mainstream which is locally called Westerdiep (Fig. VI.1).

All the lots were formerly subjected to the application of fertilizers and were used for hay production. The aftermath was grazed by cattle. Fertilizer application ceased in the year of acquisition or one year later depending on the time of acquisition, namely, spring or autumn. Only the relatively dry part of lot 1371 has been reclaimed i.e. the soil had been ploughed superficially prior to 1946. The years of acquisition of the different lots are listed in Table VI.1. Annual hay-making by the State Forestry Commission or by farmers was carried out in the period between the end of June and the beginning of August, with an occassional second cut in September and/or grazing of the aftermath.

It was expected that the restoration from grasslands under agricultural practices to species-rich grassland would depend on the substrate. Lots were, therefore, compared in the upper course (Anlo) and the middle course (Taarlo), and, where present, dry and wet areas were compared within lots. Since groups of lots, on a similar substrate, are under the hay-making without fertilization regime for different numbers of years, the importance of that period of hay-making for the succession can be studied.

It can also be expected that the pattern of succession will depend on the management practice which is adopted. Several experiments were, therefore, carried out in the upper course (Loefvledder) and in the lower course (Lammertshem) in recently acquired lots, referred to as recent reserves for convenience. As hay-making in July is the usual management practised by the State Forestry Commission it was regarded as the standard practice with which the following practices could be compared.

– Cutting late in the season could enhance late-

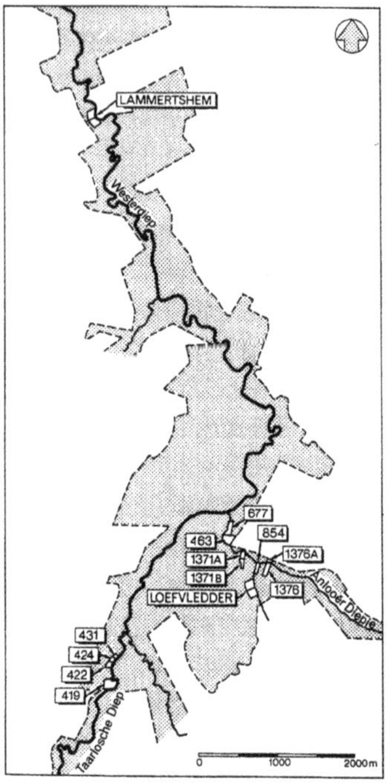

Fig. VI.1. Situation of valley grassland study areas in the reserve Drentsche A, which is indicated by the dashed line.

Table VI.1. Survey of study areas and lots in the valley grasslands, the year of acquisition and the year in which vegetation recording started.

Study area	Year of acquisition	Start vegetation recording
Anlo (upper course)		
463	1972	1974
677	1967	1972
854	1967	1972
1371A, B	1946	1974
1376	1972	1972
1376A	1972	1972
Taarlo (middle course)		
419	1965	1972
422	1972	1972
424	1967	1972
431	1972	1972
Loefvledder (upper course)		
recent reserve	1972	1973
older reserve	1967	1973
Lammertshem (lower course)		
recent reserve	1972	1973
older reserve	1967	1973

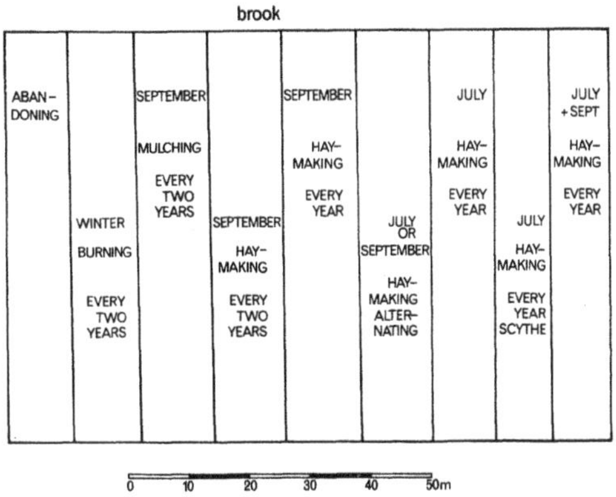

Fig. VI.2. Lay-out of the experimental management practices in the recent reserves.

flowering species, so September hay-making was started.
- Irregular hay-making could prevent the thriving of species adapted to July or September cutting, so alternating of July and September hay-making was started.
- Hay-making could be missed for a year for practical reasons, so September hay-making every second year was started.
- Sometimes the removal of the swath is impossible, so September mulching (cutting and leaving the mown material) every second year was started.
- A higher mowing frequency could enhance the removal of nutrients, so annual hay-making in both July and September was started.
- All cutting experiments are done by the State Forestry Commission with their heavy machines. Annual July cutting by scythe was started as a comparison. The cuttings were removed by hand to prevent soil compaction.
- Sometimes cutting is replaced by burning because of a lack of money and the standing crop is then removed by fire e.g. in Germany. So burning was in principle carried out every two years in the winter period if the upper layer of the litter was dry enough (Photo VI.1). The bottom layer of the litter was always moist, therefore the burning is referred to as cold fire (Schiefer, 1982).
- Finally, part of the lot was abandoned to find out if and when succession would lead to a woody vegetation.

The lay-out of the management regimes is given in Fig. VI.2.

Another series of experiments was initiated in Loefvledder and in Lammertshem in lots acquired earlier with species-richer communities under July hay-making without fertilizer applications and referred to as older reserves. The main comparison concerns the ongoing removal of nutrients with subsequent vegetational succession versus the maintainance or replenishment of nutrients after their removal with the subsequent stabilization of vegetational succession. The usual hay-making in July was again regarded as the standard practice for comparison.
- Maintenance of nutrients by leaving the cuttings, so annual mulching in July was started.
- Maintenance of nutrients by less cutting, so hay-making every second year in July was started.
- Maintenance of nutrients by less cutting and leaving the cuttings, so mulching every second year in July was started.
- Maintenance of nutrients by irregular mulching, so annual July cutting was started, but removal of the cuttings only took place every second year.
- Replenishing of nutrients by the annual application of 50 kg N · ha^{-1} (nitrogen is supposed to be the limiting factor in the production) in the spring as Ammonium Nitrogen Limestone (to prevent the pH from lowering) and annual hay-making in July. The annual nutrient removal with hay-making equalled c. 100 kg N · ha^{-1}. The difference in the nitrogen budget is replenished by precipitation.
- Replenishing of nutrients by the annual application in the spring of 50 kg N · ha^{-1} as Ureum Formaldehyde releasing nitrogen very slowly in contrast to the application of Ammonium Nitrogen Limestone and annual hay-making in July.
- Annual cutting by scythe and the removal of the cuttings by hand was started in order to make a comparison.

The lay-out of the experiments is shown in Fig. VI.3.

Site descriptions are given in section VI.B.

The effects of the usual July management were described by the year-to-year recording of perma-

Photo VI.1. Burning the litter in the Loefvledder study area.

nent plots in the Anlo and Taarlo areas (section VI.D). The effects of the different management treatments in the upper and the lower course were both compared by sequential mapping and by year-to-year recording of the permanent plots (section VI.E).

Classification of the plant communities which occurred in the Anlo and Taarlo areas was done by comparing the relevés of the 80 permanent plots from the start of recording to 1980. The classification of the plant communities found in the areas under comparative practices was done by comparing the permanent plots each year from 1972 to 1980. The syntaxonomical position of the coena and their synecology could be discussed after classification (section VI.C).

Vegetational trends were obtained from permanent plot data using the dissimilarities between each year and the first year. Species diversity and cover percentage increase of decrease in different plant communities under the management regimes were compared with findings elsewhere (section VI.E).

VI.B. Site description

VI.B.1. Soil, geomorphological and hydrological conditions

The general outline of the Drenthe Plateau of the brooks draining it are given by De Gans (1980) and the source area, upper course, middle course and lower course are detailed by Grootjans (1980) (cf. section III.B.1). More details concerning the present study area can be derived from Everts et al. (1984) and will be quoted briefly here.

The upper courses often feature narrow valleys with a sandy soil without or with only a thin layer of peat. These areas are fed by subsurface flow from the adjacent plateau, sometimes also with a base flow discharging seepage water from a deep aquifer containing mineral-rich (Ca^{2+}) groundwater. The influence of the subsurface flow is always predominant. This groundwater type is mineral-poor because of its origin from rainwater and the short length of time it is in the soil. The Anlo and Loefvledder study areas represent upper course areas in secondary incisions in the Plateau being tributaries of the main valley (Photo's III.2, 3 and 4). The study areas occur in the upper course where there is

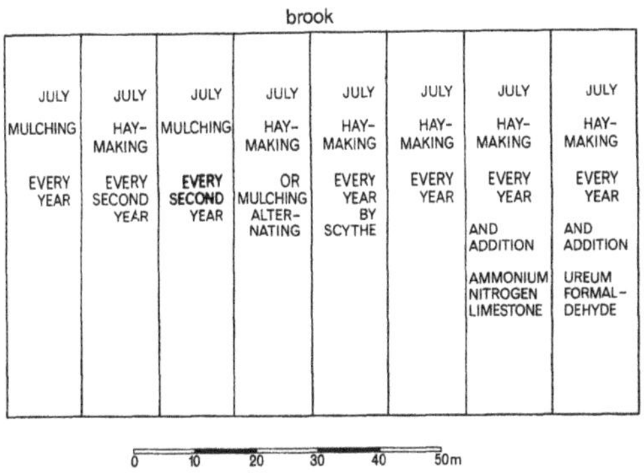

Fig. VI.3. Lay-out of the experimental management practices in the older reserves.

no impermeable boulder-clay in the subsoil. The base flow can, therfore, reach the surface and a transition to the middle course is in fact found. The study areas generally feature 'Veld' podzol soils in their high and dry parts with a layer of raw humus in the topsoil and a thick infiltration layer of colloidal humus in the subsoil. 'Beek' and 'Broek' earth soils, which have an earthened mineral humic ('Beek') on peaty ('Broek') topsoil, were found in the wet parts of the lots. Locally, more peat had accumulated and 'Made' peat soils with an earthened topsoil occurred in the wet areas. Soil nomenclature has been derived from the soil classification by De Bakker & Schelling (1966) adapted to the Drentsche A system by Van Heuveln (1980).

The middle courses are characterized by broad valleys filled with a thick peat layer (Photo III.5). No heavy loam layer called 'potklei' is found in the study area Taarlo, so discharged seepage water has free access to the surface. This base flow is so strong that a hydraulic head of 1.5 m (cf. Grootjans, 1985) is reached resulting in peat surfaces with a convex shape. The meadows in the middle course do not become inundated by the brook because of this convex shape rising above the brook bank levees. The lots at the Taarlo study area featured 'Vlier' peat soils which have an unearthened topsoil.

The lower course, including the study area Lammertshem, features a very broad valley containing thick and flat accumulations of peat. Only the central part of the valley belongs to the nature reserve of the Drentsche A. It is surrounded by polders which are in agricultural use and can only be drained artificially due to the low level with respect to Ordnance Date. The nature reserve and polder are separated by a dam and a deep ditch on the polder side of the dam (Photo III.6) (Boedeltje & Bakker, 1980). Drainage has become more intensive due to land consolidation programmes. As a result the influence of the slowly discharging deep mineral-rich groundwater flows had diminished (Bakker et al., 1987). Subsequently the nutrient-rich flood water from the river, resulting from annual inundations (Photo VI.2), becomes more important. The soil in the Lammertshem area should predominantly be assigned to 'Made' peat soil as it has a earthened topsoil and locally some 'Vlier' peat soil with unearthened topsoil.

VI.B.2. Groundwater tables

The general hydrological conditions of the study sites are given in section VI.B.1, but some details are shown in the present section. The groundwater tables were measured every two or three weeks. The groundwater dynamics have been summarized by the transformation of the groundwater table fluctuations into duration lines i.e. graphs representing the cumulative number of days with a certain groundwater group (5 cm). The graph represents the length of time the groundwater is in the soil layers (Niemann, 1963; 1973). The comparison of groundwater tables was based on records from April 1976 up until April 1977 for the Anlo and Taarlo study areas and from January to December 1977 for the Loefvledder and Lammertshem study areas.

The shape of the duration lines can provide supplementary information on groundwater dynamics. The quotient of mathematical mean ($\bar{x}$) and median (M) of the duration lines should be taken into account (Niemann, 1973). If $\bar{x}/M>1$, the higher groundwater levels are more frequent than lower ones and the duration line has a convex shape. Convex shaped duration lines are associated with a continuous supply of groundwater by a regional base flow, particularly during the winter and early spring. Concave shaped duration lines, with $\bar{x}/M<1$ are associated with the infiltration of rain-

Photo VI.2. Inundation of the lower course in the Lammertshem study area.

water in the wet seasons followed by rapidly decreasing groundwater levels, until the regional base flow level has been reached. Straight duration lines, with $\bar{x}/M \pm 1$ often point to a decreased supply of base flow water (Grootjans & Ten Klooster, 1980). Statistical analysis of duration line bundles is according to Niemann (1973) and Grootjans & Ten Klooster (1980). The Wilcoxon-Mann-Whitney test has been used to test whether or not two bundles are centered differently. All available data were used, not only the extreme duration lines.

The duration line bundles of the study areas and the comparisons are shown in Fig. VI.4A–D, whereas the mathematical mean ($\bar{x}$), median (M) and their quotients are given in Table VI.2. The upper course Anlo area could, therefore, be subdivided into a 'dry' and a 'wet' subset of groundwater tables. These subsets differ significantly ($P<0.001$) (Fig. VI.4A). The drier parts reveal an $\bar{x}/M$ of about 0.8, clearly indicating infiltration in the sandy 'Veld' podzol soils. The wetter parts on peaty soil, showing an $\bar{x}/M$ of about 1.05, indicate slightly discharging seepage water with a locally higher supply (lot 854) or locally little infiltration (lot 677) (Table VI.2).

The 'wet' localities in the upper course Loefvledder area appear similar to the Anlo 'wet' localities. Locally, some infiltration is found in a 'dry' area with significantly ($P<0.05$) lower groundwater tables in the wet season than those in the wetter area (Fig. VI.4B).

The middle course Taarlo area shows similar groundwater tables over all the lots (Fig. VI.4C). The $\bar{x}/M$ of about 0.92 indicates some infiltration, whereas a discharge of seepage water could be expected. This discrepancy might be due to the influence of a pumping station at a distance of 5 km (Everts et al., 1984; Grootjans, 1985).

The lower course Lammertshem area revealed the highest $\bar{x}/M$ quotient (Table VI.2). Some discharging deep groundwater, apparently still reaches the surface, although the high value might partly be due to inundations from the brook.

VI.B.3. The vegetation prior to acquisition

Land consolidation schemes were proposed for large areas of the Drenthe Plateau in the 1950's and 1960's. Prior to land consolidation vegetation maps were made indicating the agricultural value of all grassland lots. Although it was later decided that the grasslands bordering the Drentsche A river and its tributaries should not be incorporated into the land consolidation, these meadows were mapped in the early 1960's. Some

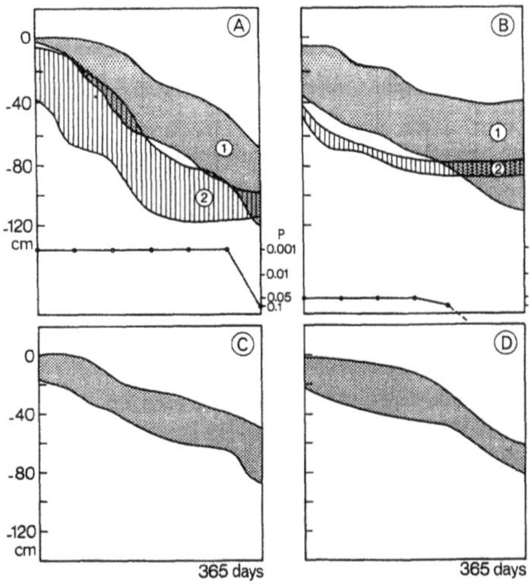

Fig. VI.4A–D. Comparison of duration line bundles A: in the Anlo study area in 1976/77 in 12 groundwater tubes in 'wet' sites (1) and 7 in 'dry' sites (2), B: in the Loefvledder study area in 1977 in 10 groundwater tubes in wetter sites (1) and 2 in drier sites (2), C: in the Taarlo study area in 1976/77 in 7 groundwater tubes and D: in the Lammertshem study area in 1977 in 12 groundwater tubes. Results of statistical analyses (abscissa) are indicated.

Table VI.2. Average values from (n) duration lines of mathematical mean (X̄), median (M) and the quotient of mathematical mean and median in the separate study areas.

Study area (n)	X̄	M	X̄/M
Anlo 'dry'			
463 (2)	−82 cm	−108 cm	0.76
677 (2)	−68 cm	−79 cm	0.86
854 (1)	−83 cm	−106 cm	0.78
1376 (2)	−66 cm	−78 cm	0.85
Anlo 'wet'			
463 (2)	−45 cm	−45 cm	1.00
677 (2)	−28 cm	−30 cm	0.93
854 (3)	−39 cm	−31 cm	1.26
1376 (1)	−31 cm	−30 cm	1.03
1371A (2)	−32 cm	−31 cm	1.03
1371B (2)	−54 cm	−52 cm	1.04
Taarlo			
419 (2)	−43 cm	−49 cm	0.88
422 (2)	−25 cm	−28 cm	0.89
431 (3)	−42 cm	−43 cm	0.98
Loefvledder			
'dry' (2)	−72 cm	−80 cm	0.90
'wet' (10)	−52 cm	−50 cm	1.04
Lammertshem (12)	−31 cm	−25 cm	1.24

information is, therefore, available on the initial agricultural value and botanical composition of the study areas (Proefstation voor Akker- en Weidebouw, 1965; 1966).

Only the dry 'Veld' podzoil soils in the Anlo study area were classified as 'intermediate' from an agricultural point of view, due to their moist character. The other soils in the Anlo area and the whole Loefvledder, Taarlo and Lammertshem areas were classified as poor or insufficient from an agricultural point of view because these were too wet.

The moist soils in the Anlo area (Photo VI.3) were characterized by the dominance of *Glyceria fluitans, Poa trivialis, Ranunculus repens* and also featured *Festuca pratensis, Alopecurus geniculatus, Polygonum hydropiper, Juncus effusus* and *Cardamine pratensis.* The wet Taarlo and Lammetshem areas were dominated by *Glyceria maxima, Poa trivialis, Alopecurus geniculatus, Agrostis stolonifera* and also contained *Poa pratensis, Festuca pratensis, Glyceria fluitans, Deschampsia cespitosa, Holcus lanatus, Carex* spp., *Ranunculus repens, Trifolium repens* and *Rumex acetosa.*

No information is available on the botanical composition of the dry areas.

Some earlier information is available from Schimmel (1955). He described all brook valleys on the Drenthe Plateau in the early 1950's, which eventually resulted in the Drentsche A brook system being designated as a nature reserve. The Anlo and Loefvledder areas were strongly influenced by agricultural practices at that time with the exception of lot 1371. This lot contained numerous species which indicated nutrient-poor soil conditions viz. *Viola palustris, Carex hostiana, C. panicea, Dactylorhiza maculata, Menyanthes trifoliata, Molinia caerulea, Juncus conglomeratus, Juncus acutiflorus, Potentilla palustris.* Somewhat north of the Taarlo study area Schimmel (1955) also described a community of *Carex acutiformis* with interesting species: *Crepis paludosa, Caltha palustris, Rhinanthus angustifolius, Dactylorhiza majalis, Lychnis flos-cuculi, Phyteuma nigrum.* Despite an annual fertilizer dressing of 300 kg ground basis slag, 300 kg Potassium and 200 kg Ammonium Nitrogen Limestone · ha^{-1}, 38 species were found in a relevé $(4 \times 4\,m)^2$. Schimmel already suggested that discharged seepage water contributed to this type of species-rich community despite the heavy fertilizer application.

VI.C. Vegetation classification, syntaxonomy and synecology

VI.C.1. Vegetation classification

VI.C.1.a. Anlo and Taarlo (Table VI.3, VI.4)

The vegetation was described in relevés $(2 \times 2\,m)^2$ of permanent plots and dissimilarities between relevés were calculated, as described in section II.C.3. Permanent plots were fixed in transects perpendicular to the brook. Each transect runs

Photo VI.3. Lot in the Anlo study area still used by a farmer and applied with fertilizers featuring a *Poo-Lolietum* with *Cardamine pratensis* along the ditches.

down a gradient from the hedge-bank bordering the brook valley and the plateau to the brook. The number of permanent plots in the different lots varied, depending on the abiotic (relief, soil, hydrological conditions) differentiation within each lot. Nineteen permanent plots were fixed in the dry parts and 35 in the wet parts of the Anlo study area, and 30 permanent plots in the Taarlo study area.

All 606 relevés from the period between 1972 and 1980 were not clustered together. Because of the large differences within the Anlo area, each lot was clustered separately, whereas the relevés of the Taarlo lots were clustered together. The clustering resulted in 23 clusters at a dissimilarity level of 50% and 79 subclusters at a dissimilarity level of 38%. A rough qualitative comparison was made of the different synoptic tables. The final synoptic table shows the frequency and cover percentages for most of the species (Table VI.3). Only the clusters are discussed syntaxonomically and synecologically as it would be too detailed to discuss all the subclusters. The position of the clusters in the lots and during the study period is shown in Table VI.4A–K. Most of the lots comprise of several clusters, but some lots belong to one cluster. This is particularly the case in the Taarlo study area where the lots do not reach the sandy valley flanks. Many (sub)clusters only represent the relevés of permanent plots over a period of one or a few years.

The species-poor clusters 8, 9 and 10 in lot 1376A are considered to be *Molinio-Arrhenatheretea* basal communities. Cluster 8, a BC *Stellaria media* [*Molinio-Arrhenatheretea*] represents all permanent plots in the first year. From 1974 onwards the gradient becomes subdivided into two parts. The two permanent plots close to the hedge-bank represent cluster 9, a BC *Dactylis glomerata* [*Molinio-Arrhenatheretea*], characterized by tall grasses and forbs. Cluster 10, a BC *Agrostis capillaris* [*Molinio-Arrhenatheretea*] comprises the two permanent plots near the brook and features *Festuca rubra* and *Agrostis capillaris*, initiating a *Festuco-Cynosuretum*. The application of fertilizers, grazing and dunging by horses until 1971 contributes to the poor development of the stands.

The species-poor clusters 15, 16 and 17 in the dry area of lot 463 can be classified as *Lolio-Cynosuretum* with *Poo-Lolietum* elements. The lot received annual fertilizer applications until 1971. Cluster 15, a *Lolium perenne/Bromus hordeaceus* community, mainly represents relevés from the first four years of two permanent plots nearest to the hedge-bank. Cluster 17 a *Lolium perenne/Anthoxanthum odoratum* community, includes the relevés from the first years of the plots at some distance from the hedge-bank. The last four years of the whole dry area are represented by cluster 16, a *Lolium perenne/Agrostis stolonifera* community.

Table VI.3. Synoptic table of the coena of the valley grasslands. The frequency of occurrence is indicated and the mean coverage (+: 1–5%, a: 6–10%, 1: 11–20%, 2: 21–30% etc.). Character and differential species of syntaxa are given: Art: Artemisietea vulgaris, PL: Poo-Lolietum, A: Arrhenatherion elatioris, MA: Molinio-Arrhenatheretea, LP: Lolio-Potentillion, LC: Lolio-Cynosuretum, FC: Festuco-Cynosuretum, CP: Calthion palustris, Ccn: Caricion curto-nigrae, JA: Juncion acutiflori. The coena 11 (Anthriscus sylvestris community) and 13 (Juncus articulatus community) are omitted because they contain only one relevé.

Syntaxonomic position		Basal communities [Molinio - Arrhenatheretea]			Lolio-Cynosuretum/ Poo-Lolietum			Lólio-Cynosuretum			Lolio-Cynosuretum/ Festuco-Cynosuretum		Lolio-Cynosuretum/ Calthion palustris/ Caricion curto-nigrae			Calthion palustris/ Filipendulion			Calthion palustris/ Caricion curto-nigrae		Caricion curto-nigrae/ Calthion palustris	Calthion palustris/ Juncion acutiflori
Coenon		8	9	10	15	16	17	12	1	2	20	21	18	19	22	5	6	7	14	23	3	4
Lot		1376A	1376A	1376A	463	463	463	854	1376	1376	677	677	463	463	677	422 431	419 422	424	854	677	1371A	1371B
Number of relevés		4	13	11	10	18	6	15	50	5	16	24	13	29	9	96	136	8	47	31	28	35
	Cirsium arvense	50/+	69/+	54/+	.	.	.	.	2/+	20/+	.	.	.	.	.	.	.	.	.	.	.	.
Art	Urtica dioica	25/+	8/+	18/+	.	22/+	.	.	.	.	.	.	.	.	.	19/+	.	.	.	.	.	.
Art	Rumex obtusifolius	75/a	13/2	72/+	20/+	39/+	50/+	.	12/+	80/+	19/+	17/+	8/+	21/+	.	9/+	.	.	.	.	.	.
	Stellaria media	75/3	62/+	9/+	90/+	22/+	.	.	.	.	.	4/+	.	3/+	6/+	13/+	.	13/+	.	.	.	.
A	Glechoma hederacea	50/+	100/+	100/+	20/+	22/+	.	53/+	.	.	.	.	.	.	.	18/+	1/+	.	.	.	.	.
A	Dactylis glomerata	.	92/+	81/+	20/+	11/+	.	100/+	.	.	25/+	25/+	.	.	.	.	.	100/+	6/+	.	.	57/+
A	Bromus hordeaceus	75/+	69/+	73/+	100/2	100/a	83/+	.	.	.	.	.	31/+	38/+	.	23/+	16/+	87/+	.	.	.	.
A	Anthriscus sylvestris	.	61/+	45/+	.	17/+	.	93/+	.	.	.	21/+	.	.	.	42/+	6/+	.	.	.	.	.
PL	Lolium perenne	25/+	54/+	81/+	90/1	100/1	83/	73/+	4/+	60/+	6/+	9/+	62/+	69/+	.	22/+	3/+	75/+	11/+	.	.	.
PL	Poa annua	.	.	.	20/+	.	83/+	.	.	.	13/+	9/+	23/+	.	33/+	.	.	.	.	3/+	.	.
PL	Plantago major	.	.	.	.	50/+	.	.	.	.	.	.	23/+	41/+	.	3/+	.	.	.	.	.	.
MA	Holcus lanatus	100/3	100/3	100/1	100/a	100/1	100/+	93/a	100/+	100/4	100/a	100/+	100/a	93/a	100/+	92/a	99/a	100/a	96/+	100/a	100/a	100/a
MA	Rumex acetosa	50/a	100/a	100/1	100/+	100/a	50/+	100/a	98/+	100/1	94/a	100/+	100/a	97/+	100/+	98/+	96/+	100/+	100/+	100/+	100/+	100/+
MA	Cardamine pratensis	25/+	8/+	18/+	50/+	89/+	83/+	93/+	98/+	100/+	50/+	63/+	100/+	100/+	77/+	95/+	85/+	63/+	98/+	97/+	93/+	97/+
MA	Cerastium fontanum	75/+	38/+	91/+	90/+	100/+	100/+	67/+	74/+	100/+	31/+	42/+	100/+	97/+	44/+	99/+	97/+	75/+	72/+	94/+	96/+	94/+
MA	Poa pratensis	100/a	100/+	100/+	80/+	100/+	67/+	100/+	82/+	80/+	100/a	96/+	54/+	72/+	55/+	39/+	21/+	38/+	51/+	35/+	68/+	13/+
MA	Plantago lanceolata	.	.	.	.	.	.	93/+	4/+	.	.	92/+	.	.	33/+	96/+	100/1	100/+	55/+	58/+	100/+	100/1
MA	Deschampsia cespitosa	.	.	.	.	.	.	.	16/+	20/+	31/+	38/a	.	.	22/+	.	1/+	.	.	6/+	.	54/+
	Taraxacum spec.	50/+	92/+	100/+	70/+	94/+	83/+	100/+	60/+	100/+	63/+	46/+	85/+	100/+	22/+	99/+	63/+	87/+	70/+	71/+	7/+	83/+
LP	Alopecurus geniculatus	50/+	.	.	50/+	50/+	100/a	.	84/+	80/+	.	13/+	100/a	79/+	.	.	.	.	11/+	19/+	.	.
	Glyceria fluitans	.	.	.	.	.	33/+	14/+	90/+	40/+	.	9/+	100/+	76/+	.	.	.	.	26/+	16/+	.	.
	Stellaria uliginosa	.	.	.	.	.	33/+	.	18/+	60/+	.	.	92/+	28/+	11/+	9/+	.	.	6/+	48/+	.	.
	Leontodon autumnalis	.	.	.	.	.	.	20/+	18/+	.	.	29/+	21/+	.	.	6/+	5/+	.	15/+	16/+	.	.
	Phleum pratense	.	.	.	30/+	11/+	33/+	14/+	.	.	.	.	8/+	10/+	.	3/+	.	.	4/+	.	.	.
LP	Agrostis stolonifera	.	8/+	.	.	83/a	.	67/+	36/+	.	.	13/+	15/+	24/+	11/+	4/+	.	.	32/+	45/+	11/+	37/+
	Sagina procumbens	.	.	.	.	6/+	16/+	.	4/+	60/+	.	.	61/+	69/+	22/+	3/+	.	.	2/+	19/+	.	.
LC	Poa trivialis	50/+	85/+	81/+	80/+	100/a	67/+	93/+	100/a	.	31/+	92/+	46/+	100/+	100/+	97/+	86/+	87/a	77/+	58/+	96/+	54/+
LC	Bellis perennis	.	.	.	20/+	56/+	67/+	.	30/+	60/+	.	.	100/+	100/+	11/+	41/+	15/+	.	21/+	6/+	.	.
LC	Cynosurus cristatus	.	.	.	20/+	78/+	83/+	40/+	58/+	20/+	13/+	42/+	85/+	100/a	44/+	51/+	43/+	75/+	57/+	100/a	100/a	86/+
LC	Ranunculus repens	25/+	77/a	91/a	80/+	100/1	83/+	100/a	100/1	100/1	100/+	96/+	100/a	100/1	88/+	91/+	94/a	100/+	100/a	94/a	93/+	35/+
LC	Trifolium repens	.	8/+	54/+	20/+	61/+	83/+	80/+	78/+	80/1	6/+	46/+	69/+	93/a	66/+	56/+	57/+	87/+	94/+	94/a	96/a	69/+
LC	Festuca pratensis	.	.	18/+	10/+	67/+	50/+	73/+	100/a	100/a	50/+	25/+	85/+	97/+	44/+	93/+	74/+	100/a	60/+	87/+	54/+	40/+
LC	Ranunculus acris	.	.	18/+	40/+	6/+	16/+	100/+	98/+	100/1	31/+	46/+	62/+	45/+	44/+	100/+	95/+	100/+	100/+	94/+	100/+	74/+
	Carex ovalis	.	.	.	.	.	.	.	20/+	20/+	.	9/+	8/+	38/+	11/+	.	.	.	.	3/+	.	29/+
MA	Trifolium pratense	.	.	.	.	.	.	33/+	16/+	.	13/+	.	15/+	62/+	.	76/+	46/+	.	43/+	13/+	100/a	29/+
	Juncus articulatus	.	.	.	.	.	.	14/+	28/+	.	.	.	69/+	69/+	.	7/+	3/+	.	28/+	19/+	18/+	.
	Mentha aquatica	.	.	.	.	.	.	4/+	.	.	.	.	31/+	35/+	.	11/+	.	.	23/+	10/+	14/+	.
	Lotus uliginosus	.	.	.	.	.	.	7/+	6/+	.	.	9/+	23/+	59/+	44/+	55/+	44/+	.	26/+	77/+	100/+	97/+
	Juncus effusus	.	.	.	.	.	.	47/+	62/+	60/+	69/+	67/+	62/+	93/+	100/+	6/+	4/+	25/+	91/+	100/+	93/+	34/+
FC	Festuca rubra	50/a	54/+	100/1	.	.	16/+	93/a	74/+	100/a	100/3	100/3	85/+	76/+	100/a	76/a	92/a	38/+	98/a	87/+	100/+	86/+
FC	Agrostis capillaris	25/+	15/+	91/a	40/+	61/+	.	.	10/+	.	31/+	100/2	69/+	17/+	.	.	.	.	.	39/+	14/+	89/a
FC	Anthoxanthum odoratum	.	15/+	.	.	61/+	100/a	100/+	94/+	60/+	100/+	100/a	92/+	100/a	100/a	89/+	100/a	100/+	100/1	97/+	100/+	100/+
FC	Trifolium dubium	.	.	.	10/+	.	.	14/+	.	.	.	9/+	8/+	31/+	.	34/+	3/+	.	.	10/+	.	9/+
FC	Luzula campestris	.	.	.	.	.	.	.	.	.	13/+	42/+	.	.	44/+	.	.	.	.	.	43/+	57/+
FC	Stellaria graminea	.	.	.	.	.	.	47/+	.	.	.	.	.	.	.	23/+	18/+	63/+	2/+	.	64/+	43/+
	Cirsium palustre	.	.	.	20/+	22/+	.	60/+	4/+	.	44/+	50/+	.	31/+	77/+	64/+	82/+	.	30/+	90/+	86/+	100/+
	Equisetum fluviatile	.	.	.	.	.	.	27/+	26/+	.	19/+	17/+	.	.	55/+	93/+	97/+	100/+	60/+	100/+	100/+	54/+
	Agrostis canina	.	.	.	.	.	.	.	.	.	.	.	.	.	66/+	.	.	.	.	3/+	4/+	20/+
	Rhinanthus angustifolius	.	.	.	.	6/+	.	73/+	48/+	.	50/+	67/+	54/+	72/a	100/+	68/+	93/+	87/+	96/a	97/a	100/a	97/a
	Equisetum palustre	.	.	.	10/+	.	33/+	27/+	68/+	40/+	.	4/+	69/+	76/+	.	51/+	46/+	63/+	94/+	.	100/+	97/+
CP	Myosotis palustris	.	.	.	.	.	.	.	30/+	20/+	.	4/+	38/+	21/+	.	74/+	7/+	87/+	43/+	48/+	25/+	17/+
CP	Lychnis flos-cuculi	.	.	.	.	.	.	47/a	.	.	.	.	92/+	62/+	22/+	91/+	88/+	50/+	89/+	74/+	86/+	74/+
CP	Ajuga reptans	.	.	.	.	.	.	.	.	.	.	.	23/+	3/+	.	49/+	90/+	75/+	4/+	29/+	79/+	91/+
CP	Caltha palustris	.	.	.	.	.	.	.	.	.	.	.	31/+	38/+	22/+	31/+	52/+	25/+	100/a	97/1	82/+	.
Ccn	Galium palustre	.	.	.	.	.	.	7/+	42/+	.	6/+	.	54/+	66/+	11/+	6/+	.	.	79/+	32/+	43/+	.
Ccn	Carex nigra	.	.	.	.	.	.	.	20/+	.	25/+	46/+	.	.	55/+	16/+	38/+	.	96/+	94/a	100/a	66/+
LC	Prunella vulgaris	.	.	.	.	.	.	.	.	.	.	17/+	.	.	11/+	1/+	2/+	.	.	58/+	18/+	77/+
F	Filipendula ulmaria	.	.	.	.	.	.	.	.	20/+	.	.	.	.	.	100/1	100/a	100/2	2/+	39/+	39/+	49/+
F	Lythrum salicaria	.	.	.	.	.	.	.	.	.	.	.	.	.	.	25/+	56/+	100/+	.	.	11/+	17/+
F	Valeriana offcinalis	.	.	.	.	.	.	.	.	.	.	.	.	.	.	23/+	10/+	100/+	4/+	.	79/+	11/+
A	Veronica chamaedrys	.	.	.	.	.	.	33/+	.	.	.	.	.	.	.	28/+	65/+	87/+	.	.	14/+	.
CP	Bromus racemosus	.	.	.	.	.	.	.	.	.	.	.	.	.	.	49/+	13/+	38/+	.	.	.	.
CP	Carex acutiformis	.	.	.	.	.	.	.	.	.	.	.	.	.	.	92/1	96/a	100/+	.	.	.	.
CP	Crepis paludosa	.	.	.	.	.	.	.	.	.	.	.	.	.	.	88/+	50/a	63/+	.	.	61/+	20/+
LC	Veronica arvensis	.	.	.	40/+	22/+	33/+	.	.	.	.	.	15/+	27/+	.	83/+	10/+	13/+	.	.	.	.
	Glyceria maxima	.	.	.	.	.	.	.	.	.	.	.	.	.	.	64/+	14/+	100/+	.	.	.	.
	Phragmites australis	.	.	.	.	.	.	.	.	.	.	.	.	.	.	6/+	40/+	13/+	.	.	.	.
A	Vicia cracca	.	.	.	.	.	.	.	.	.	.	.	.	.	.	44/+	49/+	38/+	.	.	.	.
A	Phyteuma nigrum	.	.	.	.	.	.	.	.	.	.	.	.	.	.	7/+	44/+	100/+	.	.	.	.
A	Ranunculus ficaria	.	.	.	.	.	.	.	.	.	.	.	.	.	.	5/+	4/+	.	.	.	.	.
	Rhinanthus minor	.	.	.	.	.	.	.	.	.	.	.	.	.	.	1/+	7/+	.	.	10/+	.	.
A	Leucanthemum vulgare	.	.	.	.	.	.	.	.	.	.	.	.	.	.	.	1/+	.	.	.	.	.
A	Heracleum sphondylium	.	.	.	.	.	.	.	.	.	.	.	.	.	.	.	4/+	38/+	.	.	.	.
A	Phalaris arundinacea	.	.	.	.	.	.	.	.	.	.	.	.	.	.	12/+	13/+	.	.	.	.	.
	Stellaria palustris	.	.	.	.	.	.	.	.	.	.	.	.	.	.	16/+	10/+	.	.	18/+	.	.
	Lysimachia nummularia	.	.	.	.	.	.	.	.	.	.	.	.	.	.	15/+	.	13/+	.	3/+	.	.
	Angelica sylvestris	.	.	.	.	.	.	.	.	.	.	.	.	.	.	.	1/+	.	.	19/+	.	.
	Carex disticha	.	.	.	.	.	.	.	.	.	.	.	.	.	.	3/+	8/+	.	.	.	.	.
	Achillea ptarmica	.	.	.	.	.	.	.	.	.	.	.	.	.	.	15/+	3/+	.	.	13/+	4/+	11/+
	Lysimachia vulgaris	.	.	.	.	.	.	.	.	.	.	.	.	.	.	3/+	.	25/+	.	.	15/+	37/+
CP	Dactylorhiza majalis	.	.	.	.	.	.	27/+	.	.	.	.	.	.	.	.	.	.	34/+	6/+	82/+	91/+
	Galium uliginosum	.	.	.	.	.	.	.	.	.	.	.	.	.	.	8/+	11/+	.	.	.	75/+	9/+
Ccn	Ranunculus flammula	.	.	.	.	.	.	.	40/+	40/+	.	.	46/+	24/+	.	.	.	.	53/+	23/+	14/+	.
Ccn	Eleocharis palustris	.	.	.	.	.	.	.	24/+	40/+	.	.	.	.	.	20/+	.	.	.	26/+	.	.
Ccn	Menyanthes trifoliata	.	.	.	.	.	.	.	.	.	.	.	.	.	.	.	.	.	.	.	100/3	.
Ccn	Potentilla palustris	.	.	.	.	.	.	.	.	.	.	.	.	.	.	.	.	.	.	.	96/+	.
Ccn	Viola palustris	.	.	.	.	.	.	.	.	.	.	.	.	.	.	1/+	1/+	.	.	.	100/+	23/+
Ccn	Carex echinata	.	.	.	.	.	.	.	.	.	.	.	.	.	.	.	.	.	.	.	54/+	17/+
	Dactylorhiza maculata	.	.	.	.	.	.	.	.	.	.	.	.	.	.	.	.	.	.	.	46/+	3/+
JA	Succisa pratensis	.	.	.	.	.	.	.	.	.	.	.	.	.	.	.	.	.	.	.	4/+	.
JA	Juncus acutiflorus	.	.	.	.	.	.	.	.	.	.	.	.	.	.	.	.	.	43/+	10/+	21/+	100/3
JA	Juncus conglomeratus	.	.	.	.	.	.	.	.	.	.	.	.	.	.	.	.	.	.	.	7/+	86/+
JA	Luzula multiflora	.	.	.	.	.	.	.	.	.	.	.	.	.	.	.	.	.	.	.	.	43/+
JA	Carex panicea	.	.	.	.	.	.	.	.	.	.	.	.	.	.	.	.	.	.	.	25/+	17/+

Addenda: Achillea millefolium, Anemone nemorosa, Elymus repens, Festuca arundinacea, Galium aparine, Hypericum tetrapterum, Hypochaeris radicata, Juncus bufonius, Lysimachia thyrsiflora, Montia fontana, Potentilla erecta, Polygonum hydropiper, Rumex crispus, Berula erecta, Valeriana dioica, Veronica serpyllifolia.

Photo VI.4. Lot in the Anlo study area which has been cut without fertilizer application for 7 years featuring a *Lolio-Cynosuretum*. *Ranunculus repens* and *Cardamine pratensis* occur all over the lot whereas *Juncus acutiflorus* is restricted to the ditches.

The 'dry' clusters 12 (lot 854, reserve since 1967), 1 and 2 (lot 1376, fertilizer application until 1971) are considered to represent a *Lolio-Cynosuretum* (Photo VI.4). Cluster 12, a *Holcus lanatus/Anthriscus sylvestris* community, represents relevés close to the hedge-bank and contains *Arrhenatherion elatioris* elements. Cluster 1, a *Holcus lanatus/Poa trivialis* community, occurred in the whole lot during nearly the whole study period, wereas cluster 2, a *Holcus lanatus/Ranunculus acris* community only occurred in 1974.

The clusters 20 and 21, occurring in the dry part of lot 677 (a reserve since 1967) can be classified as a transition between *Lolio-Cynosuretum* and *Festuco-Cynosuretum*. Cluster 20, a *Festuca rubra/Holcus lanatus* community represents relevés of the first three years of the permanent plots. Cluster 21, a *Festuca rubra/Agrostis capillaris* community represents the last four years of the same permanent plots.

The clusters 18 and 19 (lot 463, receiving fertilizer applications until 1971) and 22 (lot 677, reserve since 1967) are found in wet areas and are considered *Lolio-Cynosuretum* with *Calthion palustris* and *Caricion curto-nigrae* elements. Cluster 18, a *Cynosurus cristatus/Alopecurus geniculatus* community represents the relevés of the first few years, whereas cluster 19, a *Cynosurus cristatus/Trifolium pratense* community, represents the relevés of the last four years of the same permanent plots. Cluster 22, a *Cynosurus cristatus/Cirsium palustre* community represents the transition between the dry and wet area in lot 677.

The clusters 5, 6 and 7 occurred in the wet Taarlo lots and can be classified as *Calthion palustris* with *Filipendulion* elements. There is an affinity with the *Carex acutiformis* community (cf. *Angelico-Cirsietum oleracei* Tx.em. Soo 40) within the *Calthion palustris* (Photo VI.5). The convex area of peat dried superficially during the summer period causing some N-mineralization resulting in the occurrence of *Arrhenatherion elatioris* elements (Grootjans 1980). Cluster 5, a *Carex actuiformis/Crepis paludosa* community (Photo VI.6) is found in the lots 422 and 431, cluster 6, a *Carex actuiformis/Ajuga reptans* community, in the lots 419 and 421 and cluster 7, a *Carex acutiformis/Phyteuma nigrum* community in the permanent plot close to some trees in lot 424.

The 'wet' clusters 14 and 23 are considered as *Calthion palustris* with *Caricion curto-nigrae* elements. Cluster 14, a *Caltha palustris/Anthoxanthum odoratum* community occurs in lot 854 (Photo VI.7), cluster 23, a *Caltha palustris/Cynosurus cristatus* community, in lot 677. Both clusters, therefore, occur in lots that have been reserves since 1967.

Cluster 3, a *Menyanthes trifoliata* community, is found in lot 1371A and is a transition between *Caricion curto-nigrae* and *Calthion palustris*. Cluster 4 occurs in lot 1371B and is a transition between *Calthion palustris* and *Juncion acutiflori* (Photo VI.8). Both clusters occur in a lot that is a reserve over more than 30 years.

VI.C.1.b. Loefvledder (Table VI.5, 6, 7 and Fig. VI.5, 6)

Each transect runs down a gradient towards the brook and contains three permanent plots, namely, the 'high' plot, at some distance from the brook,

Table VI.4(A–K). Cluster number to which each permanent plot each year was assigned.

A: Lot 422

Years / Plot	72	73	74	76	77	78	78	80
1	5.4	5.4	5.7	5.1	5.3	5.1	5.1	5.1
2	5.4	5.4	5.7	5.1	5.1	5.1	5.1	5.1
3	5.4	5.4	5.7	5.1	5.1	5.1	5.1	5.1
4	5.4	5.5	5.7	5.1	5.1	5.1	5.1	5.1
5	5.5	5.5	5.7	5.1	5.1	5.1	5.1	5.1
6	5.4	5.4	5.7	5.1	5.1	5.1	5.1	5.1

B: Lot 431

Years / Plot	72	73	74	76	77	78	79	80
1	5.8	5.8	5.8	5.2	5.2	5.2	5.2	5.2
2	5.8	5.8	5.8	5.2	5.2	5.2	5.2	5.2
3	5.8	5.8	5.8	5.2	5.2	5.2	5.2	5.2
4	5.8	5.8	5.6	5.2	5.2	5.2	5.2	5.2
5	5.9	5.6	5.6	5.3	5.2	5.2	5.2	5.3
6	5.9	5.6	5.6	5.3	5.3	5.3	5.3	5.3

C: Lot 1376

Years / Plot	72	73	74	76	77	78	79	80
1	1.1	1.1	1.1	1.5	1.5	1.5	1.3	1.3
2	1.1	1.1	1.1	1.5	1.5	1.5	1.3	1.3
3	1.2	1.2	2.1	1.4	1.4	1.4	1.4	1.4
4	1.1	1.1	2.1	1.5	1.5	1.5	1.5	1.5
5	1.2	1.2	2.1	1.2	1.4	1.4	1.4	1.4
6	1.1	1.1	2.1	1.5	1.5	1.5	1.3	1.3
7	1.2	1.2	2.1	1.6	-	1.6	1.6	1.6

D: Lot 1371

Years / Plot	74	75	76	77	78	79	80
1	3.1	3.1	3.1	3.1	3.1	3.1	3.1
2	3.1	3.1	3.1	3.1	3.1	3.1	3.1
3	3.1	3.1	3.1	3.1	3.1	3.1	3.1
4	3.2	3.2	3.2	3.2	3.2	3.2	3.2
5	4.1	4.1	4.3	4.3	4.3	4.3	4.3
6	4.1	4.1	4.2	4.2	4.2	4.2	4.2
7	4.1	4.1	4.2	4.2	4.2	4.2	4.2
8	4.1	4.1	4.2	4.2	4.2	4.2	4.2
9	4.1	4.1	4.2	4.2	4.2	4.2	4.2

E: Lot 419

Years / Plot	72	73	74	76	77	78	79	80
1	6.1	6.2	6.2	6.2	6.2	6.2	6.2	6.2
2	6.1	6.2	6.2	6.2	6.2	6.2	6.2	6.2
3	6.1	6.2	6.2	6.2	6.2	6.2	6.2	6.2
4	6.2	6.2	6.2	6.2	6.2	6.2	6.2	6.2
5	6.2	6.2	6.2	6.2	6.2	6.2	6.2	6.2
6	6.2	6.2	6.2	6.2	6.2	6.2	6.2	6.2
7	6.2	6.2	6.2	6.2	6.2	6.2	6.2	6.2
8	6.3	6.2	6.2	6.2	6.2	6.2	6.2	6.2
9	6.3	6.2	6.3	6.3	6.3	6.3	6.3	6.3

F: Lot 424

Years / Plot	72	73	74	76	77	78	79	80
1	7.1	7.1	7.1	7.1	7.2	7.2	7.2	7.2
2	6.4	6.4	6.4	6.4	6.4	6.4	6.4	6.4
3	6.4	6.4	6.4	6.4	6.4	6.4	6.4	6.4
4	6.4	6.4	6.4	6.4	6.4	6.4	6.4	6.4
5	6.4	6.4	6.4	6.4	6.4	6.4	6.4	6.4
6	6.4	6.4	6.4	6.4	6.4	6.4	6.4	6.4
7	6.4	6.4	6.4	6.4	6.4	6.4	6.4	6.4
8	6.4	6.4	6.4	6.4	6.4	6.4	6.4	6.4
9	6.4	6.4	6.5	6.4	6.4	6.4	6.4	6.4

G: Lot 1376A

Years / Plot	72	74	76	77	78	79	80
1	8.1	9.2	9.2	9.2	9.2	9.1	9.1
2	8.2	9.2	9.2	9.2	9.2	9.1	9.1
3	8.2	9.2	10.1	10.2	10.2	10.2	10.2
4	8.2	10.3	10.2	10.2	10.2	10.2	10.2

H: Lot 854

Years / Plot	72	73	74	76	77	78	79	80
1	11.1	12.1	12.4	12.3	12.3	12.3	12.3	12.3
2	12.2	12.1	12.4	12.2	12.2	12.2	12.2	12.2
3	13.1	14.1	14.1	14.6	14.7	14.7	14.6	14.6
4	14.1	14.1	14.1	14.6	14.6	14.6	14.6	14.6
5	14.3	14.3	14.1	14.4	14.7	14.7	14.7	14.7
6	14.1	14.3	14.2	14.5	14.5	14.5	14.5	14.5
7	14.2	14.2	14.1	14.5	14.7	14.7	14.7	14.7
8	14.3	14.3	14.8	14.5	14.9	14.9	14.9	14.9

J: Lot 463

Years / Plot	72	73	74	76	77	78	79	80
1	15.1	15.3	15.1	15.3	16.1	16.1	16.1	16.1
2	15.4	15.3	15.2	15.3	16.1	16.1	16.1	16.1
3	17.1	17.3	15.2	16.1	16.1	16.1	16.1	16.1
4	17.1	17.3	15.2	16.1	16.1	16.1	16.1	16.1
5	17.2	17.3	18.1	19.1	19.1	19.1	19.2	19.2
6			18.1	19.1	19.1	19.1	19.2	19.2
7			18.2	19.3	19.3	19.3	19.3	19.2
8			18.2	18.3	19.2	19.2	19.2	19.2
9			18.2	18.3	19.2	19.2	19.2	19.2
10			18.2	18.3	19.2	19.2	19.2	19.2
11			18.2	18.3	18.3	18.4	19.2	19.2

K: Lot 677

Years / Plot	72	73	74	76	77	78	79	80
1	20.1	20.1	20.2	22.1	21.1	21.1	21.1	21.1
2	20.1	20.2	20.2	22.3	21.3	21.3	21.3	21.3
3	20.2	20.2	20.2	22.3	21.3	21.3	21.3	21.3
4	20.2	20.1	20.3	22.3	21.3	21.3	21.3	21.3
5	20.2	20.2	20.2	22.3	21.3	21.3	21.3	21.3
6	20.1	22.2	23.1	22.2	21.2	21.2	21.2	21.2
7	23.2	22.3	23.1	22.2	23.3	23.3	23.3	23.3
8	23.2	23.1	23.1	23.4	23.3	23.3	23.3	23.3
9	23.2	23.4	23.1	23.4	23.3	23.3	23.3	23.3
10	23.2	23.4	23.1	23.5	23.6	23.6	23.6	23.6
11	23.2	-	23.1	23.5	23.6	23.6	-	-

Photo VI.5. Lot in the Taarlo study area which has been cut without fertilizer application for 5 years featuring a *Carex acutiformis* community with *Lychnis flos-cuculi* and *Rhinanthis angustifolius*.

Photo VI.6. Lot in the Taarlo study area which has been cut without fertilizer application for 5 years featuring a *Carex acutiformis* community with *Crepis paludosa* and *Filipendula ulmaria*.

Photo VI.7. Lot in the Anlo study area which has been cut without fertilizer application for 15 years featuring a *Calthion palustris* community with clumps of *Juncus acutiflorus* occurring in the lot and *Cardamine pratensis* and *Ranunculus acris*.

Photo VI.8. Lot in the Anlo study area which has been cut without fertilizer application for 35 years featuring a transition between *Calthion palustris* and *Juncion acutiflori* with *Juncus acutiflorus* dominating all over the lot and *Ranunculus acris*.

nearer the brook the 'middle' plot and adjacent to the brook, the 'low' plot. Each management regime featured duplicate transects and, thus, six permanent plots. In the recent reserve involving nine management regimes 54 permanent plots were, therefore, recorded, whereas 48 plots were recorded in the older reserve since eight management practices were involved.

Permanent plot data from 1973, 1978 and 1983 were clustered (II.C.3). Synoptic tables were made for both the recent reserve and the older reserve starting from the results of these clusterings. Clusters from different years were lumped if these proved to be similar.

The dendrograms resulting from clusterings of the recent reserve are shown in Fig. VI.5 for 1973, 1978 and 1983, respectively. It is clear that all the coena found in 1973 had disappeared in 1978 and that most coena found in 1983 already occurred in 1978.

Table VI.5 shows the corresponding synoptic table including 14 coena, all of which can be reckoned among transitions between *Lolio-Plantaginion* and basal communities of the *Molinio-Arrhenatheretea* with mainly *Lolio-Cynosuretum* elements (*Agrostis/Poa* community). Many species indicating nutrient-rich soil conditions were involved. Three coena contain the majority of relevés featuring a *Festuca pratensis* group (coenon N), a *Cerastium fontanum* group (coenon P) and an intermediate coenon O. Other coena include small numbers of relevés and can be seen as minor variants of the prevailing communities characterized by one or a few differential species viz. *Festuca rubra* type (coenon E), *Heracleum sphondylium* type (coenon F), *Alopecurus geniculatus* type (coenon G), *Potentilla anserina* type (coenon K), *Ranunculus repens* type (coenon L), *Agrostis stolonifera* type (coenon M), *Elymus repens* type (coena A, B, C, D and H).

The coena B, C and D showed high dissimilarity percentages with respect to the other coena due to the cover of *Elymus repens* and *Urtica dioica*. These three coena (and coenon A) only occurred in parts of the management sites 1, 2, 3 and 4 ('abandoning', 'burning' and 'September hay-making or mulching every two years'). Since this situation already existed in 1973 prior to the start of the management practices it must be due to an abiotic phenomenon. Indeed the permanent plots concerned are found in the relatively dry area, with lower groundwater tables in the wet season, rather than in the wetter area (see section VI.B.2). These permanent plots cannot, therefore, be taken into account in the comparison of management regimes. All other permanent plots in the different management practices seemed to be distributed randomly over the clusters in 1973. In 1983, however, a relationship had arisen between management regimes and coena (Fig. VI.5). The 1978 clustering result

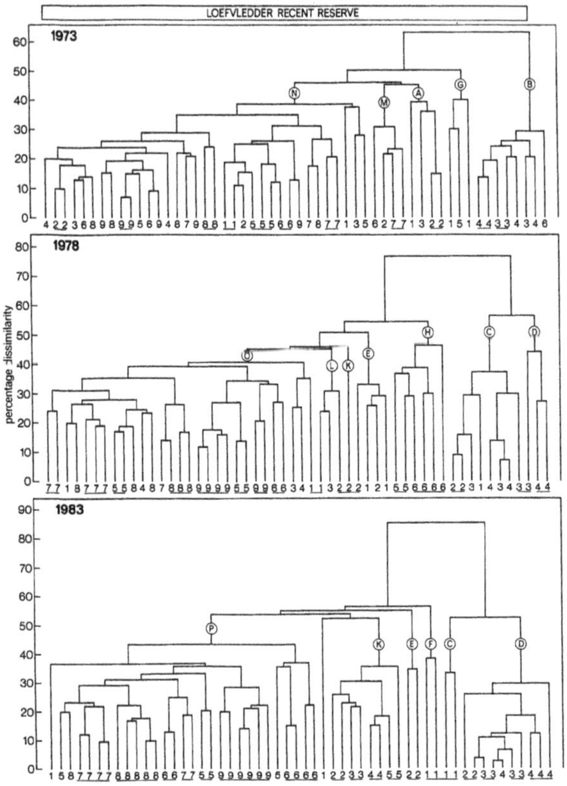

Fig. VI.5. Dendrograms of all permanent plots in the recent reserve of the Loefvledder area in 1973, 1978 and 1983, respectively, showing the coena A–P corresponding with Table VI.5 and with an indication of the management regimes (1: abandoning, 2: burning, 3: mulching in September every second year; 4: hay-making in September every second year, 5: annual hay-making in September, 6: hay-making alternating in July or in September, 7: annual hay making in July, 8: annual hay-making in July with the scythe, 9: hay-making in both July and September).

appeared intermediate. The *Heracleum sphondylium* type (coenon F) was found in the abandoned site, *Cirsium arvense* type (coenon E) in the burned site. *Potentilla anserina* (coenon K) dominated in the regimes concerning September hay-making or mulching every two years, September hay-making every year and burning (regimes 2, 3, 4 and 5).

Coenon P (*Cerastium fontanum* type) included all the regimes involving July hay-making (regimes 6, 7, 8 and 9, see Fig. VI.5). Subclusters could be distinguished at the 30% dissimilarity level within coenon P, corresponding with each of these four treatments (Table VI.6). The July–September alternating hay-making regime was characterized by the *Anthriscus sylvestris* group, July hay-making by the *Polygonum mite* group and July hay-making by scythe by the *Festuca rubra* group. Hay-making in July and September (alternating or every year) obviously favoured *Rhinanthus angustifolius*. The management practice involving hay-making twice a year featured the *Alopecurus geniculatus* group, indicating elements of a *Lolio-Cynosuretum* with a few species indicating nutrient-poor soil conditions.

The clustering of the older reserve is shown in Fig. VI.6 for 1973, 1978 and 1983, respectively.

Table VI.7 shows the corresponding synoptic table including 9 coena, all of which can be classified as *Festuco-Cynosuretum* Tx.apud Büker 1942. Species indicating both nutrient-rich and nutrient-poor conditions were involved. In 1973 at the beginning of the experiment the lot included stands with *Molinia-Arrhenatheretea* elements (coenon Q), and stands with *Caricion curto-nigrae* elements (coenon S). The permanent plots in the different management regimes seemed to be distributed randomly over the clusters in 1973. In 1978, however, July mulching every year (regime 1) was related to high cover percentages of *Ranunculus repens* and *Rumex acetosa* (coenon T). The less frequently mulched (practices 2 and 3) and the artificially fertilized plots (regimes 7 and 8) were characterized by the *Poa*

Table VI.5. Synoptic table of the coena occurring in the recent reserve of Loefvledder in the upper course during the period 1973 to 1983 and corresponding with the clusterings in Fig. VI.5. P: species indicating nutrient-poor soil conditions; R: nutrient-rich conditions; I: intermediate. The occurrence of species is given in percentages and in mean cover percentages (+: 1–5; a: 6–10; 1: 11–20; 2: 21–30% etc.).

	Coenon	B	C	D	H	A	E	F	K	L	M	G	N	O	P
	Year	1973	1978/83	1978/83	1978	1973	1978/83	1983	1978/83	1978	1973	1973	1973	1978	1983
	Number of relevés	8	10	13	5	4	6	2	9	3	4	3	35	29	29
R	Elymus repens	100/6	100/7	100/3	100/3	100/3	.	.	11/+	.	25/+	.	3/a	14/+	21/a
R	Galeopsis tetrahit	.	20/+	.	.	.	.	.	.	.	.	.	.	.	.
I	Dactylis glomerata	.	10/+	.	.	.	.	.	.	.	.	.	.	.	.
R	Urtica dioica	25/+	30/1	100/3	.	.	33/+	.	.	.	.	.	.		3/+
-	Holcus mollis	.	30/a	69/2	.	.	17/+	.	.	.	.	.	.		14/+
R	Cirsium arvense	13/+	80/+	85/1	40/+	75/+	100/1	50/+	44/+	67/+	50/+	33/+	26/+	14/+	10/+
R	Rumex obtusifolius	.	30/+	31/+	60/+	.	33/1	100/+	.	.	25/+	.	6/+	21/+	24/+
I	Festuca rubra	25/+	10/+	.	50/+	75/a	100/1	50/+	100/+	100/+	75/+	67/a	83/+	97/a	90/a
R	Senecio sylvaticus	.	.	8/+	.	.	33/+	.	11/+	.	.	.	.	.	.
I	Lotus uliginosus	.	.	.	.	.	33/+	.	11/+	.	.	.	.	7/+	.
R	Cirsium vulgare	.	.	.	.	.	17/+	.	.	.	.	.	.	.	.
I	Epilobium tetragonum	.	.	8/+	.	.	67/+	100/+	11/+	.	.	.	.	.	25/+
-	Rubus sp.	.	.	.	.	.	.	50/1	.	.	.	.	.	.	.
R	Heracleum sphondylum	.	10/+	.	.	.	.	100/2	.	.	.	.	.	.	.
R	Poa pratensis	100/a	20/a	8/+	100/a	100/ a	100/ 1	100/a	89/+	67/2	100/+	100/1	94/1	100/a	100/+
I	Holcus lanatus	88/+	70/+	38/+	100/3	100/1	100/1	100/4	100/3	100/3	100/1	100/2	100/1	100/4	100/4
R	Poa trivialis	88/a	.	15/+	100/a	100/1	17/+	.	33/+	100/+	100/+	67/+	94/1	97/a	90/a
-	Taraxacum sp.	50/+	.	.	40/+	100/+	33/+	.	11/+	.	.	100/+	49/+	31/+	34/+
R	Rumex crispus	.	10/+	.	40/+	.	17/+	50/+	22/+	.	.	33/+	11/+	7/+	10/+
R	Potentilla anserina	.	.	.	20/+	.	33/+	.	100/4	67/+	.	.	3/+	21/+	31/+
R	Ranunculus repens	75/+	10/+	8/+	40/+	75/+	100/a	100/+	100/1	100/4	100/+	67/1	89/a	100/+	100/1
R	Agrostis stolonifera	63/a	.	8/+	100/+	100/+	100/a	100/a	100/a	67/+	100/4	100/a	100/2	86/1	90/a
R	Festuca pratensis	38/+	10/+	.	.	100/+	33/+	.	33/+	33/+	100/+	67/a	100/1	86/+	87/+
R	Lolium perenne	.	.	.	.	75/+	.	.	.	.	75/+	33/1	51/a	4/+	7/+
R	Alopecurus geniculatus	.	.	.	.	25/+	.	.	.	.	75/+	100/a	83/a	10/+	17/+
I	Cardamine pratensis	13/+	.	.	40/+	100/+	.	.	.	.	.	67/+	89/+	66/+	76/+
R	Trifolium repens	.	.	.	.	25/+	.	.	.	.	50/+	33/+	29/+	10/+	14/+
I	Ranunculus acris	.	.	.	.	75/+	.	.	.	33/+	.	.	20/+	3/+	10/+
R	Phleum pratense	.	.	.	20/+	.	.	.	11/+	.	.	.	3/+	.	3/+
P	Equisetum palustre	.	.	.	.	.	.	.	.	.	25/+	.	11/+	7/+	21/+
I	Agrostis capillaris	.	.	.	20/+	.	33/+	.	44/+	33/+	.	.	.	34/+	21/+
I	Cerastium fontanum	38/+	.	.	80/+	100/+	.	.	55/+	.	50/+	33/+	54/+	59/+	83/+
I	Rumex acetosa	25/+	10/+	15/+	60/+	25/+	50/+	100/+	89/a	33/+	.	67/+	57/+	83/+	100/ 1
P	Rhinanthus angustifolius	.	.	.	.	.	.	.	22/+	.	.	.	.	.	59/a
I	Anthoxanthum odoratum	.	.	.	.	.	.	.	.	.	.	.	3/+	3/+	38/+
P	Juncus effusus	.	.	.	.	.	.	.	.	.	.	.	.	3/+	3/+
R	Anthriscus sylvestris	.	.	.	.	.	.	.	.	.	.	.	.	10/+	17/+
R	Polygonum mite	.	.	.	.	.	.	.	.	.	.	.	.	3/+	3/+
I	Cynosurus cristatus	.	.	.	.	.	.	.	.	.	.	.	.	7/+	7/+
R	Plantago major	.	.	.	.	.	.	.	.	.	.	.	3/+	.	7/+
I	Plantago lanceolata	.	.	.	.	.	.	.	11/+	.	.	.	.	.	3/+

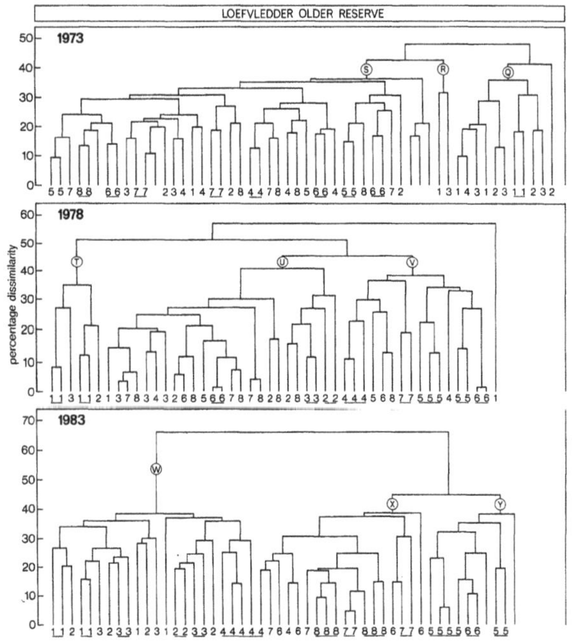

Fig. VI.6. Dendrograms of all permanent plots in the older reserve of the Loefvledder area in 1973, 1978 and 1983, respectively, showing the coena Q–Y corresponding with Table VI.7 and with an indication of the management regimes (1: annual mulching in July, 2: hay-making in July every second year, 3: mulching in July every second year, 4: hay-making or mulching alternating in July, 5: annual hay-making in July with the scythe, 6: annual hay-making in July, 7: annual hay-making in July and application of Ammonium Nitrogen Limestone (50 kg N · ha^{-1}), 8: annual hay-making in July and application of Ureum Formaldehyde (50 kg N · ha^{-1})). The permanent plots which got no figure were not taken into consideration because the management was not consistent.

pratensis group (coenon U). The regimes involving only hay-making annually or every second year (regimes 5 and 6) were not characterized by particular species (coenon V). In 1983 all practices in which mulching was included (practices 1, 2, 3 and 4) were characterized by the *Holcus lanatus* group (coenon W). The regimes involving hay-making annually without fertilizer applications (regimes 5 and partly 6) were characterized by the *Carex nigra* group (coenon Y) with many species indicating nutrient poor soil conditions. If artificial fertilizer was also involved (regimes 7, 8) the *Juncus effusus* group (coenon X) characterized the vegetation.

VI.C.1.c. Lammertshem (Table VI.8, 9 and Fig. VI.7, 8, 9)

The description of the vegetation and the lay-out of the permanent plots was similar to that of the Loefvledder area mentioned in the previous section. Due to the wet years at the end of the 1970's hay-making and mulching, according to the experimental scheme, became impossible and the practices were, therefore, terminated in 1980. Hence, only permanent plot data from 1973 and 1979 were clustered and listed in synoptic tables.

The dendrograms resulting from the recent reserve are shown in Fig. VI.7 for 1973 and 1979 coena, which can all be classified as transitions between basal communities of the *Molinio-Arrhenatheretea* and *Calthion palustris* communities. The latter might be focused on the *Senecioni-Brometum racemosi caricetosum aquatilis* (Everts et al., 1980). Moreover, *Filipendulion* elements can be traced. Despite the recent acquisition from farmers the proportion of species indicating nutrient-rich soil conditions is low, probably owing to the anaerobic conditions.

Three clusters could be discerned at the 40% dissimilarity level at the beginning of the experiment in 1973. These clusters seem to point to small abiotic differences. Coenon A (*Cerastium fontanum* type) includes all permanent plots at some distance from the brook, the 'high' plots and the permanent plots of three adjoining regimes close to the brook. All other permanent plots close to the brook, the 'low' plots form coenon H (*Agrostis stolonifera* type). Coenon F comprises all intermediate plots featuring a high cover degree of *Ranunculus repens* in the slightly lower lying central area of the lot (Fig. VI.8A).

Despite the initial abiotic differences the 1979 clusters partly represent (groups of) management regimes suggesting a prevailing influence of management above abiotic differences. Coenon B (*Ranunculus acris* type) includes both July hay-making every year and hay-making twice a year, whereas coenon C with abundant *Rumex acetosa* and *Festuca rubra* represents July cutting by scythe. Coenon G (*Holcus lanatus* type) includes the July or September hay-making alternating regime. This holds for all permanent plots irrespective of their distance from the brook. The arrangement of permanent plots into coena further depends on the position of the permanent plots in the lot (Fig. VI.8). The plots close to the brook feature two coena; alternating July or September hay-making coenon G (*Holcus lanatus* type) and the practices including abandoning, burning, September mulching or hay-making every two years and September hay-making every year: coenon J (dominating *Phalaris arundinacea*). The plots at some distance from the brook feature two other coena: abandoning, burning and September mulching every two years: coenon E (*Filipendula ulmaria* type) and September hay-making every one or two years or alternating with July hay-making: coenon D (*Phalaris arundinacea*/*Filipendula ulmaria* type). The intermediate plots show characteristics of both the 'low' and 'high' permanent plots.

Table VI.6. Synoptic table of the composition of plant communities in 1983 in different management regimes in coenon P in the recent reserve of Loefvledder as given in Table VI.5. P: species indicating nutrient-poor soil conditions; R: nutrient-rich conditions; I: intermediate. The occurrence of species is given in percentages of frequency and in mean cover percentages (+: 1–5; a: 6–10, 1: 11–20; 2: 21–30% etc.).

	Hay-making	July or September	July	July	July + September
	Management practice	Alternating	Every year	Every year by scythe	Every year
	Number of relevés	4	4	5	6
I	Cardamine pratensis	75/+	100/+	100/+	100/+
R	Festuca pratensis	100/+	100/+	100/+	100/+
I	Rumex acetosa	100/+	100/+	100/a	100/+
R	Ranunculus repens	100/2	100/1	100/2	100/1
-	Taraxacum sp.	100/+	25/+	25/+	50/+
I	Holcus lanatus	100/3	100/4	100/2	100/4
I	Cerastium fontanum	100/+	100/+	80/+	50/+
R	Anthriscus sylvestris	50/+	.	.	.
R	Rumex crispus	50/+	.	.	.
R	Rumex obtusifolius	50/+	.	.	.
R	Elymus repens	100/+	25/+	.	.
R	Poa trivialis	100/a	75/+	80/+	83/+
P	Rhinanthus angustifolius	100/+	25/+	.	100/+
I	Epilobium tetragonum	50/+	50/+	20/+	.
R	Cirsium arvense	25/+	.	25/+	.
R	Polygonum mite	.	25/+	.	.
R	Agrostis stolonifera	100/a	100/1	60/+	100/+
R	Poa pratensis	100/+	100/a	100/+	100/+
I	Festuca rubra	75/+	100/+	100/1	100/+
-	Holcus mollis	.	.	20/+	.
R	Potentilla anserina	.	.	40/+	.
I	Agrostis capillaris	.	50/+	.	67/+
I	Anthoxanthum odoratum	.	50/+	20/+	83/+
P	Equisetum palustre	.	50/+	.	83/+
R	Alopecurus geniculatus	.	.	.	83/+
I	Glyceria fluitans	.	.	.	50/+
R	Lolium perenne	.	.	.	17/+
I	Cynosurus cristatus	.	.	.	33/+
R	Trifolium repens	.	.	.	67/+
I	Leontodon autumnalis	.	.	.	17/+
I	Carex ovalis	.	.	.	17/+
P	Juncus effusus	.	.	.	17/+
I	Plantago lanceolata	.	.	.	17/+
P	Prunella vulgaris	.	.	.	17/+
I	Ranunculus acris	.	.	.	17/+

Table VI.7. Synoptic table of the coena occurring in the older reserve of Loefvledder in the upper course during the period 1973 to 1983 and corresponding with the clusterings in Fig. VI.6. P: species indicating nutrient-poor soil conditions; R: nutrient-rich conditions, I: intermediate. The occurrence of species is given in percentages of frequency and in mean cover percentages (+: 1–5; a: 6–10; 1: 11–20; 2: 21–30% etc.).

	Coenon	Q	R	S	T	U	V	W	X	Y
	Year	1973	1973	1973	1978	1978	1978	1983	1983	1983
	Number of relevés	10	2	35	6	25	16	20	17	8
R	Alopecurus geniculatus	20/+	.	3/+	.	4/+	.	.	.	.
-	Taraxacum sp.	90/+	.	26/+	.	12/+	.	.	.	.
R	Festuca pratensis	100/+	50/+	29/+	17/+	20/+	19/+	15/+	.	.
R	Poa trivialis	60/+	.	51/+	67/+	80/+	75/+	65/+	29/+	.
I	Cardamine pratensis	100/+	50/+	23/+	50/+	24/+	56/+	70/+	6/+	.
I	Ranunculus acris	100/+	50/+	71/+	17/+	20/+	13/+	15/+	.	.
I	Plantago lanceolata	50/+	.	9/+	17/+	4/+	6/+	20/+	.	.
R	Trifolium repens	60/+	50/+	63/+	.	.	.	.	.	.
I	Leontodon autumnalis	20/+	.	31/+	.	.	.	.	.	.
P	Ranunculus flammula	20/+	.	20/+	.	4/+	.	.	.	.
I	Anthoxanthum odoratum	100/a	100/1	100/1	100/+	100/+	100/a	90/+	100/+	100/+
I	Lotus uliginosus	20/+	100/+	46/+	.	.	13/+	5/+	.	.
I	Festuca rubra	100/a	100/2	100/1	100/a	100/1	100/3	80/1	100/1	100/2
I	Agrostis capillaris	100/a	50/+	100/1	100/a	100/2	100/2	95/1	100/+	100/+
P	Juncus acutiflorus	.	.	3/+	33/a	.	.	20/a	12/+	.
P	Equisetum palustre	40/+	.	11/+	.	4/+	.	20/+	.	.
I	Lychnis flos-cuculi	10/+	.	23/+	.	12/+	.	10/+	18/+	13/+
I	Galium palustre	10/+	.	23/+	33/+	20/+	25/+	30/+	12/+	25/+
P	Luzula campestris	10/+	.	23/+	.	12/+	31/+	10/+	12/+	13/+
P	Hydrocotyle vulgaris	.	50/+	37/+	.	4/+	.	.	18/+	.
P	Carex curta	.	.	11/+	.	12/+	.	.	18/+	.
R	Ranunculus repens	100/+	100/+	100/a	100/1	84/a	94/a	70/+	47/+	50/+
I	Rumex acetosa	100/a	100/+	100/a	100/1	100/a	100/a	100/2	94/+	100/+
R	Poa pratensis	100/+	100/+	100/+	100/a	84/1	81/+	95/+	53/+	75/+
I	Carex ovalis	.	50/+	57/+	17/+	76/+	44/+	35/+	29/+	50/+
I	Holcus lanatus	90/+	50/+	54/+	100/a	96/a	94/a	100/2	100/1	87/+
P	Cirsium palustre	10/+	50/+	17/+	17/+	20/+	50/+	60/+	47/+	50/+
I	Cerastium fontanum	10/+	.	.	.	4/+	.	50/+	.	.
R	Anthriscus sylvestris	.	.	.	.	.	.	10/+	.	.
P	Juncus effusus	70/+	100/+	100/a	17/+	80/a	94/a	85/a	100/1	100/+
P	Potentilla erecta	.	.	.	.	.	.	.	12/+	.
P	Carex nigra	90/+	100/+	91/+	17/+	85/+	94/+	55/+	100/+	100/1
P	Viola palustris	.	50/+	43/+	.	36/+	25/+	25/+	71/+	100/+
P	Agrostis canina	.	.	.	.	28/a	38/+	.	47/a	63/+
P	Nardus stricta	.	.	.	.	.	.	.	.	13/+
P	Juncus conglomeratus	.	.	.	.	.	.	.	.	25/+

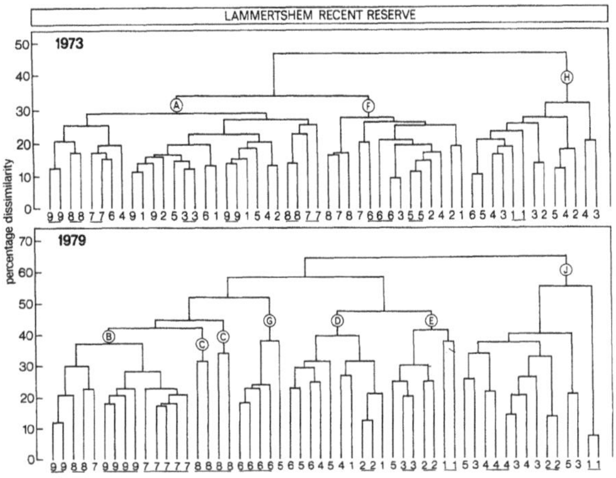

Fig. VI.7. Dendrograms of all permanent plots in the recent reserve of the Lammertshem area in 1973 and 1979, respectively, showing the coena A–J corresponding with Table VI.8 and with an indication of the management regimes (1: abandoning, 2: burning, 3: mulching in September every second year, 4: hay-making in September every second year, 5: annual hay-making in September, 6: hay-making alternating in July or in September, 7: annual hay-making in July, 8: annual hay-making in both July and September).

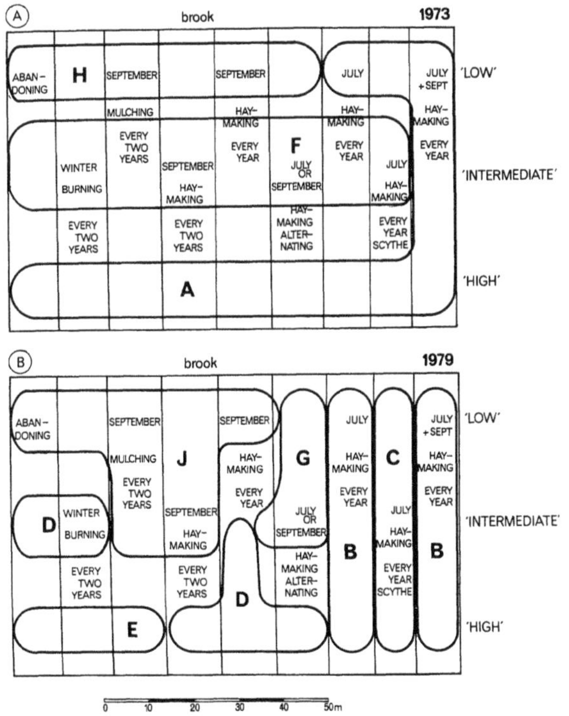

Fig. VI.8. Position of the coena A–J corresponding with Table VI.8 in 1973 and 1979, respectively, in the 'low', 'intermediate' and 'high' sites in the recent reserve of the Lammertshem area.

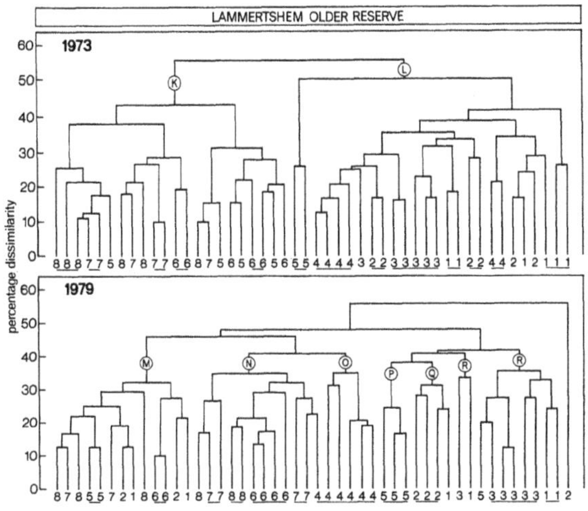

Fig. VI.9. Dendrograms of all permanent plots in the older reserve of the Lammertshem area in 1973 and 1979, respectively, showing the coena K–R corresponding with Table VI.9 and with an indication of the management regimes (1: annual mulching in July, 2; hay-making in July every second year, 3: mulching in July every second year, 4: hay-making or mulching alternating in July, 5: annual hay-making in July with the scythe, 6: annual hay-making in July, 7: annual hay-making in July and application of Ammonium Nitrogen Limestone (50 kg N · ha^{-1}), 8: annual hay-making in July and application of Ureum Formaldehyde (50 kg N · ha^{-1}).

The dendrograms resulting from the permanent plots of the older reserve are shown in Fig. VI.9 for 1973 and 1979, respectively.

Table VI.9 shows the corresponding synoptic table including 8 coena, which all tend towards the *Valeriano-Filipenduletum phalaridetosum arundinaceae* Meisel 1969 (see Everts et al., 1980). Only a few species indicating nutrient-rich or drier soil conditions are involved. A ditch subdivides the lot into two equal parts (each containing 4 practices), which are characterized by different coena: coenon K with the *Valeriana officinalis* group and coenon L with the *Ranunculus acris* group. It is not clear which factors cause the subdivision. Different agricultural or management practices prior to 1973 could be an explanation.

The increasing wetness in the late 1970's, which has previously been mentioned, resulted in a high *Calamagrostis canescens* cover degree (coenon M) in nearly all permanent plots at some distance from the brook despite different management regimes. The vegetation appears to be governed by management practices in most other permanent plots. The practices involving July hay-making every year with or without artificial fertilizers are combined in coenon N (*Holcus lanatus* type). July cutting by scythe is characterized by the nutrient-rich species *Urtica dioica* and *Poa trivialis* (coenon P). July hay-making or mulching alternating is characterized by the *Festuca pratensis* group (coe-

Table VI.8. Synoptic table of the coena occurring in the recent reserve of Lammertshem in the lower course during the period 1973 to 1979 and corresponding with the clusterings of Fig. VI.7. P: species indicating nutrient-poor soil conditions; R: nutrient-rich conditions; I: intermediate. The occurrence of species is given in percentages of frequency and in mean cover percentages (+: 1–5; a: 6–10; 1: 11–20; 2: 21–30% etc.).

	Coenon	G	C	B	A	J	D	E	F	H
	Year	1979	1979	1979	1973	1979	1979	1973	1973	1973
	Number of relevés	5	4	14	27	14	10	7	14	13
I	Holcus lanatus	100/3	100/+	100/1	100/2	71/a	60/a	71/+	100/2	92/+
R	Poa trivialis	100/1	100/a	100/1	89/+	86/a	60/+	57/+	93/+	100/+
I	Rumex acetosa	20/+	75/1	100/a	78/+	.	10/+	.	57/+	38/+
I	Festuca rubra	.	75/1	71/+	41/+	.	60/+	100/+	14/+	.
R	Alopecurus pratensis	.	25/+	.	.	.	.	.	.	.
I	Stellaria glutinosa	.	25/+	7/+	.	.	.	.	.	.
-	Lythrum salicaria	.	25/+	.	.	.	10/+	.	.	.
I	Ranunculus acris	40/+	25/+	71/+	59/+	.	20/+	.	14/+	.
R	Poa pratensis	20/+	.	71/+	37/+	14/+	20/+	.	43/+	.
I	Lysimachia nummularia	.	.	14/+	.	.	.	.	.	.
P	Rhinanthus angustifolius	.	.	36/+	.	.	.	.	.	.
I	Plantago lanceolata	.	.	7/+	.	.	.	.	.	.
I	Glyceria fluitans	.	.	.	4/+	.	.	.	.	.
I	Carex disticha	.	.	.	15/+	.	.	.	.	.
R	Glechoma hederacea	.	.	.	7/+	.	.	.	.	.
I	Ajuga reptans	.	.	.	4/+	.	.	.	.	.
I	Anthoxanthum odoratum	.	50/+	43/+	70/+	.	.	14/+	2/+	.
I	Cerastium fontanum	.	.	86/+	100/+	.	.	.	57/+	.
R	Urtica dioica	10/+	.	.	.	14/+	.	.	.	.
P	Juncus effusus	.	.	.	.	7/+	.	.	.	.
R	Phalaris arundinacea	100/1	100/2	100/1	81/1	100/6	100/5	86/a	86/+	100/1
I	Filipendula ulmaria	60/+	100/1	93/2	100/a	21/+	100/2	100/4	64/+	15/+
I	Valeriana officinalis	.	25/+	7/+	15/+	.	30/+	29/1	.	.
I	Deschampsia cespitosa	.	.	21/+	19/+	.	40/+	57/+	7/+	8/+
R	Ranunculus repens	.	100/+	100/2	100/2	42/+	20/+	29/+	100/4	100/3
I	Cardamine pratensis	60/+	75/+	100/+	100/+	7/+	40/+	29/+	93/+	62/+
R	Festuca pratensis	20/+	.	50/+	100/+	.	10/+	.	100/+	38/+
I	Lychnis flos-cuculi	.	.	.	93/+	.	.	.	79/+	8/+
-	Caltha palustris	100/1	100/+	100/1	100/a	57/+	90/a	71/+	100/1	100/a
-	Taraxacum sp.	.	25/+	71/+	100/+	7/+	20/+	57/+	86/+	23/+
I	Myosotis palustris	20/+	75/+	71/+	67/+	.	.	.	100/+	31/+
-	Carex aquatilis	100/+	100/+	93/+	11/+	57/+	60/+	57/+	64/+	69/+
R	Glyceria maxima	.	.	7/+	22/+	.	20/+	43/+	21/+	46/+
I	Senecio aquaticus	.	.	7/+	19/+	.	10/+	14/+	29/+	.
P	Carex nigra	20/+	.	7/+	22/+	14/+	20/+	14/+	29/+	.
R	Rumex crispus	.	.	.	11/+	.	.	.	14/+	.
R	Trifolium repens	.	.	7/+	20/+	.	.	.	21/+	.
P	Veronica scuttellata	.	.	.	19/+	.	.	.	14/+	.
-	Calamagrostis canescens	40/+	25/+	14/+	7/+	21/+	.	.	.	.
R	Agrostis stolonifera	60/+	.	21/+	15/+	79/1	.	29/+	100/a	100/1
I	Galium palustre	20/+	75/+	71/+	41/+	7/+	.	.	57/+	92/+
R	Alopecurus geniculatus	.	.	.	.	.	.	.	64/+	77/+
P	Stellaria palustris	.	.	14/+	.	.	.	.	7/+	31/+
R	Rorippa amphibia	.	20/+	.	.	.	.	.	.	46/+
R	Poa palustris	.	.	.	.	7/+	.	.	.	15/+

Table VI.9. Synoptic table of the coena occurring in the older reserve of Lammertshem in the lower course during the period 1973 to 1979 and corresponding with the clusterings of Fig. VI.9. P: species indicating nutrient-poor soil conditions; R: nutrient-rich conditions; I: intermediate. The occurrence of species is given in percentages of frequency and in mean cover percentages (+: 1–5; a: 6–10; 1: 11–20; 2: 21–30% etc.).

	Coenon	K	L	M	N	O	P	Q	R
	Year	1973	1973	1979	1979	1979	1979	1979	1979
	Number of relevés	24	24	13	12	5	3	4	11
P	Lysimachia vulgaris	4/+	.	.	.	.	.	.	.
I	Valeriana officinalis	100/3	100/1	100/1	100/2	100/1	100/2	100/2	91/a
I	Ranunculus acris	8/+	83/+	.	8/+	60/1	33/+	50/+	9/+
R	Glechoma hederacea	.	67/+	23/+	8/+	40/1	67/+	.	27/+
I	Plantago lanceolata	.	63/+	.	.	60/1	.	.	9/+
P	Valeriana dioica	.	67/+	8/+	.	20/1	.	25/+	18/+
-	Angelica sylvestris	.	21/+	.	.	.	.	.	.
P	Carex nigra	.	17/+	.	.	.	.	.	.
I	Leontodon autumnalis	.	4/+	.	.	.	.	.	.
I	Trifolium pratense	.	4/+	.	.	.	.	.	.
I	Senecio aquaticus	.	4/+	.	.	.	.	.	.
R	Phalaris arundinacea	92/2	83/a	100/a	100/1	80/2	100/a	100/a	91/3
R	Glyceria maxima	46/+	93/+	5/+	67/+	40/1	40/+	25/+	63/+
I	Anthoxanthum odoratum	.	63/+	.	.	60/+	.	25/+	.
-	Taraxacum sp.	13/+	67/+	54/+	50/+	80/+	33/+	50/+	27/+
-	Caltha palustris	63/+	100/+	62/+	75/+	100/+	33/+	100/+	36/+
-	Lythrum salicaria	25/+	88/+	62/+	75/+	100/+	.	50/+	81/+
I	Equisetum fluviatile	33/+	50/+	.	8/+	.	.	25/+	.
P	Juncus effusus	13/+	25/+	15/+	25/+	40/+	.	25/+	.
-	Carex aquatilis	.	42/+	15/+	33/+	60/+	33/+	50/+	27/+
I	Lysimachia nummularia	.	25/+	.	.	20/+	.	.	.
P	Achillea ptarmica	.	8/+	15/+	.	.	.	.	9/+
I	Lychnis flos-cuculi	.	17/+	.	.	.	.	25/+	9/+
P	Galium uliginosum	.	29/+	.	.	20/+	.	.	.
P	Juncus conglomeratus	.	.	8/+	8/+	.	.	.	.
R	Poa palustris	.	.	15/+	16/+	20/+	.	.	.
-	Calamagrostis canescens	38/1	17/+	100/4	58/1	20/+	33/+	.	9/+
I	Holcus lanatus	17/+	100/a	69/+	100/3	100/a	67/+	25/+	45/+
I	Deschampsia cespitosa	.	.	.	8/+	.	.	.	.
R	Anthriscus sylvestris	.	.	.	8/+	.	.	.	.
R	Festuca pratensis	.	58/+	8/+	8/+	100/+	.	25/+	.
I	Festuca rubra	.	58/+	.	.	80/+	.	25/+	25/+
I	Rumex acetosa	4/+	100/+	54/+	42/+	100/a	.	100/+	63/+
I	Galium palustre	4/+	33/+	15/+	25/+	60/+	33/+	.	.
R	Ranunculus repens	.	58/+	33/+	83/+	100/+	.	.	27/+
I	Cardamine pratensis	.	.	69/+	83/+	100/+	73/+	.	18/+
R	Poa trivialis	67/+	50/+	77/+	100/+	100/+	100/1	25/+	45/+
R	Urtica dioica	.	.	23/+	42/+	.	100/a	25/+	27/+
I	Filipendula ulmaria	96/3	100/3	100/2	100/3	100/2	100/4	100/3	100/5

non O). The large *Filipendula ulmaria* cover degree characterizes July mulching every year or second year (coenon R), whereas July hay-making every second year is not characterized by particular species (coenon Q). The abandoned permanent plots are scattered over various coena.

VI.D. Vegetational changes

VI.D.1. Standard hay-making without fertilizer application in upper- and middle course (Anlo and Taarlo)

VI.D.1.a. Succession schemes

All permanent plots in the Anlo and Taarlo study area were subjected to hay-making without fertilizer application. The plots could be subdivided into three groups. Taarlo, Anlo dry and Anlo wet as has been mentioned in section VI.C.1.a. Each annual relevé of each permanent plot was ascribed to one of the subclusters resulting from the clustering (Table VI.4A–K.). According to this analysis the vegetational change was described schematically in terms of (sub)cluster transitions (e.g. Van Noordwijk-Puijk et al., 1979) in the permanent plots as shown in Fig. VI.10A–G.

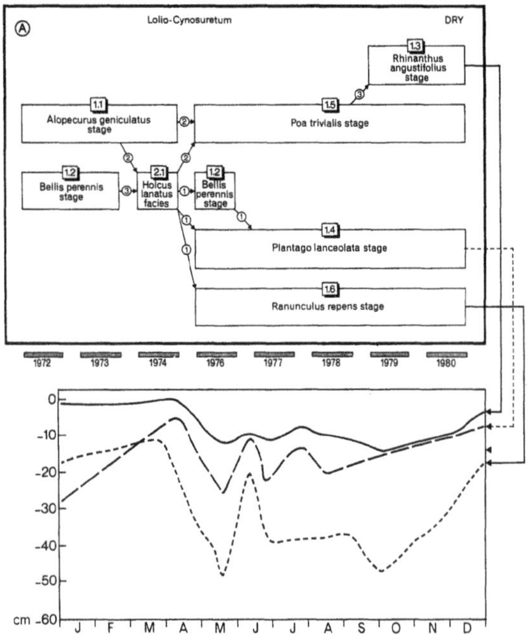

It is remarkable that most transitions between subclusters took place during the first years of the study period, whereas few transitions could be recorded during the last years. Only transitions from one subcluster to another are found in most lots. In some lots, however, the number of subclusters (temporarily) increased during the process of haymaking without fertilizer application (Fig. VI.11)

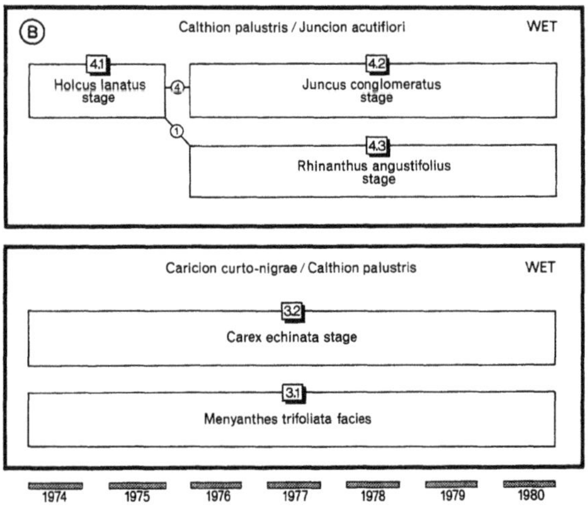

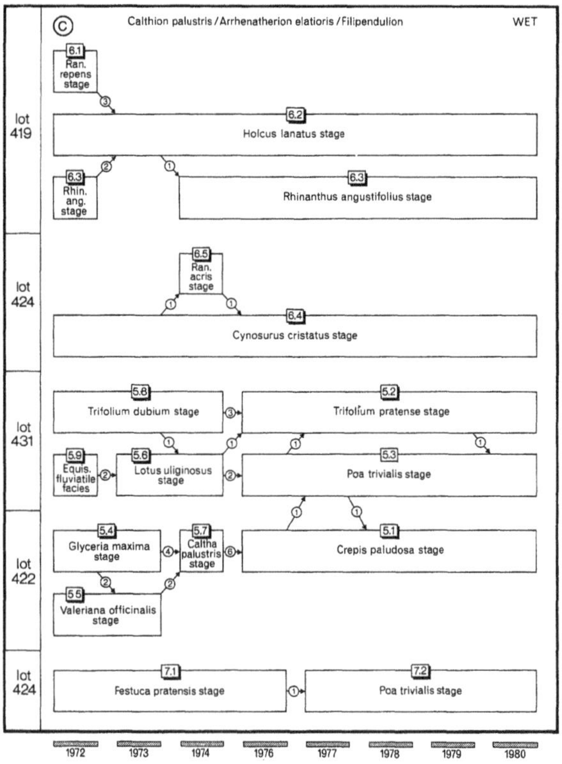

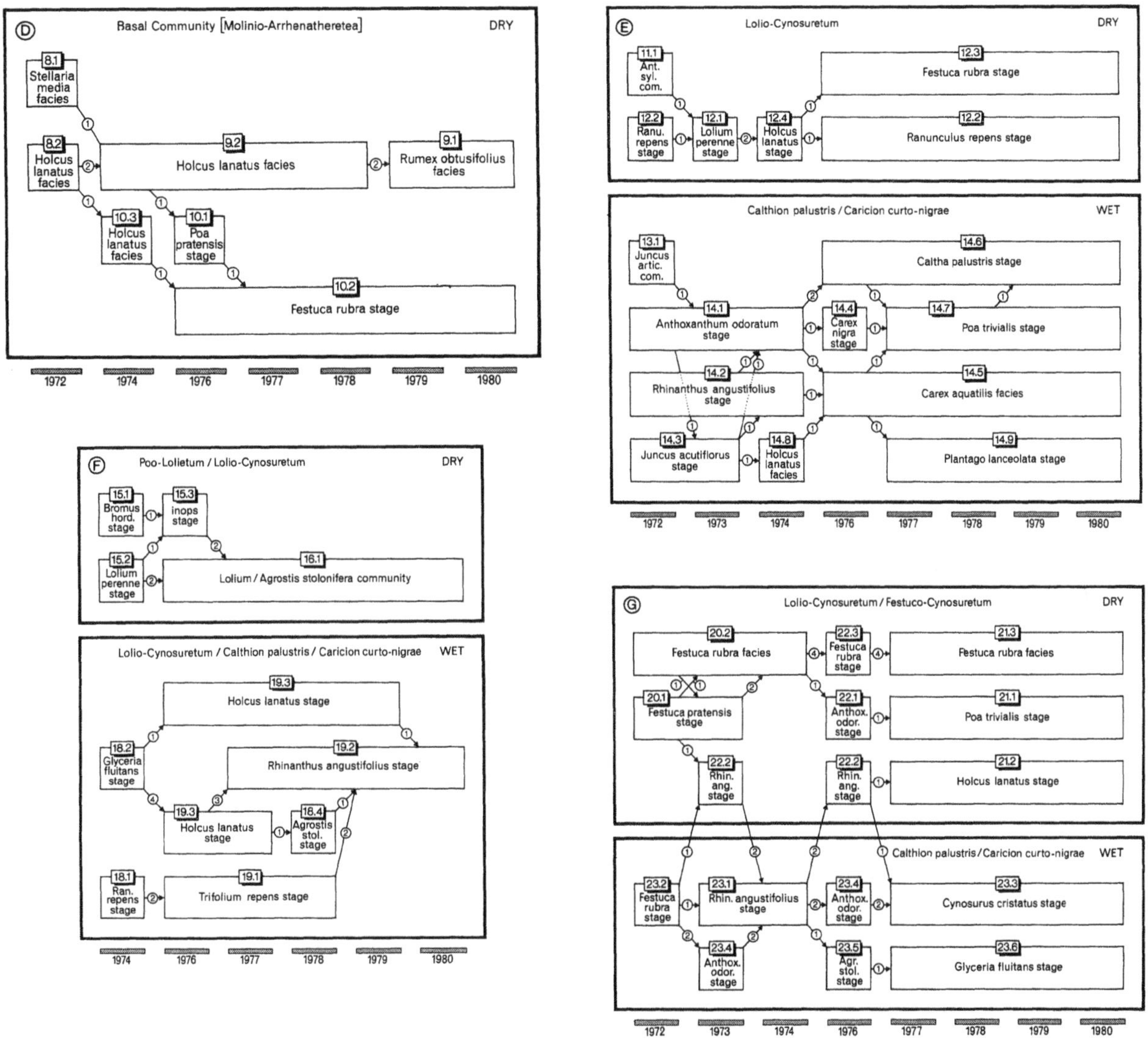

Fig. VI.10A–G. Transitions of (sub)clusters corresponding with Table VI.3 in the permanent plots in A: lot 1376 and the relation with different groundwater tables in 1979, B: lot 1371 AB, C: lots 419, 422, 424, 431, D: lot 1376A, E: lot 854, F: lot 463 and G: lot 677. The numbers in the arrows indicate the numbers of transitions (after Van Duuren et al., 1981).

viz. lot 1376 (cluster 1), lot 1376A (including clusters 8, 9, 10) and lot 463 (including clusters 18, 19). These lots have their recent acquisition in common. It is, therefore, suggested that when the presence of agricultural practices (viz. artificial fertilizer) has been withdrawn, previously masked abiotic differences can emerge again resulting in different plant communities. Particularly in lot 1376, the divergence of plant communities seems to be related to different groundwater tables (Fig. VI.10A). The decrease of community diversity in lot 463 can be attributed to the decrease of the dominant *Holcus lanatus* and *Ranunculus repens,* respectively. Apparently the coarse vegetation pattern, with local dominance of different species, changed into a more uniform stand with a high *Rhinanthus angustifolius* coverage all over the wet part of lot 463.

Since one could possibly not easily imagine the succession from the description of the coena (Table VI.3) and their rather abstract transition schemes

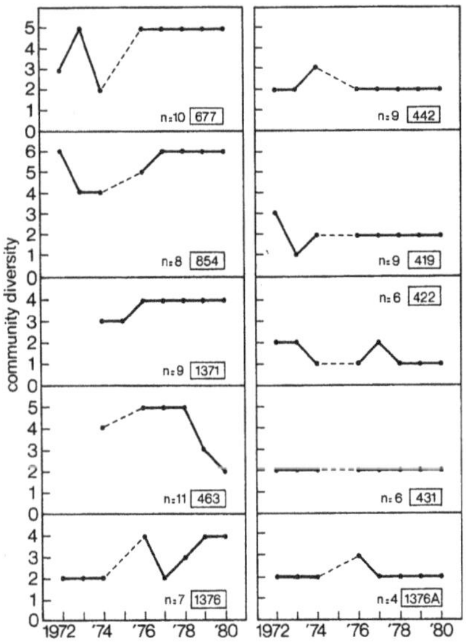

Fig. VI.11. Changes in community diversity (number of (sub) clusters) in the lots with different numbers of permanent plots under hay-making without fertilizer applications.

(Fig. VI.10A–G), information from individual permanent plots is presented in order to give examples of the main vegetational changes.

In the dry parts of the Anlo study area the transition from stands with *Poo-Lolietum* elements to stands with *Lolio-Cynosuretum* elements is characterized by the decrease of species indicating nutrient-rich soil condition viz. *Lolium perenne, Alopecurus geniculatus* and *Bromus hordeaceus* and the increase of intermediate and nutrient-poor species viz. *Anthoxanthum odoratum* and *Rhinanthus angustifolius,* but also nutrient-rich species *Agrostis stolonifera, Poa trivialis* and *Ranunculus repens* (Table VI.10). The transition from stands with *Lolio-Cynosuretum* elements, in its turn, to stands with *Festuco-Cynosuretum* elements is characterized by the decrease of *Festuca pratensis* and *Poa pratensis* and the increase of *Agrostis capillaris, Cirsium palustre, Deschampsia cespitosa* and *Hypochaeris radicata* (Table VI.11).

In the wet parts of the Anlo study area the transition from stands with *Lolio-Cynosuretum* elements to stands with *Calthion palustris* and *Caricion curto-nigrae* elements is found. It is characterized by the decrease of *Cerastium fontanum* and *Alopecurus geniculatus* and the increase of *Rhinanthus angustifolius, Galium palustre, Caltha palustris* and *Ranunculus flammula* (Table VI.12). A well developed *Caricion curto-nigrae/Calthion palustris* is characterized by a considerable number of species indicating nutrient-poor soil conditions (Table VI.13). It hardly changed during a ten year period of hay-making without fertilizer application. The appearance of *Succisa pratensis,* however, and the decrease of *Molinio-Arrhenatheretea* and *Lolio-Cynosuretum* elements point to a transition towards a fen meadow community (*Cirsio-Molinietum*), but it will certainly take a long time before such a community develops, if ever. A *Calthion palustris/Caricion curto-nigrae* can also develop towards a stand with *Juncus acutiflorus* (Table VI.14). When this community has completely developed many species indicating nutrient-poor soil conditions are included. In fact, an incomplete *Juncetum acutiflori* (*Juncion acutiflori*) developed, which hardly changed during a ten years period of hay-making without fertilizer application (Table VI.15). Again, it will take a long time before a complete *Juncetum acutiflori* has developed, if ever.

In the wet Taarlo study area the transition of stands with *Glyceria maxima* to stands with *Crepis paludosa* within a *Carex acutiformis* community is characterized by the decrease of nutrient-rich species viz. *Rumex crispus, Poa pratensis, Glyceria maxima, Urtica dioica* and *Anthriscus sylvestris* and an increase of the nutrient-poor species *Rhinanthus angustifolius* (Table VI.16). Once the *Crepis paludosa* stand had been achieved only little changes took place during a ten years period of hay-making without fertilizer application (Table VI.17), with the exception of an enormous decrease of *Carex acutiformis.*

The afore-mentioned changes in permanent plots are characterized by the decrease and increase of certain species. It should be kept in mind, however, that a lot of species seem to fluctuate irregularly around a certain equilibrium. Species featuring large year-by-year fluctuations in cover degree are *Holcus lanatus, Festuca rubra, Agrostis capillaris, Carex nigra, Ranunculus repens, Rumex acetosa, Plantago lanceolata,* and *Caltha palustris. Rhinanthus angustifolius* seems to feature a long term fluctuation of the populations involved: an increase in *Lolio-Cynosuretum* communities and a decrease in well developed *Calthion palustris/Caricion curto-nigrae* communities.

VI.D.1.b. Vegetational trends

Vegetational changes in the permanent plots were quantified and interpreted as trends or gradual deviations from the initial stand as indicated in section II.C.3. The trends are shown in Fig. VI.12A–C for Anlo dry, Anlo wet and Taarlo, respectively. Each dissimilarity graph between the first year and subsequent years includes one or more permanent plots belonging to one or more clusters as indicated in Fig. VI.4A–K dealing with the subdivision of (sub)clusters over the permanent plots.

The unidirectional character of vegetational changes was already evident from the transitions of (sub)cluster and is confirmed by the dissimilarities.

The dissimilarity between the first year and subsequent years increased significantly ($P<0.05$) in the study area Anlo dry (Fig. VI.12B) with the exception of the lots 854 and 677 which had been acquired earlier than the other lots. However, in the wet parts of the Anlo study area (except for lot 463) only lot 1371, acquired a long time ago, did not show an increasing dissimilarity.

No dissimilarity change at all could be detected in the lots in the Taarlo study area.

VI.D.1.c. Nutrient-rich and nutrient-poor species

Despite large year-to-year fluctuations unidirectional vegetational changes could also be detected by measuring the mean quotient of species indicating nutrient-poor and nutrient-rich soil conditions. The poor/rich quotients of the Anlo study area are shown in Fig. VI.13A–B for the dry and wet parts, respectively. Lot 1371, acquired a long time ago, reached the highest poor/rich quotient. The mean poor/rich quotients are significantly lower in the dry than in the wet parts of the lots ($P<0.05$ ac-

Table VI.10. Changes in cover percentage of species in a permanent plot (4 m^2) on dry sandy soil in the Anlo study area (lot 463) under hay-making without fertilizer applications after 1972. The coena correspond with those indicated in Table VI.3. P: species indicating nutrient-poor soil conditions; R: nutrient-rich conditions; I: intermediate. Character and differential species of syntaxa are given: PL: Poo-Lolietum, MA: Molinio-Arrhenatheretea, LC: Lolio-Cynosuretum, F: Filipendulion, CP: Calthion palustris, Ccn: Caricion curto-nigrae, JA: Juncion acutiflori, A: Arrhenatherion elatioris; LP: Lolio-Potentillion.

		Year	1974	76	77	78	79	80	81	82	83	84
		Coenon	15	16	16	16	16	16				
		Number of species	15	19	19	20	21	22	20	19	16	20
PL	R	Lolium perenne	40	4	4	2	1	1	1	.	.	.
LP	R	Alopecurus geniculatus	10	2	2	2	2	2	1	2	1	1
MA	I	Cerastium fontanum	4	1	1	1	1	1	1	1	1	1
A	R	Bromus hordeaceus	10	2	20	4	1	1	1	1	.	.
LC	R	Trifolium repens	.	2	2	4	2	2	1	4	.	1
MA	R	Poa pratensis	.	1	1	1	1	2	1	2	.	1
LC	R	Festuca pratensis	.	1	1	1	1	1	.	1	2	1
LP	R	Agrostis stolonifera	.	.	1	8	20	12	8	8	1	8
LC	R	Poa trivialis	.	1	8	20	12	20	4	4	4	12
LC	R	Ranunculus repens	4	2	4	8	20	30	20	12	30	20
	P	Rhinanthus angustifolius	.	.	.	.	.	1	1	8	8	12
FC	I	Anthoxanthum odoratum	.	1	1	1	2	2	4	8	8	12
LC	I	Cynosurus cristatus	2	1	1	1	1	1	1	2	4	2
	I	Glyceria fluitans	.	.	.	.	.	.	.	.	1	1
MA	I	Holcus lanatus	4	10	30	30	2	4	8	30	20	20
MA	I	Rumex acetosa	4	2	8	1	20	2	2	1	2	2
	R	Phleum pratense	1	.	.	.	.	2	1	.	.	1
FC	I	Agrostis capillaris	2	2	.	.	.	2	1	.	.	1
MA	I	Cardamine pratensis	1	1	1	1	1	1	1	2	1	1
LC	I	Bellis perennis	2	1	1	1	2	1	1	1	.	1
	–	Taraxacum sp.	2	1	1	1	1	.	1	1	.	1
	R	Rumex obtusifolius	.	1	4	2	2	.	1	.	.	.
	R	Elymus repens	.	2	.	1	1	1	.	.	.	.
PL	R	Stellaria media	.	.	1	.	1	.	.	.	.	.
PL	R	Plantago major	.	.	.	1	1	1	.	1	.	.
FC	I	Trifolium dubium	2	.	.	.	.	.	.	.	.	.
	I	Veronica arvensis	4	.	.	.	.	.	.	.	.	.
CP	I	Myosotis palustris	.	.	.	.	.	1	.	.	.	1
	I	Rhinanthus minor	.	.	.	.	.	.	.	1	.	.
CP	–	Caltha palustris	.	.	.	.	.	.	.	.	2	.
CP	I	Lychnis flos-cuculi	.	.	.	.	.	.	.	.	1	.
	I	Mentha aquatica	.	.	.	.	.	.	.	.	1	.

cording to Wilcoxon's test). This could be due to the better agricultural value of the dry areas than the wet areas prior to acquisition (see section VI.B.3). Lot 1376 is the only one in which the poor/rich quotient increased significantly ($P<0.05$), in both the dry and wet area. Agricultural practices ceased in the year preceding the start of the experiment in this lot. The poor/rich quotient decreased in the wet parts of the lots 677 and 854, which had already been under the hay-making without fertilization regime for a longer time. This discrepancy can probably be attributed to superficial drainage in order to enable hay-making with heavy machines.

The poor/rich quotient in the permanent plots near the hedge-bank (cluster 7) is lower than elsewhere in the same lot in the Taarlo study area (Fig. VI.13C). The poor/rich quotient increased significantly ($P<0.05$) in the lots in which agricultural practices ceased the year preceding the start of the experiment.

VI.D.1.d. Species diversity

The mean species diversity increased significantly ($P<0.05$) in most wet parts of the Anlo study area (Fig. VI.14A), whereas an increase was found in a minority of the dry parts of the same area (Fig. VI.14B). Species diversity is significantly lower ($P<0.05$, according to Wilcoxon's test) in the dry parts of the lots 463, 677 and also during the first

Table VI.11. Changes in cover percentage of species in a permanent plot (4 m^2) on dry sandy soil in the Anlo study area (lot 677) under hay-making without fertilizer applications after 1967. See Table VI.10 for remaining information.

		Year	1972	73	74	76	77	78	79	80	81	82	83	84
		Coenon	20	20	20	22	21	21	21	21				
		Number of species	7	11	18	14	16	16	14	15	19	18	16	16
LC	R	Festuca pratensis	2	.	2	.	.	.	.	.	.	.	.	.
MA	I	Cardamine pratensis	.	2	1	.	.	1	.	.	.	.	.	.
MA	R	Poa pratensis	2	10	10	1	1	1	1	1	1	1	1	1
LC	R	Ranunculus repens	4	4	4	1	1	1	4	1	1	2	1	1
MA	I	Holcus lanatus	10	4	10	2	8	2	1	1	1	1	4	1
FC	I	Agrostis capillaris	.	.	40	10	20	10	20	20	12	30	30	30
	P	Cirsium palustre	.	.	1	.	1	1	1	1	1	2	1	1
	I	Carex ovalis	.	.	.	1	1	1	.	1	1	1	1	1
MA	I	Plantago lanceolata	.	.	.	.	1	1	1	1	1	1	.	.
MA	I	Deschampsia cespitosa	.	1	2	1	4	4	12	12	12	4	12	4
FC	P	Hypochaeris radicata	.	.	.	.	.	1	1	1	2	1	.	.
Ccn	P	Hydrocotyle vulgaris	.	.	.	.	.	.	.	.	1	1	1	1
	I	Leontodon autumnalis	.	.	.	.	1	.	.	.	.	1	2	1
Ccn	P	Carex nigra	.	.	.	.	.	.	.	.	.	1	2	2
FC	I	Anthoxanthum odoratum	2	4	4	1	8	4	2	1	1	1	2	8
FC	I	Festuca rubra	40	10	40	10	60	50	40	30	30	20	40	30
LC	R	Poa trivialis	.	2	.	1	1	1	1	1	1	.	1	1
MA	I	Rumex acetosa	10	4	4	1	1	1	4	2	8	8	8	4
	P	Juncus effusus	.	2	2	1	1	1	.	1	1	1	2	2
FC	P	Luzula campestris	.	2	2	1	2	1	.	1	2	4	2	2
PL	R	Poa annua	.	.	2	1	.	.	.	.	.	.	.	.
	-	Taraxacum sp.	.	.	1	.	.	.	.	.	.	.	.	.
LC	I	Ranunculus acris	.	.	1	.	.	.	.	.	.	.	.	.
	-	Quercus robur	.	.	1	.	.	.	1	.	1	1	.	.
	P	Rhinanthus angustifolius	.	.	1	1	1	.	.	.	.	.	.	.
	I	Dactylis glomerata	.	.	.	.	.	.	1	.	.	.	.	.
PL	R	Lolium perenne	.	.	.	.	.	.	.	.	1	.	.	.
LC	I	Cynosurus cristatus	.	.	.	.	.	.	.	.	1	.	.	.

years in 854, than in the wet parts. No significant changes in the species diversity were found in the Taarlo study area (Fig. VI.14C).

VI.D.2. Hay-making regimes in the upper course (Loefvledder)

VI.D.2.a. Cover changes of abundant species

The recent reserve showed some abiotic differences as indicated in section VI.B.2 and a subsequent heterogeneous vegetation (see section VI.C.2.b). The permanent plots involved cannot, therefore, be taken into account in the comparison of management regimes. This was the case for four

Table VI.12. Changes in cover percentage of species in a permanent plot (4 m^2) on wet peaty soil in the Anlo study area (lot 463) under hay-making without fertilizer applications after 1972. See Table VI.10 for remaining information.

Year			1974	76	77	78	79	80	81	82	83	84
Coenon			18	18	18	18	19	19				
Number of species			23	24	22	22	23	31	30	26	27	26
PL	R	Poa annua	2	.	.	.	.	.	.	.	.	.
MA	I	Cerastium fontanum	4	1	1	1	1	1	1	1	1	1
LC	R	Trifolium repens	4	.	.	.	1	2	1	1	2	1
	I	Sagina procumbens	4	4	.	.	.	1	1	1	1	.
LP	R	Alopecurus geniculatus	10	1	1	8	1	1	1	1	1	1
LC	R	Poa trivialis	.	1	4	1	2	8	2	2	4	4
LC	R	Festuca pratensis	.	2	1	4	8	1	1	1	.	1
	P	Juncus effusus	.	1	1	1	8	4	1	1	2	4
	P	Juncus articulatus	.	1	.	1	.	1	1	1	1	2
LP	R	Agrostis stolonifera	.	.	1	10	1	1	2	2	1	1
LC	R	Ranunculus repens	4	4	12	30	20	20	12	12	30	12
	P	Rhinanthus angustifolius	.	.	.	4	40	12	20	30	8	40
Ccn	I	Galium palustre	.	.	.	1	1	1	1	1	1	1
CP	–	Caltha palustris	.	.	.	.	1	2	1	1	2	4
Ccn	P	Ranunculus flammula	.	1	.	.	.	1	1	1	1	1
MA	I	Holcus lanatus	4	20	40	2	.	2	8	4	20	4
LC	R	Poa pratensis	2	.	.	.	.	.	1	.	.	.
PL	R	Lolium perenne	4	.	.	.	1	1	1	.	.	.
FC	I	Agrostis capillaris	2	.	.	.	.	1	.	.	.	.
LC	I	Cynosurus cristatus	4	2	2	.	2	4	1	2	4	2
FC	I	Anthoxanthum odoratum	10	1	1	2	8	12	4	4	8	8
	I	Glyceria fluitans	4	2	1	1	2	8	1	1	1	1
FC	I	Festuca rubra	4	2	1	.	.	1	1	.	.	.
MA	I	Rumex acetosa	2	1	2	1	2	4	2	1	2	1
MA	I	Cardamine pratensis	2	1	1	1	1	2	1	1	1	1
LC	I	Ranunculus acris	1	.	1	.	1	1	1	1	1	1
LC	I	Bellis perennis	2	1	1	1	1	2	1	1	.	1
	–	Taraxacum sp.	2	1	1	1	1	1	1	1	1	1
	P	Equisetum palustre	2	1	1	4	2	1	1	2	2	8
	I	Stellaria glutinosa	2	1	1	1	.	1	1	1	1	1
CP	I	Lychnis flos-cuculi	1	.	1	1	1	2	1	1	1	1
	I	Mentha aquatica	.	1	1	.	1	1	1	.	1	.
CP	I	Myosotis palustris	.	.	.	1	.	1	.	1	.	.
	I	Veronica serpyllum	1	.	.	.	.	.	.	.	.	.
PL	R	Plantago major	.	1	.	.	.	.	.	.	.	.
	–	Epilobium obscurum	.	1	.	.	.	.	.	.	.	.
F	I	Filipendula ulmaria	.	1	.	.	.	.	.	.	.	.
F	–	Lythrum salicaria	.	.	1	.	.	.	.	.	.	.
	P	Epilobium palustre	.	.	.	1	.	.	.	.	1	.
	I	Polygonum hydropiper	.	.	.	.	.	.	.	.	1	.

Table VI.13. Changes in cover percentage of species in a permanent plot (4 m^2) on wet peaty soil in the Anlo study area (lot 1371A) under hay-making without fertilizer applications after 1946. See Table VI.10 for remaining information.

		Year	1974	75	76	77	78	79	80	81	82	83	84
		Coenon	3	3	3	3	3	3	3				
		Number of species	34	37	36	34	33	37	41	39	34	36	36
LC	R	Festuca pratensis	2	.	.	.	.	.	.	.	.	.	.
FC	I	Anthoxanthum odoratum	4	4	1	2	1	1	2	1	1	1	1
	P	Rhinanthus angustifolius	4	4	10	2	2	2	2	1	1	1	1
MA	I	Holcus lanatus	4	10	1	12	4	1	1	1	2	2	2
LC	I	Cynosurus cristatus	4	10	1	12	12	1	1	1	1	1	2
LC	R	Trifolium repens	10	4	10	4	4	2	2	1	.	2	1
FC	I	Trifolium pratense	10	4	4	8	8	4	1	1	1	2	4
MA	I	Plantago lanceolata	4	4	4	12	8	4	1	1	1	2	2
MA	R	Poa pratensis	2	1	1	.	4	1	1	1	.	.	.
LC	R	Poa trivialis	2	4	1	1	2	2	1	2	.	.	.
CP	I	Crepis paludosa	.	.	1	1	.	1	2	1	1	1	8
CP	I	Ajuga reptans	.	.	1	1	2	2	1	1	1	1	.
Ccn	P	Potentilla palustris	.	.	.	.	.	1	1	1	.	1	.
JA	P	Succisa pratensis	.	.	.	.	.	.	1	.	1	.	1
Ccn	P	Menyanthes trifoliata	2	2	20	12	20	20	12	40	50	60	60
	P	Agrostis canina	.	1	.	.	.	.	.	.	1	1	1
FC	I	Festuca rubra	4	4	1	4	1	1	1	1	2	2	2
Ccn	I	Galium palustre	2	2	1	1	1	1	1	1	1	2	1
FC	P	Luzula campestris	2	4	1	1	1	1	1	1	.	.	1
	P	Dactylorhiza maculata	1	2	1	1	1	1	1	1	.	.	1
JA	P	Carex panicea	2	2	1	1	2	1	1	1	1	1	1
Ccn	P	Carex echinata	4	.	1	1	1	1	2	1	2	1	1
LC	R	Ranunculus repens	2	2	1	1	4	4	1	1	2	4	8
LP	P	Dactylorhiza majalis	1	1	1	1	4	2	8	2	2	1	1
Ccn	P	Carex nigra	10	40	4	12	1	12	12	12	8	8	20
	I	Equisetum fluviatile	4	4	1	1	1	1	4	2	1	4	8
	P	Equisetum palustre	2	1	10	10	10	10	2	10	2	2	1
	P	Juncus effusus	2	2	1	1	4	2	1	1	1	1	1
F	I	Valeriana officinalis	4	2	1	1	1	1	1	1	1	2	1
MA	I	Rumex acetosa	2	4	1	1	4	2	2	1	1	1	1
	I	Lotus uliginosus	10	4	2	4	2	1	1	1	2	4	2
LC	I	Ranunculus acris	4	10	2	1	4	4	4	4	2	2	2
CP	I	Lychnis flos-cuculi	2	2	1	1	1	1	1	1	2	2	1
	P	Cirsium palustre	2	2	1	2	2	1	2	1	1	1	1
Ccn	P	Viola palustris	2	2	1	1	1	1	2	2	2	1	4
	P	Galium uliginosum	2	2	1	1	.	.	1	1	2	1	1
CP	-	Caltha palustris	1	2	1	1	1	2	1	1	1	1	1
MA	I	Cardamine pratensis	2	2	.	1	1	1	2	1	1	1	1
MA	I	Cerastium fontanum	2	2	1	1	1	1	1	1	1	1	1
	P	Stellaria graminea	.	2	1	.	.	.	.	.	1	1	1
LC	P	Prunella vulgaris	.	2	1	.	.	.	.	.	.	.	.
	P	Stellaria palustris	.	.	.	.	1	1	1	.	.	.	.
FC	I	Agrostis capillaris	.	.	.	.	.	.	1	1	.	.	.
	P	Potentilla erecta	.	1	.	.	.	.	.	.	.	.	.
JA	P	Juncus conglomeratus	.	1	.	.	.	.	.	.	.	.	.
Ccn	P	Ranunculus flammula	.	.	.	.	.	1	.	.	.	1	.
	-	Taraxacum sp.	.	.	.	.	.	.	1	.	.	.	.
	P	Juncus articulatus	.	.	.	.	.	.	.	.	1	1	.
	R	Stellaria media	.	.	.	.	.	.	.	1	.	.	.
	-	Betula pubescens	.	.	.	.	.	.	1	1	.	.	.
	-	Alnus glutinosa	.	.	.	.	.	.	.	.	.	.	1

Table VI.14. Changes in cover percentage of species in a permanent plot (4 m^2) on wet peaty soil in the Anlo study area (lot 677) under hay-making without fertilizer applications after 1967. See Table VI.10 for remaining information.

		Year	1972	73	74	76	77	78	79	80	81	82	83	84
		Coenon	23	23	23	23	23	23	23	23				
		Number of species	26	24	36	32	32	33	33	37	38	40	38	36
FC	I	Trifolium pratense	10	.	.	.	.	.	1	1	1	1	1	1
FC	I	Anthoxanthum odoratum	10	10	4	1	2	4	1	1	1	1	2	2
FC	I	Festuca rubra	10	2	4	.	.	1	.	1	.	1	.	.
LC	R	Festuca pratensis	2	2	4	1	1	1	1	1	1	1	1	2
MA	I	Cerastium fontanum	4	2	4	1	1	1	1	1	1	1	1	1
CP	I	Ajuga reptans	4	4	4	.	.	.	.	.	.	.	.	.
	I	Equisetum fluviatile	10	4	4	4	12	4	1	1	1	2	2	1
LC	I	Cynosurus cristatus	2	4	4	2	12	12	8	1	1	1	2	4
	P	Cirsium palustre	2	2	2	1	1	.	1	1	.	.	.	.
CP	I	Myosotis palustris	.	.	2	1	.	1	4	2	2	2	2	1
	I	Sagina procumbens	.	.	1	.	1	.	1	1	1	1	1	1
D	I	Filipendula ulmaria	.	.	4	1	2	2	1	1	2	2	2	2
MA	I	Plangago lanceolata	.	.	.	1	1	1	1	1	1	1	1	1
LP	R	Agrostis stolonifera	.	.	.	10	1	1	.	1	1	2	1	1
	I	Glyceria fluitans	.	.	.	1	1	1	8	8	12	8	8	4
Ccn	P	Ranunculus flammula	.	.	.	1	.	1	1	4	2	2	4	2
LP	R	Alopecurus geniculatus	.	.	.	1	1	1	1	1	1	1	1	1
	I	Leontodon autumnalis	.	.	.	.	1	1	1	1	1	.	.	1
	I	Polygonum hydropiper	.	.	.	.	.	.	.	1	1	1	1	.
LC	R	Ranunculus repens	4	4	2	2	4	4	4	12	8	20	20	20
F	-	Lythrum salicaria	.	.	.	.	.	.	.	.	1	1	1	.
LC	R	Poa trivialis	.	2	.	2	2	2	4	4	4	12	12	8
JA	P	Juncus acutiflorus	1	.	.	.	1	1	.	.	.	1	2	8
FC	P	Stellaria graminea	.	.	.	.	.	.	.	.	.	1	1	1
CP	P	Dactylorhiza majalis	.	.	.	.	.	.	.	.	.	.	1	1
MA	I	Holcus lanatus	4	2	4	4	1	1	1	1	4	4	12	4
MA	I	Rumex acetosa	2	2	4	1	1	1	2	2	1	1	2	2
LC	I	Ranunculus acris	2	2	4	1	2	1	4	4	2	2	1	1
	P	Juncus effusus	2	2	2	1	1	1	1	1	1	1	2	1
MA	I	Cardamine pratensis	2	2	4	1	1	1	2	2	1	1	1	1
Ccn	P	Carex nigra	2	2	10	2	2	4	2	4	1	4	4	4
Ccn	I	Galium palustre	2	.	2	.	1	1	1	1	1	2	2	1
LC	R	Trifolium repens	10	4	10	1	12	12	12	2	2	2	8	20
CP	-	Caltha palustris	10	4	30	10	12	20	20	20	12	8	4	8
CP	I	Lychnis flos-cuculi	1	.	1	1	1	1	1	1	1	1	1	1
	I	Lotus uliginosus	4	2	.	1	4	1	4	2	2	4	12	2
	P	Rhinanthus angustifolius	4	4	10	10	8	4	4	2	1	4	2	20
	I	Stellaria glutinosa	2	2	4	1	.	.	1	1	1	1	1	1
	P	Carex rostrata	4	.	.	.	.	1	.	.	1	.	.	.
	P	Juncus articulatus	.	2	.	.	1	1	.	1	1	1	1	1
Ccn	-	Eleocharis palustris	.	2	2	1	1	1	1	1	1	1	2	2
FC	I	Agrostis capillaris	.	.	4	1	.	.	.	.	.	.	.	.
LC	P	Prunella vulgaris	.	.	2	.	.	.	.	.	.	1	.	.
	-	Alnus glutinosa	.	.	1	1	1	.	.	.	.	.	.	.
	I	Mentha aquatica	.	.	1	.	.	.	1	1	1	4	4	2
FC	I	Trifolium dubium	.	.	1	1	1	1	.	.	.	.	.	.
PL	R	Poa annua	.	.	1	.	.	.	.	.	.	.	.	.
	-	Angelica sylvestris	1	.	1	.	.	.	.	.	.	.	.	.
	-	Taraxacum sp.	.	1	1	.	.	.	.	.	.	.	.	.
	I	Veronica serpyllum	.	.	1	.	.	.	.	.	.	.	.	.
	-	Sium erectum	.	.	1	.	.	.	.	.	.	.	.	.
	P	Lysimachia vulgaris	.	.	1	.	.	.	.	.	.	.	.	.
	P	Achillea ptarmica	.	.	.	.	1	.	.	1	.	.	.	.
	I	Rhinanthus minor	.	.	.	.	.	.	1	.	.	.	.	.
	R	Bidens tripartita	.	.	.	.	.	.	.	.	1	1	.	.
MA	R	Poa pratensis	.	.	.	.	.	.	.	.	1	.	.	.
	I	Montia fontana	.	.	.	.	.	.	.	.	.	1	.	.
	I	Carex ovalis	.	.	.	.	.	.	.	.	.	.	1	.

Table VI.15. Changes in cover percentage of species in a permanent plot (4 m^2) on wet peaty soil in the Anlo study area (lot 1371B) under hay-making without fertilizer applications after 1946. See Table VI.10 for remaining information.

Year			1974	75	76	77	78	79	80	81	82	83	84
Coenon			4	4	4	4	4	4	4				
Number of species			31	31	25	29	33	34	34	34	35	30	30
FC	I	Festuca rubra	10	4	.	1	1	1	1	1	2	1	.
MA	I	Deschampsia cespitosa	4	4	.	1	.	.	1	1	1	.	.
LC	R	Ranunculus repens	4	4	1	2	1	1	1	1	1	1	2
MA	I	Cerastium fontanum	2	2	10	1	1	1	1	1	1	1	1
FC	P	Luzula campestris	2	4	.	1	4	.	.	.	.	.	.
	P	Potentilla erecta	.	.	1	2	1	1	1	.	1	1	1
LC	I	Ranunculus acris	.	.	.	.	1	1	.	2	1	2	1
JA	P	Luzula multiflora	.	.	.	.	1	4	1	1	1	1	2
	-	Veronica chamaedrys	.	.	.	.	.	.	1	.	.	1	1
LC	R	Festuca pratensis	.	.	.	.	.	.	.	.	1	.	1
F	I	Valeriana officinalis	.	.	.	.	.	.	.	.	1	.	1
Ccn	P	Viola palustris	.	2	1	2	1	1	1	1	1	4	8
JA	P	Juncus acutiflorus	40	40	20	20	20	20	40	40	40	40	30
MA	R	Poa pratensis	2	2	.	.	.	.	1	1	.	.	1
LC	I	Cynosurus cristatus	2	4	1	2	1	1	1	2	1	1	1
FC	I	Anthoxanthum odoratum	10	4	1	2	1	1	1	2	2	4	8
MA	I	Holcus lanatus	2	4	2	20	1	2	1	2	2	2	2
FC	I	Agrostis capillaris	4	2	20	40	20	8	8	4	2	2	4
Ccn	P	Carex nigra	4	2	2	1	2	4	2	1	2	2	4
	I	Equisetum fluviatile	2	2	.	1	1	1	1	1	.	1	.
F	I	Filipendula ulmaria	2	2	.	.	.	1	.	1	.	.	.
	P	Equisetum palustre	2	2	1	1	1	1	1	1	1	1	1
CP	P	Dactylorhiza majalis	2	.	1	2	4	4	2	1	1	1	1
	P	Juncus effusus	2	1	.	.	.	.	.	1	1	1	1
JA	P	Juncus conglomeratus	2	4	1	1	8	8	12	1	1	1	1
	I	Lotus uliginosus	2	4	1	2	1	1	1	12	4	8	2
MA	I	Plantago lanceolata	40	4	10	50	30	20	8	4	12	20	8
	P	Rhinanthus angustifolius	4	4	1	2	1	8	4	1	2	4	12
MA	I	Rumex acetosa	4	4	1	2	2	2	2	12	2	2	2
	P	Cirsium palustre	2	2	1	10	8	4	2	2	1	1	1
LC	P	Prunella vulgaris	2	4	1	2	2	1	1	1	1	1	1
CP	I	Ajuga reptans	1	2	1	2	1	1	1	1	1	1	1
MA	I	Cardamine pratensis	2	2	.	1	1	1	1	1	1	.	1
LC	R	Trifolium repens	2	4	1	1	2	2	1	1	1	1	1
FC	P	Stellaria graminea	1	.	.	.	.	1	.	.	1	1	.
	-	Taraxacum spec.	1	1	1	.	1	1	1	.	1	.	1
CP	I	Lychnis flos-cuculi	.	2	1	2	1	1	1	1	1	1	1
	P	Lysimachia vulgaris	1	2	1	2	1	1	1	1	2	1	1
	I	Dactylis glomerata	.	.	.	1	1	1	1	.	1	.	.
LC	R	Poa trivialis	.	.	.	.	2	2	2	4	.	.	.
	P	Myosotis discolor	.	.	.	.	1	.	1	1	1	.	.
	R	Stellaria media	1	.	.	.	.	.	.	1	.	.	.
Ccn	P	Carex echinata	.	2	.	.	.	.	.	.	.	.	.
	-	Quercus robur	.	.	1	.	.	.	.	.	.	.	.
	I	Carex ovalis	.	.	.	.	.	.	.	1	.	.	.
	-	Betula pubescens	.	.	.	.	.	.	.	.	1	.	.

Table VI.16. Changes in cover percentage of species in a permanent plot (4 m²) on wet peaty soil in the Taarlo study area (lot 431) under hay-making without fertilizer applications after 1972. See Table VI.10 for remaining information.

Year			1972	73	74	76	77	78	79	80	81	82	83	84
Coenon			5	5	5	5	5	5	5	5				
Number of species			24	32	35	33	29	33	30	31	34	24	25	25
	I	Equisetum fluviatile	40	4	4	1	1	2	1	1	1	1	2	1
LP	R	Rumex crispus	1	1	.	.	.	.	.	.	.	.	.	.
	I	Lysimachia nummularia	1	4	.	.	.	.	.	.	.	.	.	.
	I	Silene dioica	1	1	.	.	.	.	.	.	.	.	.	.
MA	I	Cerastium fontanum	2	4	4	1	1	1	1	1	1	1	1	1
MA	I	Rumex acetosa	2	4	4	1	2	1	1	1	1	1	2	1
MA	R	Poa pratensis	1	4	2	.	.	.	.	.	.	.	.	.
	R	Glyceria maxima	10	4	2	.	.	.	.	.	.	.	.	.
	I	Veronica arvensis	2	4	4	1	1	1	1	1	1	.	.	1
MA	I	Plantago lanceolata	10	4	10	10	4	2	1	4	1	1	1	2
	P	Equisetum palustre	4	2	4	4	2	1	1	1	2	1	2	1
	R	Urtica dioica	2	2	1	1	1	1	.	.	.	.	1	.
MA	I	Holcus lanatus	4	4	4	4	20	12	1	1	.	1	2	2
LC	I	Bellis perennis	1	1	2	1	1	1	1	.	.	1	.	.
LC	R	Ranunculus repens	4	2	2	1	.	1	1	1	1	.	.	.
A	R	Anthriscus sylvestris	10	4	2	1	2	2	1	1	1	.	.	.
	I	Lotus uliginosus	.	4	40	4	8	8	8	2	20	8	8	12
FC	I	Trifolium dubium	.	.	2	1	1	2	1	1	1	.	.	1
FC	I	Festuca rubra	.	.	10	4	2	2	2	2	1	2	2	2
CP	I	Lychnis flos-cuculi	.	.	2	.	.	1	1	1	.	.	1	.
	P	Cirsium palustre	.	.	2	1	.	1	.	1	1	1	2	1
F	I	Filipendula ulmaria	10	10	10	20	20	20	12	20	12	40	20	30
	P	Rhinanthus angustifolius	.	.	.	.	1	2	8	4	1	2	4	2
FC	I	Anthoxanthum odoratum	1	2	4	1	2	12	4	8	4	2	8	2
LC	R	Festuca pratensis	1	4	2	1	2	1	2	1	1	2	2	1
LC	R	Poa trivialis	2	2	2	1	12	10	8	8	4	2	2	2
LC	I	Cynosurus cristatus	1	2	4	1	2	2	2	2	2	1	2	1
MA	I	Cardamine pratensis	.	2	4	1	1	1	1	1	1	.	1	1
LC	I	Ranunculus acris	4	2	4	1	2	8	4	2	2	2	2	2
	-	Taraxacum sp.	2	2	2	1	2	2	1	1	1	2	2	1
CP	I	Carex acutiformis	1	4	40	10	4	8	8	2	20	20	20	40
CP	I	Crepis paludosa	.	2	10	10	4	4	4	10	2	20	12	12
F	-	Lythrum salicaria	.	1	1	.	.	.	.	.	.	.	.	.
CP	I	Myosotis palustris	.	2	4	1	1	1	1	.	1	.	1	.
	R	Glechoma hederacea	.	2	1	1	1	2	1	1	1	.	.	.
	I	Vicia cracca	.	2	2	2	4	2	1	1	1	1	1	1
	-	Veronica chamaedrys	.	.	1	1	.	.	.	1	1	.	.	.
LC	R	Trifolium repens	.	.	.	1	2	1	2	1	.	.	.	.
PL	R	Lolium perenne	.	.	.	1	.	.	.	1	1	.	.	.
CP	I	Bromus racemosus	.	.	.	1	1	1	1	1	1	.	.	.
Ccn	P	Carex nigra	.	.	.	.	.	.	.	.	1	1	.	.
CP	I	Ajuga reptans	.	.	.	.	.	.	.	.	1	1	.	.
	I	Phyteuma nigrum	.	.	.	.	.	.	.	.	1	1	.	.
	I	Glyceria fluitans	.	.	.	.	.	.	.	.	1	.	.	.
Ccn	I	Galium palustre	.	.	.	.	.	.	.	.	1	.	.	.
	R	Stellaria media	.	.	.	.	.	1	.	.	.	.	.	.
	I	Mentha aquatica	.	.	.	1	.	.	.	.	.	.	.	.
	R	Phleum pratense	.	1	2	.	.	.	.	.	.	.	1	1
FC	P	Stellaria graminea	.	.	1	.	.	.	.	.	.	.	.	.
LC	P	Prunella vulgaris	.	.	.	.	.	.	.	.	.	.	.	1

Table VI.17. Changes in cover percentage of species in a permanent plot (4 m^2) on wet peaty soil in the Taarlo study area (lot 424) under hay-making without fertilizer applications after 1967). See Table VI.10 for remaining information.

		Year	1972	73	74	76	77	78	79	80	81	82	83	84
		Coenon	6	6	6	6	6	6	6	6				
		Number of species	30	27	36	31	31	32	33	33	33	30	31	32
	R	Ranunculus ficaria	4	.	.	.	.	.	.	.	.	.	.	.
	R	Glyceria maxima	1	.	.	.	.	.	.	.	.	.	.	.
CP	I	Lychnis flos-cuculi	2	2	2	1	2	8	2	1	1	1	1	1
CP	I	Carex acutiformis	40	4	10	10	4	8	8	4	4	2	2	1
	P	Rhinanthus angustifolius	4	4	4	2	2	8	4	1	2	2	2	1
LC	I	Ranunculus acris	4	4	10	4	4	8	8	4	2	2	4	2
LC	I	Bellis perennis	1	1	1	1	1	1	1	1	.	.	.	.
LC	R	Trifolium repens	.	.	2	1	2	2	4	1	1	2	3	2
FC	P	Stellaria graminea	.	.	2	1	2	2	4	1	1	2	1	1
FC	I	Trifolium pratense	.	.	4	.	.	.	2	1	2	4	4	2
	I	Lotus uliginosus	.	1	.	1	1	2	8	8	8	12	12	8
CP	I	Crepis paludosa	2	4	10	2	4	.	12	20	20	12	12	30
	P	Agrostis canina	.	.	.	.	.	.	.	.	.	1	1	1
	I	Equisetum fluviatile	4	4	4	1	1	1	4	1	1	8	12	8
MA	I	Holcus lanatus	2	2	4	1	2	1	1	1	1	2	2	2
LC	R	Festuca pratensis	2	2	.	1	1	1	2	2	1	1	2	1
FC	I	Anthoxanthum odoratum	4	4	4	1	10	2	4	2	4	2	4	2
FC	I	Festuca rubra	4	.	10	1	1	1	2	2	2	2	4	4
	I	Phragmites australis	2	2	2	1	1	1	1	1	1	1	2	2
LC	I	Cynosurus cristatus	1	4	4	2	30	20	12	1	4	1	4	2
LC	R	Poa trivialis	2	4	4	1	4	1	2	1	2	4	4	4
CP	I	Bromus racemosus	1	.	.	.	.	1	1	.	1	.	.	.
MA	R	Poa pratensis	.	.	1	.	.	.	.	.	.	.	.	.
MA	I	Plantago lanceolata	4	4	4	10	4	2	2	2	1	4	8	2
MA	I	Cerastium fontanum	2	2	2	1	.	1	1	1	1	1	1	1
MA	I	Rumex acetosa	2	4	4	1	2	1	1	4	1	2	2	2
	I	Vicia cracca	2	.	2	1	1	1	2	1	1	2	1	1
	P	Equisetum palustre	2	2	1	.	1	1	1	1	1	1	1	1
CP	-	Caltha palustris	40	10	40	2	2	8	8	12	4	4	8	12
F		Lyhrum salicaria	1	2	2	1	1	1	2	1	1	2	2	1
F	I	Filipendula ulmaria	4	4	40	4	1	1	2	2	2	12	4	8
LC	R	Ranunculus repens	2	2	1	.	2	1	1	1	2	1	2	2
CP	I	Ajuga reptans	2	2	2	.	1	1	1	2	1	2	2	1
	P	Cirsium palustre	1	2	2	2	4	4	2	4	2	2	2	1
Ccn	P	Carex nigra	2	.	4	1	1	1	2	1	.	1	2	2
A	I	Phyteuma nigrum	1	.	.	1	.	1	1	2	1	.	.	.
CP	I	Myosotis palustris	.	2	.	.	.	.	.	.	.	.	.	.
	P	Stellaria palustris	.	1	.	.	.	1	.	.	.	.	.	.
MA	I	Cardamine pratensis	.	2	2	1	1	1	1	1	1	1	1	1
	P	Juncus effusus	.	.	1	.	.	.	.	.	.	.	.	.
	-	Taraxacum sp.	.	.	2	.	.	.	.	.	.	.	.	.
	-	Veronica chamaedrys	.	.	2	1	.	.	.	1	1	.	.	1
	P	Juncus articulatus	.	.	1	.	.	.	.	.	.	.	.	.
FC	P	Luzula campestris	.	.	2	1	1	.	.	.	.	.	.	.
	P	Carex rostrata	.	.	1	.	.	.	.	.	.	.	.	.
	P	Galium uliginosum	.	.	.	1	.	.	.	.	.	.	.	.
	I	Achillea millifolium	.	.	.	.	1	.	.	.	.	.	.	.
	I	Veronica arvensis	.	.	.	.	.	.	.	.	1	.	.	.

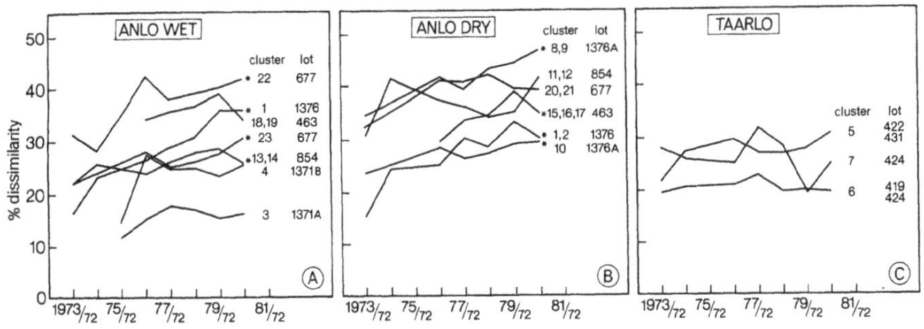

Fig. VI.12A–C. Changes of percentages dissimilarity between the first year and subsequent years under hay-making without fertilizer applications in different lots in A: Anlo 'wet' area, B: Anlo 'dry' area, C: Taarlo area. * Significant change according to Spearman's rank correlation coefficient ($P<0.05$).

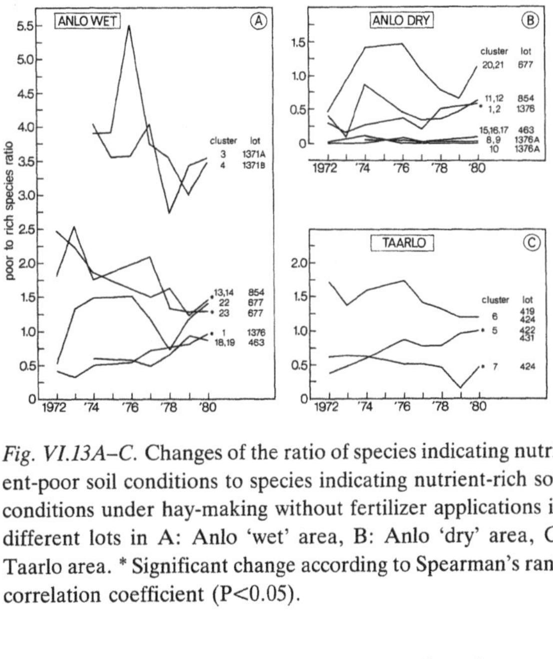

Fig. VI.13A–C. Changes of the ratio of species indicating nutrient-poor soil conditions to species indicating nutrient-rich soil conditions under hay-making without fertilizer applications in different lots in A: Anlo 'wet' area, B: Anlo 'dry' area, C: Taarlo area. * Significant change according to Spearman's rank correlation coefficient ($P<0.05$).

out of the six available permanent plots in some treatments. Therefore, in each management site two permanent plots, with an initially similar vegetation, have been taken for comparison.

The changes in mean cover percentages of the principal species in two permanent plots at each management site since 1973 are shown in Fig. VI.15A–J. Species indicating nutrient-rich soil conditions decreased rapidly at all management sites viz. *Lolium perenne, Festuca pratensis, Alopecurus geniculatus. Holcus lanatus* increased rapidly at all management sites. It suffered two temporary setbacks, a minor one in the very dry summer of 1976 and a major one in 1979 after a severe winter and wet spring followed by two wet winters. It increased again in 1983. The rise of *Ranunculus repens* and, to a much lesser degree, of *Agrostis stolonifera* often coincided with the *Holcus lanatus* decline. This suggests complementary fluctuations caused by the invasion of the gaps in the thinner *Holcus* stands by the two stoloniferous species.

The changes of the individual management regimes will now be considered. The abandoned site was characterized by the temporary occurrence of *Cirsium arvense* (Photo's VI.9, 10 and 11), *Rumex obtusifolius* and the recent development of *Heracleum sphondylium* (Fig. VI.15A). The latter might indicate a transitional phase to a community with tall forbs. The *Cirsium arvense* cover percentage fluctuated in the burned site (Fig. VI.15B). *Festuca rubra* developed better at the burning site than at all other management experiments. *Potentilla anserina* appeared and maintained itself.

The two September cutting regimes that took place every two years showed great resemblances (Fig. VI.15C–D). *Potentilla anserina* succeeded *Ranunculus repens* as the dominating species. *Festuca rubra* only appeared when the cuttings were removed, whereas *Cirsium arvense* appeared at the mulching regime.

The annual September hay-making regime featured the appearance of *Potentilla anserina,* but the species did not reach high cover percentages (Fig. VI.15E). *Poa trivialis* appeared in all annual September hay-making regimes. The presence of *Elymus repens* and *Rumex obtusifolius* and the absence of *Potentilla anserina* distinguished the al-

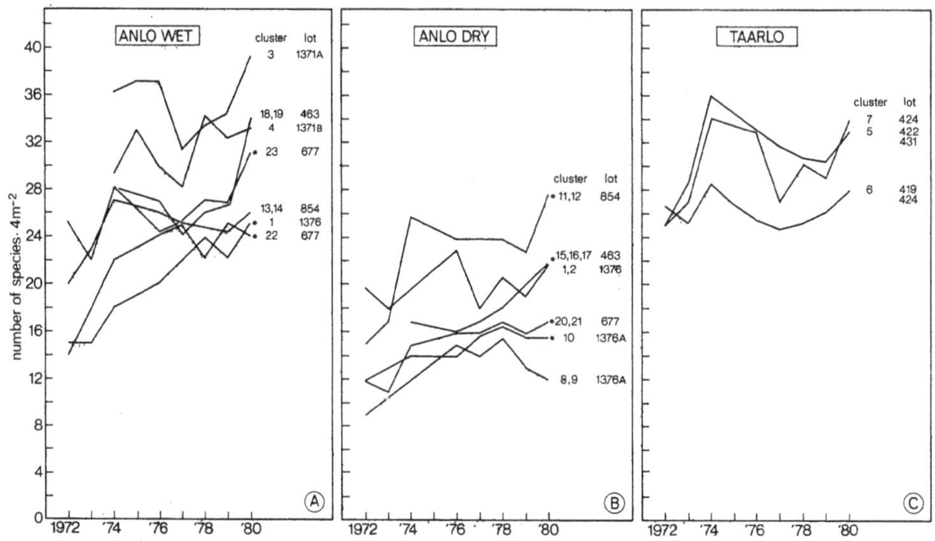

Fig. VI.14A–C. Changes of numbers of species · 4 m^{-2} under hay-making without fertilizer applications in different lots in A: Anlo 'wet' area, B: Anlo 'dry' area, C: Taarlo area. Although the clustering showed that nearly all permanent plots in lot 1376 belonged to cluster 1, the groundwater tables showed clear differences and therefore the plots were assigned to both 'wet' and 'dry' areas. * Significant change according to Spearman's rank correlation coefficient (P<0.05).

ternating July/September hay-making regime from the other two regimes.

Ranunculus repens increased with July hay-making, whereas *Agrostis stolonifera* increased with July hay-making by scythe. Moreover, the occurrence of *Festuca rubra* characterized the hay-making regime by scythe (Fig. VI.15H).

Vegetation mapping according to combinations of dominant species revealed that the successional changes in the permanent plots occurred over larger areas (Fig. VI.16). Some changes were not recorded within permanent plots viz. the local dominance of *Rubus* sp. at the abandoned site and the local occurrence of *Dactylis glomerata* with several management regimes. Moreover, *Urtica dioica* took over large areas of the slightly elevated ground.

The older reserve showed few abiotic differences. Some irregularities were, nevertheless, found in the clustering. This means that two permanent plots close to the hedgebank or the brook sometimes deviated from the other permanent plots in each treatment. The two permanent plots in the middle of each site were, therefore, compared.

The changes in mean cover percentages of the abundant species in two permanent plots at each management site since 1973 are shown in Fig. VI.17A–H. The site with mulching every year showed a decrease of *Festuca rubra* and *Anthoxanthum odoratum,* a rapid increase of *Rumex acetosa* followed by a set-back after ten years, a temporary (four years) increase of *Ranunculus repens* and a large dominance of *Holcus lanatus* during the last two years (Fig. VI.17A). *Festuca rubra* maintained itself longer with July hay-making every second year than with mulching every second year, but its cover percentage fluctuated regularly with peaks in the years after hay-making. This was also more or less the case with *Ranunculus repens* and *Agrostis capillaris*. *Holcus lanatus* reached abundance after twelve years (Fig. VI.17B). Mulching every second year resulted in a similar picture to hay-making every second year, with the exception of the last few years when only *Rumex acetosa* and *Holcus lanatus* reached higher cover percentages (Fig. VI.17C).

Alternating hay-making and mulching resulted in the maintenance of *Festuca rubra,* whereas *Holcus lanatus* did not become dominant (Fig. VI.17D). The patterns of fluctuations are irregular

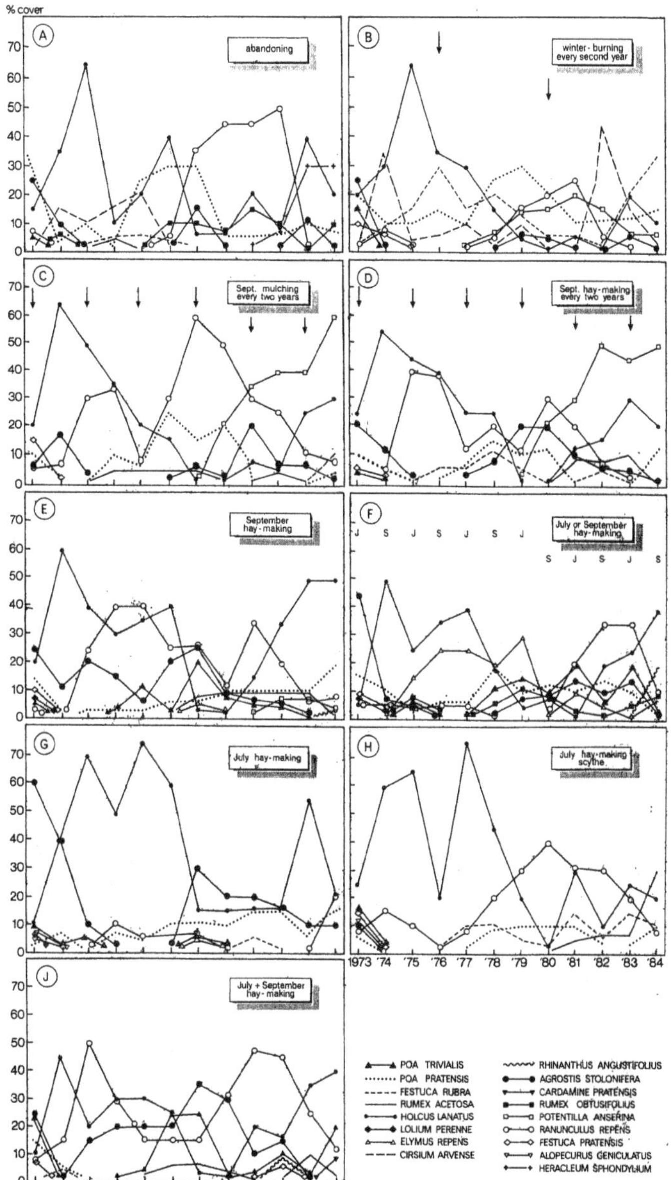

Fig. VI.15A–J. Changes of cover percentages of abundant (coverage ⩾4%) species in the *Agrostis stolonifera/Poa pratensis* community in the recent reserve of the Loefvledder area under different management regimes.

despite hay-making in the even and mulching in the odd years. The only regularity seems to be the peaks of *Rumex acetosa* alternating with the peaks of other species.

Annual hay-making by scythe resulted in a dominance of *Festuca rubra* at the cost of *Juncus effusus* (Fig. VI.17E), whereas *Juncus* maintained itself better with annual hay-making with machines (Fig. VI.17F).

The application of artificial fertilizer showed an increase of *Agrostis canina* (Fig. VI.17G) and *Agrostis capillaris* (Fig. VI.17H) with Ammonium Nitrogen Limestone and Ureum Formaldehyde, respectively. *Juncus effusus* maintained itself in both regimes and the high cover degree of *Rumex acetosa* is remarkable.

Vegetation mapping revealed that the succesion-

Photo VI.9. View of the vegetation with much *Ranunculus repens* of the Loefvledder study area before the start of abandoning.

al changes occurred over larger areas (Fig. VI.18). The only exceptions are the small patches with the dominance of *Juncus acutiflorus* and *Plantago lanceolata.*

VI.D.2.b. Vegetational trends

The vegetational trends in the recent reserve are shown in Fig. VI.19. The high dissimilarity of the second year compared to the initial vegetation (1973) is striking in all management regimes, since this initial value often amounts to only 20–40% (see Fig. VI.12A–C for the Anlo and Taarlo study areas). All the treatments apparently had such an influence that the vegetation already deviated strongly from the initial vegetation from the second year onwards. In later years fluctuations in dissimilarity percentages were found which correspond with the fluctuations in the dominant species mentioned in the previous section. Only July hay-making by scythe showed a significant ($P<0.05$) increase in the dissimilarity percentage.

The trends in the older reserve are shown in Fig. VI.20. The continuation of the standard hay-making regime in July showed the lowest deviation from the initial vegetation. All annual hay-making regimes or these alternating with mulching only, showed fluctuations. The only exception is the add-

Photo VI.10. The same situation as Photo VI.9 after three years of abandoning. Note the accumulation of litter and the dominance of *Cirsium arvense* in the background.

Photo VI.11. The same situation as Photo VI.9 after 15 years of abandoning. *Heracleum sphondylium* reached dominance.

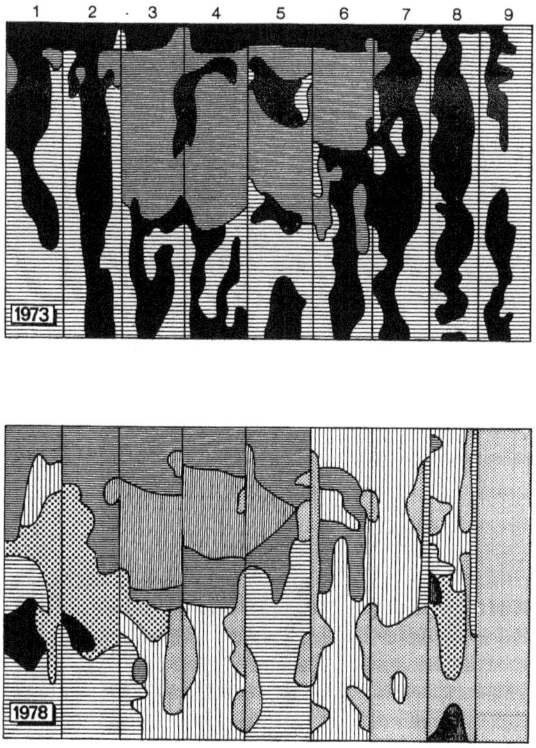

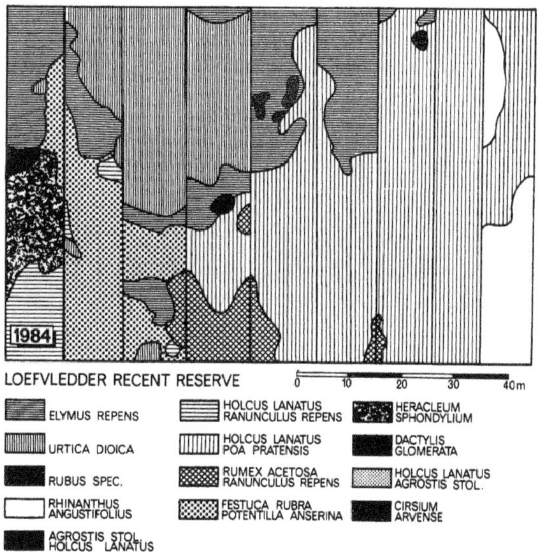

Fig. VI.16. Simplified vegetation map of the recent reserve of the Loefvledder area in 1973, 1978 and 1984, respectively. The management regimes imposed are 1: abandoning, 2: burning, 3: mulching in September every second year, 4: hay-making in September every second year, 5: annual hay-making in September, 6: hay-making alternating in July or in September, 7: annual hay-making in July, 8: annual hay-making in July with the scythe, 9: hay-making in both July and September (after Bakker & de Vries, 1985A).

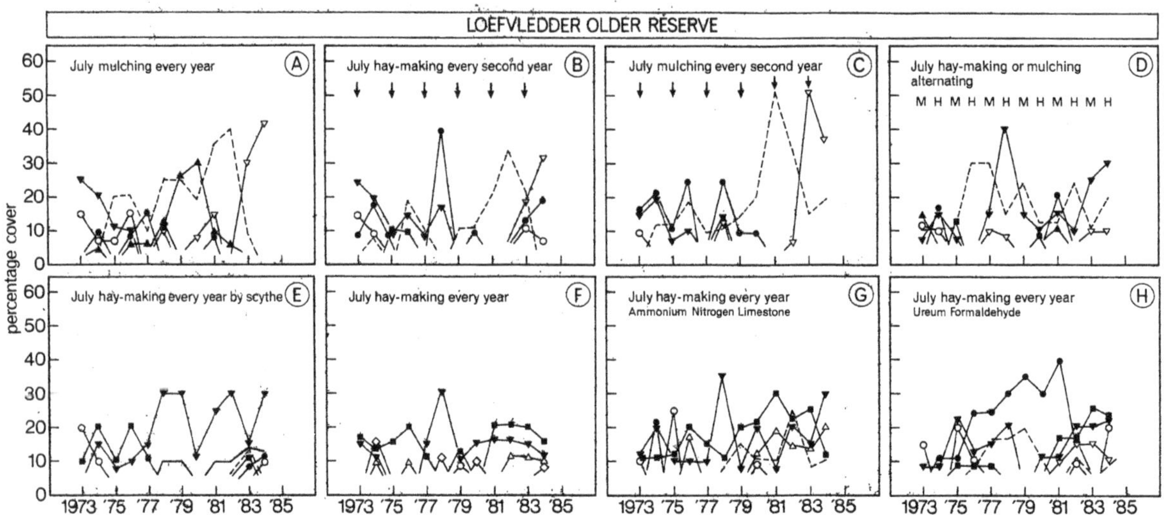

Fig. VI.17A–H. Changes of cover percentages of abundant (coverage ≥8%) species in the *Festuca rubra/Agrostis capillaris* community in the older reserve of the Loefvledder area under different management regimes. Festuca rubra ▼——▼, Anthoxanthum odoratum ○——○, Holcus lanatus ▽——▽, Rumex acetosa ----, Ranunculus repens ▲——▲, Agrostis capillaris ●——●, Juncus effusus ■——■, Agrostis canina △——△, Viola palustris ———, Carex nigra ◇——◇.

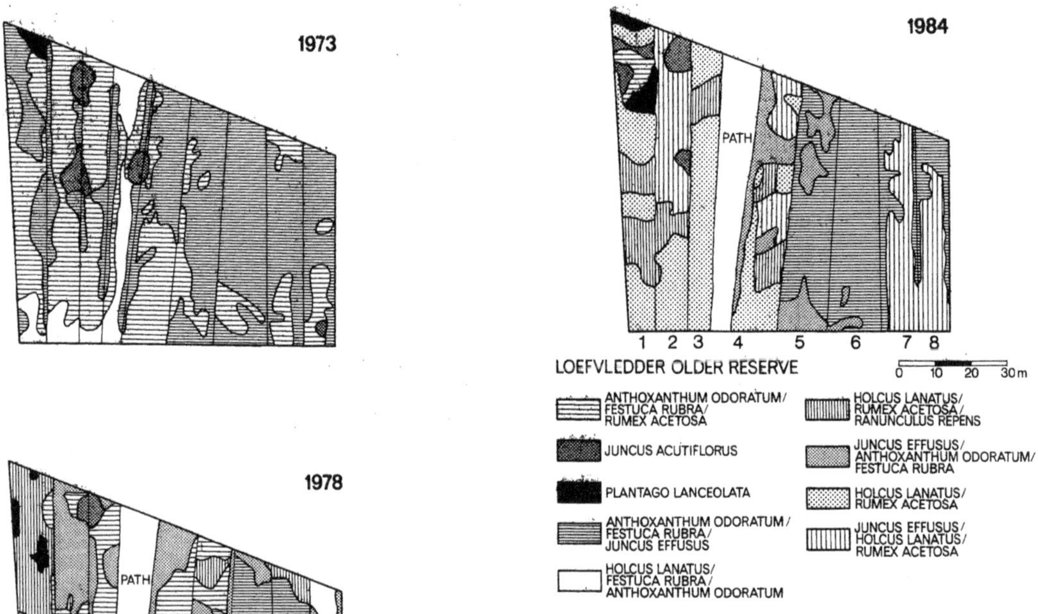

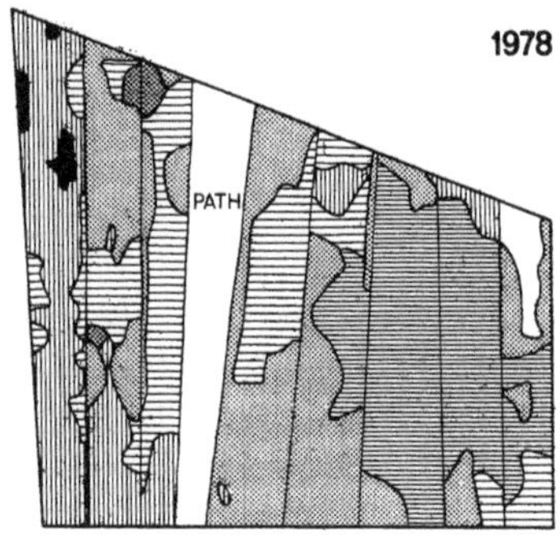

Fig. VI.18. Simplified vegetation map of the older reserve of the Loefvledder area in 1973, 1978 and 1984, respectively. The management regimes imposed are 1: annual mulching in July, 2: hay-making in July every second year, 3: mulching in July every second year, 4: hay-making or mulching alternating in July, 5: annual hay-making in July with the scythe, 6: annual hay-making in July, 7: annual hay-making in July and application of Ammonium Nitrogen Limestone (50 kg $N \cdot ha^{-1}$), 8: annual hay-making in July and application of Ureum Formaldehyde (50 kg $N \cdot ha^{-1}$).

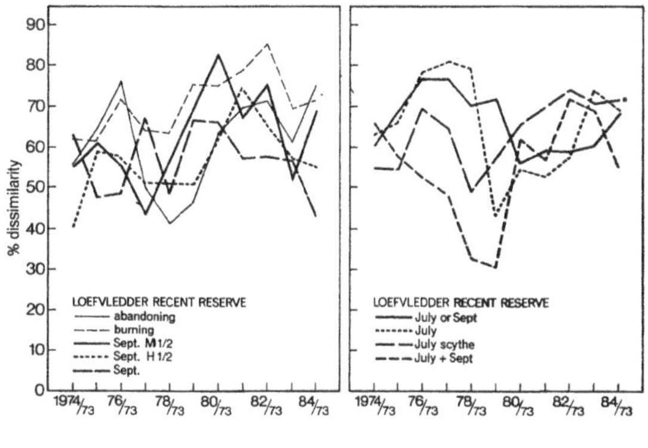

Fig. VI.19. Changes of percentages dissimilarity between the first year and subsequent years in the *Agrostis stolonifera/Poa pratensis* community in the recent reserve of the Loefvledder area under different management practices. * Significant change according to Spearman's rank correlation coefficient (P<0.05).

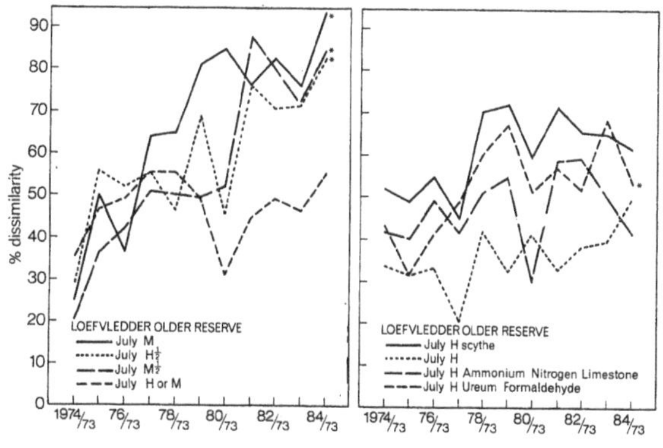

Fig. VI.20. Changes of percentages dissimilarity between the first year and subsequent years in the *Festuca rubra/Agrostis capillaris* community in the older reserve of the Loefvledder area under different management practices. * Significant change according to Spearman's rank correlation coefficient (P<0.05).

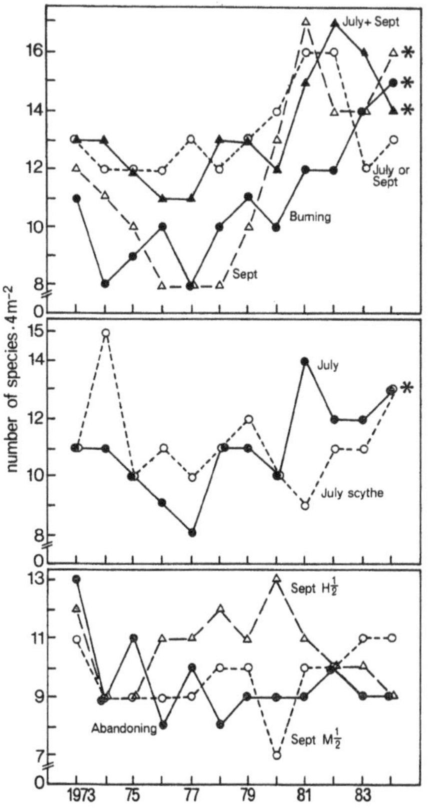

Fig. VI.21. Changes of number of species $\cdot$ 4 m^{-2} in the *Agrostis stolonifera/Poa pratensis* community in the recent reserve of the Loefvledder area under different management practices. * Significant change according to Spearman's rank correlation coefficient (P<0.05) (after Bakker & de Vries, 1985B).

Photo VI.12. View of the vegetation, with a great deal of *Ranunculus repens,* of the Loefvledder study area before the start of hay-making in both July and September.

Photo VI.13. The same situation as Photo VI.12 after three years of twice hay-making. *Holcus lanatus* became very dominant.

Photo VI.14. The same situation as Photo VI.12 after 15 years of twice hay-making. The coverage of *Holcus lanatus* decreased, whereas *Rhinanthus angustifolius* appeared.

ing of Ureum Formaldehyde fertilizer resulting in a significant increase in dissimilarity. The regimes including mulching or alternating with abandoning also showed a significant increasing divergence ($P<0.05$) resulting in the highest deviation from the initial vegetation.

VI.D.2.c. Species diversity

The mean species diversity in the recent reserve is shown in Fig. VI.21. It increased significantly ($P<0.05$) in the treatments July + September hay-making (Photo's VI.12, 13 and 14), September hay-making, July hay-making and burning. This was, however, in most cases after a temporary set-back. Species diversity only tended to decrease with abandoning and September mulching every second year.

In the older reserve (Fig. VI.22) the species diversity decreased significantly ($P<0.05$) in the regimes including mulching every year, every second year and hay-making every year by scythe and Ureum Formaldehyde fertilizer application. Species diversity only fluctuated in the other regimes.

VI.D.3. Hay-making regimes in the lower course (Lammertshem)

VI.D.3.a. Cover changes of abundant species

The recent reserve showed some abiotic differences and, as a consequence, heterogeneous vegetation (see section VI.C.1.c). All permanent plots involved cannot, therefore, be taken into account in the comparison of management regimes. Only the plots at some distance from the brook featured similar vegetation at the start of the experiment. Only these two permanent plots have, therefore, been used for comparisons.

The changes in mean cover percentages of the principal species in the two permanent plots at each management site since 1973 are shown in Fig. VI.23A–J. *Festuca pratensis,* indicating nutrient-rich soil conditions, decreased rapidly at all management sites. *Holcus lanatus* suffered a temporary set-back in most regimes in 1976 and again in 1979. The temporary abundancy of *Taraxacum* sp., *Poa trivialis* and *Cardamine pratensis* coincided with the decline in *Holcus lanatus* in 1976. The *Holcus lanatus* fluctuations were small in the plots with hay-making in July + September.

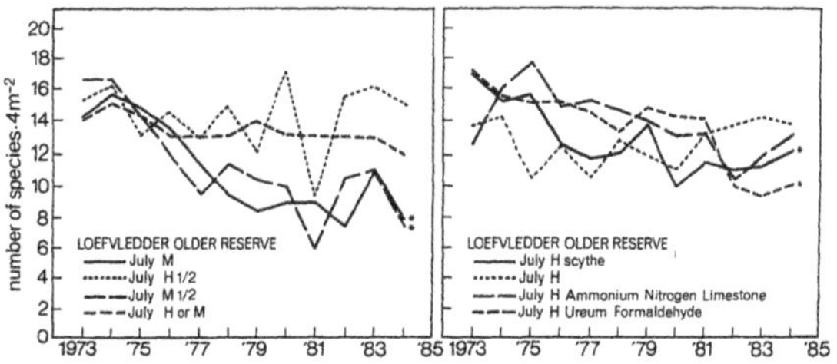

Fig. VI.22. Changes of number of species $\cdot 4\,m^{-2}$ in the *Festuca rubra/Agrostis capillaris* community in the older reserve of the Loefvledder area under different management practices. * Significant change according to Spearman's rank correlation coefficient ($P<0.05$).

All treatments in which July hay-making was involved showed fairly similar cover percentages of *Holcus lanatus* during the first three years (Fig. VI.23). *Ranunculus repens* maintained itself during the whole or most of the study period (Photo's VI.15 and 16) with the exception of when hay-making by scythe was involved. It was then replaced by *Phalaris arundinacea. Filipendula ulmaria* did not reach very high cover percentages.

Holcus lanatus increased sharply in the second year with the five other treatments (Fig. 23), whereas *Ranunculus repens* and *Caltha palustris* almost disappeared. Later, *Filipendula ulmaria* became the dominating species. *Valeriana officinalis* only reached considerable cover percentages in the abandoned site in the last year of the study period (Photo's VI.17 and 18). *Poa trivialis* showed a remarkable dominance during a few years with September hay-making, as with July or September hay-making.

Vegetation mapping revealed that succession occurred everywhere in the lot, but in 1978 it did not yet lead to a pattern which was entirely determined by management regimes (Fig. VI.24). The former *Agrostis stolonifera/Caltha palustris* community had changed into a *Holcus lanatus/Phalaris arundinacea* community when annual hay-making was involved. In the management practices with (temporary) litter accumulation, however, *Phalaris arundinacea* became the only dominant species. In some management regimes *Rumex acetosa* became co-dominant close to the brook.

Photo VI.15. View of the vegetation in the management regime hay-making in both July and September in the Lammertshem study area in the first year without fertilizer application.

Photo VI.16. The same situation as Photo VI.15 after seven years of hay-making twice a year.

Photo VI.17. The Lammertshem study area in the first year without fertilizer application.

Photo VI.18. The same situation as Photo VI.17 after seven years of abandoning. Note the tall forbs *Filipendula ulmaria* and *Valeriana officinalis*.

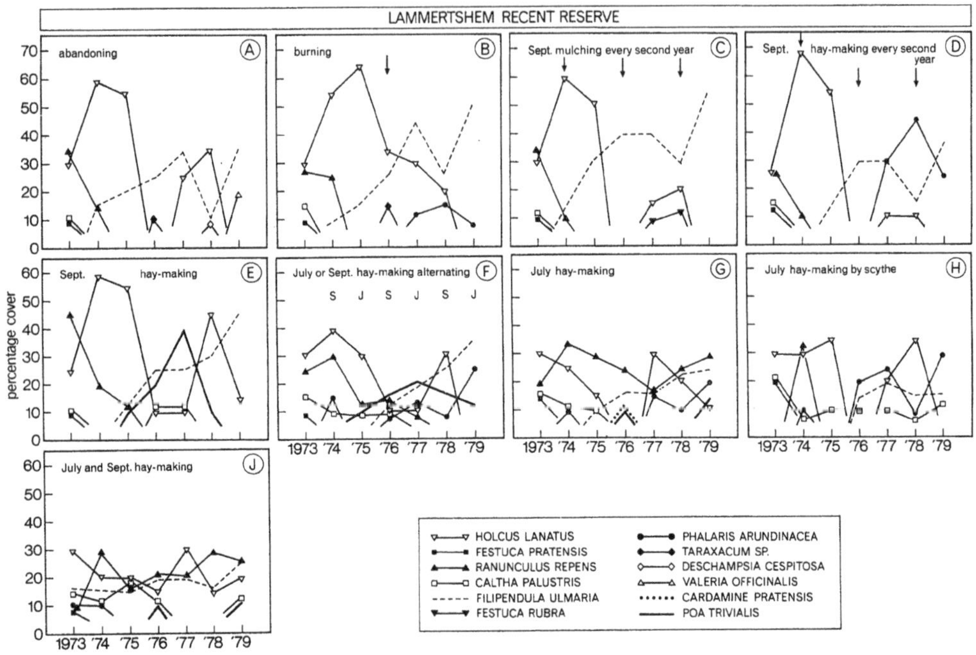

Fig. VI.23A–J. Changes of cover percentages of abundant (coverage ⩾8%) species in the *Ranunculus repens/Caltha palustris* community in the recent reserve of the Lammertshem area under different management regimes.

Some changes took place in all management regimes, namely the increase of *Phalaris arundinacea* and *Glyceria maxima*. The first species indicates large changes in the groundwater table which can be attributed to the rather recent heightening of the level of the brook as a reaction to the measured lowering of the groundwater table over 50 cm in the whole valley area (Streefkerk & Van Hoorn, 1985). The second species indicates eutrophication which can be the result of increased and prolonged inundation. *Lychnis flos-cuculi* and *Myosotis palustris,* character species of the *Calthion palustris,* decreased at the same time. This can be attributed to a diminished proportion of Ca^{2+}-rich seepage water (Everts et al., 1984).

The older reserve showed few abiotic differences. Nevertheless, some irregularities were found in the clustering (see section VI.C.1.c), which meant that permanent plots, either close to or a long way from the brook took deviating positions. The two permanent plots in the middle of each treatment were taken for comparison starting from initial similar stands. The changes in mean cover percentages of the abundant species, since 1973, in two permanent plots at each management site are shown in Fig. VI.25A–H. *Phalaris arundinacea* and *Holcus lanatus* appeared in the site with mulching every year and showed a large increase when *Filipendula ulmaria* suffered a sharp decrease (Fig. VI.25A). A similar development was found with July hay-making every second year with only slightly smaller cover percentages for *Filipendula ulmaria* and *Valeriana officinalis* and *Holcus lanatus* reaching a higher cover degree (Fig. VI.25B). Mulching every second year resulted in a different picture from hay-making every second year, since *Holcus lanatus* increased in the years after mulching, whereas *Valeriana officinalis* and *Phalaris arundinacea* increased in the intermediate years (Fig. VI.25E).

Alternating hay-making or mulching resulted in irregularly fluctuating patterns of dominance, despite hay-making in the odd years and mulching in the even years (Fig. VI.25G).

Annual hay-making with a scythe eventually resulted in an increase in *Valeriana officinalis, Phalaris arundinacea* and *Poa trivialis* (Fig. VI.25C),

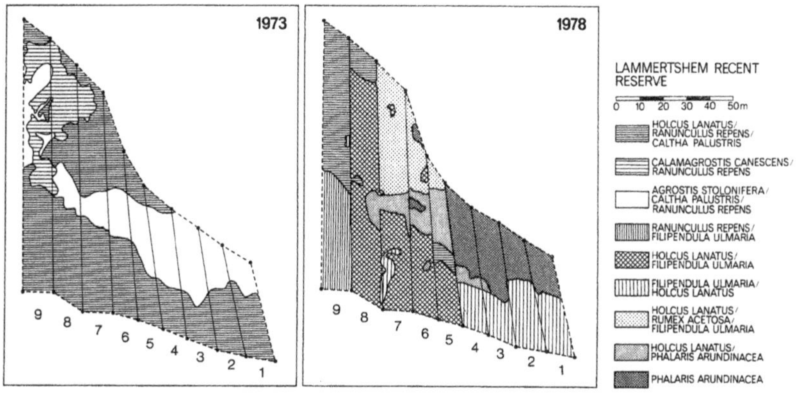

Fig. VI.24. Simplified vegetation map of the recent reserve of the Lammertshem area in 1973 and 1978, respectively. The management regimes imposed are 1: abandoning, 2: burning, 3: mulching in September every second year, 4: hay-making in September every second year, 5: annual hay-making in September, 6: hay-making alternating in July or in September, 7: annual hay-making in July, 8: annual hay-making in July with the scythe, 9: hay-making in both July and September.

whereas *Holcus lanatus* became co-dominant with annual hay-making with machines (Fig. VI.25D).

The application of artificial fertilizer showed a decrease of *Phalaris arundinacea* and an increase of *Holcus lanatus* (Fig. VI.25F) with Ammonium Nitrogen Limestone, and of *Calamagrostis canescens* (Fig. VI.25H) with Ureum Formaldehyde.

Vegetation mapping revealed that the successional changes occurred over larger areas (Fig. VI.26). The only exception was the wet zone with *Phalaris arundinacea* and *Calamagrostis canescens* at some distance from the brook.

VI.D.3.b. Vegetational trends

The vegetational trends in the recent reserve are shown in Fig. VI.27. The dissimilarity percentages of the treatments involving hay-making in July or in July + September hardly exceeded 60%, whereas

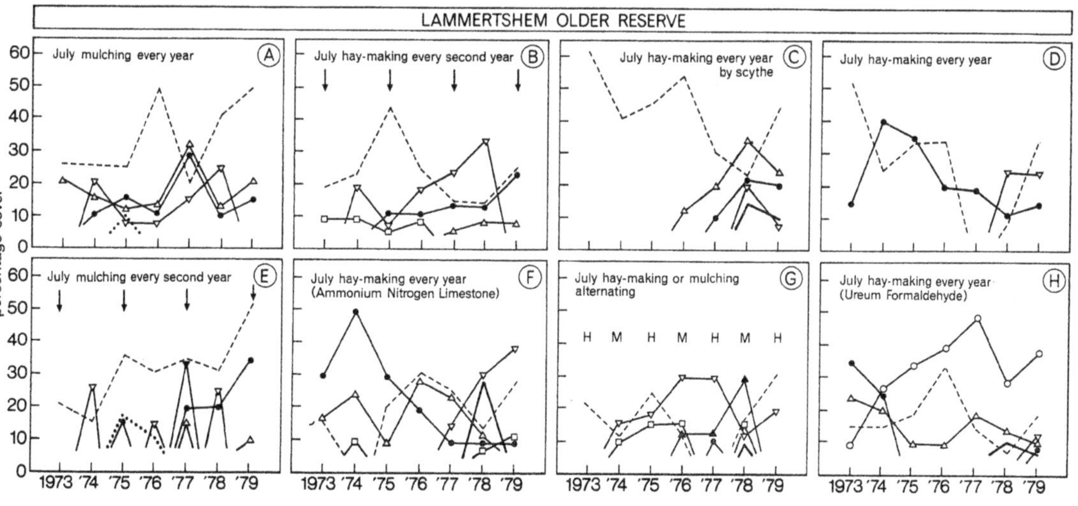

Fig. VI.25A–H. Changes of cover percentages of abundant (coverage ⩾8%) species in the *Valeriana officinalis/Phalaris arundinacea* community in the older reserve of the Lammertshem area under different management regimes. Phalaris arundinacea ●——●, Calamagrostis canescens ○——○, Poa trivialis ———, Filipendula ulmaria ----, Valeriana officinalis △——△, Holcus lanatus ▽——▽, Caltha palustris □——□, Ranunculus repens ▲——▲, Glyceria maxima · · · ·, Glechoma hederacea ▼——▼.

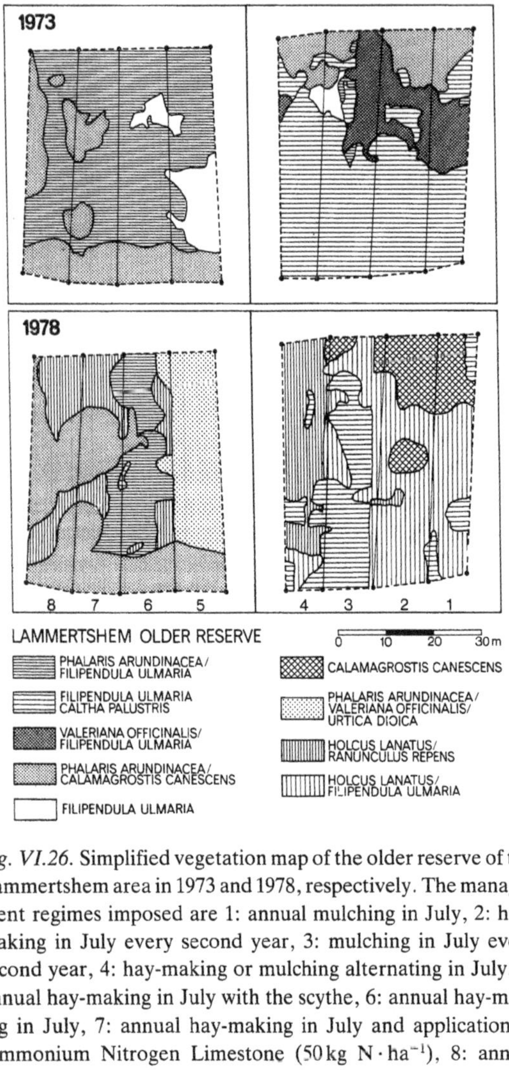

Fig. VI.26. Simplified vegetation map of the older reserve of the Lammertshem area in 1973 and 1978, respectively. The management regimes imposed are 1: annual mulching in July, 2: hay-making in July every second year, 3: mulching in July every second year, 4: hay-making or mulching alternating in July, 5: annual hay-making in July with the scythe, 6: annual hay-making in July, 7: annual hay-making in July and application of Ammonium Nitrogen Limestone (50 kg $N \cdot ha^{-1}$), 8: annual hay-making in July and application of Ureum Formaldehyde (50 kg $N \cdot ha^{-1}$).

all the other regimes eventually reached values between 80 and 90%. Perhaps the first group of treatments resembled the former agricultural practices to such an extent that only relatively small deviations from the initial vegetation were found. Burning, September hay-making every second year, September hay-making every year and July or September hay-making showed a significant ($P<0.05$) increase of the dissimilarity percentage.

The trends in the older reserve are shown in Fig. VI.28. July mulching every year showed the lowest deviation from the initial vegetation. None of the

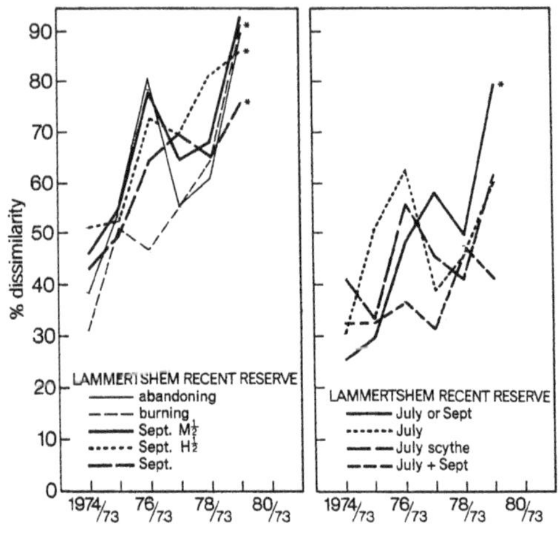

Fig. VI.27. Changes of percentages dissimilarity between the first year and subsequent years in the *Ranunculus repens/Caltha palustris* community in the recent reserve of the Lammertshem area under different management practices. * Significant change according to Spearman's rank correlation coefficient ($P<0.05$).

dissimilarity percentages of the practices increased significantly over the whole study period because of the frequent fluctuations.

VI.D.3.c. Species diversity

The mean species diversity in the recent reserve is shown in Fig. VI.29. It decreased significantly

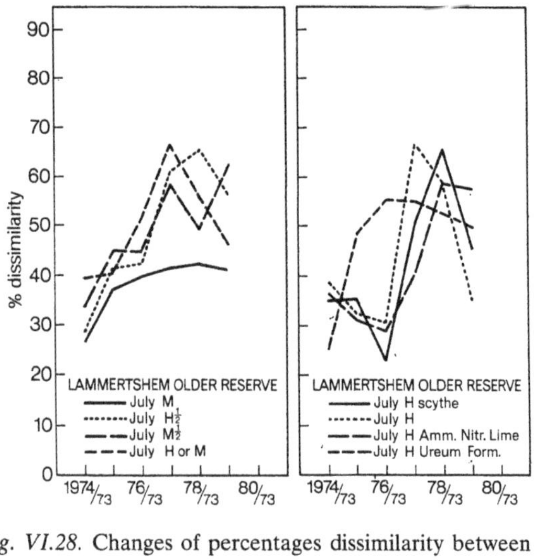

Fig. VI.28. Changes of percentages dissimilarity between the first year and subsequent years in the *Valeriana officinalis/ Phalaris arundinacea* community in the older reserve of the Lammertshem area under different management practices.

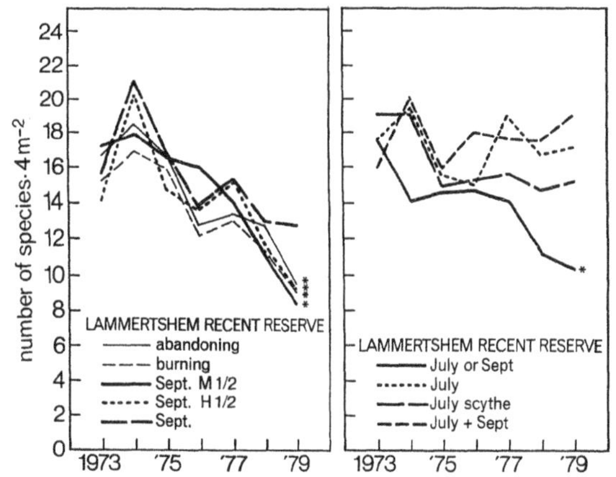

Fig. VI.29. Changes of number of species $\cdot$ 4 m^{-2} in the *Ranunculus repens/Caltha palustris* community in the recent reserve of the Lammertshem area under different management practices. * Significant change according to Spearman's rank correlation coefficient (P<0.05).

(P<0.05) in most management regimes apart from September hay-making every year and the July hay-making regimes. The species diversity tended to increase in the July + September hay-making regime.

In the older reserve (Fig. VI.30) the comparison of species diversity in all regimes is difficult due to the different initial situations. It can, however, be concluded that a significant (P<0.05) increase is found with the addition of 50 kg N $\cdot$ ha^{-1} in Ammonium Nitrogen Limestone and a significant decrease with July mulching every second year.

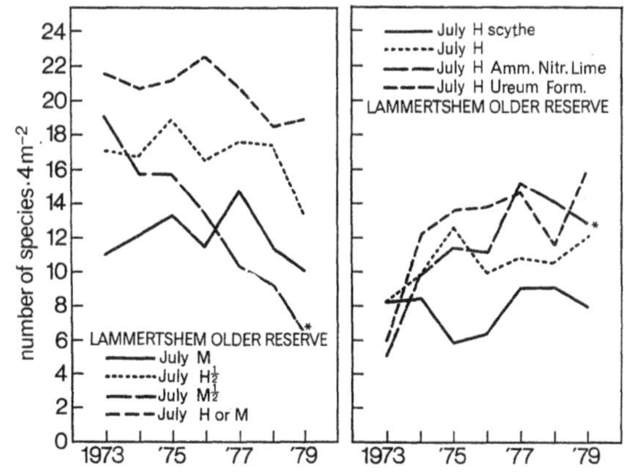

Fig. VI.30. Changes of number of species $\cdot$ 4 m^{-2} in the *Valeriana officinalis/Phalaris arundinacea* community in the older reserve of the Lammertshem area. * Significant change according to Spearman's rank correlation coefficient (P<0.05).

VI.D.4. Species changes

The cover changes of abundant species in the upper course Loefvledder area and in the lower course Lammertshem area have been discussed in the sections VI.D.2 and 3. This section summarizes the changes of all the species in the management practices.

The successional response of each species was analyzed according to the procedure mentioned in section II.C.3. The large changes are in agreement with the information provided in the sections VI.D.2 and 3.

Groups of plants species can be discerned in all the reserves (Table VI.18A–B and VI.19A–B). The groups are subdivided into species reacting in a similar way in nearly all management practices viz. constant species (group A), decreasers with different regimes (group B), increasers with different regimes (group C), decreasers with regimes involving litter accumulation including mulching, hay-making every second year, burning and abandoning (group D), increasers with regimes involving litter accumulation (group E), decreasers with regimes involving annual hay-making including hay-making in July, in September, in both July and September, alternating in July or September (group F), increasers with regimes involving annual hay-making (group G).

A comparison between the management regimes showed that the response of some individual species was similar in different plant communities. In the *Agrostis/Poa* and *Festuca/Agrostis* communities, occurring in the upper course, *Rumex acetosa* increased and *Agrostis stolonifera* decreased with all regimes. *Festuca pratensis, Poa trivialis* and *Trifolium repens* decreased, but *Lotus uliginosus* increased with regimes involving litter accumulation (Table VI.20). In the *Ranunculus/Caltha* and the *Valeriana/Phalaris* communities in the lower course *Phalaris arundinacea* and *Valeriana officinalis* increased, but *Holcus lanatus* decreased with regimes involving litter accumulation (Table VI.20).

Other species, however, showed different re-

Table VI.18A. Cover changes of all species occurring in ⩾50% of the comparable permanent plots in each management regime in the recent reserve of Loefvledder in the upper course. Changes are related to the initial cover percentages in each of the plots; small changes indicate cover changes <10%, large changes ⩾10% cover. +: small increase; ++: large increase; −: small decrease; −−: large decrease; C: constant; Λ: temporarily increase. P: species indicating nutrient-poor conditions; R: nutrient-rich conditions; I: intermediate.

LOEFVLEDDER Recent reserve Agrostis stolonifera/ Poa pratensis community			July + September hay-making	July hay-making	July hay-making scythe	July or September hay-making alternating	September hay-making	September hay-making every two years	September mulching every two years	Winter burning every two years	Abandoning
A	Constant with all regimes										
	I	Agrostis capillaris	±	C	C	C	C	Λ	C	C	C
B	Decrease with different regimes										
	R	Lolium perenne	±	-	-	C	- -	C	C	-	-
	R	Alopecurus geniculatus	C	-	C	-	C	C	-	-	-
	R	Agrostis stolonifera	C	C	-	C	C	- -	C	- -	C
C	Increase with different regimes										
	I	Rumex acetosa	+	+	+ +	+	+ +	+	±	+	+
	R	Potentilla anserina	+	+	+	Λ	+	+ +	+ +	+ +	+
	R	Rumex crispus	Λ	Λ	Λ	Λ	C	.	Λ	Λ	Λ
	R	Ranunculus repens	C	+ +	+ +	+ +	Λ	Λ	±	Λ	Λ
	R	Cirsium arvense	±	Λ	Λ	C	Λ	.	C	Λ	Λ
	-	Holcus mollis	+	+	+	.	+	.	.	+	.
	I	Festuca rubra	C	C	+ +	C	+	Λ	C	+ +	Λ
	R	Poa pratensis	C	Λ	±	C	C	C	Λ	Λ	C
	I	Epilobium tetragonum	+	+	+	+	C	.	.	+	+
	R	Polygonum mite	C	+	.	.	C	.	.	+	.
	I	Carex ovalis	+	.	.	.	C	.	.	.	Λ
	R	Urtica dioica	.	.	.	.	+	.	.	+	.
	P	Juncus effusus	+	.	.	.	.	.	.	.	+
D	Decrease with regimes involving litter accumulation										
	I	Cardamine pratensis	C	C	C	C	C	-	-	-	-
	R	Festuca pratensis	C	C	C	C	C	-	- -	- -	-
	R	Poa trivialis	C	C	C	±	Λ	C	-	- -	C
	I	Ranunculus acris	C	Λ	C	Λ	C	-	C	-	-
	R	Trifolium repens	+	C	±	-	Λ	C	.	-	-
	-	Taraxacum sp.	±	±	C	C	C	C	C	Λ	-
E	Increase with regimes involving litter accumulation										
	I	Holcus lanatus	C	C	±	C	C	Λ	Λ	Λ	C
	R	Polygonum persicaria	.	C	C	C	Λ	Λ	Λ	Λ	Λ
	R	Rumex obtusifolius	±	Λ	Λ	Λ	±	.	.	±	+ +
	I	Lotus uliginosus	.	.	.	.	.	.	+	Λ	+
	R	Senecio sylvaticus	.	.	.	.	.	.	.	+	+
	R	Phleum pratense	.	C	.	C	.	Λ	.	.	.
	I	Galium palustre	.	.	C	.	.	.	.	.	Λ
	R	Galium aparine	.	.	.	.	.	.	.	+	.
	R	Cirsium vulgare	.	.	.	.	.	.	.	C	.
	-	Rubus spec.	.	.	.	.	.	.	.	.	+
	R	Heracleum sphondylium	.	.	.	Λ	.	.	.	.	+ +
G	Increase with regimes involving annual hay-making										
	P	Rhinanthus angustifolius	+ +	+	.	+ +	+	Λ	.	.	.
	I	Anthoxanthum odoratum	+	+	C	+	.	.	.	.	.
	R	Plantago major	+	+	C	.	.	.	.	.	.
	I	Glyceria fluitans	+	+	.	.	.	.	.	.	Λ
	I	Cynosurus cristatus	C	Λ	C	.	C	.	.	.	.
	I	Plantago lanceolata	C	.	.	.	C	.	.	.	.
	I	Sagina procumbens	.	+	Λ	.	.	.	.	.	.
	-	Peplis portula	.	.	Λ	Λ	C	.	.	.	.
	I	Stellaria uliginosa	±	.	Λ	.	.	.	.	.	.
	I	Juncus bufonius	+	.	Λ	Λ	.	.	.	.	.
	I	Montia fontana	Λ	.	.	.	.	Λ	.	.	.
	R	Bromus hordeaceus	Λ	.	.	Λ	.	.	.	.	.
	P	Equisetum palustre	C	C	+	.	.	.	.	.	.
	R	Poa annua	C	.	.	.	.	.	.	.	.
	I	Cerastium fontanum	±	C	C	+	+	Λ	C	-	Λ
	P	Equisetum arvense	C	C	.	.	.	.	.	.	.
	R	Anthriscus sylvestris	Λ	Λ	.	+	.	.	.	.	.
	P	Stellaria palustris	Λ	.	.	.	.	.	.	.	.
	R	Stellaria media	C	Λ	C	Λ	.	.	.	.	.
	P	Prunella vulgaris	C	.	.	.	.	.	.	.	.
	R	Alopecurus pratensis	C	.	.	+	.	.	.	.	.
	I	Veronica serpyllum	Λ	.	.	.	.	.	.	.	.
	I	Polygonum hydropiper	Λ	.	+	.	.	.	.	.	.
	I	Leontodon autumnalis	+	.	.	.	.	.	.	.	.
	P	Stellaria graminea	.	C	C	.	.	.	.	.	.
	R	Elymus repens	.	Λ	.	Λ	Λ	.	.	.	.
	I	Dactylis glomerata	.	Λ	.	.	.	.	.	.	.
	P	Agrostis canina	.	.	.	C	.	.	.	.	.
	I	Valeriana officinalis	.	.	.	C	.	.	.	.	.
	R	Polygonum aviculare	.	.	.	+	.	.	.	.	.

Table VI.18B. Cover changes of all species occurring in ⩾50% of the comparable permanent plots in each management regime in the older reserve of Loefvledder in the upper course. Changes are related to the initial cover percentages in each of the plots; small changes indicate cover changes <10%, large changes ⩾10% cover. +: small increase; ++: large increase; −: small decrease; − −: large decrease; C: constant; Λ: temporarily increase. P: species indicating nutrient-poor conditions; R: nutrient-rich conditions; I: intermediate.

LOEFVLEDDER Older reserve Festuca rubra/ Agrostis capillaris community	July mulching every year	July hay-making every second year	July mulching every second year	July hay-making or mulching alternating	July hay-making scythe	July hay-making Ammonium Nitrogen Limestone	July hay-making Ureum Formaldehyde	July hay-making
A Constant with all regimes								
- Taraxacum spec.	C	C	C	C	C	C	C	C
I Glyceria fluitans	.	C	C	C	C	.	C	.
P Nardus stricta	.	C	.	.	.	.	.	C
B Decrease with different regimes								
R Agrostis stolonifera	-	-	C	C	-	C	C	-
I Anthoxanthum odoratum	-	C	- -	C	C	.	C	±
I Ranunculus acris	C	±	-	-	C	-	±	C
I Plantago lanceolata	C	C	±	-	.	.	.	-
P Hydrocotyle vulgaris	.	-	.	C	C	+	-	C
C Increase with different regimes								
I Rumex acetosa	+ +	C	+ +	C	+	+ +	Λ	+
P Cirsium palustre	±	±	C	+	±	±	Λ	C
I Cerastium fontanum	C	+	C	+	C	.	+	+
P Juncus acutiflorus	+	C	-	C	C	.	+	+
P Viola palustris	-	+	C	+	+ +	C	C	+
D Decrease with regimes involving litter accumulation								
F Festuca pratensis	-	-	-	±	C	C	+	C
I Festuca rubra	- -	±	-	C	+	C	C	C
R Poa trivialis	C	-	C	Λ	-	Λ	Λ	±
R Trifolium repens	C	-	-	-	C	C	C	C
P Luzula campestris	-	C	-	C	C	C	-	+
R Glyceria maxima	C	C	C	-	.	.	.	.
P Equisetum palustre	-	C	-	C	-	.	.	.
P Ranunculus flammula	C	C	.	-	C	C	C	C
I Stellaria uliginosa	C	C	-	C	C	C	C	.
I Bellis perennis	.	.	.	-	.	.	.	.
E Increase with regimes involving litter accumulation								
R Poa pratensis	Λ	Λ	Λ	Λ	±	C	C	C
R Anthriscus sylvestris	+	.	.	.	.	.	.	.
R Alopecurus geniculatus	C	Λ	.	.	.	C	.	.
I Cynosurus cristatus	C	.	.	.	.	.	.	.
I Galium palustre	C	+	+	C	±	±	-	±
P Dactylorhiza majalis	C	.	.	.	.	.	.	.
I Trifolium pratense	C	.	.	C	.	.	.	.
P Stellaria graminea	+	C	.	.	C	.	.	.
P Prunella vulgaris	.	C	C	.	.	.	.	.
P Equisetum arvense	.	+	.	.	.	.	.	.
R Phleum pratense	.	.	.	C	.	.	.	.
P Rhinanthus angustifolius	.	.	.	+	.	.	.	.
I Lysimachia nummularia	.	.	.	C	.	.	.	.
I Equisetum fluviatile	.	.	.	C	.	.	.	.
F Decrease with regimes involving annual hay-making								
I Lotus uliginosus	C	C	C	C	-	-	-	-
R Ranunculus repens	Λ	C	C	C	-	-	-	-
I Cardamine pratensis	C	C	C	C	Λ	±	-	-
I Leontodon autumnalis	C	C	C	C	C	C	-	C
G Increase with regimes involving annual hay-making								
I Holcus lanatus	C	C	C	C	±	+	+	+
P Juncus effusus	C	C	C	C	- -	+ +	+ +	+
I Carex ovalis	C	C	±	±	±	±	Λ	Λ
P Agrostis canina	.	.	.	C	+	C	C	+
P Carex nigra	C	±	C	C	+	C	C	C
I Agrostis capillaris	C	C	C	C	C	C	Λ	+
I Lychnis flos-cuculi	C	C	.	C	-	C	C	+
P Juncus conglomeratus	.	.	.	.	+	+	.	+
P Luzula multiflora	.	.	.	.	.	.	C	+
P Carex curta	.	.	.	.	.	+	.	Λ
P Carex panicea	.	.	.	.	.	.	.	Λ
P Potentilla erecta	.	.	.	.	.	.	.	+
P Carex echinata	.	.	.	.	.	C	.	.
- Galium mollugo	.	.	.	.	.	Λ	.	.
R Stellaria media	.	.	.	.	.	C	.	.
R Alopecurus pratensis	.	.	.	.	.	.	C	.
R Cirsium arvense	.	.	.	.	.	.	C	.

Table VI.19A. Cover changes of all species occurring in ⩾50% of the comparable permanent plots in each management regime in the recent reserve of Lammertshem of in the lower course. Changes are related to the initial cover percentages in each of the plots; small changes indicate cover changes <10%, large changes ⩾10% cover. +: small increase; ++: large increase; −: small decrease; −−: large decrease; C: constant; Λ: temporarily increase. P: species indicating nutrient-poor conditions; R: nutrient-rich conditions; I: intermediate.

LAMMERTSHEM Recent reserve Ranunculus repens/ Caltha palustris community		July + September hay-making	July hay-making	July hay-making scythe	July or September hay-making alternating	September hay-making	September hay-making every two years	September mulching every two years	Winter burning every two years	Abandoning
A	Constant with all regimes									
R	Poa pratensis	±	C	C	C	C	.	C	.	C
R	Agrostis stolonifera	.	C	C	.	C	C	C	.	C
I	Leontodon autumnalis	C	C	C	C	.	.	.	C	.
-	Lythrum salicaria	.	.	C	.	.	.	.	C	.
I	Veronica serpyllum	.	.	.	C	.	.	C	C	.
R	Stellaria media	.	.	.		C	C	.	.	.
R	Rumex crispus	C	.	C	.	.	.	.	.	C
B	Decrease with different regimes									
R	Festuca pratensis	-	C	- -	C	- -	- -	- -	- -	- -
-	Caltha palustris	C	- -	C	- -	C	- -	C	C	C
R	Ranunculus repens	C	C	C	- -	- -	- -	- -	- -	- -
I	Myosotis palustris	C	C	-	C	-	C	-	C	.
I	Ranunculus acris	C	C'	-	-	C	Λ	C	Λ	-
I	Senecio aquaticus	C	C	C	-	C	-	.	C	C
I	Lychnis flos-cuculi	C	C	-	C	-	C	-	-	-
C	Increase with different regimes									
I	Filipendula ulmaria	+	+ +	+ +	+ +	+ +	+ +	+ +	+ +	+ +
I	Festuca rubra	+	Λ	+	Λ	C	Λ	C	C	+
R	Glyceria maxima	C	C	Λ	Λ	C	C	C	.	Λ
I	Galium palustre	Λ	+	C	C	Λ	C	Λ	C	C
I	Deschampsia cespitosa	C	Λ	C	+	C	+	.	+	+
P	Carex nigra	.	C	C	Λ	C	Λ	Λ	C	C
D	Decrease with regimes involving litter accumulation									
I	Cardamine pratensis	+	Λ	Λ	Λ	C	-	Λ	-	-
I	Holcus lanatus	C	C	C	C	C	- -	- -	Λ	-
I	Cerastium fontanum	C	C	C	C	C	C	C	-	-
E	Increase with regimes involving litter accumulation									
R	Phalaris arundinacea	-	C	C	C	C	+ +	+	+ +	+
I	Rumex acetosa	C	C	C	±	C	C	Λ	Λ	C
I	Valeriana officinalis	-	.	C	.	C	C	C	.	+
R	Urtica doica	.	.	.	.	.	+	.	.	.
I	Lysimachia nummularia	.	.	.	.	.	.	Λ	C	Λ
P	Juncus effusus	.	.	.	.	.	.	.	Λ	.
R	Ranunculus ficaria	.	.	.	.	.	.	.	C	.
F	Decrease with regimes involving annual hay-making									
-	Taraxacum spec.	-	-	C	C	-	C	C	Λ	C
R	Trifolium repens	-	C	C	Λ	C	C	C	.	C
G	Increase with regimes involving annual hay-making									
I	Anthoxanthum odoratum	+	C	C	Λ	C	C	-	-	C
R	Poa trivialis	+ +	Λ	C	Λ	Λ	C	C	-	C
I	Agrostis capillaris	C	Λ	.	.	.	.	.	.	.
I	Plantago lanceolata	+	.	.	.	.	.	.	.	.
-	Carex aquatilis	+	C	+	+	+	+	±	.	C
P	Rhinanthus angustifolius	C	C	.	.	.	.	.	.	.
P	Carex rostrata	.	C	Λ	C	.	.	.	.	.
P	Galium uliginosum	.	.	C	.	.	.	.	.	.
R	Poa palustris	.	Λ	.	.	.	.	.	.	.
P	Stellaria palustris	.	+	.	.	.	.	.	.	.
I	Stellaria uliginosa	.	C	.	.	.	.	.	.	.
I	Ajuga reptans	.	C	.	.	.	.	.	.	.
-	Myosotis cespitosa	.	C	.	.	.	.	.	.	.
R	Rorippa amphibia	.	C	.	.	.	.	.	.	.
P	Stellaria graminea	.	C	.	.	.	.	.	.	.
I	Glyceria fluitans		.	.	C	.	.	.	.	.
P	Juncus articulatus	.	.	.	.	C	.	.	.	.

Table VI.19B. Cover changes of all species occurring in ≥50% of the comparable permanent plots in each management regime in the older reserve of Lammertshem in the lower course. Changes are related to the initial cover percentages in each of the plots; small changes indicate cover changes <10%, large changes ≥10% cover. +: small increase; ++: large increase; −: small decrease; −−: large decrease; C: constant; Λ: temporarily increase. P: species indicating nutrient-poor conditions; R: nutrient-rich conditions; I: intermediate.

		LAMMERTSHEM Older reserve Valeriana officinalis/ Phalaris arundinacea community	July mulching every year	July hay-making every second year	July mulching every second year	July hay-making or mulching alternating	July hay-making scythe	July hay-making Ammonium Nitrogen Limestone	July hay-making Ureum Formaldehyde	July hay-making
A		Constant with all regimes								
	R	Poa trivialis	C	C	C	C	C	C	C	C
	I	Filipendula ulmaria	C	C	C	C	C	C	C	C
	-	Lythrum salicaria	C	C	C	C	C	±	C	C
	I	Cardamine pratensis	C	C	C	C	C	C	C	C
	R	Poa palustris	±	C	C	C	.	.	C	C
	R	Poa pratensis	C	C	.	.	.	C	C	.
	I	Ranunculus flammula	.	.	.	C	.	.	C	.
	I	Festuca rubra	C	C	C	C	.	.	C	.
	R	Festuca pratensis	C	C	C	C	.	C	.	.
	I	Deschampsia cespitosa	.	C	.	.	.	C	.	C
C		Increase with different regimes								
	-	Calamagrostis canescens	+	+	.	C	+	+	±	±
	R	Glyceria maxima	C	-	Λ	Λ	Λ	±	Λ	C
	R	Urtica dioica	C	+	+	.	+	Λ	Λ	C
	I	Myosotis palustris	.	Λ	Λ	Λ	Λ	Λ	C	.
	P	Juncus effusus	C	-	Λ	±	.	.	Λ	C
	R	Glechoma hederacea	C	C	Λ	Λ	+	.	Λ	C
D		Decrease with regimes involving litter accumulation								
	-	Caltha palustris	-	- -	-	C	C	C	C	C
	P	Valeriana dioica	C	C	-	-	.	.	C	.
	I	Equisetum fluviatile	-	C	C	-	C	C	C	C
	I	Ranunculus acris	C	C	-	C	C	C	C	C
	I	Anthoxanthum odoratum	C	-	-	C	.	.	.	.
	P	Carex nigra	.	C	-	-	.	.	.	.
	I	Rumex acetosa	Λ	C	-	C	C	Λ	±	C
	I	Lychnis flos-cuculi	C	C	-	Λ	.	C	C	Λ
	P	Galium uliginosum	Λ	C	-	C	.	.	C	Λ
	I	Plantago lanceolata	C	-	-	Λ	.	.	C	.
E		Increase with regimes involving litter accumulation								
	R	Phalaris arundinacea	C	+ +	+ +	C	+ +	- -	C	C
	I	Valeriana officinalis	C	±	Λ	+	C	C	C	C
	-	Angelica sylvestris	Λ	Λ	±	±	C	C	C	C
	I	Veronica serpyllum	.	.	.	C	.	.	.	.
	I	Agrostis capillaris	.	.	.	C	.	.	.	.
	I	Trifolium pratense	.	.	.	C	.	.	.	.
	I	Cerastium fontanum	.	.	.	C	.	.	.	.
	I	Senecio aquaticus	.	C	.	C	.	.	.	.
	P	Prunella vulgaris	.	.	.	C	.	.	.	.
	P	Stellaria palustris	.	.	.	±	.	.	.	.
	P	Achillea ptarmica	C	C	Λ	C	.	.	.	.
	P	Carex rostrata	.	C	C	.	.	.	.	.
	R	Ranunculus ficaria	.	.	C	.	.	.	.	.
	I	Leontodon autumnalis	.	.	C	.	.	.	.	.
	R	Rumex obtusifolius	.	.	C	.	.	.	.	.
	I	Ajuga reptans	.	Λ	.	.	.	.	.	.
	R	Agrostis stolonifera	.	C	.	.	.	.	.	.
F		Decrease with regimes involving hay-making								
		Lysimachia vulgaris	C	C	C	C	-	C	C	C
G		Increase with regimes involving annual hay-making								
	R	Ranunculus repens	C	C	C	C	C	+	C	+
	I	Holcus lanatus	C	C	C	Λ	C	+ +	±	+ +
	-	Taraxacum spec.	C	C	-	C	+	+	C	+
	R	Anthriscus sylvestris	.	.	.	.	.	Λ	C	.
	R	Rumex crispus	.	.	C	C	.	.	Λ	.
	I	Galium palustre	.	C	C	C	C	C	C	Λ
	-	Carex aquaticus	C	C	±	C	Λ	±	±	+
	-	Epilobium obscurum	.	.	.	.	C	C	Λ	C
	R	Stellaria media	C	C	.	.	Λ	.	Λ	.
	I	Lysimachia nummularia	C	C	±	C	Λ	.	C	.
	P	Juncus conglomeratus	.	.	.	.	.	±	.	C
	R	Chamerion angustifolium	.	.	.	.	C	.	.	Λ
	R	Polygonum amphibium	.	.	.	.	.	.	.	Λ
	R	Alopecurus pratensis	.	.	.	.	+	.	.	.
	R	Cirsium arvense	.	.	.	.	Λ	.	.	.
	P	Lathyrus palustris	.	.	.	.	.	.	C	.

sponses in different plant communities with similar management practices (Table VI.20). *Holcus lanatus* and *Rumex acetosa*, for example, were favoured by litter accumulation in the *Agrostis/Poa* community and became the dominating species during a certain period, but were immediately suppressed by the dominating *Phalaris arundinacea* in the *Valeriana/Phalaris* community.

Particularly 'intermediate' and 'poor' species decreased and increased in the *Valeriana/Phalaris* community in regimes including litter accumulation, whereas the proportion of 'rich' species was higher in the other three communities (Table VI.21). 'Rich' species, in particular, increased with annual hay-making in the *Valeriana/Phalaris* community, but the proportion of 'poor' species was higher in the other communities. It is remarkable that similar results with respect to poor and rich

Table VI.20. Response of individual species on various management practices in four different plant communities. C: constant; +: increase; −: decrease.

	Agrostis stolonifera/ Poa pratensis community	Festuca rubra/ Agrostis capillaris community	Ranunculus repens/ Caltha palustris community	Valeriana officinalis/ Phalaris arundinacea community
Agrostis capillaris	C	+ hay-making		
Anthoxanthum odoratum	+ hay-making	− all regimes		
Cerastium fontanum	+ hay-making	− all regimes		
Festuca rubra	+ all regimes	− litter accumulation		
Agrostis stolonifera	− all regimes	− all regimes		
Festuca pratensis	− litter accumulation	− litter accumulation		
Trifolium repens	− litter accumulation	− litter accumulation		
Lotus uliginosus	+ litter accumulation	+ litter accumulation		
Holcus lanatus	+ litter accumulation	+ hay-making	− litter accumulation	− litter accumulation
Ranunculus repens	+ all regimes	− hay-making	− all regimes	+ hay-making
Cardamine pratensis	− litter accumulation	− hay-making	− litter accumulation	C
Rumex acetosa	+ all regimes	+ all regimes	+ litter accumulation	− litter accumulation
Poa trivialis	− litter accumulation	− litter accumulation	+ hay-making	C
Phalaris arundinacea			+ litter accumulation	+ litter accumulation
Valeriana officinalis			+ litter accumulation	+ litter accumulation

Table VI.21. Percentages of species indicating rich (R), intermediate (I) and poor (P) soil conditions involved in various management practices in different plant communities.

	Agrostis stolonifera/ Poa pratensis community R / I / P	Festuca rubra/ Agrostis capillaris community R / I / P	Ranunculus repens/ Caltha palustris community R / I /P	Valeriana officinalis/ Phalaris arundinacea community R / I /P
Constant with all regimes	–/ 4/ –	–/ 5/ –	29/ 9/ –	21/25/ –
Decrease with different regimes	11/ –/ –	8/14/ 5	14/18/ –	–/ –/ –
Increase with different regimes	25/17/12	–/10/14	7/18/13	16/ 5/ 8
Decrease with litter accumulation	11/ 7/ –	31/24/13	–/14/ –	–/15/46
Increase with litter accumulation	25/13/ –	31/24/24	21/14/13	21/40/31
Decrease with annual hay-making	–/ –/ –	8/14/ –	7/ –/ –	–/ –/ –
Increase with annual hay-making	29/58/88	23/19/43	21/43/75	42/15/15
Total number of species	28/24/ 8	13/21/21	14/22/ 8	19/20/13

species were found in both the recent and older reserve in the Loefvledder study area, irrespective of the number of years without fertilizer applications. It seems that the response of individual species not only depends on factors influencing the nutrient status of the soil, but also on the structure of the sward, in particular, the absence or presence of litter.

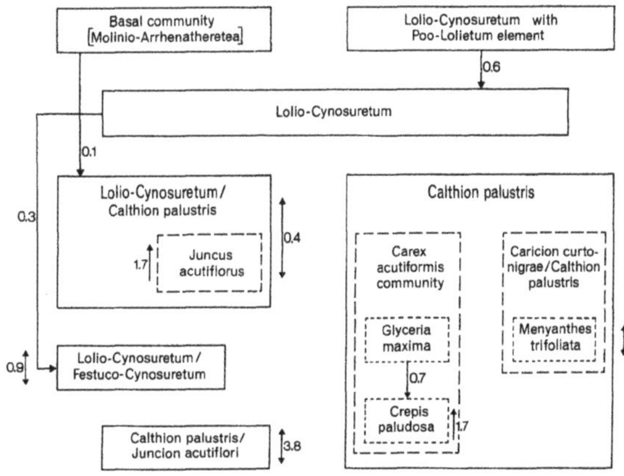

Fig. VI.31. Principal succession observed in valley grasslands. The direction of the arrow indicates the direction of the succession, arrows going up and down indicate fluctuations within a community. Figures besides the arrows indicate the mean ratio of species indicating nutrient-poor soil conditions to species indicating nutrient-rich soil conditions during the period of observation. Everts et al. (1984) suggest successional lines from the top to the bottom of the figure during the process of hay-making or grazing without fertilizer applications.

VI.E. Succession and species diversity

VI.E.1. Principal succession

Everts et al. (1984) compiled a scheme showing the principal succession series in the Drentsche A valley with hay-making or grazing without fertilizer application. Their successional lines are, however, hypothetical, since they are based on the comparison of different plant communities which occur simultaneously at one particular moment in the landscape. This study investigates whether the observed vegetational changes are in agreement with the hypothetical succession scheme on dry, sandy soil and wet, peaty soil (Fig. VI.31). Moreover, the changes of the poor/rich quotients were calculated in order to indicate the soil fertility status. This is discussed in detail in chapter X.

On dry, sandy soil stands with *Poo-Lolietum* elements can succeed into stands with *Lolio-Cynosuretum* elements, ten years after the cessation of fertilizer application (cf. also Table VI.10). The succession from stands with *Lolio-Cynosuretum* elements to stands with *Festuco-Cynosuretum* elements was also observed in the same period (cf. also Table VI.11). In another case, however, no succession was observed over a period of ten years starting with a *Lolio-Cynosuretum* stand with *Calthion palustris* elements. This was despite the fact that the fertilizer application had been discontinued five years before the start of the observations. The stands with *Festuco-Cynosuretum* elements also did not change during ten years of observation. Indeed, most successional steps on dry soil expected by Everts et al. (1984) were found, but the succession does not always take place at regular time intervals. Sometimes a phase can maintain itself for at least twelve years despite continual hay-making without fertilizer application. It is the initial composition of the sward that determines the place in the successional series and not the number of years of hay-making without fertilizer application. If succession proceeds, it corresponds more or less with the increasing poor/rich quotients.

Only the succession from stands with *Lolio-Cynosuretum* elements to stands with *Calthion palustris* and *Caricion curto-nigrae* elements were observed on wet, peaty soil over a period of ten years (cf. also Table VI.12). The community of *Carex acutiformis* (*Calthion palustris*) with *Glyceria maxima* succeeded into the variant with *Crepis paludosa* in ten years (cf. Table VI.16). All other communities maintained themselves on wet soil during the ten years, namely *Caricion curto-nigrae/Calthion palustris* (Table VI.13), *Juncion acutiflori* (Table VI.15), community of *Carex acutiformis* with *Crepis paludosa* (Table VI.17). If the hypothetical succession (Everts et al. 1984) occurs, it is generally attended with increasing poor/rich ratio's. It is, however, remarkable that, despite continual hay-making without fertilizer application, the poor/rich quotient decreased in some plant communities. This seemingly contradiction will be discussed in chapter X.

VI.E.2. Species diversity

VI.E.2.a. Abandoning

The cessation of all agricultural practices or management activities results in litter accumulation and a decrease in species diversity according to the literature (Table VI.22). In this study similar phenomena were found in semi-natural inland communities and the findings of other authors will be discussed. Nearly all authors who have compared different management practices, mentioned a decreased number of species after 5–10 years of abandoning. This is in agreement with our results. The only exceptions were found by Schiefer (1981) and Schreiber & Schiefer (1985) in a *Gentiano-Koelerietum,* a *Carex acutiformis* community and a *Calthion/Filipendulion* community and by Dierschke (1985) in a *Mesobromion* community. All these communities were species-rich. Rosenthal et al. (1985), however, found an increase in a species-poor community.

Abandoned valley fen meadows showed a considerable decrease in species-richness after three years (Wheeler, 1983). Species-rich fen vegetation is becoming increasingly uncommon in Broadland, United Kingdom, due to large standing crops and litter masses as a result of abandoning (Wheeler & Giller, 1982). Regnèll (1980) reached similar conclusions for calcareaous mire communities in Sweden after the cessation of grazing or haymaking.

Ellenberg (1978) suggested a diverging development on an abandoned pasture and meadow. Shrubs were already present in these pastures because of marginally agricultural use prior to abandoning. Former pastures can, therefore, develop into woodland after three or four decades, whereas former hayfields remain without trees because they cannot emerge in the accumulated layer of litter.

Runge (1985) did observe a successional series in an abandoned *Lolio-Cynosuretum* in Germany after a grazing period of

Table VI.22. Changes in number of species after different periods of abandoning in various plant communities with various management prior to the abandoning.

Country	Management prior to the experiment	Plant community	Management regime: Number of years	Management regime: Practice	Changes in number of species per unit area (+ : increase; = : constant; − : decrease)	Assessment
The Netherlands	hayfield	Ranunculus/Caltha community	6	abandoning	− 17 → 10 . 4 m^{-2}	this study
	hay-pasture	Agrostis / Poa community	11	abandoning	− 13 → 9 . 4 m^{-2}	this study
	pasture	Arrhenatheretum elatioris	11	abandoning	− 37 → 33 . 10 m^{-2}	Willems 1983 A
	hay-pasture	Arrhenatherion elatioris	8	abandoning	− 52 → 38 . 100 m^{-2}	Oomes & Mooi 1981
Germany	hayfield	Arrhenatherion elatioris	4(8)	abandoning	+ 33 → 38(42).25 m^{-2}	Schiefer 1981;Schreiber & Schiefer 1985
	unfertilized pasture	Gentiano-Koelerietum	4(9)	abandoning	= 42 → 42(42). 25 m^{-2}	,,
	hayfield	Mesobrometum	4(9)	abandoning	− 50 → 45 (34).25 m^{-2}	,,
	unfertilized hayfield	Arrhenatherion elatioris	3(7)	abandoning	− 39 → 31 (25).25 m^{-2}	,,
	5 years abandoned	Festuco-Brometea	6(10)	abandoning	− 23 → 23(17). 25 m^{-2}	,,
	pasture	Cynosurion cristati	4(9)	abandoning	− 40 → 36(32). 25 m^{-2}	,,
	hayfield	Mol.Arrh. Bromus erectus	4(6)	abandoning	− 39 → 27(24). 25 m^{-2}	,,
	hayfield	Mol.Arrh. Carex acutiformis	4(6)	abandoning	= 28 → 28(28). 25 m^{-2}	,,
	unfertilized hayfield	Calthion/Filipendulion	4(18)	abandoning	= 40 → 40(38). 25 m^{-2}	,,
	hayfield	Calthion/Juncion acutiflori	4(9)	abandoning	− 24 → 28(20). 25 m^{-2}	,,
	pasture	Lolio - Cynosuretum	5	abandoning	+ 7 → 14. 5 m^{-2}	Rosenthal et al. 1985
	16 years abandoned	Molinietum	5	abandoning	− 42 → 29. 25 m^{-2}	Briemle 1985
	abandoned	Mesobromion	12	abandoning	+ 21 → 24. 4 m^{-2}	Dierschke 1985
United Kingdom	abandoned	Mesobromion	c.12	abandoning	= 6 → 6 *)	Green 1980

* Point quadrat analysis.

Table VI.23. Changes in number of species after different periods of burning (1: every year; $^1/_2$: every second year) in various plant communities with various management prior to the burning experiments.

Country	Management prior to the experiment	Plant community	Management regime: Number of years	Practice	Frequency	Period	Changes in number of species per unit area (+ : increase; = : constant; – : decrease)	Assessment
The Netherlands	hayfield	Ranunculus/Caltha community	6	burning	½	winter	– 13 → 9 : 4 m^{-2}	this study
	hay-pasture	Agrostis/Poa community	11	burning	½	winter	+ 11 → 15 . 4 m^{-2}	this study
Germany	hayfield	Arrhenatherion elatioris	4	burning	1	winter	= 39 → 39 . 25 m^{-2}	Schiefer 1981,1982;Schreiber & Schiefer 1985
	unfertilized pasture	Gentiano-Koelerietum	4	burning	1	winter	= 50 → 48 . 25 m^{-2}	,,
	unfertilized hayfield	Arrhenatherion elatioris	3	burning	1	winter	= 35 → 35 . 25 m^{-2}	,,
	5 years abandoned	Festuco-Brometea	6	burning	1	winter	+ 26 → 34 . 25 m^{-2}	,,
	unfertilized hayfield	Calthion / Filipendulion	4	burning	1	winter	– 37 → 24 . 25 m^{-2}	,,
	16 years abandoned	Molinietum	5	burning	1	winter	– 32 → 26 . 25 m^{-2}	Briemle 1985
	16 years abandoned	Filipenduletum	5	burning	1	winter	– 31 → 26 . 25 m^{-2}	,,

Table VI.24. Changes in number of species after different periods of mulching every second year in various plant communities with various management prior to the mulching experiments.

Country	Management prior to the experiment	Plant community	Management regime: Number of years	Practice	Frequency	Period	Changes in number of species per unit area (+ : increase; = : constant; – : decrease)	Assessment
The Netherlands	hayfield	Ranunculus/Caltha community	6	mulching	½	September	– 17 → 9 . 4 m^{-2}	this study
	hayfield	Valeriana/Phalaris community	6	mulching	½	July	– 19 → 6 . 4 m^{-2}	,,
	hay-pasture	Agrostis / Poa community	11	mulching	½	September	= 11 → 11 . 4 m^{-2}	,,
	unfertilized hay-pasture	Festuca/Agrostis community	11	mulching	½	July	– 17 → 8 . 4 m^{-2}	,,
Germany	unfertilized pasture	Gentiana Koelerietum	4	mulching	½	August	= 48 → 51 . 25 m^{-2}	Schiefer 1981,1983;Schreiber & Schiefer 1985
	unfertilized hayfield	Arrhenatherion elatioris	3	mulching	½	August	– 33 → 21 . 25 m^{-2}	,,
	5 years abandoned	Festuco - Brometea	6	mulching	½	August	= 24 → 27 . 25 m^{-2}	,,
	pasture	Cynosurion cristati	4	mulching	½	August	= 38 → 38 . 25 m^{-2}	,,
	hayfield	Mol. Arrh. Bromus erectus	4	mulching	½	August	– 37 → 33 . 25 m^{-2}	,,
	hayfield	Mol. Arrh. Carex acutiformis	4	mulching	½	August	= 26 → 25 . 25 m^{-2}	,,
	unfertilized hayfield	Calthion/Filipendulion	4	mulching	½	August	– 40 → 35 . 25 m^{-2}	,,
	hayfield	Calthion/Juncion acutiflori	4	mulching	½	August	= 28 → 27 . 25 m^{-2}	,,
	16 years abandoned	Molinietum	5	mulching	½	August	= 34 → 32 . 25 m^{-2}	Briemle 1985

about twenty years. *Trifolium repens, Cynosurus cristatus* and *Lychnis flos-cuculi* had disappeared after three years and *Cardamine pratensis, Poa pratensis, Plantago lanceolata, Anthoxanthum odoratum, Festuca pratensis, Lotus uliginosus* and *Ranunculus acris* after ten years. *Deschampsia cespitosa* became the dominant species from that period onwards whereas in the case of *Alnus glutinosa* this was after twelve years. Not a single tree seedling was, however, observed in a similar initial sward after 17 years of abandoning so it seems to be difficult to predict the emergence of trees. This community developed into a *Festuco-Cynosuretum* accompanied with tall forbs viz. *Cirsium arvense, Rumex acetosa* and *R. obtusifolius*. In the plot with *Alnus glutinosa* emergence which was studied little litter was found whereas a heavy layer was observed in the plot without the invasion of trees. Schreiber & Schiefer (1985) found that shrubs invaded relatively rapidly in abandoned communities with quick litter decomposition. Hardly any tree seedlings emerged in other communities with a permanent litter layer or accumulation of litter during the first ten years of abandoning.

Stöcklin & Gisi (1985) suggest that two factors contribute to the litter accumulation in hay-fields in Switzerland after the cessation of agricultural practices: (i) dead plant material is no longer removed, and (ii) a shift in species composition from easily decomposable herbs in cultivated hay-fields to grasses with a higher content of cellulose and lignin in fallows favouring the accumulation of a thick layer of litter.

In this study some species decreased or disappeared after abandoning and this is in agreement with the results of Runge (1985) and Schreiber & Schiefer (1985). The same species did not, however, always respond in a similar way. This can probably be ascribed to different responses of particular species in initially different plant communities (see section VI.D.4). Hülbusch (1986) also paid attention to this phenomenon and suggested that the 'behaviour' of a species can be much influenced by the composition and hence the interference of the surrounding vegetation. It seems relevant to consider life forms and growth forms rather than species. Schreiber & Schiefer (1985) concluded from a large number of fallow experiments in Germany that tall-growing hemicryptophytes and rhizome-plants proved to be able to compete and to dominate in the new set of species, strongly suppressing stolon- and rosette hemicryptophytes. Willems (1985) reached similar conclusions and emphasized the character of the growth form of the dominant species in determining species diversity. It can be concluded that growth forms show a more consistent successional response than do species.

VI.E.2.b. Burning

In the present study the species diversity increased in the moist *Agrostis stolonifera/Poa pratensis* community and decreased in the wet *Ranunculus repens/ Caltha palustris* community after burning for 11 and 6 years, respectively. Briemle (1985) also found a decrease in the wet *Filipenduletum* and a similar result was recorded by Schreiber & Schiefer (1985) in a *Calthion/Filipendulion* community, whereas the species diversity was found to increase in dry communities (Table VI.23).

The method adopted for burning might also be important for the species diversity. Only dead standing material and, locally, the upper litter layer was burned in our study area, because most of the litter was too wet at the end of the winter period. This type of burning is called 'cold fire' (Schiefer, 1982). Burning, therefore, might result in litter accumulation in our study area.

Krüsi (1981) found that winter burning favoured some tall-growing species and was disadvantageous for one small-sized species in *Mesobromion* communities in Switzerland. Briemle (1985) reported a decreasing species diversity in a *Filipenduletum* in Germany after annual burning. This was ascribed to the fact that despite the application of 'hot-fire' the tall-growing species were not hampered, but many small-sized species were killed. This can be prevented by burning every second year (Schiefer, 1981).

Schreiber (1980) and Schiefer (1982) concluded, from the results of diverging plant communities in Germany, that winter burning alone could not prevent the vegetation from succeeding into the direction similar to that of the laissez-faire management. The changes only proceed much slower, but small species disappear in favour of tall-growing species with rhizomes and below-ground tillers. Controlled burning alone cannot, therefore, maintain semi-natural grassland communities. It must be combined with a hay-making or mulching regime (Schiefer, 1981). Green (1983) reached a similar conclusion for herbaceous vegetation in the United Kingdom. He also recommended controlled burning as a management tool for the reclamation of unmanaged grassland.

Table VI.25. Changes in number of species after different periods of mulching once a year in various plant communities with various management prior to the mulching experiments.

Country	Management prior to the experiment	Plant Community	Management regime: Number of years	Practice	Frequency	Period	Changes in number of species per unit area (+ : increase, = : constant, − : decrease)	Assessment
The Netherlands	unfertilized hayfield	Valeriana/Phalaris community	6	mulching	1	July	− 12 → 10 . 4 m^{-2}	this study
	unfertilized pasture	Festuca / Agrostis community	11	mulching	1	July	− 14 → 8 . 4 m^{-2}	,,
Germany	hayfield	Arrhenatherion elatioris	4	mulching	1	August	+ 36 → 41 . 25 m^{-2}	Schiefer 1981,1983;Schreiber & Schiefer 1985
	unfertilized pasture	Gentiano-Koelerietum	4	mulching	1	August	+ 55 → 60 . 25 m^{-2}	,,
	hayfield	Mesobrometum	4	mulching	1	August	+ 53 → 61 . 25 m^{-2}	,,
	unfertilized hayfield	Arrhenatherion elatioris	3	mulching	1	August	= 41 → 37 . 25 m^{-2}	,,
	5 years abandoned	Festuco - Brometea	6	mulching	1	August	+ 21 → 24 . 25 m^{-2}	,,
	hayfield	Mol. Arrh. Bromus erectus	3	mulching	1	August	= 35 → 36 . 25 m^{-2}	,,
	hayfield	Mol. Arrh. Carex acutiformis	3	mulching	1	August	= 31 → 28 . 25 m^{-2}	,,
	unfertilized hayfield	Calthion / Filipendulion	3	mulching	1	August	+ 35 → 41 . 25 m^{-2}	,,
	hayfield	Calthion / Juncion acutiflori	4	mulching	1	August	= 33 → 31 . 25 m^{-2}	,,
	pasture	Lolio - Cynosuretum	5	mulching	1	August	+ 8 → 12 . 5 m^{-2}	Rosenthal 1984
	16 years abandoned	Molinietum	5	mulching	1	August	= 37 → 34 . 25 m^{-2}	Briemle 1985
	16 years abandoned	Filipenduletum	5	mulching	1	August	+ 27 → 34 . 25 m^{-2}	,,
	abandoned	Mesobromion	12	mulching	1	August	+ 21 → 28 . 4 m^{-2}	Dierschke 1985

VI.E.2.c. Mulching

The species diversity remained constant in the *Agrostis stolonifera/Poa pratensis* community, but decreased in the other communities in our study area with mulching every second year. Other authors also found no change in species diversity (Briemle, 1985; Schreiber & Schiefer, 1985), or a decrease (Table VI.24).

Mulching every year resulted in a lowering of the species diversity in this study. This contradicts the findings of other authors who only mentioned a minimal decrease in a minority of cases studied (Briemle, 1985; Schreiber & Schiefer, 1985) and an increase in most communities studied (Table VI.25). This is probably due to the fact that no litter accumulation was observed, despite mulching. It was apparently mineralized quickly, because of the high autumn temperatures in southern Germany. Only if mulching was done later than mid-August was the breakdown of litter hampered in the subalpine zone because of low temperatures (Schiefer, 1981). In that case litter persisted until the early summer of the following year. Moreover, mulching implied finely divided cutting, whereas the returned cuttings were left as a swathe in this study.

The mineralization of the mulched material was even stronger if mulching took place twice a year (Schiefer, 1981; Schreiber & Schiefer, 1985). This management practice resulted in an increase of species diversity in all the communities studied (table VI.26). Particularly species of the *Arrhenatherion, Cynosurion* and *Mesobromion* increased, being small species and species indicating agricultural practices. This was attributed to the higher frequency of management practices since most lots were cut or mulched only once a year and late in summer before the start of the experiments. Mulching once a year, therefore, implied more or less a continuation of former agricultural practices with subsequently little vegetational changes. The period of mulching, however, proved to be very important depending on the community type. Mulching in August resulted in an increase of small species in low productive dry *Mesobromion* communities and a decrease in the highly productive moist *Arrhenatheretum elatioris*. These species did increase with a June mulching regime in the *Arrhenateretum* since the dominance of tall-growing grasses was prevented. Mulching in August resulted in the highest species-diversity in wet *Molinio-Arrhenatheretea* communities due to the late start of the growing season.

The rapid vegetational changes in *Arrhenatheretum* communities between two mulching periods resulted in a vegetation similar to laissez-faire with mulching every second year. This regime, however, resulted in the persistance of *Festuco-Brometea* and *Nardo-Cullunetea* communities, as does mulching every year (Schiefer, 1981).

VI.E.2.d. Hay-making

Hay-making twice a year resulted in an increase of the species diversity in this study. This is in agreement with the findings of many other authors. Kapfer (1987) mentioned a constant number of species. Oomes & Mooi (1981; 1985) also reported a constant number of species in cases where the first cut was done relatively early in the season (Table VI.27).

The results in hay-making three times a year seem to be contradictory (Table VI.28). Kapfer (1987) found a constant number of species starting from a fertilized hayfield which implies more or less continuation of the former agricultural practices. Dierschke (1985), however, found an increasing species diversity, but he started with an abandoned hayfield.

Hay-making once a year in July resulted in a stabilization or an increase of the species diversity in our study areas. This is in agreement with the results of other authors (Table VI.28).

If the annual cut for hay-making was performed later than mid-August the number of species also stabilized or increased in the communities in this study. Most other authors also found an increase or stabilization of the species diversity, with the exception of Kapfer (1987) in the *Angelico-Cirsietum* and the *Lolio-Cynosuretum* (Table VI.28).

It seems that the timing of the cutting is important. In the relatively low productive Loefvledder area (400 gdw $\cdot$ m^{-2}) the site which was cut in September contained more species than that cut in July, whereas the reverse was true in the high productive Lammertshem area (800 gdw $\cdot$ m^{-2}). A large standing crop in the summer period apparently hampers many species. Rowell et al. (1985) also found that the species-richness increased within three years with July cutting in a very productive stand of *Calamagrostis canescens* and *Filipendula ulmaria* in the United Kingdom, whereas it did not change with September cutting. This was attributed to the strong decline of *Phalaris arundinacea* in the site which was cut in July.

The species diversity decreased in most of the communities studied with hay-making every second year and remained constant in the *Festuca/Agrostis capillaris* community. Oomes & Mooi

Table VI.26. Changes in number of species after different periods of mulching twice a year in various plant communities with various management prior to the mulching experiments.

Country	Management prior to the experiment	Plant community	Management regime				Changes in number of species per unit area + : increase = : constant – : decrease	Assessment
			Number of years	Practice	Frequency	Period		
Germany	hayfield	Arrhenatherion elatioris	4 (8)	mulching	2	June/August	+ 44 → 44(50). 25 m^{-2}	Schiefer 1981,1983; Schreiber & Schiefer 1985
	unfertilized pasture	Gentiano-Koelerietum	4 (8)	mulching	2	June/August	+ 43 → 49(54). 25 m^{-2}	,,
	hayfield	Mesobrometum	4 (8)	mulching	2	June/August	+ 53 → 54(59). 25 m^{-2}	,,
	unfertilized hayfield	Arrhenatherion	3 (7)	mulching	2	June/August	+ 38 → 40(48). 25 m^{-2}	,,
	5 years abandoned	Festuco-Brometea	6 (10)	mulching	2	June/Augsut	+ 24 → 32(34). 25 m^{-2}	,,
	pasture	Cynosurion cristati	4 (8)	mulching	2	June/August	= 39 → 42(42). 25 m^{-2}	,,
	hayfield	Mol. Arrh. Bromus erectus	4	mulching	2	June/August	+ 33 → 38 . 25 m^{-2}	,,
	hayfield	Mol. Arrh. Carex acutiformis	4	mulching	2	June/August	= 35 → 36 . 25 m^{-2}	,,
	unfertilized hayfield	Calthion/Filipendulion	4 (8)	mulching	2	June/August	+ 37 → 37(41). 25 m^{-2}	,,
	hayfield	Calthion/Juncion acutiflori	4 (9)	mulching	2	June/August	= 27 → 32(29). 25 m^{-2}	,,
	pasture	Lolio-Cynosuretum	5	mulching	2	July/October	+ 9 → 16 . 5 m^{-2}	Rosenthal et al. 1985

(1981) also observed a decrease, but Dierschke (1985) mentioned an increase (Table VI.28). The latter started with an abandoned hayfield.

VI.E.2.e. Comparison of the management regimes
All or nearly all practices caused an increase in the number of species in some communities (Table VI.29). The management experiment described by Schmidt (1981, 1984) started with the bare soil of a sterilized arable field. Any practice would, therefore, result in an increased number of species. Rosenthal (1984) and Oomes & Mooi (1985) started with a hay pasture where fertilizer applications had been carried out. The *Mesobromion* communities studied by Dierschke (1985), Green (1980), Wells (1980) and the *Festuco-Brometea* (Schiefer, 1981) were all abandoned shortly before or a little longer before the start of the management experiments. All other plant communities revealed different reactions upon the different management regimes imposed with respect to the increase, maintenance or decrease of species diversity.

It is apparent that the plant community involved and/or the management practices prior to the start of the experiment determines whether a particular management regime results in an increase of the species diversity, maintenance of the number of species or even a decrease in the number of species. Compare, for example, the results of August mulching in an *Arrhenatherion elatioris* (Schiefer, 1981), in a *Molinietum* (Briemle, 1985) and in a *Calthion/Filipendulion* (Schiefer, 1981).

A problem in ranking the different management practices with respect to their impact on species diversity is not only the diverging initial plant community, but also the different numbers of years the results of the experiments have been studied. Unfortunately, some experiments were only carried out for a period of a few years. It is obvious that different conclusions can be drawn after 8 years than after 4 years (cf. Schiefer, 1981; Schreiber & Schiefer, 1985). This is in agreement with the results found in the present study. Moreover, not all management practices were carried out in all the plant communities studied. Unfortunately no comparison was made between return and removal of the cuttings in the initial management experiment done by Schiefer (1981), but this omission was recently rectified (Schiefer, 1983; Schreiber & Schiefer, 1985).

A general trend can nevertheless be seen in the sequence of the management regimes in all the plant communities with respect to an increasing number of species. This sequence is hay-making twice>hay-making once>mulching twice>mulching once>hay-making every second year>mulching every second year> abandoning. Only the effects of burning vary too widely to give this management practice a position in the sequence.

VI.E.2.f. Cutting regimes and species composition
Although the different initial plant communities hamper the comparison of the effects of cutting regimes on the species composition, it is possible to draw some general conclusions. In our study area July hay-making every second year in a way resembled mulching. This was because litter could accumulate every second year when no hay-making took place. This promoted the high cover degree of *Holcus lanatus, Valeriana officinalis* and *Phalaris arundinacea* in the *Valeriana officinalis/Phalaris arundinacea* community and *Festuca rubra, Agrostis capillaris* and *Holcus lanatus* in the *Festuca/Agrostis capillaris* community. Oomes & Mooi (1981) obtained similar results since *Filipendula ulmaria* increased with June hay-making every second year in *Arrhenatherion elatioris* in the Netherlands, whereas the small species *Anthoxanthum odoratum, Plantago lanceolata, Cardamine pratensis* and *Ranunculus acris* decreased. Krüsi (1981), however, found no differences between June hay-making every year or every second year in *Mesobrometum* communities in Switzerland. Dierschke (1985) mentioned no differences between hay-making every second year and mulching every year in highly productive *Mesobromion* communities in Germany. Hay-making every second year, however, resembled hay-making every year in low productive communities.

Oomes & Mooi (1981) studied different cutting regimes in an *Arrhenatherion elatioris* in the Netherlands for a period of eight years. In comparing cutting in June and in September they conclud-

Table VI.27. Changes in number of species after different periods of hay-making twice a year (2) or three times a year (3) in various plant communities with various management prior to the hay-making experiments.

Country	Management prior to the experiment	Plant community	Management regime: Number of years	Practice	Frequency	Period	Changes in number of species per unit area (+ : increase; = : constant; − : decrease)	Assessment
The Netherlands	hayfield	Ranunculus/Caltha community	6	hay-making	2	July/September	+ 16 → 19 . 4 m^{-2}	this study
	hay-pasture	Agrostis/Poa community	11	hay-making	2	July/September	= 13 → 14 . 4 m^{-2}	,,
	hay-pasture	Arrhenatherion elatioris	8	hay-making	2	May/September	+ 52 → 55 . 100 m^{-2}	Oomes & Mooi 1981
	hay-pasture	Arrhenatherion elatioris	8	hay-making	2	June/September	= 52 → 52 . 100 m^{-2}	,,
	pasture	Poo-Lolietum	12	hay-making	2	June/October	+ 24 → 36 . 400 m^{-2}	Oomes & Mooi 1985
	pasture	Poo-Lolietum	12	hay-making	2	July/October	= 31 → 31 . 400 m^{-2}	,,
Germany	pasture	Lolio-Cynosuretum	5	hay-making	2	July/October	+ 9 → 20 . 5 m^{-2}	Rosenthal et al. 1985
	abandoned	Mesobromion	12	hay-making	2	June/September	+ 19 → 32 . 4 m^{-2}	Dierschke 1985
	hayfield	Lolio-Cynosuretum	4	hay-making	2	July/October	= 29 → 29 . 16 m^{-2}	Kapfer 1987
United Kingdom	abandoned	Mesobromion	c. 12	hay-making	2	winter	+ 6 → c.35 *)	Green 1980
	abandoned	Mesobromion	c. 12	hay-making	2	summer	+ 6 → c.40 *)	,,
Germany	hayfield	Angelico-Cirsietum	4	hay-making	3	July/Aug/Oct	= 31 → 30 . 20 m^{-2}	Kapfer 1987
	hayfield	Lolio-Cynosuretum	4	hay-making	3	June/Aug/Oct	= 32 → 30 . 20 m^{-2}	,,
	abandoned	Mesobromion	12	hay-making	3	June/July/Sept	+ 19 → 31 . 4 m^{-2}	Dierschke 1985

* Point quadrat analysis.

Table VI.28. Changes in number of species after different periods of hay-making before and after August (1) or every second year (½) in various plant communities with various management prior to the hay-making experiments.

Country	Management prior to the experiment	Plant community	Management regime: Number of years	Management regime: Practice	Management regime: Frequency	Management regime: Period	Changes in number of species per unit area (+ : increase; = : constant; – : decrease)	Assessment
The Netherlands	hayfield	Ranunculus/Caltha community	6	hay-making	1	July	= 17 → 17 . 4 m^{-2}	this study
	unfertilized hayfield	Valeriana/Phalaris community	6	hay-making	1	July	+ 8 → 12 . 4 m^{-2}	,,
	hay-pasture	Agrostis/Poa community	11	hay-making	1	July	+ 11 → 13 . 4 m^{-2}	,,
	unfertilized hay-pasture	Festuca/Agrostis community	11	hay-making	1	July	= 13 → 14 . 4 m^{-2}	,,
	hay-pasture	Arrhenatherion elatioris	8	hay-making	1	May	= 52 → 55 . 100 m^{-2}	Oomes & Mooi 1981
	hay-pasture	Arrhenatherion elatioris	8	hay-making	1	June	= 52 → 53. 100 m^{-2}	,,
	pasture	Poo-Lolietum	12	hay-making	1	June	+ 27 → 31 . 400 m^{-2}	Oomes & Mooi 1985
	pasture	Poo-Lolietum	12	hay-making	1	July	= 28 → 29 . 400 m^{-2}	,,
Germany	pasture	Lolio-Cynosuretum	5	hay-making	1	July	+ 8 → 21 . 5 m^{-2}	Rosenthal 1984
	abandoned	Mesobromion	12	hay-making	1	June	+ 22 → 30 . 4 m^{-2}	Dierschke 1985
The Netherlands	hayfield	Ranunculus/Caltha community	6	hay-making	1	September	= 15 → 13 . 4 m^{-2}	this study
	hay-pasture	Agrostis/Poa community	11	hay-making	1	September	+ 12 → 16. 4 m^{-2}	,,
	pasture	Arrhenatheretum elatioris	11	hay-making	1	August	+ 28 → 40 . 2 m^{-2}	Willems 1983A
	hay-pasture	Arrhenatherion elatioris	8	hay-making	1	August	= 52 → 53 . 100 m^{-2}	Oomes & Mooi 1981
	pasture	Poo-Lolietum	12	hay-making	1	October	+ 29 → 35 . 400 m^{-2}	Oomes & Mooi 1985
Germany	16 years abandoned	Molinietum	5	hay-making	1	September	= 34 → 35 . 25 m^{-2}	Briemle 1985
	16 years abandoned	Filipenduletum	5	hay-making	1	September	+ 31 → 40 . 25 m^{-2}	,,
	hayfield	Angelico-Cirsietum	4	hay-making	1	October	– 29 → 25 . 20 m^{-2}	Kapfer 1987
	hayfield	Lolio-Cynosuretum	4	hay-making	1	October	– 36 → 28 . 20 m^{-2}	,,
	hayfield	Lolio-Cynosuretum	4	hay-making	1	October	+ 35 → 45. 16 m^{-2}	,,
	unfertilized hayfield	Molinietum	4	hay-making	1	August	= 31 → 30 . 4 m^{-2}	Kapfer & Pfadenhauer 1986
	hayfield	Molinietum	4	hay-making	1	October	= 25 → 23 . 4 m^{-2}	,,
United Kingdom	abandoned	Mesobromion	c. 12	hay-making	1	winter	+ 6 → c. 40 *	Green 1980
The Netherlands	hayfield	Ranunculus/Caltha community	6	hay-making	½	September	– 14 → 9 . 4 m^{-2}	this study
	unfertilized hayfield	Valeriana/Phalaris community	6	hay-making	½	July	– 17 → 13 . 4 m^{-2}	,,
	hay-pasture	Agrostis/Poa community	11	hay-making	½	September	– 12 → 9 . 4 m^{-2}	,,
	unfertilized hay-pasture	Festuca/Agrostis community	11	hay-making	½	July	= 15 → 16 . 4 m^{-2}	,,
	hay-pasture	Arrhenatherion elatioris	8	hay-making	½	June	= 52 → 49 . 100 m^{-2}	Oomes & Mooi 1981
Germany	abandoned	Mesobromion	12	hay-making	½	August	+ 22 → 32 . 4 m^{-2}	Dierschke 1985

* Point quadrat analysis.

Table VI.29. Ranking of management practices on a similar substrate with respect to the changes in the number of species. Management practice H: hay-making, M: mulching, B: burning, A: abandoned. Frequency of practice 1: once a year, 2: twice a year, 3: three times a year, $^1/_2$: every second year. The months in which management practices were carried out are only indicated if the particular practice was carried out in different periods of the year.

Management prior to the experiment	Plant community	Number of species: Increase	Constant	Decrease	Assessment
Sterilized arable field	Molinio-Arrhenatheretea	Oct H 1 > May H 1 > M 1 > H 2 > A			Schmidt 1981, 1985
Pasture	Lolio - Cynosuretum	H 1 > H 2 > M 2 > A > M 1			Rosenthal 1984, 1985
Abandoned	Mesobromion	H 2 > H 3 > H ½ > H 1 / M 1 > A			Dierschke 1985
Pasture	Poo - Lolietum	June + Oct H 2 > Oct H 1 > June H 1 >	July H 1 / July + Oct H 2		Oomes & Mooi 1985
Abandoned	Mesobromion	winter H 2 > summer H 2 >	A		Green 1980
Abandoned	Mesobromion		May H 1/June H 1/July H 1/May M 1/June M 1/July M 1		Wells 1971
5 years abandoned	Festuco-Brometea	M 2 > B > M 1 > M ½ >	A		Schiefer 1981, Schreiber & Schiefer 1985
Unfertilized pasture	Gentiano-Koelerietum	M 2 / M 1 >	M ½ / A / B		,, ,,
Hayfield	Calthion/Juncion acutiflori	M 2 / A >	M 1 / M ½		,, ,,
16 years abandoned	Filipenduletum	H 1 > M 1 >	>	B	Briemle 1985
Hay-pasture	Agrostis/Poa community	Sept H 1 > B > July H 1 / H 2 >	M ½ >	H ½	This study
Hay-pasture	Arrhenatheretum elatioris	H 1 >	>	A	Willems 1983A
Hay-pasture	Arrhenatherion elatoris	May H 1 / June H 1 / May + Sept H 2 >	> H 1 / June + Sept. H 2 >	H ½ > A	Oomes & Mooi 1985
Hayfield	Mesobrometum	M 1 >	> M 2 >	A	Schiefer 1981, Schreiber & Schiefer 1985
Pasture	Cynosurion cristati	M 2 >	> M ½ >	A	,, ,,
Hayfield	Mol. Arrh. Bromus erectus	M 2 >	> M 1 >	M ½ > A	,, ,,
Unfertilized hayfield	Calthion / Filipendulion	M 1 >	> M 2 / A >	M ½ > B	,, ,,
Hayfield	Lolio-Cynosuretum		H 2 >	H 1	Kapfer 1987
Hayfield	Lolio-Cynosuretum		H 2 >	H 1	,,
Hayfield	Angelico-Cirsietum		H 3 >	H 1	,,
Unfertilized hayfield	Valeriana/Phalaris community	H 1 >	>	M > H ½ > M ½	This study
Hayfield	Ranunculus/Caltha community	H 2 >	> July H 1 >	Sept H 1 > B/H ½ > M ½/A	,,
Unfertilized hay-pasture	Festuca/Agrostis community		H 1 / H ½ >	M 1 > M ½	,,
Hayfield	Mol. Arrh. Carex acutiformis		M 2 / A / M ½ >	M 1	Schiefer 1981, Schreiber & Schiefer 1985
16 years abandoned	Molinietum	H 1 > M 1 >	>	B	Briemle 1985
Unfertilized hayfield	Arrhenatherion elatioris		M 2 / B >	M 1 > A / M ½	Schiefer 1981, Schreiber & Schiefer 1985
Hayfield	Arrhenatherion elatioris		M 2 / B >	M 1 > A	,, ,,

ed that some lower growing species only persisted or spread in the June mowing regime (*Plantago lanceolata, Lolium perenne, Bellis perennis, Ranunculus repens, Taraxacum officinale*). *Plantago lanceolata, Bromus hordeaceus* and *Lychnis flos-cuculi* increased markedly in the regime including cutting twice. *Filipendula ulmaria* increased markedly during the first three years with September hay-making.

In order to regenerate litter fen communities in south Germany, Kapfer (1987) studied cutting regimes involving hay-making in June, August and October and a single cut in October. He started with an *Angelico-Cirsietum*/*Lolio-Cynosuretum* and a poor *Lolio-Cynosuretum*. Species of the *Lolio-Cynosuretum* decreased in both management regimes and species indicating less nutrient-rich soil conditions increased. Tall growing forbs like *Cirsium oleraceum, Alopecurus pratensis* increased particularly in the October cut regimes as did *Festuca rubra*.

Krüsi (1981) compared different cutting regimes in *Mesobrometum* communities in Switzerland over a three years period. The June regime seemed to be advantageous for medium-sized species but unfavourable for tall-growing species. The October regime did not show these differences. Egloff (1986) also found in *Molinion* communities in Switzerland that tall forbs were favoured in September cut sites as compared with cuts in June or in July after two years.

It is apparent from the afore-mentioned data that late cutting regimes differentiate themselves from earlier cutting regimes by some increase of tall-growing species. The effects on species diversity are not always the same and certainly depend on the community in which the management practices are carried out. The same is the case for the frequency of hay-making.

CHAPTER VII

The salt marsh vegetation

VII.A. Introduction

This chapter covers the succession in salt marsh plant communities subjected to different management practices viz. renewed cattle-grazing (with mowing as a comparison) thirteen years after abandoning, continued abandoning, ceasing grazing and continued grazing. The questions to be answered are (i) how does species diversity and equitability change after the resumption of grazing and after abandoning, (ii) can certain vegetational trends be quantified and (iii) can a successional scheme be defined after abandoning or renewed grazing of salt marsh plant communities including a time-scale?

In 1972 grazing was resumed in part of the study area thirteen years after the cessation (see section IV.C). The effects of renewed grazing and continued abandoning were compared using both sequential mapping in (1971 before resumption of grazing, 1976 and 1981; section D), and by annual recording of permanent plots (section C). Moreover, a number of cutting regimes (cutting once in June or August and cutting twice in both June and August) were applied to different plant communities.

Since the effects of ceasing grazing in 1958 had not been monitored, in the autumn of 1973 two exclosures were established in the continuously grazed part of the study area. The succession was recorded by sequential mapping in 1973, 1978 and 1985 and by annual recording of permanent plots.

Classification of the plant communities was done by clustering nearly 1000 relevés. The syntaxonomical position of the coena and their synecology could be discussed after classification (section B).

Vegetational trends were derived from permanent plot data and quantified by the dissimilarities between each year and the first year (section C). Species diversity and cover percentage increase or decrease under grazing, mowing and abandoning, in different plant communities, were compared with findings elsewhere (section E).

VII.B. Vegetation classification, syntaxonomy and synecology

VII.B.1. Vegetation classification

Nearly 1000 relevés from the Dutch Wadden Sea area were used to classify the plant communities of the salt marshes and their low dunes (Boersma, 1988). The relevés were not only derived from Schiermonnikoog but also from other areas in order to get a degree of generalization. C. 760 relevés were used from Schiermonnikoog, parts of which had previously been published in internal reports (Van Der Duim, 1973; Prins, 1974; Schwab & Schwab-Vos, 1974; Norder & Ruyter, 1977; Kasteleijn, 1985). C. 120 relevés were used from the grazed salt marsh on the island of Terschelling (Schwab & Schwab-Vos, 1974; Heinemeijer, 1982) and c. 120 relevés were used from the salt marshes and land reclamation fields along the Groningen mainland coast (Dijkema, 1975).

The results are presented in the dendrogram in Fig. VII.1. Three main clusters could easily be recognised at a dissimilarity level of c. 90% representing the lower salt marsh (coena 1.1 to 1.8), the mid- and upper salt marsh (coena 2.1 to 2.16) and the low dunes (coena 3.1 to 3.3). The ranking of the coenon numbers approximately reflects their positions relative to MHW. This is shown in detail in Tables VII.1–4. Only the *Elymus repens* dominated stands (coenon 2.12) and the prolonged inundated *Scirpus maritimus* community (coenon 1.9) apparently indicate such deviating circumstances that these cannot be assigned to the main clusters.

Twelve clusters of the coena found in the study area can be recognized at the dissimilarity level of 63%. The lower salt marsh main cluster is divided into three clusters, the mid- and upper salt marsh main cluster contains seven clusters and the low dune main cluster is not subdivided. This dissimilarity level is chosen for practical reasons to represent vegetation maps at a scale of 1 : 7500 (see Fig. VII.13). More detailed information could be incorporated during field surveys which resulted in vegetation maps, at a scale of 1 : 1000 or larger, namely at the dissimilarity level of c. 55%. More details can, therefore,be depicted in the exclosure maps at a scale of 1 : 500 in Fig. VII.16, and in the transition schemes in Table VII.11 and 12. Some clusters were lumped because these could not be discerned in the field situation, despite their higher dissimilarity level. Most clusters containing one relevé were not taken as a coenon, because these were not recognized during vegetation mapping.

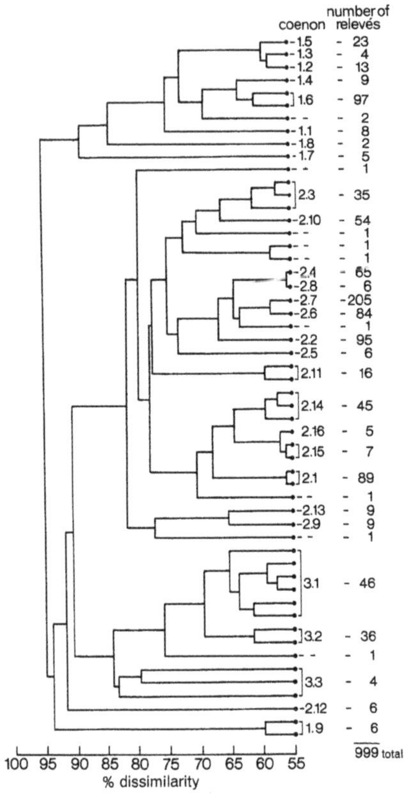

Fig. VII.1. Simplified (all dissimilarity percentages below 55% are not distinguished) dendrogram of salt marsh relevés from Schiermonnikoog and additional relevés from Terschelling and mainland salt marshes. (after Boersma, 1988).

VII.B.2. *Mud flats and lower salt marsh (Table VII.1)*

The coena 1.1 and 1.2 can be assigned to the *Salicornietum strictae* (Table VII.1). Coenon 1.1 features a facies of *Spartina anglica* and occurs on the tidal flats, whereas coenon 1.2 is a *Salicornietum strictae* and is found within the range of daily inundations and sometimes in wet depressions in the lower salt marsh.

According to Westhoff & Den Held (1969) the coena 1.4 and 1.5 can be considered as *Puccinellietum maritimae*, variant with *Aster tripolium* and variant typicum, respectively. The variant typicum can be seen as an initial phase of the *Puccinellietum maritimae* with a high annual frequency of inundations at the transition from higher tidal flats to the lower salt marsh, whereas the variant with *Aster tripolium* occurs on the lower salt marsh. Coenon 1.3 represents the grazed counterpart of coenon 1.4 and should be classified as a BC *Puccinellia maritima* [*Puccinellion maritimae*] (Basal Geselschaft sensu Kopecký & Hejný (1974, 1978)). It can be observed at the sides of creeks and in small creeks which are regularly flooded and continuously accreting sediments.

Coenon 1.6 can be classified as a *Plantagini- Limonietum* and is mostly found in the lower salt marsh depressions behind the cliffed levee bordering the tidal flats. In some ungrazed areas *Halimione portulacoides* can reach an abundance. Although Beeftink (1965) and Westhoff & Den Held (1969) mention this coenon characteristic for ungrazed salt marsh, it is often found in grazed areas in the Schiermonnikoog salt marsh, possibly being a transitional phase after the resumption of grazing.

Coenon 1.7 is assigned to the *Halimionetum portulacoidis*. It can be observed on well drained sides of creeks and in small silted creeks in the low salt marsh and is restricted to ungrazed areas.

Coenon 1.8 should be classified as a BC *Puccinellia maritima* [*Puccinellio-Spergularion*] due to the alliance character species *Spergularia salina*. It is found several dm higher above MHW than the other lower salt marsh communities in depressions cut off from direct tidal influences. Stagnant water during the winter months facilitates the destruction of the turf during the summer periods in grazed areas.

Coenon 1.9 can be classified as a *Halo-Scirpetum maritimi*, being incomplete, possibly due to grazing. The abundance of *Scirpus maritimus*, in particular, is lower than described by Beeftink (1977A). It is found in depressions cut off from the direct tidal influences and is, therefore, inundated with fresh or salt water for long periods of the year. This coenon builds mosaics with coenon 1.8 mentioned above.

VII.B.3. *Higher salt marsh (Table VII.2)*

The higher salt marsh coena can all be assigned to the *Armerion maritimae* with the exception of coenon 2.1 belonging to the *Angelicion litoralis* (Table VII.2).

Coenon 2.1 is classified as a *Atriplici-Agropyretum pungentis* (*Agropyron pungens* = *Elymus pycnanthus*). It is found particularly along creeks, but also somewhat higher on the salt marsh and is restricted to ungrazed areas. The dense canopy of *Elymus pycnanthus* explains the deviating position in the clustering with respect to the remaining higher salt marsh coena. It certainly decreases species diversity by competitive exclusion and prevents the establishment of a number of species. If grazed, other species can co-occur with *Elymus pycnanthus* and the species is found in a number of coena (2.8, 2.15, 2.16, 3.1, 3.3). This range of communities with a high *Elymus pycnanthus* coverage can be attributed to the remnants of the communities which *Elymus pycnanthus* invades after the deposition of drift material (Beeftink, 1965) or by vegetative spread by stolons once they were established.

Coenon 2.2 can be assigned to the *Artemisietum maritimae*. *Juncus gerardii* and *Triglochin maritima* are found in the grazed areas while the *Plantago maritima* cover is larger in the ungrazed area and *Artemisia maritima* can become dominant compared to the grazed areas. The coenon is found on the levee along the tidal flats.

The coena 2.3 and 2.4 variant with *Juncus maritimus* which are assigned to the lower occurring *Juncetum gerardii* typicum, find their counterparts in the coena 2.6 variant with *Juncus gerardii* and 2.8 variant with *Juncus maritimus*, respectively, in the somewhat higher occurring *Juncetum gerardii leontodontetosum autumnalis*. Although *Juncus maritimus* is the dominating species the variants with *Juncus maritimus* are not classified as a *Juncetum maritimi* Bilik 1956 (Runge, 1980) since other character species viz. *Oenanthe lachenalii* and *Apium graveolens* are absent. Sykora (1983) mentions that *Juncus maritimus* is by no means characteristic for any of the *Lolio-Potentillion anserinae* communities and that the *Juncus maritimus* dominated vegetation belongs to the *Armerion maritimae*.

According to Runge (1980) and Sykora (1983) *Juncus maritimus* dominated stands are hardly grazed or are not grazed at all. Although *Juncus maritimus* forms some kind of protection for a number of species thriving within the coarse *Juncus* bushes, the stands are certainly grazed which is in contradiction to the impressions of the afore-mentioned authors. This is probably due to the high stocking rate (1.6 animals $\cdot$ ha^{-1}) in the Schiermonnikoog salt marsh (Looyen, 1983). *Festuca rubra*, *Juncus gerardii* and *Agrostis stolonifera* within the *Juncus* bushes and to a lesser degree *Juncus maritimus* itself are grazed, which, combined with trampling, results in thin *Juncus maritimus* communities.

The *Juncus maritimus* variant of the *Juncetum gerardii* typicum is observed in salt marsh depressions and along creek bank levees, whereas the variant of the *Juncetum gerardii leontodontetosum autumnalis* is found on the gradient to the low dunes (Photo VII.1). The coena 2.3, 2.6 and 2.7 can occur in relatively low parts of the higher salt marsh between the low dunes or in the higher salt marsh at brackish transitions to the dunes, but the preponderance of *Glauco-Puccinellietalia* and *Armerion maritimae* species, like *Festuca rubra, Juncus gerardii, Glaux maritima, Plantago maritima, Armeria maritima* and *Triglochin maritima* justifies their classification into the *Juncetum gerardii*.

According to Sykora (1982; 1983) coenon 2.9 can be assigned to the *Ononido-Caricetum distantis* Runge 1966 within the alliance *Armerion maritimae*. Sykora (1983) states that this association is assigned to the alliance *Armerion* and not to the *Lolio-Potentillion* because of the presence of only a few *Lolio-Potentillion* species and this is confirmed by our data.

VII.B.4. Upper salt marsh and feet of low dunes (Table VII.3)

Coenon 2.11 cannot be classified easily due to the lack of abundantly occurring character species of lower syntaxonomical levels. The occurrence of the *Plantaginetalia* character species *Potentilla anserina, Agrostis stolonifera, Lolium perenne* and *Leontodon autumnalis* perhaps justifies a classification as a BC *Festuca rubra* [*Plantaginetalia*]. This community is found in the

Table VII.1. Synoptic table of the coena of the lower salt marsh. The frequency of occurrence is indicated and the mean coverage (+: 1–5%, a: 6–10%, 1: 11–20%, 2: 21–30% etc.).

Class	Thero-Salicornietea		Asteretea tripolii						
Order	Thero-Salicornietalia		Glauco-Puccinellietalia						
Alliance	Thero-Salicornion		Puccinellion maritimae					Puccinellio-Spergularion salinae	Halo-Scirpion
Association	Salicornietum strictae		Puccinellietum maritimae			Plantagini-Limonietum	Halimionetum portulacoidis		Halo-Scirpetum maritimi
Lower syntaxa/coena	Spartina facies	typicum	typicum	variant with Aster tripolium	Basal community				
Coenon	1.1	1.2	1.5	1.4	1.3	1.6	1.7	1.8	1.9
Number of relevés	8	13	23	9	4	97	5	2	6
Situation above MHW cm	< -15	-35-0	0	25	25	25-35	25	55-65	55-65
Annual frequency of inundations	> 500	350-600	350	130	130	85-130	130	30-40	30-40
Grazing	-	-	-	-	+	+	-	+	+
Suaeda maritima	88/+	15/+	70/+	100/4	100/+	86/+	60/+	100/+	.
Salicornia europaea	88/+	100/a	100/a	89/+	100/+	76/+	60/+	100/+	.
Puccinellia maritima	75/1	62/+	100/a	100/1	100/+	100/4	100/+	100/+	.
Spartinetum townsendii									
Spartina anglica	100/3	69/+	61/+	89/+	.	30/+	20/+	.	.
BC Pucc.mar. [Puccinellion]									
Glaux maritima	.	.	9/+	.	75/+	49/a	.	50/+	83/+
Puccinellietum maritimae and Plantagini-Limonietum									
Aster tripolium	50/+	.	43/+	89/+	.	82/+	20/+	.	.
Spergularia maritima	.	.	39/+	56/+	.	46/+	.	.	.
Plantagini-Limonietum									
Limonium vulgare	.	8/+	35/+	11/+	.	49/a	.	.	.
Plantago maritima	.	.	9/+	11/+	.	49/+	.	.	.
Halimionetum portulacoidis									
Halimione portulacoides	25/+	.	13/+	56/+	.	48/+	100/9	.	.
Festuca rubra	.	.	.	.	.	27/+	60/+	.	.
Puccinellio-Spergularion salinae and Halo-Scirpetum maritimi									
Scirpus maritimus	.	.	.	.	.	.	.	100/+	100/a
Juncus gerardii	.	.	9/+	.	25/+	14/+	.	100/+	67/+
Agrostis stolonifera	.	.	.	.	.	2/+	.	100/+	83/+
Puccinellio-Spergularion salinae									
Spergularia salina	.	.	26/+	11/+	.	33/+	.	100/1	.
Juncus bufonius	.	.	.	.	.	.	.	100/a	.
Atriplex prostrata	.	.	.	44/+	.	22/+	20/+	100/+	.
Other taxa									
Triglochin maritima	.	.	22/+	22/+	.	52/+	.	50/+	17/+
Artemisia maritima	.	.	.	11/+	.	37/+	40/+	.	.

Addenda: taxa occurring in one coenon in less than 20% of the relevés:
1.7 Cochlearia anglica, Juncus maritimus, Armeria maritima, Poa pratensis, Parapholis strigosa; 1.8 Elymus pycnanthus; 1.9 Potentilla anserina, Phragmites australis.

Table VII.2. Synoptic table of the coena of the higher salt marsh. The frequency of occurrence is indicated and the mean coverage (+: 1–5%, a: 6–10%, 1: 1: 11–20%, 2: 21–30% etc.).

Class	Artemisietea vulgaris	Asteretea tripolii						
Order	Convolvuletalia sepium	Glauco-Puccinellietalia						
Alliance	Angelicion litoralis	Armerion maritimae						
Association	Atriplici-Agropyretum pungentis	Artemisietum maritimae	Juncetum gerardii					Ononidó-Caricetum distantis
Subassociation			Juncetum gerardii typicum		Juncetum gerardii leontodontetosum			
				variant with Juncus maritimus	variant with Juncus gerardii		variant with Juncus maritimus	
Coenon	2.1	2.2	2.3	2.4	2.6	2.7	2.8	2.9
Number of relevés	89	95	35	65	84	205	6	9
Situation above MHW cm	55-75	35-65	55	45-85	75.95	65-120	65-95	55-80
Annual frequency of inundations	20-40	30-160	40	15-60	12-20	5-30	12-30	18-40
Grazing	-	+/-	+	+/-	+	+	+/-	+
Festuca rubra	85/1	100/1	63/+	98/1	100/2	100/4	100/1	100/1
Juncus gerardii	9/+	49/+	100/3	42/+	83/a	42/+	50/+	44/+
Glaux maritima	16/+	38/+	100/+	85/a	94/+	36/+	100/+	.
Plantago maritima	19/+	61/+	63/a	51/+	94/1	82/a	67/+	22/+
Agrostis stolonifera	54/+	33/+	66/+	91/a	63/+	89/1	67/a	100/1
Atriplici-Agropyretum and Juncetum gerardii variants with Elymus pycnanthus								
Elymus pycnanthus	100/3	8/+	11/+	12/+	1/+	22/+	83/+	22/+
Artemisietum maritimae								
Artemisia maritima	57/+	100/3	37/+	55/+	82/+	52/+	17/+	.
Juncetum gerardii leontodontetosum								
Armeria maritima	15/+	16/+	9/+	11/+	57/+	78/+	17/+	11/+
Juncetum gerardii variants with Juncus gerardii								
Triglochin maritima	6/+	49/+	94/+	51/+	85/+	20/+	17/+	11/+
Puccinellia maritima	.	24/+	83/a	14/+	54/a	8/+	.	.
Juncetum gerardii variants with Juncus maritimus								
Juncus maritimus	54/+	9/+	9/+	100/1	1/+	3/+	100/1	.
Juncetum gerardii variants								
Sedum acre	.	.	.	.	.	1/+	.	.
Festuca arundinacea	.	.	.	.	.	2/+	33/+	.
Leontodon autumnalis	9/+	.	.	5/+	1/+	6/+	50/+	.
Ononido-Caricetum distantis and Agrostio-Trifolietum fragiferi								
Leontodon saxatilis	.	.	.	.	.	.	.	67/+
Trifolium fragiferum	.	.	.	.	.	4/+	.	33/+
Sagina maritima	.	.	.	.	.		.	56/+
Phragmites australis	.	.	.	2/+	.	.	.	78/+
Odontites verna serotina	1/+	.	.	.	1/+	3/+	.	33/+
Centaurium littorale	.	.	.	.	.	1/+	.	78/+
Ononido-Caricetum distantis								
Plantago coronopus	7/+	.	.	.	.	22/+	.	100/+
Trifolium repens	19/+	.	9/+	2/+	2/+	56/+	50/+	89/a
Ononis spinosa	.	.	.	.	.	.	.	100/2
Lotus corniculatus	8/+	.	.	3/+	.	9/+	33/+	44/+
Carex distans	10/+	.	.	.	6/+	27/+	.	89/+
Trifolium campestre	.	.	.	.	.	.	.	11/+
Sagina nodosa	.	.	.	.	.	.	.	22/+
Juncus alpino articulatus atricapillus	.	.	.	.	.	.	.	33/+
Juncus articulatus	.	.	.	.	.	.	.	11/+
Carex trinervis	.	.	.	.	.	.	.	22/+
Carex flacca	.	.	.	.	.	.	.	78/+
Carex nigra	.	.	.	.	.	.	.	33/+
Salix repens	.	.	.	.	.	.	.	100/a
Trifolium pratense	2/+	.	.	.	.	1/+	.	78/a
Holcus lanatus	.	.	.	.	.	1/+	.	89/a
Other taxa								
Potentilla anserina	19/+	1/+	20/+	38/+	2/+	28/+	100/1	100/a
Poa pratensis	24/+	2/+	.	23/+	6/+	46/+	100/a	.
Atriplex prostrata	73/+	85/+	69/+	95/+	50/+	31/+	33/+	.
Spartina anglica	.	1/+	.	.	5/+	.	.	.
Polygonum aviculare	.	1/+	.	.	1/+	.	.	.
Atriplex littoralis	2/+	3/+	.	.	1/+	.	.	.
Agrostis capillaris	2/+	.	.	.	.	1/+	.	.
Cerastium semidecandrum	.	.	.	3/+	.	2/+	.	22/+
Stellaria graminea	2/+	.	.	.	.	2/+	33/+	.
Taraxacum spec.	1/+	.	.	.	.	3/+	.	.
Trifolium arvense	1/+	.	.	.	.	1/+	.	.
Bupleurum tenuissimum	3/+	.	.	.	.	4/+	.	.
Plantago major	.	.	.	.	.	1/+	.	.
Poa annua	.	.	.	.	.	5/+	.	.
Carex arenaria	.	.	.	.	.	1/+	.	22/+
Cirsium arvense	6/+	.	.	.	.	2/+	.	.
Carex extensa	.	.	.	.	1/+	1/+	.	.
Aster tripolium	6/+	30/+	6/+	12/+	44/+	18/+	.	.
Limonium vulgare	8/+	55/+	20/+	12/+	63/+	24/+	.	.
Spergularia maritima	10/+	27/+	14/+	9/+	35/+	19/+	.	.
Rumex acetosella	1/+	.	.	.	.	1/+	.	.
Cerastium fontanum	12/+	.	.	.	.	8/+	.	56/+
Cochlearia anglica	1/+	5/+	.	3/+	12/+	9/+	.	.
Lolium perenne	4/+	15/+	.	.	.	21/+	33/+	.
Spergularia salina	2/+	15/+	3/+	5/+	29/+	4/+	.	.
Elymus repens	1/+	.	.	2/+	.	3/+	.	.
Cochlearia danica	2/+	.	.	3/+	6/+	6/+	.	22/+
Halimione postulacoides	6/+	15/+	3/+	3/+	6/+	6/+	.	.
Salicornia europaea	3/+	13/+	9/+	5/+	25/+	5/+	.	.
Suaeda maritima	4/+	17/+	3/+	2/+	19/+	6/+	.	.

Addenda: taxa occurring in one coenon in less than 20% of the relevés: 2.2 Hordeum secalinum; 2.3 Holcus lanatus, Parapholis strigosa, Rumex crispus, Sonchus oleraceus, Trifolium pratense; 2.7 Aira praecox, Calamagrostis epigejos, Carex oederi pulchella, Festuca ovina, Galium palustre, Hydrocotyle vulgaris, Juncus bufonius, Radiola linoides, Ranunculus flammula, Scirpus rufus, Trifolium dubium, Triglochin palustris; 2.9 Ranunculus baudotii, Triglochin palustris, Eleocharis uniglumis, Puccinellia maritima.

Photo VII.1. The grazed salt marsh features stands with *Juncus gerardii* and *Festuca rubra* in the lower parts, and locally *Juncus maritimus* (dark), *Armeria maritima* at the transition to the lower dunes, which are covered by *Ammophila arenaria* (right).

oldest part of the Schiermonnikoog salt marsh protected by the Kooiduinen (Fig. IV.9) and through this a *Agrostio-Trifolietum fragiferi* community could have developed. This part of the Schiermonnikoog salt marsh is, however, treated with fertilizers (100–200 kg Nitrogen Phosphorus, often followed by a second application of 50 kg Ammonium Nitrogen Limestone $\cdot$ ha^{-1}) (Bakker, 1983). This probably causes the dominance of *Lolium perenne* in coenon 2.11 and the occurrence of *Cirsium arvense* and *C. vulgare.*

Coenon 2.12 can be assigned to the *Agropyretum repentis* Nordhagen 1940 in the alliance *Agropyro-Rumicion crispi* Nordhagen 1940 in the order *Chenopodietalia* Braun Blanquet 1936. The ungrazed stands are found in the cliffed upper salt marsh at the mainland coast (Dijkema, 1983). The grazed stands are found in the afore-mentioned salt marsh area which has fertilizers applied which probably cause the large abundance of *Elymus repens*. The co-occurrence of grazed and ungrazed parts of coenon 2.12 is due to floristical and not to structural ressemblance.

Coenon 2.13 is assigned to the *Sagino-maritimae Cochlaerietum danicae.* It is found at the feet of low dunes in mosaics (see also Westhoff & Den Held, 1969) with the *Ononido-Caricetum distantis* (coenon 2.9) mentioned in section VII.B.3.

The coena 2.14, variant inops, 2.15, variant with *Elymus pycnanthus,* and 2.16 variant with *Ammophila arenaria* can be classified as transitions between the *Armerion maritimae* salt marsh communities and the *Galio-Koelerion* dune communities. The absence of character species does not justify a more detailed classification. These communities are found at the feet and on the slopes of higher dunes and on top of lower dunes.

VII.B.5. Dunes (Table VII.4)

The coena 3.1, variant with *Ammophila arenaria,* 3.2 variant with *Festuca ovina,* and 3.3 inops are assigned to transitions of *Saginion maritimae* and *Galio-Koelerion* (Table VII.4). Due to the absence of character species a more detailed classification does not seem to be justified. The variant with *Ammophila arenaria* (coenon 3.1) is found in relatively low areas and can, therefore, be inundated once or twice a year. The other two variants are not inundated annually because of their high situation above MHW.

VII.C. Permanent plots

VII.C.1. Lay-out of permanent plots under hay-making and grazing

Most permanent plots are situated in couples at either side of a fence in order to compare abandoned and renewed grazed areas in detail. Moreover, a number of cutting regimes have been applied to areas of 5 × 8 m from 1971 or 1972 onwards (Photo VII.2). Areas which were considered to feature homogeneous plant communities were selected and cut yearly with a scythe. The swath was removed immediately after cutting.

Table VII.3. Synoptic table of the coena of the upper salt marsh and the feet of low dunes. The frequency of occurrence is indicated and the mean coverage (+: 1–5%, a: 6–10%, 1: 11–20%, 2: 21–30% etc.).

Class	Plantaginetea	Rudereto-Secalietea	Saginetea maritimae			
Order	Plantaginetalia majoris	Chenopodietalia	Saginetalia maritimae			
Alliance		Agropyro-Rumicion crispi	Saginion maritimae	Armeria maritima / Carex arenaria community		
Association		Agropyretum repentis maritimum	Sagino maritimae-Cochlearietum danicae			
Coenon	2.11	2.12	2.13	2.14	2.15	2.16
Number of relevés	16	6	9	45	7	5
Situation above MHW cm	50 - 130	75 - 95	40 - 65	85 - 125	75 - 85	85 - 105
Annual frequency of inundations	7 - 60	12- 20	30 - 80	5 - 15	15 - 20	8 - 15
Grazing	+	+/-	+	+	+	+
Agrostis stolonifera	75/1	17/+	33/+	33/+	100/+	.
Potentilla anserina	50/+	33/+	33/+	18/+	.	.
Trifolium repens	94/1		89/a	16/+	57/+	.
Festuca rubra	100/2	83/a	100/2	100/1	100/+	100/+
BC Festuca rubra - [Plantaginetalia] and Agropyretum repentis						
Cirsium arvense	50/+	33/+	.	.	.	.
Cirsium vulgare	25/+	33/+	.	.	.	.
BC Festuca rubra - [Plantaginetalia]						
Lolium perenne	100/3	17/+	.	2/+	.	20/+
Leontodon autumnalis	56/+	.	.	2/+	14/+	20/+
Agropyretum repentis						
Rumex crispus	.	50/+	.	7/+	.	.
Elymus repens	6/+	100/7	11/+	.	14/+	20/+
Atriplex prostrata	31/+	67/+	.	53/+	.	20/+
Sagina maritimae - Cochlearietum danicae						
Cochlearia danica	6/+	.	89/+	9/+	29/+	.
Sagina maritima	.	.	56/+	.	.	.
Ononis spinosa	.	.	44/+	2/+	.	.
Holcus lanatus	.	.	100/2	11/+	.	.
Trifolium pratense	.	.	89/+	2/+	.	.
Salix repens	.	.	78/+	.	.	.
Festuca ovina	.	.	78/+	.	.	.
Rumex acetosella	.	.	100/+	7/+	29/+	20/+
Cerastium semidecandrum	.	.	100/+	27/+	14/+	.
Luzula campestris	.	.	22/+	.	.	.
Leontodon saxatilis	.	.	100/+	.	14/+	.
Carex trinervis	.	.	33/+	.	.	.
Danthonia decumbens	.	.	33/+	.	.	.
Sagina nodosa	.	.	56/+	.	.	.
Aira praecox	.	.	33/+	.	.	.
Trifolium campestre	.	.	50/+	.	.	.
Trifolium arvense	.	.	89/+	.	.	.
Juncus gerardii	.	.	67/+	9/+	.	.
Sagina maritimae-Cochlearietum danicae and Armeria maritima/ Carex arenaria community						
Poa pratensis	6/+	.	100/a	89/a	100/+	100/+
Lotus corniculatus	.	.	100/a	69/+	100/+	80/+
Plantago lanceolata	.	.	33/+	7/+	.	40/+
Plantago coronopus	.	.	44/+	9/+	57/+	40/+
Carex arenaria	.	.	100/+	20/+	29/+	60/+
Armeria maritima / Carex arenaria community						
Armeria maritima	33/+	.	32/+	67/+	86/+	100/+
Plantago maritima	6/+	17/+	22/+	76/+	86/+	100/+
Variants						
Elymus pycnanthus	.	.	11/+	89/a	100/+	100/+
Ammophila arenaria	.	.	22/+	47/+	.	80/+
Other taxa						
Artemisia maritima	19/+	50/+	11/+	24/+	29/+	20/+
Cerastium fontanum	44/+	50/+	11/+	4/+	57/+	.
Bromus hordeaceus	44/+	.	.	9/+	71/+	60/+
Spergularia maritima	13/+	17/+	.	4/+	.	.
Triglochin maritima	6/+	.	.	.	.	.
Suaeda maritima	6/+	.	.	2/+	.	.
Sonchus arvensis maritima	.	17/+	.	4/+	.	.
Stellaria graminea	.	.	.	11/+	14/+	.
Carex distans	.	.	.	13/+	14/+	.
Agrostis capillaris	.	.	.	9/+	14/+	.
Stellaria media	.	.	.	4/+	.	20/+
Sagina procumbens	.	.	.	2/+	14/+	+

Addenda: taxa occurring in one coenon in less than 20% of the relevés: 2.11 Bromus hordeaceus hordeaceus, Festuca arundinacea, Hordeum marinum, Ammophila arenaria, Plantago lanceolata, Sagina procumbens, Sonchus arvensis maritima, Stellaria media, Glaux maritima; 2.12 Achillea millefolium, Bellis perennis, Polygonum aviculare, Ranunculus repens, Taraxacum spec., Hordeum secalinum, Trifolium fragiferum, Odontites verna serotina, Poa annua, Matricaria recurtita, Aster tripolium; 2.14 Atriplex littoralis, Matricaria maritima; 2.15 Eleocharis uniglumis, Empetrum nigrum, Genista tinctoria, Luzula campestris, Schoenus nigricans, Danthonia decumbens; 2.16 Centaurium littorale, Crepis capillaris, Linum catharticum, Trifolium dubium, Vicia sativa nigra.

Table VII.4. Synoptic table of the coena of the low dunes. The frequency of occurrence is indicated and the mean coverage (+: 1–5%, a: 6–10%, 1: 11–20%, 2: 21–30% etc.).

Class / Order / Alliance	Koelerio-Corynephoretea / Festuco- Sedetalia / [Galio-Koelerion]		
Lower syntaxa/coena	with Ammophila arenaria	with Festuca ovina	with Hieracium umbellatum
Coenon	3.1	3.2	3.3
Number of relevés	46	36	4
Situation above MHW cm	165 - 205	>205	>205
Annual frequency of inundations	0.5 - 2	<0.5	<0.5
Grazing	+	+	+
Carex arenaria	85/a	97/+	75/+
Festuca rubra	85/+	72/+	100/+
Trifolium arvense	39/+	64/+	25/+
Plantago lanceolata	74/+	86/+	75/+
Elymus pycnanthus	72/+	36/+	75/a
Agrostis capillaris	59/a	89/+	75/+
Galium verum maritimum	37/+	67/+	75/+
Lotus corniculatus	87/+	97/+	25/+
Poa pratensis	93/a	100/+	50/+
Cerastium fontanum	33/+	61/+	100/+
Differential species			
Trifolium campestre	33/+	7/+	.
Holcus lanatus	83/+	81/+	.
Stellaria graminea	70/+	47/+	.
Rumex acetosella	80/+	100/+	.
Bromus hordeaceus	39/+	14/+	.
Cerastium semidecandrum	28/+	50/+	.
Agrostis stolonifera	24/+	.	.
Armeria maritima	67/+	28/+	.
Ammophila arenaria	100/a	86/a	.
Aira praecox	15/+	47/+	.
Vicia sativa nigra	.	53/+	.
Vicia lathyroides	.	25/+	.
Cochlearia danica	2/+	44/+	.
Plantago coronopus	17/+	75/+	.
Trifolium repens	35/+	75/+	25/+
Festuca ovina	15/+	100/2	.
Luzula campestris	26/+	69/+	25/+
Hypochaeris radicata	7/+	53/+	.
Trifolium dubium	.	67/+	25/+
Carex trinervis	.	28/+	.
Danthonia decumbens	.	22/+	.
Sagina nodosa	.	25/+	.
Salix repens	.	28/+	.
Hieracium umbellatum	.	11/+	50/+
Sedum acre	7/+	14/+	50/+
Other taxa			
Scleranthus perennis	4/+	6/+	.
Trifolium pratense	7/+	3/+	.
Lolium perenne	11/+	11/+	25/+
Achillea millefolium	2/+	3/+	25/1
Taraxacum spec.	13/+	39/+	50/+
Poa annua	2/+	.	25/+
Leontodon autumnalis	13/+	8/+	.
Leontodon saxatilis	7/+	8/+	.
Senecio jacobaea	2/+	3/+	.
Senecio vulgaris	.	3/+	25/+
Viola curtisii	.	8/+	25/+

Addenda: taxa occurring in one coenon in less than 20% of the relevés: 3.1 Atriplex prostrata, Cirsium arvense, Rumex crispus, Glaux maritima, Hippophaë rhamnoides, Juncus gerardii, Teesdalia nudicaulis, Silene otites, Glaux maritima, Vicia cracca, Koeleria macrantha, Plantago maritima; 3.2. Jasione montana, Polygala vulgaris dunense, Corynephorus canescens, Sagina maritima, Anthoxanthum odoratum, Calamagrostis epigejos, Calluna vulgaris, Empetrum nigrum, Erica tetralix, Genista tinctoria, Linaria vulgaris, Linum catharticum, Polypodium vulgare, Potentilla anserina, Sagina procumbens, Viola canina.

Photo VII.2. Part of the salt marsh study area featuring grazed and abandoned areas and mown sites along the fence.

Since, in salt marshes, the peak standing crop or maximum above-ground biomass is found at the end of the summer period (Joenje & Wolff, 1979), the August cutting regime will remove a maximum amount of standing crop. Moreover, such a regime enables full development and seed setting of most species involved. In order to prevent full development a June hay-making regime was imposed and finally a June + August hay-making regime was imposed to prevent regrowth of the species. The latter more or less simulates the continuous set-back of the turf by grazing animals. All the cutting regimes were compared on either side of the fence in order to find out if turf which had been cut once was kept short by the grazers.

An advantage of the location of the permanent plots is that these may be considered similar in their abiotic environmental conditions. A disadvantage is, however, the small distance between the areas under different experimental regimes. For instance, there is a chance that cattle prefer mown plots, thus resulting in an increased grazing intensity as compared with the uncut plots, or that they are attracted by exclosures resulting in heavier grazing adjacent to the exclosures.

The permanent plots were laid in different plant communities shown in Table VII.5, which also gives the management regimes involved. The permanent plots which were cut in June were recorded during the first half of June, whereas all the other permanent plots were recorded annually during the period between the end of July and the end of August.

Couples of permanent plots were also laid at either side of the exclosure fence erected in 1973 in order to compare continually grazed and recently abandoned areas. The plant communities involved are shown in Table VII.5 the BC *Puccinellia maritima* [*Puccinellion maritimae*] (coenon 1.3) was called 'bare soil' for convenience. The permanent plots were recorded annually during the period between the end of August and the beginning of October.

VII.C.2. Cover changes of abundant species

This section gives a detailed description of a number of transitions between coena indicated in section VII.D.2 in the grazed salt marsh and in section VII.D.3 in the abandoned salt marsh and, therefore, completes the transition schemes. Full details of the initial situation and species composition and

Table VII.5. Number of permanent plots in different regimes in the various plant communities.

Coenon	Regime									
	Continued abandoned	Grazing	June hay-making	June + August hay-making	August hay-making	June hay-making + grazing	June + August hay-making + grazing	August hay-making + grazing	Continued grazing	Exclosure
Puccinellia maritima 'bare soil' (1.3)	–	–	–	–	–	–	–	–	1	1
Puccinellia maritima (1.4)	2	2	–	–	–	–	–	–	1	1
Elymus pycnanthus (2.1)	2	2	1	1	2	1	1	2	–	–
Artemisia maritima (2.2)	2	2	1	1	2	1	1	2	1	1
Juncus gerardii (2.3)	2	2	–	–	1	–	–	1	–	–
Juncus maritimus (2.4)	2	2	1	1	1	1	1	1	2	2
Festuca rubra/ Limonium vulgare (2.6)	1	1	1	1	1	1	1	1	2	2
Festuca rubra/ Armeria maritima (2.7)	1	1	1	1	1	1	1	1	2	2
Ammophila arenaria (3.1)	–	–	–	–	–	–	–	–	1	1

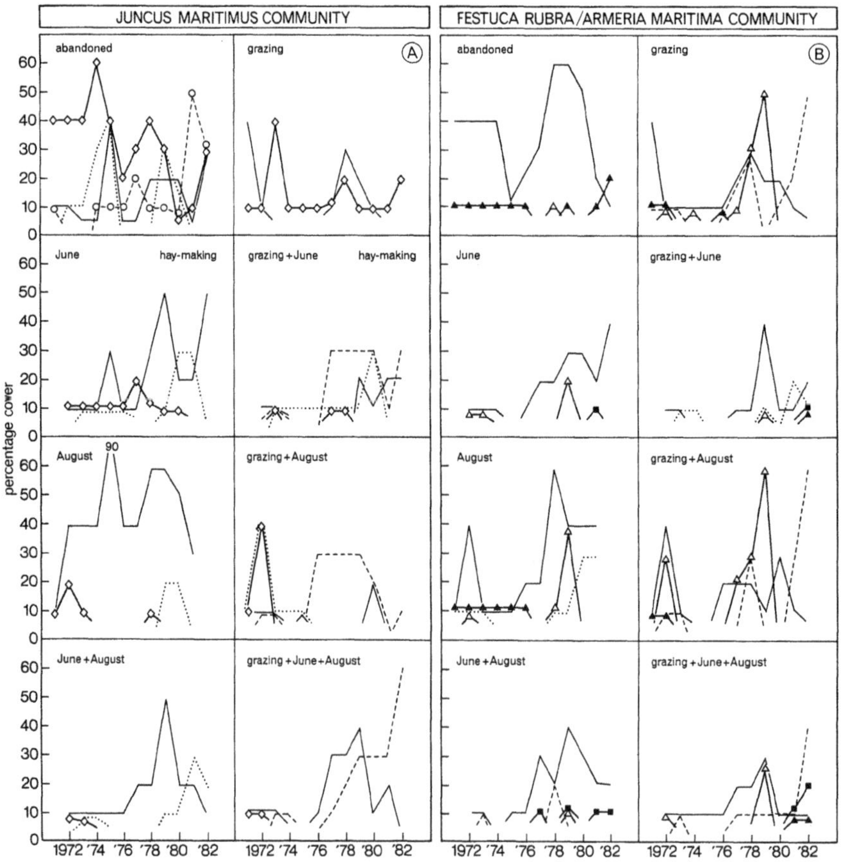

Fig. VII.2A–B. Changes of cover percentages of abundant (coverage≥10%) species under different management regimes in various plant communities.

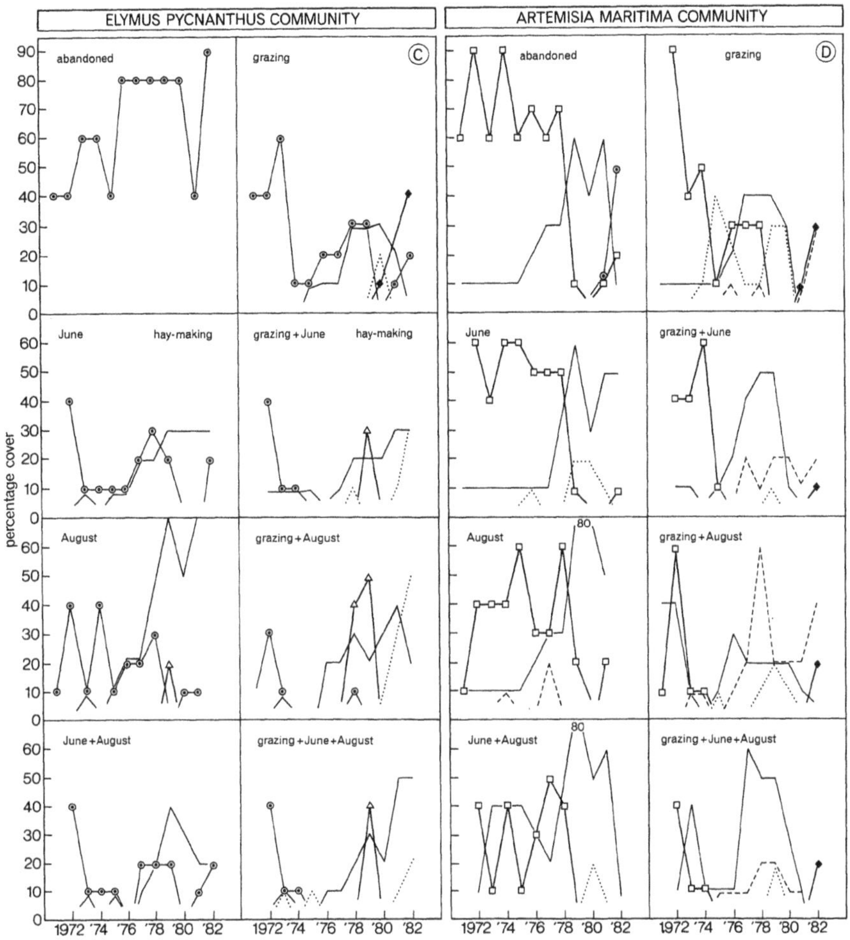

Fig. VII.2C–D. Changes of cover percentages of abundant (coverage≥10%) species under different management regimes in various plant communities.

abundance after ten years are given by Bakker (1985) for all the plant communities which have been discussed. The year-to-year changes in the *Elymus pycnanthus* community have also been published by Bakker (1984).

In the *Juncus maritimus* community (Fig. VII.2A), *Juncus maritimus* maintained itself best in the abandoned situation and next under grazing and June hay-making. *Atriplex prostrata* only reached considerable cover percentages in the abandoned community. All hay-making regimes resulted in a high *Festuca rubra* abundance (Photo VII.3) (transition of coenon 2.4 into coenon 2.7), whereas *Juncus gerardii* became abundant in all hay-making regimes combined with grazing (transition of coenon 2.4 into coenon 2.3).

Festuca rubra fluctuated strongly in all the plots in the *Festuca rubra*/*Armeria maritima* community (Fig. VII.2B), whereas *Armeria maritima* only reached abundance in the plots in which June mowing was involved. *Trifolium repens* fluctuated strongly in all regimes, but became most abundant in the grazed plots. *Plantago maritima* and *Juncus gerardii* initially occurred in all plots with low cover percentages. *Plantago maritima,* however, became abundant in the abandoned plot (few large rosettes) and in both grazed and mown plots (many small rosettes), whereas *Juncus gerardii* became particularly abundant in most grazing regimes (Photo's VII.4 and 5) and in the twice cut regime.

Elymus pycnanthus only maintained dominance in the abandoned plot in the *Elymus pycnanthus*

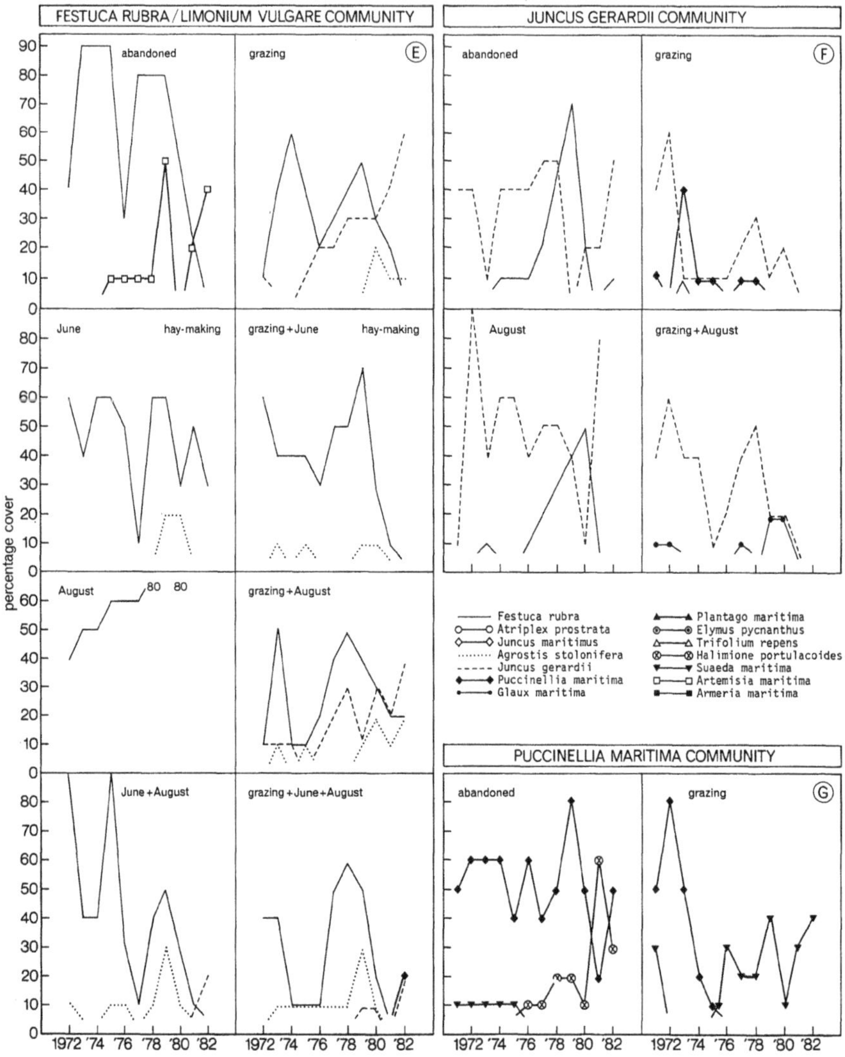

Fig. VII.2E–G. Changes of cover percentages of abundant (coverage ≥10%) species under different management regimes in various plant communities.

community (Fig. VII.2C) and nearly disappeared when grazing and hay-making were combined. *Festuca rubra* only became dominant in all the hay-making regimes (Photo's VII.6 and 7), whereas both *Festuca rubra* and *Agrostis stolonifera* were involved in all grazing experiments (transitions of coenon 2.1 to coenon 2.7). *Trifolium repens* also became temporarily abundant in the grazed plots. *Puccinellia maritima* was restricted to the purely grazing regime.

Artemisia maritima initially maintained itself best in the abandoned situation in the *Artemisia maritima* community (Fig. VII.2D). Its position was taken over by *Festuca rubra* and later by *Elymus pycnanthus* (transition of coenon 2.2 into coenon 2.1). Although its coverage fluctuated markedly, *Artemisia maritima* declined more slowly in the hay-making experiments than in the grazed plots (Photo's VII.8 and 9). *Festuca rubra* dominated most in all hay-making plots (transition of coenon 2.2 into coenon 2.6/2.7), whereas *Juncus gerardii* particularly favoured by the grazing management regimes (transition of coenon 2.2 into coenon 2.6/1.4). *Agrostis stolonifera* only favoured pure graz-

Photo VII.3. Juncus maritimus community, untouched (left) and cut in August for three years (right, behind the fence). *Festuca rubra* reached a high coverage in the cut area.

ing temporarily. In fact almost identical fluctuations of *Agrostis stolonifera* were observed in all plots of both the *Artemisia maritima* and *Festuca rubra/Limonium vulgare* communities. Since these fluctuations are not fully synchronous they cannot be attributed to climatological influences. It seemed to be a general trend that *Agrostis stolonifera* only reached high abundancies after the decrease of *Artemisia maritima* and if *Festuca rubra* coverages were not too high. *Puccinellia maritima* was restricted to the grazed plots.

Artemisia maritima took the position of *Festuca rubra* in the abandoned plot in the *Festuca rubra/ Limonium vulgare* community (Fig. VII.2E) (transition of coenon 2.7 into coenon 2.2) (Photo's VII.10 and 11). *Agrostis stolonifera* was found in all grazing regimes (transitions of coenon 2.7 into coenon 2.6 in the only grazed plot) and in the two hay-making regimes with a relatively small *Festuca rubra* cover percentage. *Juncus gerardii* became abundant in most grazing regimes and in the plot which was mown twice (transition of coenon 2.7 into coenon 1.6). *Puccinellia maritima* only appeared in the grazed site which was also mown twice. *Festuca rubra* fluctuated strongly except with the August mown regime.

In the *Juncus gerardii* community *Juncus gerardii* and *Festuca rubra* fluctuated in opposite ways in the abandoned plot and in that mown in August (Fig. VII.2F). *Juncus gerardii* decreased in the grazing regimes, whereas *Puccinellia maritima* was temporarily abundant in the grazed and *Glaux maritima* in the grazed and mown plot. *Halimione portulacoides* dominated *Puccinellia maritima* in the abandoned *Puccinellia maritima* community (Fig. VII.2G) (transition of coenon 1.6 into coenon 1.7). *Puccinellia maritima* disappeared in the grazing regime and *Suaeda maritima* became abundant (transition of coenon 1.6 into coenon 1.3).

In all plant communities most management practices have the often large year-to-year fluctuations in cover percentage of the abundant species in common. The next section deals with the question of whether the vegetational changes found in the permanent plots reveal trends, taking into account the whole species composition in the permanent plots.

Photo VII.4. Festuca rubra/Armeria maritima community untouched (right) and cut in August for three years (left). *Festuca rubra* remains low in the grazed area (foreground).

Photo VII.5. The same situation as in Photo VII.4 after fifteen years of cutting in August (left), which results in a short turf dominated by *Festuca rubra*. The grazed area is very short (foreground).

Photo VII.6. Elymus pycnanthus community untouched (right) and cut in August for three years (left, behind the fence) with a high coverage of *Festuca rubra.*

Photo VII.7. The same situation as in Photo VII.6 after fifteen years of cutting in August (right), featuring some *Artemisia maritima. Juncus maritimus* spread in the grazed area (left, before the fence).

Photo VII.8. Artemisia maritima community untouched (left) and cut in August for three years (right, behind the fence) with a high coverage of *Festuca rubra*.

Photo VII.9. The same situation as in Photo VII.8 after fifteen years of cutting in August (right) in which regime *Artemisia maritima* maintains itself, whereas it is overgrown by *Elymus pycnanthus* in the untouched area (left).

Photo VII.10. Festuca rubra/Limonium vulgare community untouched (left) and cut in August for three years (right, behind the fence), in which regime *Limonium vulgare* maintains itself best.

Photo VII.11. The same situation as in Photo VII.10 after fifteen years of cutting in August (right) in which regime the original community maintains itself, whereas it is overgrown by *Artemisia maritima* in the untouched site (left).

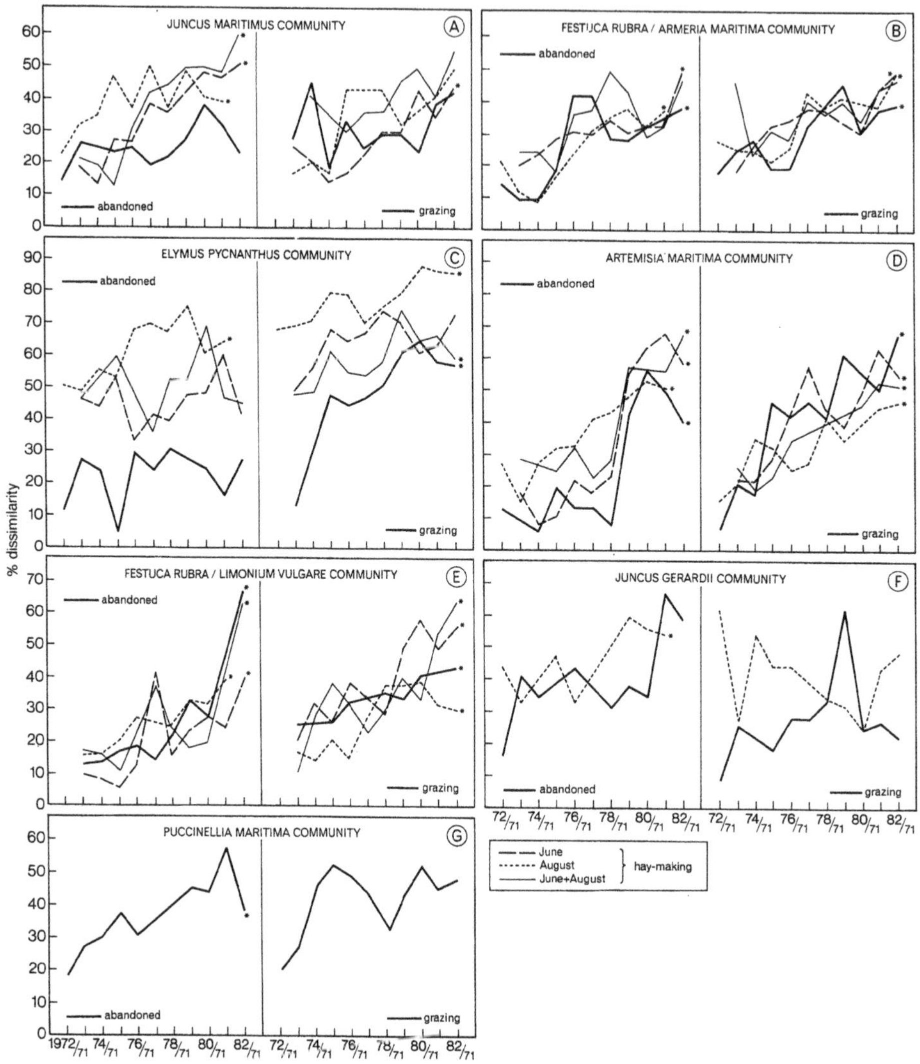

Fig. VII.3A–G. Changes of percentages dissimilarity between the first year and subsequent years under different management practices in various plant communities. * Significant change according to Spearman's rank correlation test ($P<0.05$) (after Braaksma, 1984).

VII.C.3. Vegetational trends

Vegetational changes in the permanent plots were quantified by comparing the dissimilarities for each permanent plot between each year and the initial year. The cover percentages of all the species which occurred were taken into account and these quantitative dissimilarities were calculated as indicated in section II.C.3. The increase, decrease or fluctuations of dissimilarities are shown in Fig. VII.3A–G.

The vegetation in the *Juncus maritimus* community (Fig. VI.3A) did not change in the abandoned plot, but changed significantly in all mown plots and in the plot which was grazed and mown in June.

Both the grazed and ungrazed plots which were mown in June and August revealed no change in the *Festuca rubra/Armeria maritima* community (Fig. VII.3B). The vegetation deviated more and more from the initial stand in all other plots, even in the abandoned situation.

The dissimilarity in the abandoned *Elymus pycnanthus* community remained very low, indicating little change in the stand (Fig. VII.3C). All mowing regimes revealed a sudden large deviation from the initial stand often with subsequent large fluctuations. Only the grazed and ungrazed plots which were mown in August and those which were grazed and cut in June and August revealed a gradual deviation from the initial stand. The plots which were only grazed showed a gradual vegetational change.

The vegetational change was large but gradual in all *Artemisia maritima* plots (Fig. VII.3D) and in all *Festuca rubra/Limonium vulgare* plots (Fig. VII.3E) even in the abandoned situations.

Only the plot mown in August showed a gradual deviation from the initial stand in the *Juncus gerardii* community (Fig. VII.3F). The dissimilarity in all other plots fluctuated strongly.

The abandoned *Puccinellia maritima* community (Fig. VII.3G) revealed a gradual deviation from the initial stand, whereas the grazed plot fluctuated.

An increasing dissimilarity percentage could be seen in many plant communities (Table VII.6), indicating gradual deviations from the initial stands both in abandoned and experimental plots. It was obvious that fluctuations had been superimposed on the vegetational trends and were often so large that no gradual change could be found.

The patterns of vegetational change after the imposition of new management regimes can also be demonstrated by the dissimilarities between consecutive years for each permanent plot. The range of year-to-year changes in the *Artemisia maritima* community (Fig. VII.4A) is representative for most communities. Only the *Elymus pycnanthus* community showed a pattern of small changes in the abandoned plot, an initially small change during the first years, when all the litter had not yet been removed, and an initially large change after hay-making which gradually decreased (Fig. VII.4B).

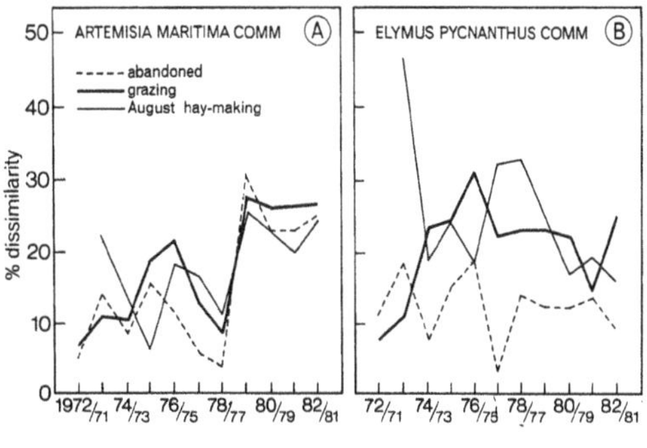

Fig. VII.4A–B. Changes of percentages dissimilarity between consecutive years under different management practices in various plant communities (after Braaksma, 1984).

VII.C.4. Species diversity

The change in number of species · 4 m^{-2} in the studied plant communities with the management re-

Table VII.6. Survey of changes in percentage dissimilarity between each year and the initial year in various plant communities with different management regimes (+: significant (P<0.05) increase, (+): increase not significant, 0: fluctuating, ·: not measured).

Abandoned	X							
Grazing					X	X	X	X
Hay-making June		X		X		X		X
Hay-making August			X	X			X	X
Juncus maritimus community	0	+	+	+	0	+	(+)	(+)
Festuca rubra/Armeria maritima community	+	+	+	(+)	+	+	+	0
Elymus pycnanthus community	(+)	(+)	+	(+)	+	(+)	+	+
Artemisia maritima community	+	+	+	+	+	+	+	+
Festuca rubra/Limonium vulgare community	+	+	+	+	+	+	+	+
Juncus gerardii community	(+)	·	+	·	0	·	0	·
Puccinellia maritima community	+	·	·	·	(+)	·	·	·

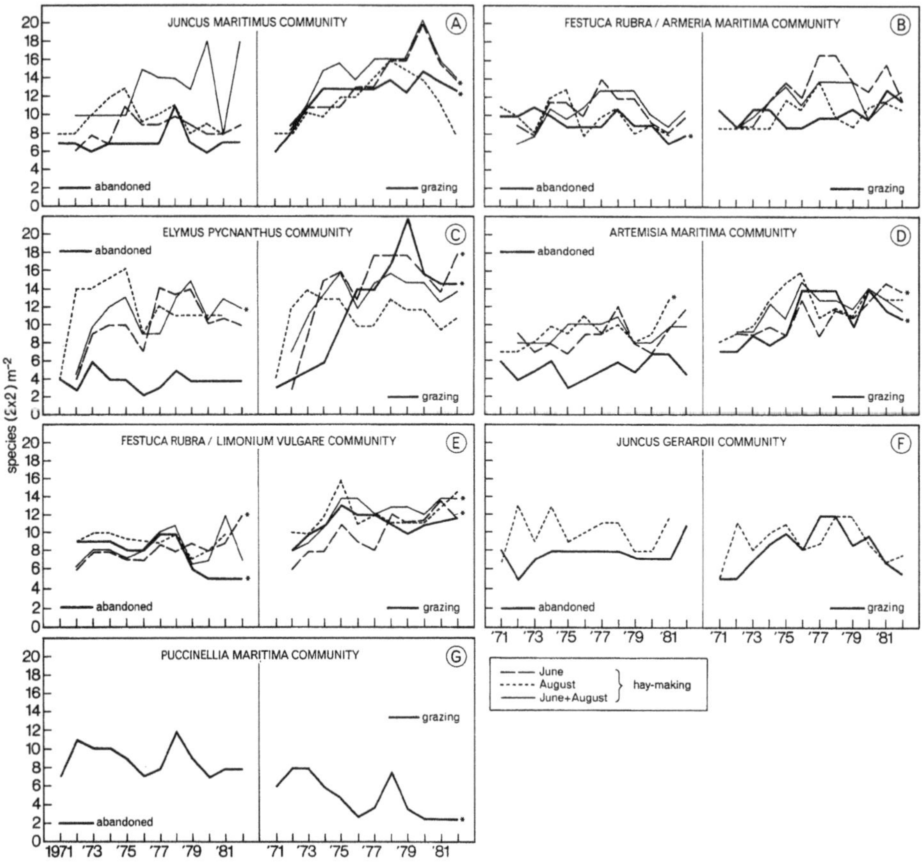

Fig. VII.5A–G. Changes of number of species · 4 m^{-2} under different management practices in various plant communities. * Significant change according to Spearman's rank correlation test (P<0.05) (after Braaksma, 1984).

gimes will be discussed in this section. Whether changes in species diversity were to be considered trends was decided in the way described in section VII.C.3 for vegetational changes, and this is indicated in Fig. VII.5A–G. Differences among the regimes are quantified by applying the sign test (P<0.05) resulting in a sequence of species diversity for the regimes in each of the plant communities (Table VII.7).

In the *Juncus maritimus* community species diversity became highest in the plots which had been grazed and those which had been cut twice (Fig. VII.5A). The plots which had been mown once, either in June or in August, revealed less species and the abandoned plot the lowest species diversity. The number of species only increased in the grazed plot and the one which had been grazed and mown in June.

The *Festuca rubra*/*Armeria maritima* community revealed little differences in species diversity for the management regimes. The species diversity of the abandoned plot decreased and was smaller than those of all other regimes (Fig. VII.5B). Only the grazed plot which had been cut twice and the grazed plot which had been cut in June showed a larger species diversity than the other grazed and/or mown regimes.

The species diversity in the abandoned *Elymus pycnanthus* regime remained very low, whereas all other regimes became species-richer (Fig. VII.5C). It increased in the plot which was mown in June and August, the grazed plot and the grazed plot which was cut in June and August. The differences among these regimes did not seem significant (Table VII.7). It should be mentioned, however, that all cutting regimes revealed a sudden increase of the species diversity, whereas the species number increased gradually in the plot which had only been grazed.

In the *Artemisia maritima* community a clear difference of species diversity levels was found in the different management regimes (Fig. VII.5D). Grazing combined with cutting regimes revealed the highest species diversity, followed by only grazing. All cutting regimes, together, showed the next lowest species diversity, whereas the abandoned plot had the lowest number of species. Species diversity only increased under the August mowing regime, as well in the grazed plot and in the plot which was grazed and cut in June.

Species diversity in the *Festuca rubra*/*Limonium vulgare* community became highest in all grazed plots combined with cutting regimes (Fig. VII.5E). The plot which was only grazed showed the next lowest species diversity. Little differences were found among the cutting regimes and the abandoned plot with the lowest species diversity. The number of species decreased in the abandoned plot, but increased in the grazed plot and those mown in June, and June + August.

In the *Juncus gerardii* community both cutting regimes showed higher species diversities than the abandoned plot, whereas the grazing regimes equaled all other regimes (Fig. VII.5F). The species diversity did not change.

The *Puccinellia maritima* community was the only community with a higher species diversity in the abandoned plot than in the grazed plot (Fig. VII.5G). This phenomenon was due to the decreasing species diversity in the grazed plot. The high stocking rate resulted in such a degree of destruction of the weak soil that only a few species could maintain themselves (Photo's VII.12 and 13).

A survey of the changes in species diversity in various plant communities with different management regimes is given in Table VII.8.

VII.C.5. Equitability

Relative equitability is a measure of the concentration of dominance in the group of species of a relevé. The relative equitability was higher in grazed than in mown plots in most communities (Fig. VII.6A–F), which could often be attributed to the increasing dominance of *Festuca rubra*.

VII.C.6. Exclosure

VII.C.6.a. Cover changes of abundant species

The abandoned 'bare soil community' revealed a succession of *Salicornia europaea* through *Suaeda*

Table VII.7. Sequence of species diversity with different management regimes per community according to the sign test ($P<0.05$) different symbols indicate significant differences. Ab: abandoned, J: June hay-making; A: August hay-making; JA: June + August hay-making; G: grazed; GJ: grazed + June hay-making; GA: grazed + August hay-making; GJA: grazed + June + August hay-making. J/A: both regimes have equal species diversities; GJA (= GJ): GJA has larger species diversity than all other regimes, but equals GJ, whereas GJ is not larger than all other regimes.

Juncus maritimus community		Festuca/ Armeria community		Elymus pycnanthus community		Artemisia maritima community		Festuca/ Limonium community		Juncus gerardii community		Puccinellia maritima community	
GJA	a	GJA	a	GJ	a	GJA	a	GJA	a	A	a	Ab	a
GJ	ab	GJ	ab	A	ade	GA	ab	GA	ab	GA	b	G	b
G	bc	GA	ac	G	abce	GJ	bc	GJ	c	Ab	c		
JA	bc	G	c	GJA	b	G	cd	G	bcd	G	abc		
GA	c	JA	c	GA	d	J	e	A	ce				
A	d	J	c	JA	de	A	ce	JA	ef				
J	de	A	cd	J	e	JA	e	Ab	f				
Ab	f	Ab	d	Ab	f	Ab	f	J	ceg				

Photo VII.12. Lower salt marsh community with *Triglochin maritima, Aster tripolium, Limonium vulgare* and *Puccinellia maritima* bordered by a higher salt marsh community with *Artemisia maritima* at the abandoned marsh.

Photo VII.13. The same area as in Photo VII.12 five years after the resumption of grazing, which resulted in bare soil and *Suaeda maritima* at the lower salt marsh and a short turf of *Festuca rubra* at the higher salt marsh.

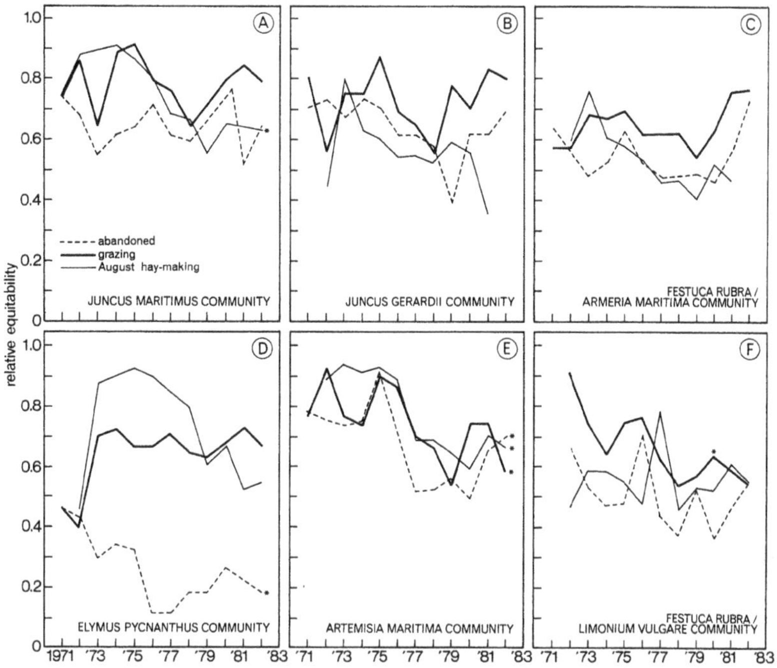

Fig. VII.6A–F. Changes of the relative equitability under different management practices in various plant communities. * Significant change according to Spearman's rank correlation test (P<0.05) (after Braaksma, 1984).

maritima to *Halimione portulacoides* (Fig. VII.7A).

After the decline of *Puccinellia maritima*, the species *Aster tripolium, Plantago maritima,* and *Glaux maritima* temporarily spread in the abandoned *Puccinellia maritima* community, followed by an increase of *Festuca rubra* and particularly *Artemisia maritima* (Fig. VII.7B).

Limonium vulgare and to a lesser extent *Aster tripolium* and *Glaux maritima* spread temporarily after cattle exclusion in the *Festuca rubra/Limonium vulgare* community (Fig. VII.7C). These species and *Juncus gerardii* declined quickly in the dense *Festuca rubra* stand which was succeeded by *Artemisia maritima.*

Although large fluctuations were observed, *Ar-*

Table VII.8. Survey of changes in species diversity from year-to-year in various plant communities with different management regimes (+: significant (P<0.05) increase, −: significant decrease, (+) increase not significant, 0: fluctuating, ·: not measured).

Abandoned	X							
Grazing					X	X	X	X
Hay-making June		X		X		X		X
Hay-making August			X	X			X	X
Juncus maritimus community	0	0	0	(+)	+	(+)	0	(+)
Festuca rubra/Armeria maritima community	−	0	0	0	(+)	(+)	0	0
Elymus pycnanthus community	0	(+)	(+)	+	+	+	(+)	(+)
Artemisia maritima community	0	(+)	+	(+)	+	+	(+)	(+)
Festuca rubra/Limonium vulgare community	−	+	0	0	(+)	+	(+)	+
Juncus gerardii community	0	·	0	·	0	·	0	·
Puccinellia maritima community	0	·	·	·	−	·	·	·

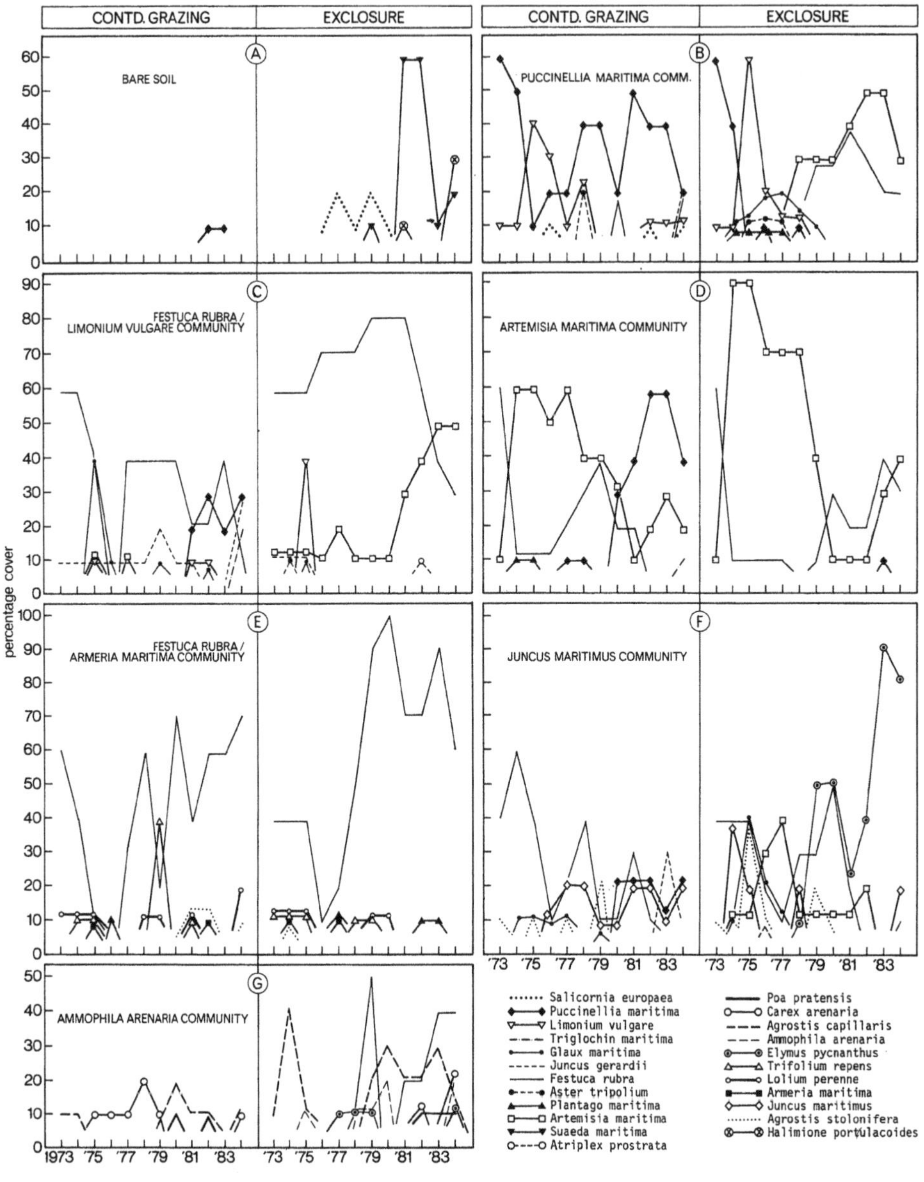

Fig. VII.7A–G. Changes of cover percentages of abundant (coverage ⩾10%) species under continued grazing and the cessation of grazing in various plant communities.

temisia maritima and *Festuca rubra* maintained themselves as the dominant species in the abandoned *Artemisia maritima* community (Fig. VII. 7D).

Festuca rubra spread largely in the *Festuca rubra/ Armeria maritima* community after exclusion (Fig. VII.7E). *Glaux maritima, Artemisia maritima* and *Festuca rubra* successively reached high cover percentages before *Elymus pycnanthus* became the dominant species in the *Juncus maritimus* community (Fig. VII.7F).

In the *Ammophila arenaria* community (Fig. VII.7G) *Agrostis capillaris, Festuca rubra, Ammophila arenaria, Elymus pycnanthus* and *Carex arenaria* increased temporarily but none of these species became really dominant after exclusion.

Some plant communities under continued grazing only revealed cover percentage fluctuations of

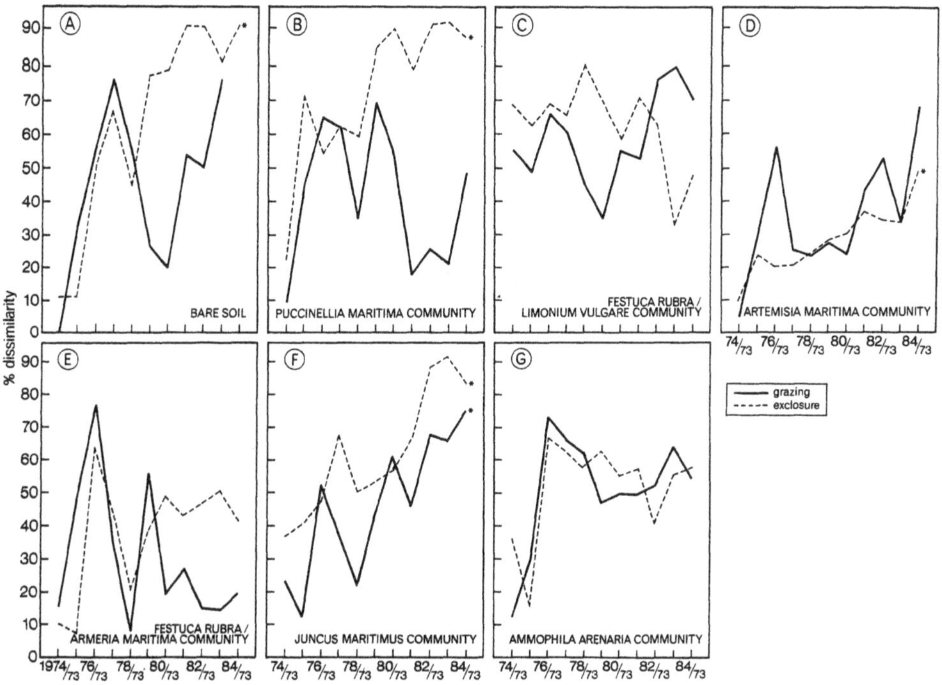

Fig. VII.8A–G. Changes of percentages dissimilarity between the first year and subsequent years under continued grazing and the cessation of grazing in various plant communities. * Significant change according to Spearman's rank correlation test ($P<0.05$).

initially dominant species. Others showed an increase of lower- or mid salt marsh species, e.g. *Puccinellia maritima, Triglochin maritima,* and *Juncus gerardii* in the *Festuca rubra/Limonium vulgare-*, the *Artemisia maritima-* and the *Juncus maritimus* community. This phenomenon must be attributed to the increased grazing intensity in the areas adjacent to the exclosures.

VII.C.6.b. Vegetational trends

The percentages dissimilarity of the continuously grazed areas showed a large increase after the establishment of the exclosures (Fig. VII.8A–G). This can be attributed to the effects of increased grazing intensities in the immediate vicinity of the exclosures. The vegetation, therefore, deviates from the initial situation in both the continuously grazed and recently ungrazed sward.

Vegetational trends in the exclosures were observed on bare soil (Fig. VII.8A), and in the *Puccinellia maritima-* (Fig. VII.8B), *Artemisia maritima-* (Fig. VII.8D), and *Juncus maritimus* community (Fig. VII.8F). The *Festuca rubra/Limonium vulgare* community (Fig. VII.8C) deviated strongly from the initial vegetation from the first year of exclusion onwards. The same held for the *Ammophila arenaria* community (Fig. VII.8G) from the exceptionally dry year of 1976 onwards. Only the *Festuca rubra/Armeria maritima* community showed irregular fluctuations (Fig. VII.8E).

VII.C.6.c. Species diversity

Species diversity declined significantly in the excluded *Festuca rubra/Limonium vulgare-* (Fig. VII.9C) and *Festuca rubra/Armeria maritima* community (Fig. VII.9E). The decrease was not significant in the *Artemisia maritima-* (Fig. VII.9D) and the *Juncus maritimus* community (Fig. VII.9F). In the *Puccinellia maritima-* (Fig. VII.9B) and the *Ammophila arenaria* community (Fig. VII.9G) species diversity fluctuated, whereas it increased on very open bare soil (Fig. VII.9A).

Fluctuations of species diversity were found in most continuously grazed plots. It only increased in the bare soil (Fig. VII.9A) and in the *Juncus mar-*

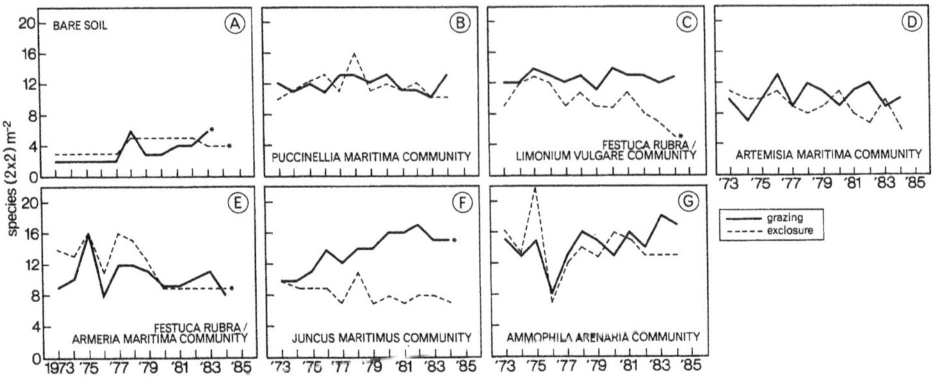

Fig. VII.9A–G. Changes of number of species · 4 m^{-2} under continued grazing and the cessation of grazing in various plant communities. * Significant change according to Spearman's rank correlation test (P<0.05).

itimus community (Fig. VII.9F). Again, the increased grazing intensity along the exclosure hampers the comparison of continuously grazed areas and exclosures.

VII.C.6.d. Equitability

The relative equitability decreased significantly in the excluded *Puccinellia maritima* community due to the dominance of *Artemisia maritima* (Fig. VII.10B), in the *Juncus maritimus* community with *Elymus pycnanthus* dominance (Fig. VII.10F), in the *Ammophila arenaria* community with also a dominance of *Festuca rubra* (Fig. VII.10G) and in the *Festuca rubra/Armeria maritima* community due to *Festuca rubra* dominance (Fig. VII.10E). Note that the species diversity only declined significantly in the latter community (Fig. VII.9E) and in the *Festuca rubra/Limonium vulgare* community (Fig. VII.9C).

In the continuously grazed area a decrease of the relative equitability was found in the *Festuca rubra/Armeria maritima* community due to a dominance of *Festuca rubra* and in the bare soil with *Puccinellia maritima* dominance (Fig. VII.10A).

VII.C.7. Comparison of management regimes

The cover changes of abundant species have been discussed in sections VII.C.2 and VII.C.6.a. The present section will summarize the changes of all species in the experiments (Table VII.9A–C).

Some quantification was added by indicating small or large increases or decreases of cover percentages, the large changes are in agreement with the information provided in sections VII.C.2 and VII.C.6.a. All grazed plots including the combinations with cutting regimes were considered together. Moreover, both plots grazed less than 13 years and permanently grazed plant communities were compared.

Nearly all species showed a decreasing cover percentage in the *Puccinellia maritima* community within 13 years of grazing regimes (Table VII.9A), whereas in all other plant communities many species increased permanently or temporarily. The *Juncus maritimus*- and the *Festuca rubra/Armeria maritima* community showed the highest number of species which maintained constant cover percentages. It was striking that most lower salt marsh species increased their cover in nearly all recently grazed plant communities. The dune species only increased in the plant communities highest above MHW. These species were also present and capable of maintaining themselves in the lower lying permanently grazed plots of the *Juncus maritimus* community, and to a lesser extent in the *Festuca rubra/Armeria maritima* community. Time will learn whether the lower salt marsh species which appeared in all recently grazed plots will persist. The increase of lower salt marsh species in the long-grazed plots must be due to the increased grazing intensity adjacent to the exclosure.

The *Juncus gerardii* community showed many species with constant cover percentages under

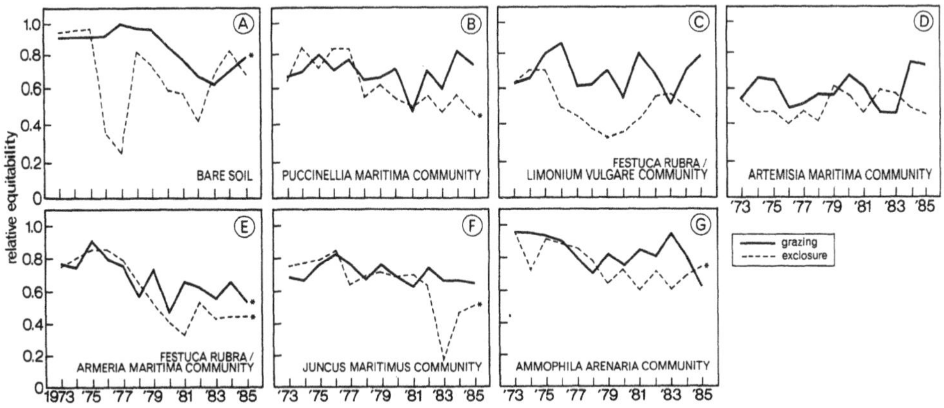

Fig. VII.10A–G. Changes of the relative equitability under continued grazing and the cessation of grazing in various plant communities. * Significant change according to Spearman's rank correlation test (P<0.05).

mowing regimes (Table VII.9B). This could be due to the fact that only an August cutting regime was practised in this plant community. A number of species showed a cover increase under the June or the August cutting regime in the *Artemisia maritima* community. The regime with cutting only in June caused a cover increase of some species in the *Festuca rubra/Limonium vulgare*- and the *Festuca rubra/Armeria maritima* community, whereas the same held for the regime where cutting was only done in August in the *Elymus pycnanthus* community. The combined June + August mowing regime, however, caused a cover increase of the highest number of species in most plant communities.

The cover changes in permanent plots abandoned for less than 13 years and more than 13 years are summarized in Table VII.9C. Since different plots are involved caution is needed in comparing the older and younger permanent plots in the plant communities. Nevertheless, it seems that the bare soil plot could represent a stage preceding that of the older plot in the *Puccinellia maritima* community. Also, the young *Festuca rubra/Limonium vulgare* plot could have preceded the older plot in the *Artemisia maritima* community. Finally, the *Elymus pycnanthus* plot could represent a stage preceding that of the young plot in the *Juncus maritimus* community. The older *Artemisia maritima* plot is on its way to becoming an *Elymus pycnanthus* community. All these changes fit in the successional schemes provided in section VII.D.3.

Many species showed decreased cover percentages during the first period of exclosure. Several lower salt marsh species and a few mid- and upper salt marsh species temporarily increased their cover percentage shortly after the cessation of grazing. These species, however, sometimes already reached lower cover percentages than the initial values or decreased in the longer abandoned plots. This phenomenon should be attributed to the large spread of a few species, namely, *Halimione portulacoides, Artemisia maritima, Festuca rubra* and *Elymus pycnanthus* with increasing height relative to MHW. Some other species increased in the *Puccinellia maritima*- and *Juncus gerardii* community.

A comparison between the regimes showed that few species appeared or increased their cover percentage in the longer abandoned salt marsh area. In the mown, and still more in the grazed salt marsh areas, many species appeared or increased their cover percentages, in accordance with the often highest species diversity in the grazed salt marsh (see section VII.C.4). It is striking that, particularly in the grazed areas, more than in the mown areas, lower salt marsh species appeared at heights with respect to MHW where these did not and do not occur in the abandoned areas. Moreover, these species already appeared in the grazed areas during the first five years after the beginning of the experiment, but were often still absent in the mown permanent plots (Bakker, 1978). The ratios between lower salt marsh-, mid- and upper salt marsh- and dune species in the plant communities at the beginning of the experiment and after ten years of re-

Table VII.9A. Cover changes of all species in permanent plots in plant communities with increasing height with respect to MHW in continuously (>13 years) and recently (<13 years) grazed areas. +: increase between 5 and 10%, ++: increase >10%, −: decrease between 5 and 10%, −−: decrease >10%, C: constant, Λ: (−) temporarily increase (with lower new cover percentage), L: lower salt marsh species, M: mid- and upper salt marsh species, D: dune species (according to Bakker & Ruyter, 1981).

		Increasing height with respect to M H W →												
		Bare soil	Puccinellia maritima community		Festuca/ Limonium community		Juncus gerardii community	Artemisia maritima community		Juncus maritimus community		Festuca/ Armeria community		Elymus pycnanthus community
G R A Z I N G		>13yr	<13	>13yr	<13	>13yr	<13yr	<13	>13yr	<13	>13yr	<13	>13yr	<13yr
L	Spartina anglica	+	-	.	.	C	.	.	.	.	.	.	.	.
L	Halimione portulacoides	.	-	.	.	.	Λ	+	.	.	Λ	.	.	.
L	Aster tripolium	.	--	C	+	.	.	+	.	.	C	.	.	Λ
L	Limonium vulgare	+	+	C	-	C	Λ	C	C	.	C	.	.	+
?	Cochlearia anglica	.	.	.	.	.	.	Λ	.	.	.	.	.	.
L	Puccinellia maritima	+	--	C	++	++	+	++	++	+	++	.	.	+
L	Triglochin maritima	.	C	C	+	C	+	+	C	C	C	+	.	Λ
L	Salicornia europaea	C	-	C	+	+	+	+	Λ	+	+	+	.	+
L	Suaeda maritima	+	-	C	+	+	+	+	Λ	+	Λ	+	.	+
L	Spergularia maritima	+	-	C	+	C	+	+	C	+	+	Λ	.	+
L/M	Spergularia salina	.	Λ	Λ	+	.	+	+	Λ	+	Λ	+	.	+
L/M	Juncus gerardii	.	.	C	++	C	C	++	C	++	+	++	Λ	+
L/M	Glaux maritima	.	Λ	C	+	.	+	+	C	C	-	+	.	+
M	Artemisia maritima	.	-	C	-	C	Λ	--	Λ	C	C	Λ	.	+
M	Festuca rubra	.	.	+	--	--	C	+	--	-	--	--	C	++
M	Atriplex prostrata	.	.	.	+	.	-	C	Λ	C	.	C	.	C
M	Plantago maritima	.	-	C	+	C	+	C	C	+	+	C	C	+
M	Agrostis stolonifera	.	.	.	+	.	C	+	-	-	-	C	C	++
M	Armeria maritima	.	.	C	C	C	Λ	+	.	+	Λ	C	C	+
M	Carex distans	.	.	.	.	.	.	.	.	Λ	.	Λ	.	+
M	Potentilla anserina	.	.	.	.	.	Λ	.	.	-	.	-	.	Λ
M	Juncus maritimus	.	.	.	.	.	.	.	.	C	+	.	.	+
M/D	Trifolium repens	.	.	.	.	.	.	.	.	Λ	.	C	C	+
M/D	Poa pratensis	.	.	.	Λ	.	.	.	.	Λ	.	-	-	Λ
?	Sagina maritima	.	.	.	.	.	.	.	.	+	.	+	.	+
M/D	Plantago coronopus	.	.	.	Λ	.	.	Λ	.	Λ	.	Λ	C	+
?	Parapholis strigosa	.	.	.	.	.	Λ	+	.	Λ	.	.	.	+
M/D	Elymus pycnanthus	.	.	.	.	.	Λ	.	.	.	+	.	Λ	--
?	Lolium perenne	.	.	.	.	.	.	.	.	.	.	.	C	.
D	Cochlearia danica	.	.	.	.	.	.	.	.	+	.	Λ	+	+
D	Cerastium fontanum	.	.	.	.	.	.	.	.	.	.	Λ	-	Λ
D	Lotus corniculatus	.	.	.	.	.	.	.	.	.	.	Λ	Λ	Λ
D	Bromus hordeaceus	.	.	.	.	.	.	.	.	.	.	Λ	.	Λ
?	Centaurium littorale	.	.	.	.	.	.	.	.	.	.	Λ	.	Λ
?	Bupleurum tenuissimum	.	.	.	.	.	.	.	.	.	.	.	.	+
D	Rumex crispus	.	.	.	.	.	.	.	.	.	.	.	.	Λ
?	Trifolium pratense	.	.	.	.	.	.	.	.	.	.	.	.	Λ

Table VII.9B. Cover changes of all species in permanent plots in plant communities with increasing height with respect to MHW in hay-making areas. +: increase between 5 and 10%, ++: increase >10%, −: decrease between 5 and 10%, −−: decrease >10%, C: constant, Λ: (−) temporarily increase (with lower new cover percentage), J: cover change particularly in June mown plot; A: cover change particularly in August mown plot, * mown in August only, L: lower salt marsh species, M: mid- and upper salt marsh species, D: dune species (according to Bakker & Ruyter, 1981).

Increasing height with respect to MHW →

H A Y - M A K I N G		Festuca/ Limonium community	Juncus gerardii community *	Artemisia maritima community	Juncus maritimus community	Festuca/ Armeria community	Elymus pycnanthus community
L	Halimione portulacoides	Λ	Λ	+ J	.	.	.
L	Aster tripolium	+ JA	.	.	.	.	.
L	Limonium vulgare	+	.	C	.	.	+ JA
?	Cochlearia anglica	.	.	Λ A	.	.	.
?	Atriplex litoralis	.	.	+	.	.	.
L	Puccinellia maritima	+	-	+ A	+ JA	+ JA	Λ JA
L	Triglochin maritima	-	C	-	C	Λ A	.
L	Salicornia europaea	+ JA	Λ	+	+ JA	.	.
L	Suaeda maritima	+ JA	Λ	+	+ JA	+ JA	Λ A
L	Spergularia maritima	+ JA	.	+ J	+ JA	.	+ JA
L/M	Spergularia salina	+ JA	Λ	+ A	.	+ JA	.
L/M	Juncus geradii	+ JA	++	-	-	C	Λ A
L/M	Glaux maritima	-	-	-	-	.	Λ A
M	Artemisia maritima	+ J	C	--	+	+ JA	+
M	Festuca rubra	++ A	C	++	++	+	++
M	Atriplex prostrata	+	C	C	C	+ JA	C
M	Plantago maritima	C	+	-	+ JA	+ J	+ A
M	Agrostis stolonifera	+ J	C	++	++	++	+
M	Armeria maritima	C	.	+ A	+	+	+
M	Carex distans	Λ J	.	.	.	+ J	+ A
M	Potentilla anserina	.	-	.	+	-	Λ
M	Juncus maritimus	.	.	.	-	.	C
M/D	Trifolium repens	+	.	.	.	+ A	+
M/D	Poa pratensis	Λ J	.	.	Λ	-	Λ
?	Sagina maritima	Λ JA	.	.	+ JA	+ JA	+ JA
M/D	Plantago coronopus	+ JA	.	.	+ JA	+	+ JA
?	Parapholis strigosa	.	.	.	+	.	+ JA
M/D	Elymus pycnanthus	.	.	+	.	+ J	--
D	Cochlearia danica	.	.	Λ J	.	.	+
D	Cerastium fontanum	.	.	Λ JA	Λ	Λ JA	+ A
D	Lotus corniculatus	.	.	.	.	+ J	.
D	Bromus hordeaceus	.	.	.	.	Λ	Λ
?	Bupleurum tenuissimum	.	.	.	.	.	Λ
D	Rumex crispus	.	.	.	.	.	Λ
?	Trifolium pratense	.	.	.	.	.	Λ

Table VII.9C. Cover changes of all species in permanent plots in plant communities with increasing height with respect to MHW in recently (<13 years) and longer (>13 years) abandoned areas. +: increase between 5 and 10%, ++: increase >10%, −: decrease between 5 and 10%, −−: decrease >10%, C: constant, Λ: (−) temporarily increase (with lower new cover percentage), L: lower salt marsh species, M: mid- and upper salt marsh species, D: dune species (according to Bakker & Ruyter, 1981).

Increasing height with respect to MHW →

A B A N D O N E D		Bare soil	Puccinellia maritima community		Festuca/ Limonium community		Juncus gerardii community	Artemisia maritima community		Juncus maritimus community		Festuca/ Armeria community		Elymus pycnanthus community
		<13yr	<13	>13yr	<13	>13yr	>13yr	<13	>13yr	<13	>13yr	<13	>13yr	>13yr
L	Spartina anglica	.	.	-	.	.	.	.	.	.	.	.	.	.
L	Halimione portulacoides	++	Λ	++	.	.	.	+	.	.	.	.	.	.
L	Aster tripolium	.	Λ	--	-	.	.	C	.	-	.	.	.	.
L	Limonium vulgare	.	Λ	.	Λ (-)	C	+	C	-	-	.	.	.	.
?	Cochlearia anglica	.	.	Λ	.	.	.	.	.	.	.	.	.	.
?	Atriplex litoralis	.	.	Λ	.	.	.	.	+	.	.	.	.	.
L	Puccinellia maritima	++	--	+	Λ	+	C	C	.	.	.	.	.	.
L	Triglochin maritima	.	Λ (-)	C	-	-	C	-	-	-	C	.	Λ	.
L	Salicornia europaea	Λ	C	-	Λ	.	+	.	.	.	.	.	.	.
L	Suaeda maritima	Λ	Λ	--	Λ	.	+	Λ	.	.	.	.	+	.
L	Spergularia maritima	.	C	-	-	.	.	C	.	.	.	.	.	.
L/M	Spergularia salina	.	.	Λ	.	.	+	.	.	.	.	.	.	.
L/M	Juncus gerardii	.	.	.	-	-	C	-	-	-	C	-	-	.
L/M	Glaux maritima	.	Λ (-)	.	Λ (-)	-	-	-	-	-	-	.	-	.
M	Artemiasia maritima	.	++	+	++	++	C	++	--	Λ	-	.	Λ	Λ
M	Festuca rubra	.	++	+	--	--	+	--	++	--	C	++	--	C
M	Atriplex prostrata	.	Λ	+	Λ	C	+	Λ	+	Λ (+)	++	Λ	-	C
M	Plantago maritima	.	Λ (-)	+	-	-	+	-	-	.	.	Λ	C	.
M	Agrostis stolonifera	.	.	.	.	-	C	C	C	-	C	-	-	-
M	Armeria maritima	.	C	.	-	.	.	.	Λ	.	.	-	C	.
M	Carex distans	.	.	.	.	.	.	.	.	.	.	.	Λ	.
M	Potentilla anserina	.	.	.	.	.	.	.	.	.	.	.	-	.
M	Juncus maritimus	.	.	.	.	.	.	.	.	Λ (-)	--	.	.	C
M/D	Poa pratensis	.	.	.	.	.	.	C	.	.	.	-	Λ	.
M/D	Elymus pycnanthus	.	.	.	.	.	.	.	++	++	+	+	.	++
M/D	Trifolium repens	.	.	.	.	.	.	.	.	.	.	-	.	.
?	Lolium perenne	.	.	.	.	.	.	.	.	.	.	-	.	.
?	Leontodon autumnalis	.	.	.	.	.	.	.	.	-	.	-	.	.
?	Sagina nodosa	.	.	.	.	.	.	.	.	.	.	-	.	.
M/D	Plantago coronopus	.	.	.	.	.	.	.	.	.	.	-	.	.
D	Cerastium fontanum	.	.	.	.	.	..	.	.	.	.	-	.	.
D	Bromus hordeaceus	.	.	.	.	.	.	.	.	.	.	-	.	.
D	Trifolium arvense	.	.	.	.	.	.	.	.	.	.	Λ	.	.
D	Lotus corniculatus	.	.	.	.	.	.	.	.	.	.	-	Λ	.

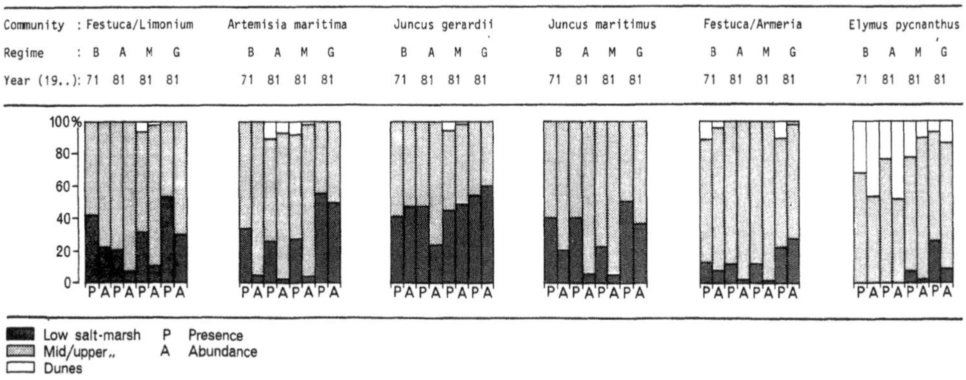

Fig. VII.11. Assignment of species to lower salt marsh group <35 cm + MHW, 35 cm + MHW<mid- and upper salt marsh group <85 cm + MHW, dune group >85 cm + MHW in permanent plots (2 × 2) m^2 from 1971 (Beginning = mean of three plots) until 1981 under continued Abandoning, Mowing in August, and Grazing in various plant communities (after Bakker, 1985).

sumed grazing revealed the highest lower salt marsh proportion in all grazed plant communities both for presence and abundance (Fig. VII.11).

Dune species also appeared in the grazed and mown areas but these remained restricted to the highest parts of the salt marsh. These phenomena will be discussed in detail in chapter X.

Changes of species diversity have been summarized in Fig. VII.12A–B. The August hay-making regimes were used in the comparison in order to involve the *Juncus gerardii* community. The average species diversity in the grazed salt marsh was reached five years after resumed grazing (Fig. VII.12A). Species diversity initially increased quicker in the mown than in the grazed plots but dropped below the grazed level. It took over five years before the species diversity in a recent exclosure dropped below that of the continuously grazed communities, but not significantly according to the Mann-Whitney test, and it had not yet reached the level of longer abandoned communities thirteen years after exclusion (Fig. VII.12B).

Permanent plot data give detailed information about succession in a limited area because the number of permanent plots is restricted. Information about the whole study area can be gained from vegetation mapping which will be discussed in section VII.D.

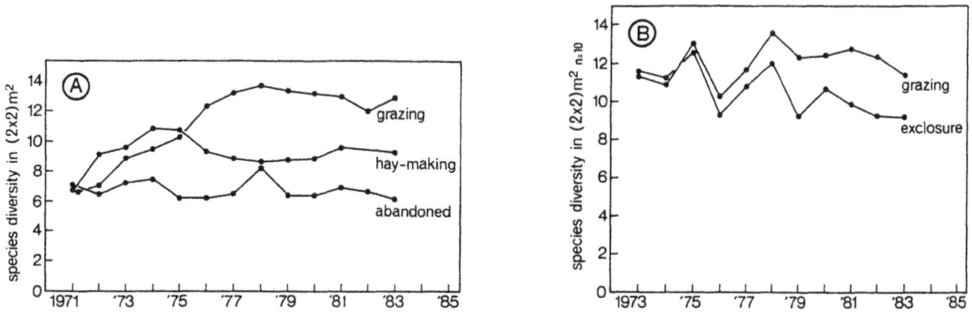

Fig. VII.12A–B. Changes in mean number of species · $4\,m^{-2}$ A: under grazing, mowing and continued abandoning (n = 7) and B: under continued grazing and recent abandoning (n = 8) in various plant communities at the mid and upper salt marsh. Differences are not significant according to the Mann-Whitney-test.

Photo VII.14. Juncus maritimus (dark) and particularly *Elymus pycnanthus* dominated the abandoned higher salt marsh.

Photo VII.15. The same area as in Photo VII.14 five years after resumption of grazing. The clumps of *Juncus maritimus* maintained themselves and 'protect' *Artemisia maritima* and *Elymus pycnanthus*. The latter species was removed elsewhere and replaced by a short turf of *Festuca rubra,* whereas the micro-relief became apparent.

VII.D. Vegetation mapping

VII.D.1. Comparison of vegetation maps

Although mapping of vegetation complexes should be avoided, in the present study one complex of communities was mapped as a mosaic in 1971, before grazing was resumed. *Juncus maritimus* and particularly *Elymus pycnanthus* dominated the varied stand on a small-scaled micro-relief. After the resumption of grazing the dominant species was largely suppressed enabling the minor species to increase and the vegetation pattern to become more manifest in 1976 (Photos VII.14 and 15). The classification of the legend units in the study area was based on c. 300 relevés made for the 1971 map (Van Der Duim, 1973; Prins, 1974) and c. 140 relevés (Norder & Ruyter, 1977) made for the 1976 map. Moreover, those who mapped in 1971 verified the description of the legend units used in 1976. In 1981 c. 100 relevés were made (Looijen, Looijen & Zomer, 1984) to control the classification and again those who mapped in 1976 verified the description of the legend units in 1981.

For a comparison of the different maps, the 1976 map was overlaid with a transparancy of the 1971 map. The distribution of the communities in 1971 and 1976 was worked out for each separate point by

Table VII.10A. Areas occupied by the coena in 1971, 1976 and 1981 in the area grazed since 1972. Comparison is made by a grid of which 400 points = 1 ha. The percentage of unchanged points is related to the first year of each study period viz. 1971–1976, and 1976–1981.

	GRAZED						1971 - 1976 Unchanged points		1976 - 1981 Unchanged points	
Coenon	1971		1976		1981			Related to 1971		Related to 1976
	Points	Ha	Points	Ha	Points	Ha	Absolute	%	Absolute	%
1.2	.	.	.	.	.	.	.	.	.	.
1.3	112	0.28	113	0.28	104	0.29	69	62	78	69
1.4	106	0.27	66	0.17	50	0.19	14	13	25	38
1.6	423	1.06	503	1.26	508	0.87	277	65	139	27
1.7	.	.	.	.	1	0.01	.	.	.	.
2.1	1233	3.08	913	2.28	836	1.01	349	28	184	20
2.2	1197	2.99	818	2.05	802	1.17	696	58	367	45
2.3	357	0.89	395	0.99	410	2.70	272	76	336	85
2.4	1762	4.41	2015	5.04	2053	5.86	1398	79	1845	92
2.6	256	0.64	415	1.04	429	0.75	65	25	201	48
2.7	1735	4.34	1965	4.91	1677	4.55	1060	61	1133	58
2.14	.	.	120	0.30	115	0.29	.	.	21	18
2.15	.	.	1375	3.44	1359	3.44	.	.	1113	81
2.16	1536	3.84	794	1.99	845	2.10	565	37	668	84
3.1	575	1.44	437	1.09	431	0.84	283	49	292	67
Mosaic	659	1.65	.	.	.	.	.	.	.	.
Bare soil	20	0.05	40	0.10	30	0.05	14	70	5	13
Water	44	0.11	46	0.11	51	0.15	33	75	46	100
Total	10015	25.05	10015	25.05	9701	24.27	5095	51	6453	64

means of a dot grid with a dot spacing of 0.5 cm, corresponding to a distance of 5 m. The 1976 and 1981 maps were compared in the same way. This enabled a study of changes in communities and the areas of the different communities to be calculated (400 points equals 1 ha at the scale 1 : 1000). The areas occupied by the various communities are shown in Table VII.10A–B. Despite a nearly similar area for several communities during the study period, the percentage of unchanged points is often rather small. Each community can undergo a transition into another community and vice versa, suggesting larger vegetation dynamics than could be derived from their area alone. The percentage of unchanged points is related to the first year of a study period. As for coenon 3.1, for instance, only 49% of the 1971 points remained unchanged in the grazed area in 1976 (Table VII.10A).

Each community mapped in 1971, for example, could have been unchanged or may have changed into several other communities in 1976. This is quantified in the transition matrices of Table VII.11 and 12. Each community in 1976 on the other hand can partly originate from several other communities from 1971 which is quantified in Table VII.11 and 12, too.

An area of c. 2 ha (the *Juncus maritimus*/*Elymus pycnanthus* mosaic in 1971) was not included in the

Table VII.10B. Areas occupied by the coena in 1971, 1976 and 1981 in the continuously abandoned area. Comparison is made by a grid of which 400 points = 1 ha. The percentage of unchanged points is related to the first year of each study period viz. 1971–1976 and 1976–1981.

	ABANDONED						1971 - 1976		1976 - 1981	
							Unchanged points		Unchanged points	
Coenon	1971		1976		1981			Related to 1971		Related to 1976
	Points	Ha	Points	Ha	Points	Ha	Absolute	%	Absolute	%
1.2	199	0.50	96	0.24	3	0.01	96	48	.	.
1.3	154	0.39	41	0.10	9	0.02	15	10	3	7
1.4	35	0.09	132	0.33	160	0.40	6	17	79	60
1.6	.	.	33	0.08	55	0.14	.	.	21	64
1.7	.	.	61	0.15	47	0.12	.	.	18	30
2.1	804	2.01	1220	3.05	1567	3.92	518	64	846	69
2.2	987	2.47	1089	2.72	456	1.14	802	82	416	38
2.3	85	0.21	76	0.19	52	0.13	62	72	52	68
2.4	820	2.05	609	1.52	828	2.07	392	48	410	67
2.6	.	.	18	0.05	6	0.02	.	.	1	5
2.7	371	0.93	297	0.74	376	0.94	188	51	194	65
2.14	.	.	2	0.01	86	0.22	.	.	2	100
2.15	.	.	28	0.07	.	.	.	.	.	.
2.16	176	0.44	126	0.32	55	0.14	78	44	39	31
3.1	92	0.23	59	0.15	58	0.15	47	51	46	78
Mosaic	147	0.37	.	.	.	.	.	.	.	.
Bare soil	.	.	1	0.01	4	0.01	.	.	.	.
Water	79	0.20	61	0.16	155	0.39	0	0	61	100
Total	3949	9.89	3949	9.89	3917	9.82	2204	56	2188	55

calculations because it had not been mapped in sufficient detail in 1971.

Since a vegetation map at scale 1 : 1000 is too large to represent, simplified maps at scale 1 : 7500 are depicted. Some coena featuring in the Tables VII.1–4 are lumped to the simplified legend units of the vegetation maps in Fig. VII.13: *Puccinellia maritima* community (coena 1.2, 1.3, 1.4, 1.6, 1.7); *Elymus pycnanthus* community (coenon 2.1); *Artemisia maritima* community (coenon 2.2); *Juncus gerardii* community (coenon 2.3); *Juncus maritimus* community (coenon 2.4); *Festuca rubra* community (coena 2.6, 2.7, 2.14); *Armeria maritima* community (coenon 2.15); *Ammophila arenaria* community (coena 2.16, 3.1). The simplified units are characterized by dominant species.

VII.D.2. Changes after the resumption of grazing

A comparison of the generalized vegetation maps (Fig. VII.13) and the areas under the different communities in Table VII.10A clearly shows that in the grazed area the *Elymus pycnanthus*- (coenon 2.1), *Ammophila arenaria*- (coena 2.16 and 3.1) and *Artemisia maritima* (coenon 2.2) communities decreased during the study period. By contrast the *Juncus gerardii*- (coenon 2.3), and the *Juncus maritimus* community (coenon 2.4) spread). The emergence of the *Armeria maritima* community (coenon 2.15) is strikingly.

All complex transitions between the different communities involving both an increase and decrease of the area of the communities are shown in details in Table VII.11A–B. The main transitions are summarized in Fig. VII.14A–B for the periods 1971–1976 and 1976–1981, respectively. During the first period the *Artemisia maritima* community (coenon 2.2) lost area in favour of the *Festuca rubra* community (coena 2.6, 2.7 and 2.14), which on its turn lost area to the *Puccinellia maritima* community (coena 1.3, 1.4, 1.6) and the *Armeria maritima* community (coenon 2.15). Moreover, the *Armeria maritima* community originated from the *Ammophila areanaria* community (coena 2.16, 3.1) and the *Elymus pycnanthus* community (coenon 2.1). The *Elymus pycnanthus* community, in turn, also decreased in favour of the *Festuca rubra* – and the

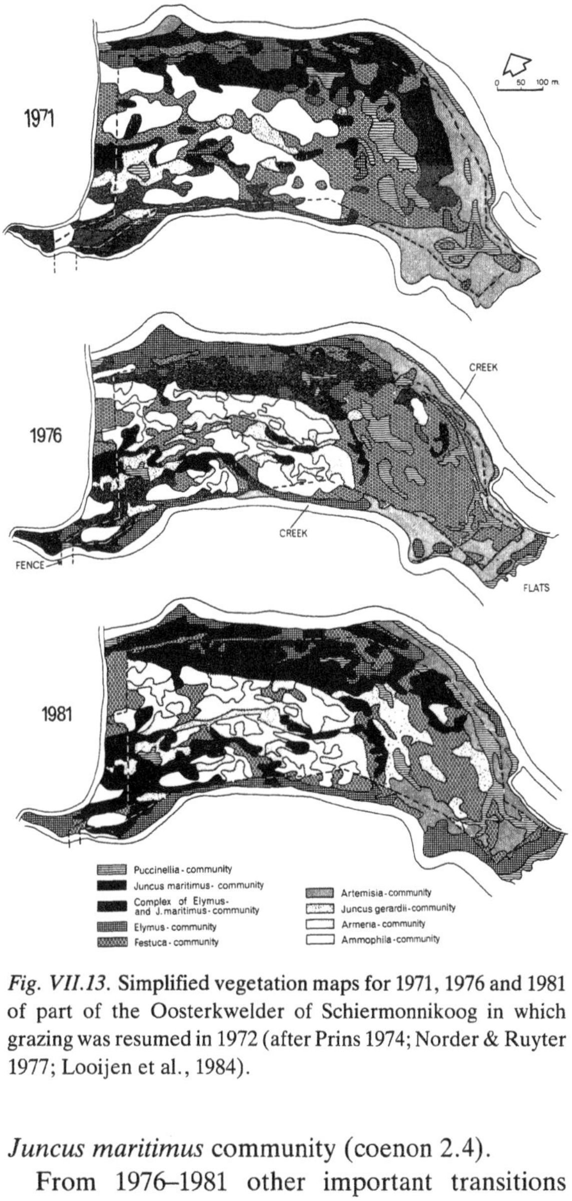

Fig. VII.13. Simplified vegetation maps for 1971, 1976 and 1981 of part of the Oosterkwelder of Schiermonnikoog in which grazing was resumed in 1972 (after Prins 1974; Norder & Ruyter 1977; Looijen et al., 1984).

Juncus maritimus community (coenon 2.4).

From 1976–1981 other important transitions took place than in the preceding period. The *Juncus maritimus* community still originated from the *Elymus pycnanthus* community, but it lost area to the *Festuca rubra* community. The latter decreased in turn, in favour of the *Juncus gerardii* community (coenon 2.3), and of the *Artemisia maritima* community. The *Juncus gerardii* community also originated from the *Puccinellia maritima*- and the *Artemisia maritima* community.

The main changes took place in the *Festuca rubra* zones and areas higher above MHW during the first period, whereas this held for the *Festuca rubra*

Table VII.11A. Transitions between the coena in the area grazed from 1972 onwards during the period between 1971 and 1976. The species composition of the coena is given in the Tables VII.1, 2, 3 and 4.

		1976							GRAZED										
		1.2	1.3	1.4	1.6	1.7	2.1	2.2	2.3	2.4	2.6	2.7	2.14	2.15	2.16	3.1	Bare soil	Water	Total
1971	1.2	.	.	.	.	.	.	.	.	.	.	.	.	.	.	.	.	.	.
	1.3	.	69	10	22	.	.	2	6	.	.	.	.	.	.	.	3	.	112
	1.4	.	9	14	50	.	.	4	.	.	.	22	.	.	.	.	7	.	106
	1.6	.	15	23	277	.	3	7	.	8	3	77	.	.	.	.	.		423
	1.7	.	.	.	.	.	.	.	.	.	.	.	.	.	.	.	.	.	.
	2.1	.	1	.	1	.	349	1	10	279	30	89	40	417	.	13	1	2	1233
	2.2	.	6	3	102	.	17	696	.	17	.	350	.	.	1	.	5	.	1197
	2.3	.	4	.	1	.	4	.	272	18	33	17	.	7	1	.	.	.	357
	2.4	.	1	5	3	.	172	68	11	1398	17	47	13	8	10	1	7	1	1762
	2.6	.	.	.	6	.	1	.	50	23	65	87	.	24	.	.	.	.	256
	2.7	.	1	9	33	.	63	25	34	98	218	1060	15	147	22	10	.	.	1735
	2.14	.	.	.	.	.	.	.	.	.	.	.	.	.	.	.	.	.	.
	2.15	.	.	.	.	.	.	.	.	.	.	.	.	.	.	.	.	.	.
	2.16	.	1	.	.	.	17	.	9	43	40	74	22	635	565	130	.	.	1536
	3.1	.	.	.	.	.	5	.	1	12	4	6	.	74	190	283	.	.	575
Mosaic		.	.	2	5	.	280	13	.	118	5	136	30	63	5	.	2	.	659
Bare soil		.	2	.	3	.	.	1	.	.	.	.	.	.	.	.	14	.	20
Water		.	4	.	.	.	2	1	2	1	.	.	.	.	.	.	1	33	44
Total		.	113	66	503	.	913	818	395	2015	415	1965	120	1375	794	437	40	46	10015

Table VII.11B. Transitions between the coena in the area grazed from 1972 onwards between 1976 and 1981. The species composition of the coena is given in the Tables VII.1, 2, 3 and 4.

		1981							GRAZED										
		1.2	1.3	1.4	1.6	1.7	2.1	2.2	2.3	2.4	2.6	2.7	2.14	2.15	2.16	3.1	Bare soil	Water	Total
1976	1.2	.	.	.	.	.	.	.	.	.	.	.	.	.	.	.	.	.	.
	1.3	.	78	11	6	.	.	.	1	.	.	2	.	.	.		4	2	104
	1.4	.	6	25	13	.	.	.	.	3	.	3	.	.	.	.	.	.	50
	1.6	.	13	28	139	.	.	18	296	.	.	14	.	.	.	.	.	.	508
	1.7	.	.	.	1	.	.	.	.	.	.	.	.	.	.	.	.	.	1
	2.1	.	.	.	1	.	184	.	3	390	3	217	12	20	.	2	4	.	836
	2.2	.	.	.	128	.	5	367	117	78	.	102	.	.	.	.	.	.	802
	2.3	.	4	.	6	.	.	.	336	5	44	12	.	1	.	.	2	.	410
	2.4	.	1	1	2	.	182	1	1	1845	.	11	.	3	1	1	2	2	2053
	2.6	.	.	.	.	.	.	3	46	12	201	158	.	8	.	.	1	.	429
	2.7	.	2	3	52	.	12	76	275	5	42	1133	13	60	3	1	.	.	1677
	2.14	.	.	.	.	.	3	.	.	1	1	42	21	44	3	.	.	.	115
	2.15	.	.	.	.	.	16	.	2	3	8	119	59	1113	37	2	.	.	1359
	2.16	.	.	.	.	.	1	.	.	1	2	5	7	125	668	36	.	.	845
	3.1	.	.	.	.	.	.	.	.	1	.	3	2	3	127	292	3	.	431
Bare soil		.	10	3	1	.	1	.	4	1	.	.	.	.	.	.	5	5	30
Water		.	.	.	.	.	.	1	.	.	.	.	.	.	.	.	.	50	51
Total		.	115	75	349	.	404	466	1081	2345	301	1821	114	1377	839	334	21	59	9701

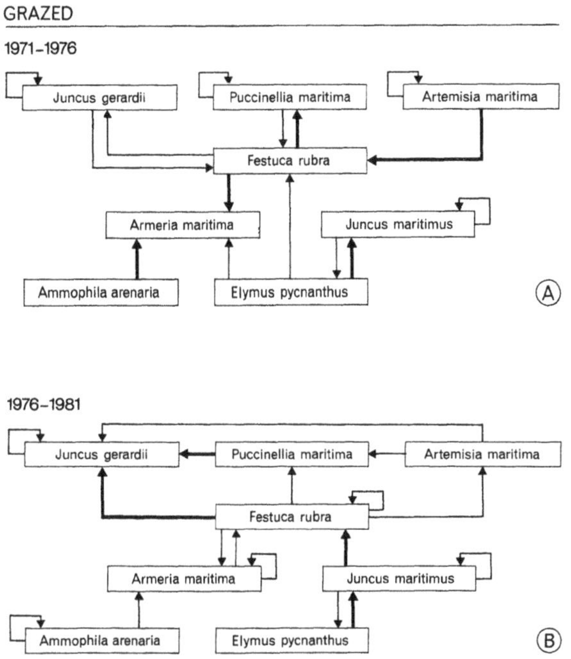

Fig. VII.14A–B. Simplified successional schemes of A: the period between 1971 and 1976 and B: the period between 1976 and 1981, after the resumption of grazing in 1972 of the salt marsh which had not been grazed since 1958. Thick arrows indicate transitions concerning >20% of the initial area, thin arrows between 10 and 20%, closed circuits indicate a steady state (>50% of the initial area remained) (after Norder & Ruyter, 1977; Looijen et al., 1984).

zone and lower situated areas during the second period. The *Festuca rubra* community situated at the transition of higher (fresher) and lower (salter) salt marsh seems very unstable as can be seen by its changes towards and from other communities. This is partly reflected in the percentage of unchanged points, in comparing the vegetation maps. (Table VII.10A). Particularly the coena 2.6 (*Juncetum gerardii leontodontetosum* variant with *Juncus gerardii*) and 2.14 (*Ononido-Caricetum distantis*) showed low percentages of unchanged points.

The *Elymus pycnanthus* community (coenon 2.1) also showed a low percentage of unchanged points, indicating little stability. This could be due to the dominance of *Elymus pycnanthus* in a number of different communities which became apparent again after the disappearance of *Elymus pycnanthus*. The sometimes high percentage of changed points in some coena of the *Puccinellia maritima* community is probably due to the large fluctuations of the annual species. The communities with low percentages of unchanged points that decreased in area can emerge at places previously occupied by other communities. Their extinction cannot, therefore, be predicted. The communities of *Artemisia maritima* (coenon 2.2) and *Ammophila arenaria* (coena 2.17 and 3.1) with a decreasing area but high percentage of unchanged points apparently cannot shift to other places and hence seem bound to disappear under continued grazing with the present stocking rate (1.6 animals $\cdot$ ha^{-1}).

Some transitions between coena in this and the next section are described in detail in section VII.C.2.

VII.D.3. Changes during continued abandoning

The *Puccinellia maritima*- (coena 1.2, 1.3, 1.4, 1.6, 1.7) and the *Festuca rubra* community (coena 2.6, 2.7, 2.14) showed little change, whereas the *Elymus pycnanthus* community (coenon 2.1) showed a large area increase in comparing the generalized vegetation maps (Fig. VII.13) and the areas under the different community (Table VII.10B). Both the *Artemisia maritima*- (coenon 2.2) and the *Ammophila arenaria* community (coena 2.16 and 3.1) showed a decrease.

The complexity of transitions between the different communities involving both increase and decrease of their area are shown in details in Table VII.12A–B. The main transitions are summarized in Fig. VII.15A–B for the periods 1971 to 1976 and 1976 to 1981, respectively. During the first period the *Elymus pycnanthus* community (coenon 2.1) originated from the *Artemisia maritima*- (coenon 2.2), the *Festuca rubra* – (coena 2.6, 2.7, 2.14) – and the *Juncus maritimus* community (coenon 2.4). This process went on partly from 1976 to 1981, but the *Juncus maritimus* community also partly originated from the *Elymus pycnanthus* community, and the *Festuca rubra* community. The differences with respect to the 1971 to 1976 period are due to local hay-making outside the fence along the second creek.

The percentages of unchanged points of the *Ely-*

Table VII.12A. Transitions between the coena in the continuously abandoned area during the period between 1971 and 1976. The species composition of the coena is given in the Tables VII.1, 2, 3 and 4.

	1976						ABANDONED											
	1.2	1.3	1.4	1.6	1.7	2.1	2.2	2.3	2.4	2.6	2.7	2.14	2.15	2.16	3.1	Bare soil	Water	Total
1971 1.2	96	8	37	30	.	4	14	.	.	.	10	.	.	.	.	.	.	199
1.3	.	15	73	.	20	1	.	.	.	.	27	.	.	.	.	.	18	154
1.4	.	.	6	2	7	3	12	.	4	.	.	.	.	.	.	.	1	35
1.6	.	.	.	.	.	.	.	.	.	.	.	.	.	.	.	.	.	.
1.7	.	.	.	.	.	.	.	.	.	.	.	.	.	.	.	.	.	.
2.1	.	.	.	.	1	518	68	1	141	.	14	.	24	20	6	1	10	804
2.2	.	2	8	1	18	135	802	6	1	.	4	.	.	.	.	.	10	987
2.3	.	.	.	.	.	4	.	62	5	9	5	.	.	.	.	.	.	85
2.4	.	2	.	.	.	228	153	6	392	.	16	.	.	4	1	.	18	820
2.6	.	.	.	.	.	.	.	.	.	.	.	.	.	.	.	.	.	.
2.7	.	.	.	.	.	114	21	.	31	3	188	.	2	8	2	.	2	371
2.14	.	.	.	.	.	.	.	.	.	.	.	.	.	.	.	.	.	.
2.15	.	.	.	.	.	.	.	.	.	.	.	.	.	.	.	.	.	.
2.16	.	.	.	.	.	68	2	.	4	2	19	.	.	78	3	.	.	176
3.1	.	.	.	.	.	17	1	.	4	4	3	.	2	15	47	.	.	92
Mosaic	.	.	.	.	.	113	3	1	14	.	11	2	.	1	.	.	2	147
Bare soil	.	.	.	.	.	.	.	.	.	.	.	.	.	.	.	.	.	.
Water	.	14	8	.	15	15	14	.	13	.	.	.	.	.	.	.	.	79
Total	96	41	132	33	61	1220	1089	76	609	18	297	2	28	126	59	1	61	3949

Table VII.12B. Transitions between the coena in the continuously abandoned area between 1971 and 1976. The species composition of the coena is given in the Tables VII.1, 2, 3 and 4.

	1981						ABANDONED											
	1.2	1.3	1.4	1.6	1.7	2.1	2.2	2.3	2.4	2.6	2.7	2.14	2.15	2.16	3.1	Bare soil	Water	Total
1976 1.2	.	2	11	3	.	.	.	.	.	.	.	.	.	.	.	.	6	22
1.3	1	3	23	1	.	1	.	.	.	.	.	.	.	.	.	.	5	34
1.4	2	4	79	6	20	1	10	.	.	.	4	.	.	.	.	2	3	131
1.6	.	.	3	21	4	1	2	.	.	.	9	.	.	.	.	.	.	40
1.7	.	.	14	4	18	2	5	.	1	.	2	.	.	.	.	.	8	54
2.1	.	.	6	1	.	846	2	.	314	.	62	40	.	6	1	.	12	1290
2.2	.	.	9	19	4	419	416	.	91	.	96	.	.	.	.	2	12	1068
2.3	.	.	.	.	.	.	.	52	2	.	.	.	.	.	.	.	.	54
2.4	.	..	.	.	.	155	2	.	410	.	2	1	.	.	.	.	6	576
2.6	.	.	.	.	.	6	.	.	.	1	4	.	.	.	.	.	.	11
2.7	.	.	.	.	.	42	18	.	4	5	194	23	.	.	.	.	.	286
2.14	.	.	.	.	.	.	.	.	.	.	.	2	.	.	.	.	.	2
2.15	.	.	.	.	.	2	.	.	.	.	1	19	.	.	.	.	.	22
2.16	.	.	.	.	.	91	.	.	.	.	2	1	.	39	11	.	.	144
3.1	.	.	.	.	.	.	.	.	.	.	.	.	.	10	46	.	.	56
Bare soil	.	.	.	.	.	.	.	.	.	.	.	.	.	.	.	.	.	.
Water	.	.	15	.	1	1	1	.	6	.	.	.	.	.	.	.	103	127
Total	3	9	160	55	47	1567	456	52	828	6	376	86	.	55	58	4	155	3917

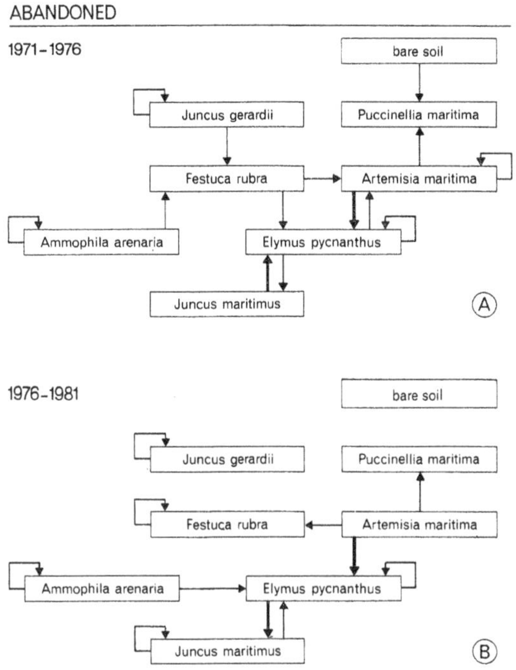

Fig. VII.15A–B. Simplified successional schemes of A: the period between 1971 and 1976 and B: the period between 1976 and 1981, of the continuously abandoned salt marsh which had not been grazed since 1958. Thick arrows indicate transitions concerning >20% of the initial area, thin arrows between 10 and 20%, closed circuits indicate a steady state (>50% of the initial area remained) (after Norder & Ruyter, 1977; Looijen et al., 1984).

mus pycnanthus community are higher than in the grazed area, because *Elymus pycnanthus* mainly spread from existing clones.

VII.D.4. Exclosures

Two exclosures were built (c. 60 × 10 m) in the continuously grazed part of the salt marsh in the autumn of 1973 (area A in Fig. IV.17) in order to illustrate the effects of ceasing grazing. One exclosure was situated at the gradient from dune to upper salt marsh featuring the zonation from *Ammophila arenaria* – via *Festuca rubra* – towards a *Juncus maritimus* community (Fig. VII.16) (Photo VII.16). Although *Ammophila arenaria* and *Festuca rubra* became more vigorous in the exclosure their communities did not spread. Initially the *Festuca rubra* community, with some *Elymus pycnanthus,* spread lower down the gradient and later

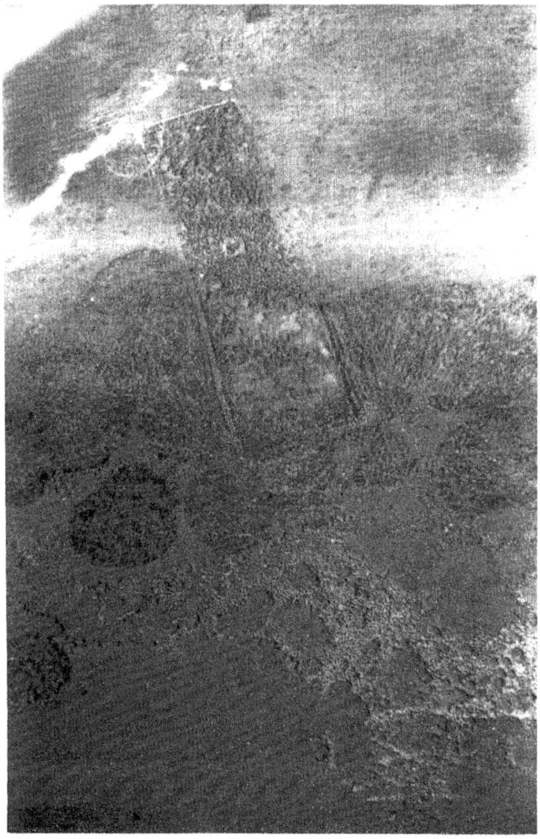

Photo VII.16. Exclosure at the higher *Festuca rubra* salt marsh towards the low dunes (white) with a *Juncus maritimus* zone (dark) at the gradient to the lower *Puccinellia maritima* salt marsh.

Elymus became so dominant that the *Festuca rubra* community changed into the *Elymus pycnanthus* community in the exclosure. The *Juncus maritimus* community became completely overgrown by *Elymus pycnanthus* (Photo's VII.17, 18 and 19) which also spread higher at the gradient. It may be expected that, except for the *Ammophila arenaria* community, the whole gradient would be covered by the *Elymus pycnanthus* community in the exclosure.

In the continuously grazed area the *Ammophila arenaria* community and the area of *Elymus pycnanthus* in the highest part of the gradient decreased. The *Juncus maritimus* community was replaced locally by the *Juncus gerardii* community. All these changes were possibly due to the increase in trampling/grazing along the fence since 1973.

Photo VII.17. The gradient of the *Festuca rubra* community (left) to the *Juncus maritimus* community with *Elymus pycnanthus* (right) before the construction of the exclosure.

Photo VII.18. The same situation as in Photo VII.17 one year after the construction of the exclosure. Note the high flowering *Juncus maritimus* tillers.

Photo VII.19. The same situation as in Photo VII.17 two years after the construction of the exclosure. Note the dominance of *Elymus pycnanthus* in the *Juncus maritimus* community, flowering tillers of *Juncus maritimus* are only visible adjacent to the fence. *Festuca rubra* dominates higher at the gradient.

Photo VII.20. Exclosure at the lower salt marsh with creeks of different size. The round clumps of *Juncus maritimus* indicate the gradient to the higher salt marsh.

Photo VII.21. A small creek with bare soil immediately after the construction of the exclosure.

Photo VII.22. The same situation as in Photo VII.21 one year after the construction of the exclosure.

Photo VII.23. The same situation as in Photo VII.21 four years after the construction of the exclosure. Note the increase of *Suaeda maritima* and *Limonium vulgare* at the lower parts and of *Artemisia maritima* at the higher parts.

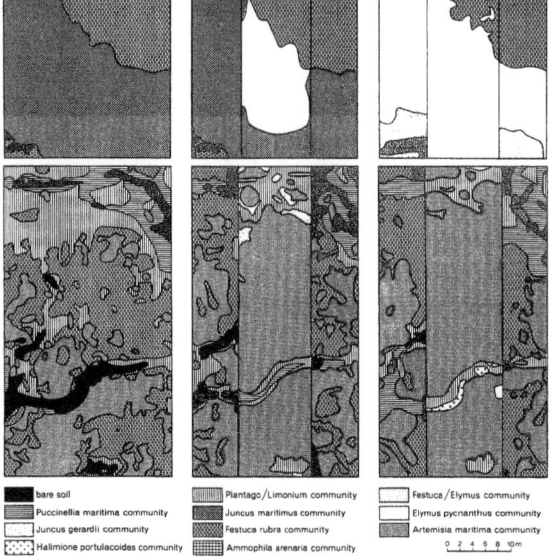

Fig. VII.16A–C. Simplified vegetation maps for A: 1973, B: 1978 and C: 1985 of two exclosures in the Oosterkwelder of Schiermonnikoog established in the autumn of 1973 (after Schwab & Schwab-Vos, 1974). The upper exclosure is situated at the transition of dune to higher salt marsh, the lower one lies at the lower salt-marsh.

←

The transformation of the *Juncus maritimus* community to the *Elymus pycnanthus* community is remarkable but seems related to the wet summer of 1985 which enhanced the production of *Elymus pycnanthus* and it was moreover protected by *Juncus maritimus*. Moreover, the *Elymus pycnanthus* coverage was much lower than inside the exclosure.

The second exclosure was built in the higher and mid salt marsh with *Festuca rubra*- and *Artemisia maritima* communities alternating with the *Puccinellia maritima* community including *Plantagini-Limonietum* and bare soil due to trampling the weak bottom (Fig. VII.16) (Photo VII.20). The area of bare soil disappeared completely in the exclosure and transformed into the *Puccinellia maritima* community of which mainly *Plantagini-Limonietum*, but locally *Halimionetum portulacoi-*

dis established in 1985 (Photos VII.21, 22 and 23). In 1978 the initial *Puccinellia maritima* community was replaced by the *Juncus gerardii-* and the *Artemisia maritima* community locally. The latter overgrew the newly emerged *Juncus gerardii* community and the whole *Festuca rubra* community inside the exclosure and the same held locally for the continuously grazed area, which might be attributed to silting up. It should be noted that the *Elymus pycnanthus* community was found inside the exclosure in 1985. The transformation of *Plantagini-Limonietum* to *Puccinellia maritima* community in the continuously grazed area could be due to heavier trampling/grazing along the fence.

The main transitions are summarized in Fig. VII.17A–B for the periods 1973 to 1978 and 1978 to 1985, respectively in the continuously grazed area and in Fig. VII.18A–B in the ungrazed area since 1973.

VII.D.5. Comparison of grazed and ungrazed salt marsh

Some structural characteristics (height of the canopy, cover and number of species) differed considerably between the grazed and ungrazed salt marsh. Difference can be seen between 1976 grazed and ungrazed relevés as compared with 1971 relevés (Fig. VII.19). Although it would be more correct to use permanent plot information, some general structural changes can be discerned in the same communities in which permanent plots were situated after the resumption of grazing. There was a decrease in mean height, in mean herb cover, and probably in mean litter cover resulting in a more open canopy and an increase of species diversity. Only the *Puccinellia maritima* community showed a decrease of species diversity which may be due to the destruction of the turf. Structural changes not only favoured the species already present but also

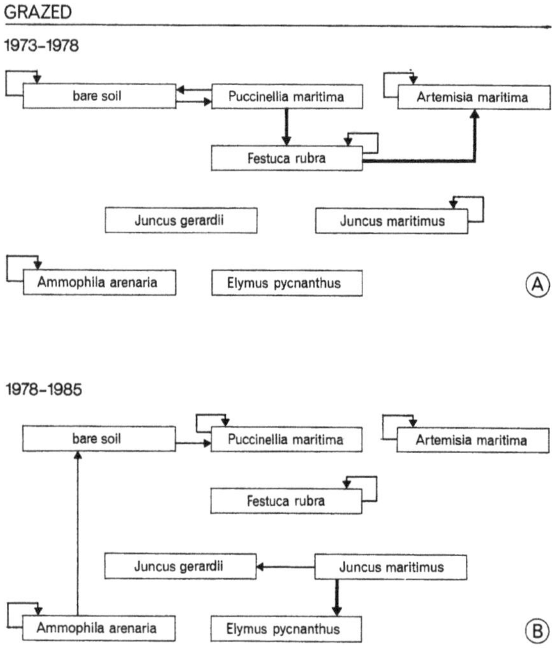

Fig. VII.17A–B. Simplified successional schemes of A: the period between 1973 and 1978 and B: the period between 1978 and 1985 of the continuously grazed salt marsh. Thick arrows indicate transitions concerning >20% of the initial area, thin arrows between 10 and 20%, closed circuits indicate a steady state (>50% of the initial area remained).

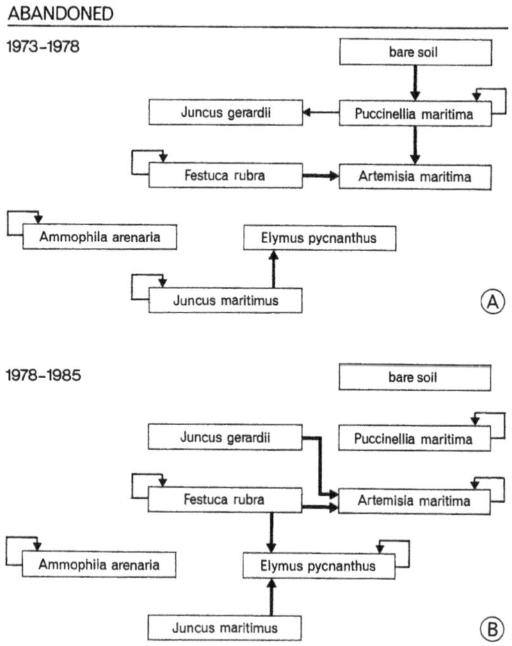

Fig. VII.18A–B. Simplified successional schemes of A: the period between 1973 and 1978 and B: the period between 1978 and 1985, after the cessation of grazing in the autumn of 1973. Thick arrows indicate transitions concerning >20% of the initial area, thin arrows between 10 and 20%, closed circuits indicate a steady state (>50% of the initial area remained).

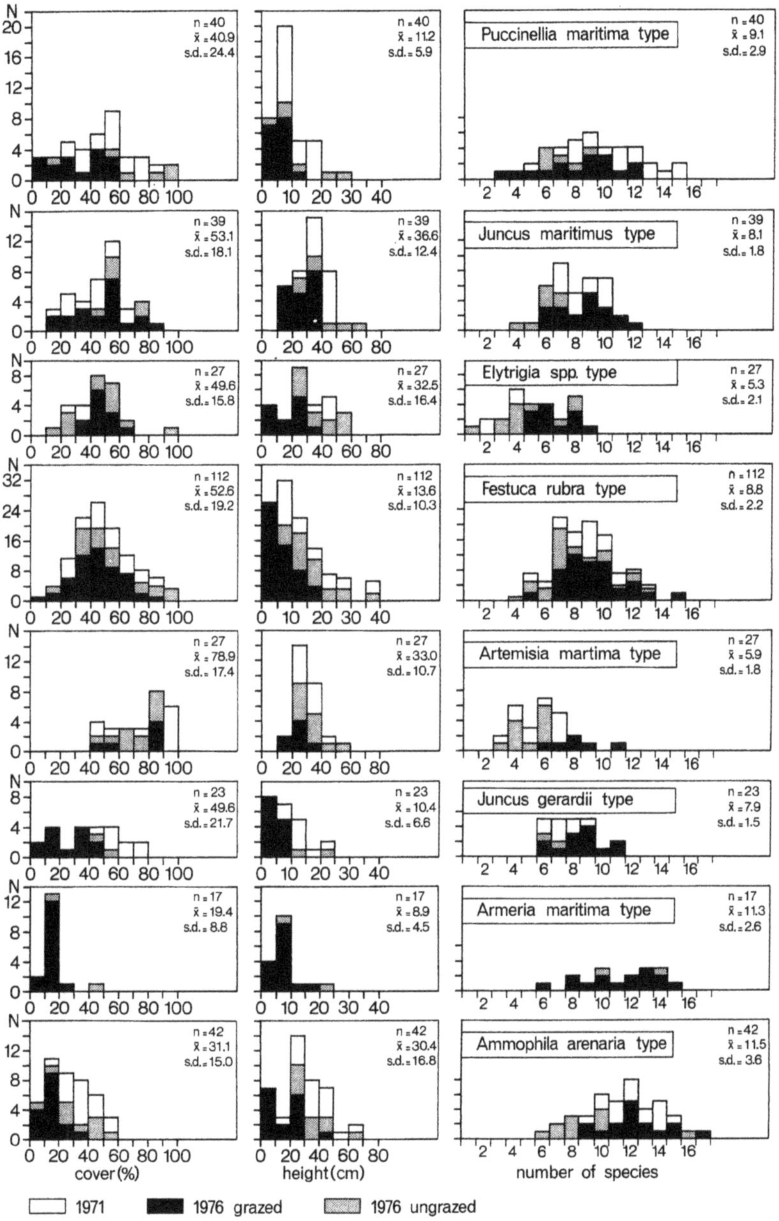

Fig. VII.19. Structural characteristics (height, cover and number of species) in 1971, 1976 grazed and 1976 continuously abandoned. n = number of relevés, x̄ = mean, s.d. = standard deviation (after Bakker & Ruyter, 1981).

facilitated the establishment of new species. These aspects were already discussed in section VII.C.

Comparison of the sequential vegetation maps revealed that, in the gradient from depressions to low dunes, changes in the zonation of the coena 2.3/2.6/2.7 differed between the grazed and ungrazed area as can be seen from the transitions between coena in Table VII.11A–B and 12A–B (Fig. VII.20). In the grazed area relatively (%) more changes were towards coena occurring previously lower down the gradient than in the ungrazed area, both for the periods 1971 to 1976 and 1976 to

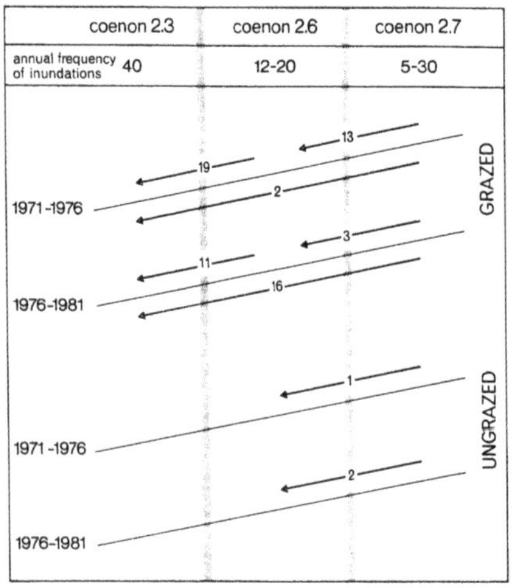

Fig. VII.20. Changes between communities at the transition from depressions in the salt marsh (coenon 2.3) to the higher salt marsh (coenon 2.7) from the periods between 1971 and 1976 and 1976 to 1981, respectively, in grazed and continuously abandoned areas. Arrows indicate the direction of change and its percentage of the initial community area (after Norder & Ruyter, 1977; Looijen et al., 1984).

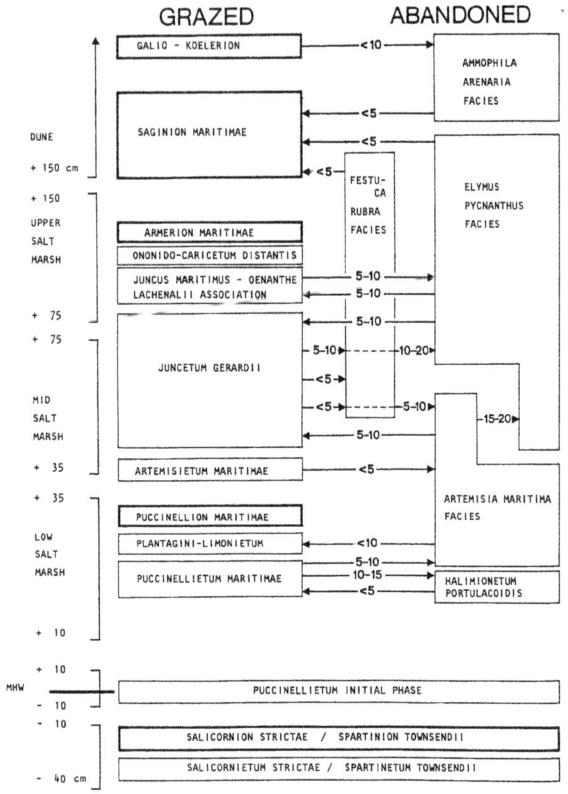

Fig. VII.21. The position of syntaxa with respect to MHW and the number of years it takes to change after abandoning or the resumption of heavy grazing (1.6 heifers · ha^{-1}).

1981. Since the zoning of *Juncetum gerardii* inops variant with *Triglochin maritima* (coenon 2.3)/*Juncetum gerardii armerietosum* variant with *Triglochin maritima* (coenon 2.6)/*Juncetum gerardii armerietosum* variant inops (coenon 2.7) corresponds with a successional series according to Westhoff (1947) and for the first two coena also with the annual frequency of inundations, 'climbing up' the higher salt marsh of lower salt marsh communities seemed to take place. On average 15% of the involved communities changed into lower situated or 'salter' communities in the grazed area, whereas this percentage amounted to 1% in the ungrazed area during the period 1971–1976. These figures were 17% and 2%, respectively, during the 1976 to 1981 period. This phenomenon can be called retrogressive succession (Beeftink, 1965). Progressive succession was, however, found on a smaller scale. In the ungrazed area only progressive succession apparently took place. The same phenomena were found in the gradient from the lower salt marsh to creekbank levees.

VII.E. Comparison with other areas and discussion

VII.E.1. Principal succession

Dijkema (1983; 1984) constructed a scheme revealing the principal succession series on salt marshes in the Dutch, German and Danish Wadden Sea at the syntaxonomical level of associations. Beeftink (1977A) published similar schemes for the south-western part of the Netherlands. Working in a small area a finer-scaled system of coena was needed for this study which fits very well in Beeftink's and Dijkema's schemes after lumping the coena to the association level (Fig. VII.21). Both authors do indicate grazing in their schemes, but they cannot give any details with respect to the time involved in the transitions. As a result of the comparison of grazed and ungrazed areas with the number of years after abandoning kept in mind, and particularly from sequential mapping and permanent plots, a time scale in the succession schemes could be introduced. Analogous to Dijkema's scheme, the associations occurring in the Oosterkwelder salt marsh area were ranged according to their position relative to MHW and their reaction upon sudden grazing or abandoning (Fig. VII.21). The communities with an *Elymus pycnanthus* facies after abandoning occurred in the mid and upper salt marsh and the dunes, whereas Dijkema (1983) observed an *Elymus repens* facies between 50 cm and 150 cm MHW. *Elymus repens* is found particularly in the mainland salt marshes which contain much clay and are subjected to artificial fertilizers.

It is obvious that the community diversity drops after ceasing grazing. Several changes towards the *Elymus pycnanthus*- or *Artemisia maritima* facies pass through the *Festuca rubra* facies, whereas some miss out on this intermediate stage. The *Festuca rubra* facies, however, can also be a terminal stage. Indications can be derived from section VII.D that cyclic changes are possible after a longer period of abandoning between *Elymus pycnanthus*- and *Artemisia maritima* facies on the one hand and their 'grazed' counterparts on the other. Cyclic patterns are known for the *Halimionetum portulacoidis* when the dominant species died back after severe frost (Beeftink, 1977A). In the long term the *Ammophila arenaria* facies can be succeeded by *Hippophaë rhamnoides* scrub (Dijkema, 1980). It can be noted that with a change in the management regime it often takes ten to twenty years after ceasing the grazing regime to reach a new equilibrium. The return to the initial situation, after the resumption of grazing, generally takes five to ten years and when combined with a mowing regime it takes less than five years.

It should be kept in mind that the number of years needed for such a change after renewed grazing holds for a stocking rate of 1.3–1.7 yearlings/ha. This rate resembles Beeftink's grazing for domestic purpose (Beeftink, 1977B) and Dijkema's moderate to heavy grazing (Dijkema, 1984). At lower stocking rates it seems likely that changes will take place more slowly and/or more locally, due to grazing patterns developed. This last point will be discussed in chapter VIII.

Westhoff (1985) warns against comparing grazed and ungrazed salt marshes. Since nearly all west European salt marshes are either grazed by domestic stock or only recently abandoned, it is hardly known how the salt marsh ecosystem would be structured under natural conditions. One of the salt marshes which has never been grazed by domestic stock is the Boschplaat, a reserve of about 4000 ha developed after 1932 on the Dutch Frisian Island of Terschelling. Roozen & Westhoff (1985) described zonation and succession on the Boschplaat salt marsh. *Salicornietum strictae* is the pioneer community. The *Plantagini-Limonietum*, gradually succeeding into the *Halimionetum portulacoidis*, covers the lower salt marsh. *Halimionetum portulacoidis* and *Artemisietum maritimae* occur on creek bank levees, locally replaced by the *Atriplici-Agropyretum pungentis*. The *Junco-Caricetum extensae* covers the higher sandy salt marsh, whereas the *Juncetum gerardii leontodontetosum* is found on silt. It should be stressed, that on the lower salt marsh, graminoids, particularly grasses, hardly contribute to vegetation cover, as it might be expected (Westhoff, 1985).

Westhoff (1987) published recently a complete zonation and succession scheme of the Boschplaat salt marsh for the period 1946 to 1986. He mentioned a community of *Limonium vulgare* and *Halimione portulacoides* which might change into a *Halimionetum portulacoidis* in the depressions of the lower salt marsh and an *Atriplici-Agropyretum pungentis* which has already emerged on the banks. An *Artemisietum maritimae* is found on the middle high salt marsh which might change into an *Atriplici-Agropyretum pungentis*. On the high salt marsh a *Junco-Caricetum extensae*, maybe as variant with *Juncus maritimus*, has developed and might change into a *Festuca rubra* facies of the *Juncetum gerardii*. On driftline material on the upper salt marsh an *Atriplici-Agropyretum pungentis* has developed. This scheme is not fully in agreement with the principal succession in this study (Fig. VII.21). In particular the transitions to the *Atriplici-Agropyretum pungentis* on the middle high salt marsh and to the *Festuca rubra* facies of the *Juncetum gerardii* are only supposed and not (yet) observed, whereas a transition of the latter into the *Atriplici-Agropyretum pungentis* is even not expected. These differences are perhaps due to the fact that the Boschplaat salt marsh has only developed during 50 years, whereas that of the Oosterkwelder already over 100 years.

Another salt marsh (300 ha) which is partly never grazed, is found on the peninsula of Skallingen, Denmark. Only the lower salt marsh is ungrazed and here a succession took place from a community dominated by *Salicornia europaea* and *Puccinellia maritima* into a community dominated by *Halimione portulacoides* between 1933 and 1978 (Jensen, 1985). Lower salt marshes in the United Kingdom which have never been grazed feature a mosaic of *Puccinellietum maritimae* and *Halimionetum portulacoidis*. In the mid marsh zone *Halimionetum portulacoidis* and *Plantagini-Limonietum* are well-represented. The *Atriplici-Agropyretum pungentis* is wide spread in the south-eastern upper marshes, and the *Agropyretum repentis maritimum* in the western upper marshes (Adam 1978; 1981). Our study area mostly resembles the south-eastern salt marshes in the United Kingdom.

Comparing the information about succession gained from sequential mapping and from permanent plots (see sections II.C and V.F.2) it can be concluded, in agreement with Londo (1974), that a limited number of permanent plots can give a relatively large amount of information. Nevertheless the permanent plot data revealed only 50% of the transitions obtained from the mapping data in the period 1971 to 1976 in the study area (Norder & Ruyter, 1977). It can be concluded that mapping is necessary to reveal all successional changes in the whole study area and that permanent plot data reveal changes of all occurring species in restricted areas.

VII.E.2. Exclusion of grazing animals

Although most of the existing information about salt marsh grazing is based on comparisons, on a single occasion, between grazed and ungrazed salt marshes, some evidence can be found concerning the reaction of the vegetation after excluding animals. In the grazed part of the Boschplaat salt marsh at Terschelling, the Netherlands, exclosures were established in the mid- and upper salt marsh and low dunes in 1972 and their effect after seven years upon the vegetation was described by Heinemeyer (1982). Species of the *Puccinellion maritimae, Armerion maritimae, Saginion maritimae* and to a lesser extent the *Nanocyperion flavescentis* decreased or disappeared from the *Juncetum gerardii leontodontetosum, Scirpetum rufi* and *Lolio-Potentillion anserinae* communities. In particular *Juncus gerardii, Glaux maritima* and *Plantago maritima* decreased, *Festuca rubra* formed a facies in nearly all communities, moreover *Phragmites australis, Elymus pycnanthus, E. repens* and *Calamagrostis epigejos* spread to a large extent. *Saginion maritimae* and *Galio-Koelerion* species decreased in the low dune communities. Heinemeyer stressed that halophytes occurring up to the feet of low dunes in the grazed areas with closed turf were set back to lower areas with respect to MHW in the plots after excluding domestic stock.

Festuca rubra dominance after ceasing grazing was also reported in the United Kingdom. In a decade the exclusion of grazing animals resulted in a change of a heavily grazed *Puccinellia maritima* marsh to a *Festuca rubra* or an *Elymus pycnanthus* marsh in Bridgwater Bay (Ranwell, 1968). After the exclusion of rabbits an exposed *Armeria maritima* turf in the Pembrokeshire Islands transformed, sometimes within a few months, sometimes after several years, into a dense *Festuca rubra* stand. These *Festuca rubra* stands were regarded as the natural climax in the strongly exposed situations (Gillham, 1955). Disappearance of rabbits after a myxomatosis epidemic in 1954 caused a large increase of *Festuca rubra* until 1959 in the vegetation of Blakeney Point (White, 1961). The initial vegetation resembled our *Ammophila arenaria* community.

Hansen (1982) established exclosures in the very heavily grazed mainland salt marshes of the German North Sea coast. *Suaeda maritima* and *Salicornia europaea* decreased, whereas *Armeria maritima, Aster tripolium, Triglochin maritima* and *Artemisia maritima* increased in the *Puccinellietum maritimae.* In the *Juncetum gerardii leontodontetosum Festuca rubra, Spergularia maritima* and *Limonium vulgare* increased within three years, whereas *Glaux maritima, Suaeda maritima* and *Salicornia europaea* decreased.

At the Skallingen salt marsh, on the North Sea coast of Denmark, all lower salt marsh species increased after six years of exclusion from grazing in a formerly very open *Puccinellietum maritimae* (Jensen, 1985). *Halimione portulacoides* and *Festuca rubra* spread and *Artemisia maritima* appeared, whereas lower salt marsh species decreased or disappeared in an initially closed *Puccinellietum maritimae* stand. *Artemisia maritima* developed a facies in the *Artemisietum maritimae* (Jensen, 1978). The *Juncetum gerardii leontodontetosum* and *Lolio-Potentillion anserinae* communities were covered with *Phragmites australis* as in the Boschplaat salt marsh in the Netherlands (Heinemeyer, 1982).

The spread of *Phragmites australis* in exclosures is found in salt marshes adjacent to large dunes providing fresh seepage water like at the Boschplaat and Skallingen or in marshes in mesohaline zones where *Spartina anglica* marshes can be transformed into *Phragmites* marshes in the absence of grazing (Ranwell, 1968). This phenomenon is often found in the mesohaline Baltic. In grazed brackish salt marshes in eastern Denmark the zonation features *Puccinellia maritima-, Juncus gerardii/Festuca rubra-,* and *Festuca rubra* dominated communities (Vestergaard, 1978). Within ten years after the ceasing of grazing, *Phragmites australis* reached dominance in all communities with a co-dominance of *Scirpus maritimus* in the lowest and *Elymus repens* and *Artemisia maritima* in the highest part of the gradient.

Somewhat further in the Baltic Schmeisky (1977) described the grazed salt marsh zonation near Heiligenhafen, Federal Republic of Germany, featuring *Puccinellietum maritimae, Plantago maritima* community, *Juncetum gerardii* typicum with *Juncus gerardii* dominance, *Juncetum gerardii* typicum with *Festuca rubra* dominance, *Juncetum gerardii leontodontetosum* and *Lolio-Potentillion anserinae* at the sand barrier. The counterpart which had been ungrazed for twenty years revealed, at a single occasion comparison, the zonation *Phragmites australis* community, *Juncus maritimus-Oenanthe lachenalii* association, *Juncetum gerardii* typicum with or without *Festuca rubra* dominance and *Agropyretum repentis maritimum* at the sand barrier. Three years after the establishment of exclosures species only grew taller at the lower salt marsh, whereas *Festuca rubra* increased in the *Juncetum gerardii* and *Lolium perenne* was replaced by *Elymus repens* at the sand barrier (Schmeisky, 1974). After nine years *Plantago maritima* and *Festuca rubra* spread at the cost of *Juncus gerardii* and *Glaux maritima* in the *Juncetum gerardii* typicum, whereas no changes took place in the *Festuca rubra* dominated *Juncetum gerardii* (Scmeisky, 1978).

Christiansen (1937) recorded permanent plots during the first ten years of excluding domestic stock on an island in the mouth of the Schlei, Germany. The *Puccinellia maritima* community transformed into a *Juncus gerardii* community, the *Juncus maritimus* community into an *Agrostis stolonifera* community, the *Agrostis stolonifera* community into a *Festuca rubra* community, the *Juncus gerardii* community into a *Festuca rubra* community, the *Festuca rubra* community into a *Festuca rubra* facies. *Phragmites australis* was not found on the small island. Summarizing, the lower occurring communities were excluded by the higher ones and thus lower salt marsh species were pushed aside down the gradient, just the reverse of what was found in this study after the resumption of grazing.

In some of the excluding experiments which have been described species diversity decreased, but the observation period was often too short to measure such an effect. The relative equitability, however, diminished in all experiments which agrees with our results (see also Bakker, 1985) due to the dominance of a single species. Bobbink & Willems (1987) also found that initially the equitability and later the number of species declined.

Resumption of grazing and/or hay-making after ceasing these practices is often recommended, but is rarely practised and hardly monitored. The only known example of grazing an always ungrazed salt marsh is found on the Boschplaat, Terschelling in the Netherlands, where a *Halimonietum portulacoidis* transformed into a *Suaeda maritima* community. *Agrostis stolonifera, Juncus gerardii* and *Glaux maritima* increased in a *Festuca rubra* dominated stand, and an *Elymus pycnanthus* community transformed into a *Festuca rubra* turf (pers. comm. Oosterveld). Our results are in agreement with Oosterveld's finding that in both the latter communities lower salt marsh species also invaded.

Some inconsistencies may seem to occur in the afore-mentioned survey of the effects of excluding animals from a previously grazed salt marsh. The anomaly can be explained by taking the local grazing intensity into account prior to the establishment of exclosures. In the often bare soil of heavily grazed areas lower salt marsh species always occur. They can initially spread after the cessation of grazing, and later become overgrown by species like *Festuca rubra*. In the closed turf of a moderately grazed *Festuca rubra* sward, only a few lower salt marsh species are found. After the cessation of grazing they decrease immediately by competitive exclusion by the vigorous *Festuca rubra*.

VII.E.3. Comparison of grazing and cutting regimes

Although hay-making is mentioned as an agricultural practice in salt marshes (Siira 1970; Beeftink 1977B; Dijkema 1984) hardly any data are available concerning the species composition and the abundance of longlasting cutting regimes. Todeskino (in Härdtle, 1984) compared hay-making once a year without fertilizer applications and hay-making twice a year with fertilizer applications in northern Germany (North Sea coast). The management practices had been imposed for twenty years. *Aster tripolium, Armeria maritima, Triglochin maritima, Plantago maritima* and *Juncus gerardii* hardly occurred on the areas which were cut twice and were replaced by *Atriplex prostrata* and *Elymus repens*.

A comparison of salt marshes which have been grazed and mown for several decades is only possible in a *Juncetum gerardii* community in northern Germany (North Sea Coast) (Wetzel, 1966). The regimes have a large abundance of *Juncus gerardii* and little *Aster tripolium, Atriplex littoralis* and *Armeria maritima* in common. The grazing regime is characterized by much more *Puccinellia maritima* and some more *Festuca rubra* than the mowing regime, whereas *Trifolium fragiferum* and *Glaux maritima* are only found in the grazed area. *Agrostis stolonifera* and particularly *Triglochin maritima* and *Plantago maritima* are more abundant in the cut than in the grazed area. The thriving of *Puccinellia maritima* and *Glaux maritima* in the grazing regime is in agreement with our observations, but we did not find the abundance of *Triglochin maritima* and *Plantago maritima* in the mown areas. Perhaps the performance of mowing regimes for only thirteen years is too short to reach the mentioned sward composition.

VII.E.4. Cutting an abandoned salt marsh

Some more experiments concerning mowing an abandoned stand or one which has never been touched are known, probably because it is easier to initiate small-scaled cutting treatments than large-scaled grazing regimes. In the southwest part of the Netherlands *Puccinellietum maritimae* and *Hali-*

mionetum portulacoidis which had been ungrazed for 10 years were subjected to autumn hay-making. *Puccinellia maritima* revealed a temporary set-back during the first five years of the experiment in the *Puccinellietum maritimae. Halimione portulacoides* decreased whereas *Triglochin maritima* increased (Beeftink et al., 1978A). *Halimione portulacoides* dropped from 100% to 10% after the first cut in the *Halimionetum portulacoidis* and was succeeded by *Suaeda maritima* in the third year and by *Puccinellia maritima* in the fifth year (Beeftink et al., 1978A; 1978B).

A fifteen year old salt marsh near Anglesey, United Kingdom, was invaded by *Juncus maritimus* which was supposed to overgrow other species. It was expected that the rate of spread could be effectively reduced (Packham & Liddle, 1970), based upon a unique cutting experiment reducing the aerial biomass and preventing seed formation.

A seventy years old salt marsh in eastern Denmark which had never been exploited, was subjected to an August cutting regime from 1979 onwards (Vestergaard, 1985). The dense *Festuca rubra* stand, including *Phragmites australis,* remained unchanged after four years and only *Juncus gerardii* increased. Lower down the gradient *Scirpus maritimus* stands maintained themselves, whereas high *Phragmites australis* stands became less vigorous and *Agrostis stolonifera* and *Spergularia salina* increased and *Puccinellia maritima* and *Triglochin maritima* appeared in the community lower down the gradient. It seems that lower salt marsh species invade more easily in brackish environments subjected to mowing than in salt conditions of this study.

VII.E.5. Grazing intensity

Grazing favours many halophytes, particularly lower salt marsh species, as Jensen (1985) mentioned when summarizing the findings of 26 authors, which were mainly in agreement with our results. These authors did not, however, always agree in their judgment about the positive or negative effects of grazing on salt marsh species. Beeftink (1977A) mentioned both favourable and harmful effects upon the same species. These disagreements can probably be explained by the particular area which was being studied or the variety of particular plant communities. In an earlier paper (Bakker & Ruyter, 1981) the local disparities were introduced, thus indicating local spread or cover increase and area or cover decrease of species in other parts of the study area. These local disparities have now been worked out thus, revealing the different reactions of plant species subjected to grazing and cutting regimes in Table VII.9A–B. In fact the situation is still more complex when different grazing intensities are considered (Beeftink, 1977A).

The phenomenon of locally increasing and decreasing plant communities in our study area (section VII.D.1) is in agreement with the observations on Terschelling for the *Plantagini-Limonietum*, which was thought only to be found in ungrazed areas, decreasing locally under a grazing regime as expected, but increasing at other sites with an overall hardly changing area (Westhoff, 1985). Although on another level of integration, this seems analogous to the dynamic mosaic patterns of species described by Grubb (1984).

VII.E.6. Progressive or retrogressive succession?

The shift of plant communities will be considered as well as the invading of lower salt marsh species with resumed grazing and their disappearance after the cessation of grazing with respect to progressive and retrogressive succession. Braun-Blanquet & Pavillard (1930) called succession progressive if it went from a pioneer to a climax phase, whereas, retrogressive succession was the opposite. Beeftink (1965) considered changes in the vegetation involving an increasing number of species and increasing complexity of vegetational structure to be progressive succession for example the *Salicornietum strictae- Puccinellietum maritimae- Juncetum gerardii- Lolio-Potentillion anserinae* series. The opposite series is called retrogressive succession and can be caused by both natural (e.g. erosion) and man-made factors (e.g. grazing). Grazing can at least delay progressive succession, as was pointed out by Joenje (1985) who studied waterfowl grazing on newly embanked sandflats, by Westhoff

& Sykora (1979) who described cattle grazing on flats, which had been embanked for 33 years and by Jensen (1985) who referred to cattle and sheep which had been grazing on a young salt marsh for more than 30 years.

Succession on young developing salt marshes is progressive according to both afore-mentioned definitions. Ongoing succession can, however, often result in species-poor plant communities with huge litter accumulation of e.g. *Elymus pycnanthus, E. repens* or *Phragmites australis,* and is progressive according to Braun-Blanquet & Pavillard (1930), but retrogressive according to Beeftink (1965) because of the decreasing number of species and complexity of the vegetational structure.

The introduction of grazing in the *Atriplici-Agropyretum* would cause retrogressive succession according to Braun-Blanquet & Pavillard (1930), but progressive succession according to Beeftink (1965) because of an increase in the number of species and complexity of the vegetational structure. Grazing the *Plantagini-Limonietum* would, however, cause retrogressive succession according to both definitions.

When it is desirable to define succession as either progressive or retrogressive, the successional series from pioneer to climax phase should be taken as a measure. Species diversity and vegetational structure should not be used and perhaps better not even be related to this problem at all, because of the involvement of arbitrarily chosen levels. Beeftink's (1965) definitions of progressive and retrogressive succession are, therefore, confusing.

PART FOUR

Ecosystem dynamics

CHAPTER VIII

Grazing

VIII.A. Introduction

In this chapter grazing intensities and patterns in the vegetation in salt marsh and inland mixed grassland communities are discussed. The questions to be answered are (i) which method(s) can be adopted when measuring grazing intensity?, (ii) which part of the herbage accumulated is utilized by the grazers?, (iii) is there any selection for forage quality and/or maximizing the energy intake rate whilst foraging (optimal foraging)?, (iv) on what scale can patterns in the vegetation develop and maintain themselves and to what extent can this process be influenced experimentally?

The study area Westerholt was originally heathland but was partly ploughed to a depth of 40 cm and sown to grass in the 1950's. It comprised 3 ha of *Calluna vulgaris*/*Erica tetralix*, 2 ha of deciduous woodland (*Betula* spp. and *Salix* spp.) and 6 ha of pasture land (see section V.A.). In October 1972 the area was fenced in and sheep of the Schoonebeeker breed were introduced at a stocking rate of 3 sheep ha^{-1}. Twenty-five sheep grazed during the period July to December and 40 from January to July. In spring of 1980 the flock was replaced and since then the area has received a winter rest period of two months from grazing.

Only 50 ha of the Oosterkwelder salt marsh has always been grazed. From 1958 to 1972 the other parts were abandoned. In 1972 grazing was resumed in 30 ha and in 1975 in another 20 ha. From that year onwards 160 yearlings grazed in an area of about 100 ha from the end of May until October. The area comprised lower-, mid- and upper salt marsh and some small dunes.

Both study areas were grazed at low stocking rate i.e. only part of the annual herbage accumulation is utilized by the grazers. This type of grazing management aims at the development of mosaics of short grazed turf and lightly grazed patches, provided the animals create and maintain by their grazing constant patterns in the vegetation. If such patterns do develop, grazing at low stocking rates creates extra structural diversity in the sward. This can not be achieved by cutting.

To distinguish between the specific use of the word grazing (which refers only to the defoliation process) and its more general use (which carries with it implications of the associated effects of treading and of dung and urine return), the term 'terrain-use' or 'occupancy' is preferred for grazing in general, and 'foraging' for grazing denoting consumption (see also Hodgson, 1979).

VIII.B. Grazing intensity implying terrain-use or occupancy

VIII.B.1. Estimation of occupancy from direct observation

The number of sheep and their activities in different sections of the Westerholt area were recorded every 30 min. during six periods of 24-hours and two daylight periods (half an hour before sunrise to half an hour after sunset) for six months. Moreover, the activities of one sheep (always the same animal) were noted down every minute and its position every 10 min. Daylight observations underestimate the time spent resting (on average 34%) as compared to periods of 24-hours observation (on average 50%) (Table VIII.1).

After converting the direct observations of one sheep into a theorethical flock equivalent and comparing it with the observations on the whole flock, it appeared that the total foraging and resting time and the time at which these activities took place were similar. With regard to the spatial pattern of foraging and-or resting, one sheep does not, however represent the whole flock (Fig. VIII.1). Measuring occupancy by direct observation is, therefore, only valuable when the activities of all the sheep are recorded in relation to their location in the terrain and for periods of 24-hours of observation.

On the Schiermonnikoog salt marsh study area a continued observation (with recordings every 30 min.) was carried out of the terrain use by cattle for five periods of 24-hours. It transpired that at least three daylight periods of observation were required in order to get a representative picture of the occupancy over the full observation time of five periods of 24-hours (Allersma, 1977). Similar studies were not performed for sheep in the Westerholt study area. Studies on the grazed salt marsh of the Island of Terschelling (The Netherlands) using cattle wearing a tachygraph showed that daylight observations underestimate the time spent resting (42% on average (n = 6)) as compared to periods of 24-hours (57% on average) (Otto, 1973). Observations of individual yearlings on the Schiermonnikoog salt marsh showed similar underestimations. They demonstrated that the total foraging time, the total resting time and their periodicity were similar for one yearling and for the whole flock. One yearling could not represent the whole flock of 160 animals with regard to the location and the foraging and/or resting time spent there, since the flock was nearly always subdivided into several groups consisting of 30 to 80 animals.

Table VIII.1. Activities of sheep expressed as percentages foraging and resting during periods of 24 hours and daylight observations in various periods of the year (1974) on the Westerholt study area (after De Bie, 1976).

Date	Period of observation	Activity	
		Foraging (%)	Resting (%)
9/10 July	24 hours	47	53
	daylight	60	40
21/22 August	24 hours	45	55
	daylight	55	45
3/4 September	24 hours	48	52
	daylight	60	40
18/19 September	24 hours	52	48
	daylight	67	33
30 September/	24 hours	52	48
1 October	daylight	73	27
15/16 October	24 hours	56	44
	daylight	79	21
average	24 hours	50	50
	daylight	66	34

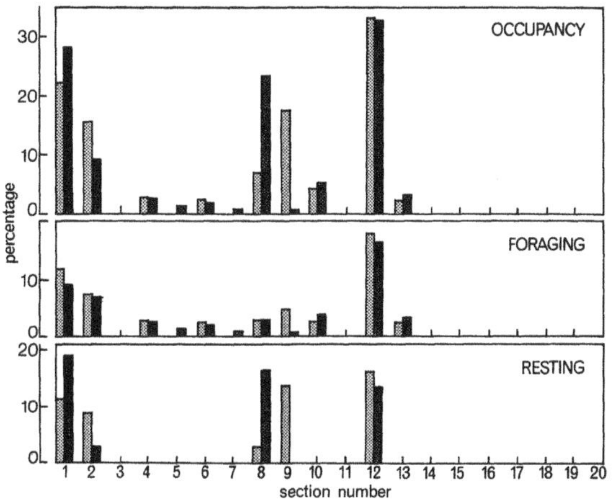

Fig. VIII.1. Total occupancy and activities of sheep expressed as percentages in each section spent by the theoretical flock (dotted bar) and the real flock (filled bar), 3–4 September 1974 (after Bakker et al., 1983).

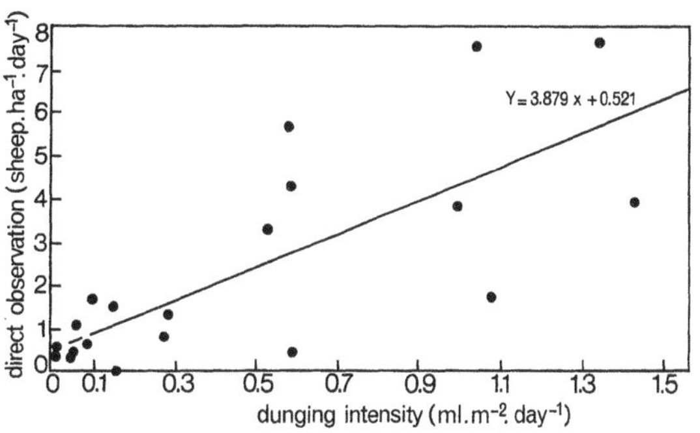

Fig. VIII.2. Regression line of the correlation (r = 0.684, P<0.001) between dunging intensity and direct observation for sheep in the Westerholt study area (after Bakker et al., 1983).

Since a great deal of direct observations appear to be necessary in order to get representative estimations, a time consuming method like this should be avoided and other methods found for the estimation of occupancy.

VIII.B.2. Estimation of occupancy from dunging intensity

The estimation of occupancy from the amount of dung voided has been adopted for several kinds of grazers e.g. geese (Ebbinge et al., 1975), deer (Bailey & Putnam, 1981), sheep (Welch, 1984), ponies and cattle (Pratt et al., 1986).

In ten plots in the Westerholt study area dung was sampled in 4 m^2 plots every two weeks in each pasture section and every month in 16–32 m^2 plots each heathland section. Figure VIII.2 shows from direct observation in the same sections that the average amount of dung and the average number of sheep were positively correlated (r = 0.684, P<0.001) for all sections together from July to October 1974. The positive correlation between direct observation and dunging intensity for the whole study area, could have been determined even if only three dung plots out of ten per section had been sampled (Bakker et al., 1983). This positive correlation applied to all heavily occupied grassland sections but not to the less occupied heathland sections, probably because of heterogeneity in space and time of dung voided. Since the study focused on the heavily occupied pasture sections, it was, in later years, sufficient to carry out dung measurements every 14 days in five sample plots of 4 m^2 each in those sections.

The relative preference – or avoidance – of the pasture area or sections of it were compared with the remaining heathland-woodland area using the dung measurements. For this purpose comparative dunging intensity values (CDI) were calculated. For a given section CDI was calculated from (cf. Hunter, 1954):

$$\mathrm{CDI} = \frac{A}{a} \times \frac{d(t_2 - t_1)}{D(t_2 - t_1)} - 1$$

where A = entire fenced-in area (m^2); a = sampled area in section a (m^2); d = amount of dung (ml) in the sampled area a, accumulated over the period ($t_2 - t_1$); D = amount of dung (ml) in the entire study area, accumulated over the period ($t_2 - t_1$). Only from July to October did the sheep clearly prefer the total grassland area, as opposed to the heathland-woodland area (Fig. VIII.3). Fig. VIII.4 shows that sheep preferred the pasture sections in which grasses reached a high coverage rather than the sections with a high *Juncus effusus* coverage. Even in the 'grass' sections of the pasture area, however, great differences in preference for individual sections were seen through the year.

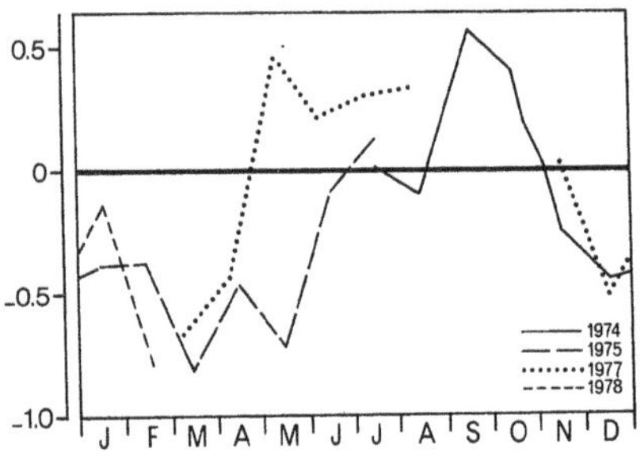

Fig. VIII.3. Preference of the sheep for the pasture land (as compared with the heathland-woodland areas) over the year, calculated from the Comparative Dunging Intensity (CDI). Positive CDI-values indicate a relative pasture 'preference', negative values an 'avoidance' and hence a relative heathland-woodland preference (after Bakker et al., 1983).

Concentrated dung accumulation probably resulted from resting. Spread pellets, however, seemed to coincide with foraging activities. Oosterveld (1976) also stressed the necessity of distinguishing between different activities. In his study area ponies, in particular, foraged in one section and defaecated in other sections. Edwards & Hollis (1982) demonstrated that cattle, ponies and deer voided large amounts of dung in latrines in the New Forest (United Kingdom) leading to transport of nutrients and the development of areas with a tall ruderal vegetation.

Ten dung plots (25 × 25) m^2 in four plant communities in the Oosterkwalder salt marsh study area were sampled every three weeks. Figure VIII.5 shows that the number of cattle recordings (direct observations) and dung patches recordings (indirect observations) were correlated positively ($r = 0.998$, $P<0.001$). Again, a discrepancy was found between foraging activities and dunging intensity. It was established from direct observations that the dominating species *Artemisia maritima* is not eaten by cattle (it contains oils that are harmful to the bacterial flora in the rumen (Nagy & Tengerdy, 1968)), but the animals spent a lot of time resting in the *Artemisia maritima* community, as indicated from the number of dung patches. Comparison of the dunging intensities in the observed plant communities revealed a relative preference for the *Artemisia maritima* community >*Festuca rubra* community and *Juncus gerardii* community >*Juncus maritimus* community (Allersma, 1977).

It can be concluded that the amount of dung voided is a reliable indication for the time spent in

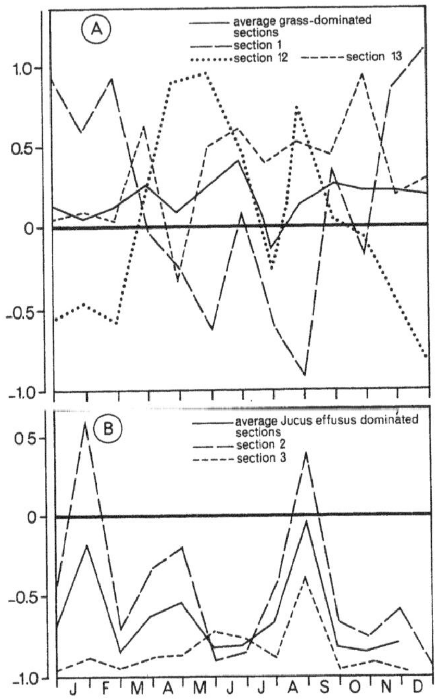

Fig. VIII.4A–B. 'Preference' of the sheep for A: the grassland sections (as compared with the *Juncus effusus* sections) and 'avoidance' of the sheep for B: the *Juncus effusus* sections (as compared with the grassland sections) calculated from the Comparative Dunging Intensity (after Bakker et al., 1983).

an area of sward (terrain-use or occupancy) in the Westerholt and Oosterkwelder study areas. It does, however, provide only little information on the activities e.g. foraging of the animals. Grazing patterns deduced from dunging intensities will be discussed more in detail in section VIII.D.

VIII.C. Grazing intensity implying foraging

VIII.C.1. Introduction

Foraging intensities can be measured in different ways. Zeevalking & Fresco (1977) estimated the foraging intensity visually from the vegetation using an index indicating the degree to which the plant species present had been grazed during the summer period. Hutchinson & King (1980) considered the area of short turf in a sward as a measure of foraging intensity. In the Westerholt study area

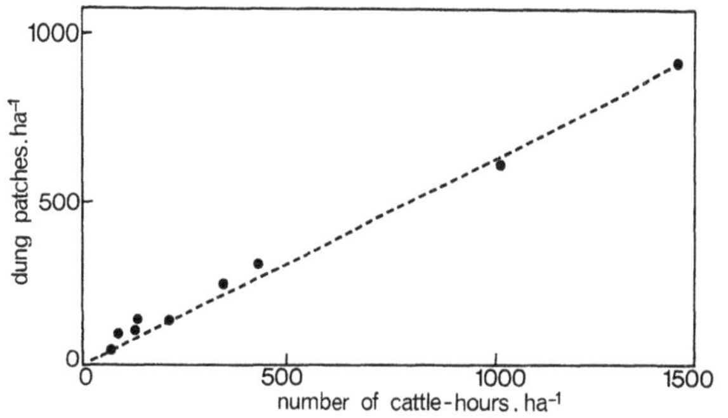

Fig. VIII.5. Relation between direct observation (number of cattle-hours · ha^{-1}) and dunging intensity (number of dung patches · ha^{-1}) on the salt marsh study area (after Bakker & Allersma, 1979).

indeed a relationship was found between grazing intensity and the proportion of short turf areas (section V.E.2). Job & Taylor (1978) found a negative correlation between the amount of grazed biomass and litterfall. The more sophisticated measurement of herbage utilized is used in estimating foraging intensity in addition to the rough indications for foraging intensity, like the area of short turf, the height of the canopy and the amount of litter.

Since grazing always implies foraging, treading and manuring, it seems important to study the effects of all these types of impact on the vegetation. Treading without foraging takes place if the animals walk from one foraging area to another or to a drinking place. There is often a fixed pattern of movement resulting in tracks. Manuring without foraging takes place if the animals defaecate in latrines. Both tracks and latrines are areas of heavy local influence, resulting in the emergence of trampling indicators and nitrophilous plant communities, respectively. It is questionable whether foraging activities also have such a heavy local impact or whether these are spread more evenly. This not only depends on the type and number of grazing animals, but also on the amount of above-ground standing crop available. Since the increase of species diversity is a major objective in conservation management, a large standing crop and litter accumulation should be prevented (Grime, 1979) and a short turf should be aimed at, at least locally (Job & Taylor, 1978). Only part of the production should be consumed and this percentage should vary in space, since foraging at different intensities is considered an advantage for the establishment of vegetation patterns with respect to canopy structure, i.e. a mosaic of short turf, tall turf and low woody vegetation (Klapp, 1965; Oosterveld, 1975; Harper, 1977).

The *utilization factor* (Esselink et al., 1989) is the herbage utilization ('t Mannetje, 1978) per unit area per unit time expressed as a proportion of the net herbage accumulation (Hodgson, 1979) per unit area per unit time, when herbage utilization is the sum of (i) the herbage consumed by the animals, (ii) subsidiary losses of herbage due to spoiling by trampling and fouling and (iii) decomposition. Herbage accumulation is the primary production minus decomposition and consumption by invertebrates and small vertebrates measured in excluded areas. Hence the term 'utilization factor' is analogous to Hodgson's 'efficiency of grazing' (Hodgson, 1979) being the herbage consumed expressed as a proportion of the net herbage accumulation (see also De Leeuw & Bakker, 1986). Foraging patterns can be quantified by attempting to measure the utilization factor of different grassland communities. Net herbage accumulation was measured at intervals of 4–6 weeks using movable cages to exclude grazing between the first and second harvests. Herbage consumption was assumed to equate with the difference in the green matter between adjacent grazed and caged sample plots after 4–6 weeks i.e. at the second harvest.

VIII.C.2. Utilization factor in the reclaimed grassland vegetation (Westerholt, grazed by sheep)

Cages were placed all over the grassland study area in the communities of *Leontodon autumnalis, Holcus lanatus* and *Agrostis capillaris* between June and December 1981, ten cages in each community. The communities of *Agrostis stolonifera* and of *Juncus acutiflorus* were too small to permit the placement of ten cages. Since very little dung was found and it was known from direct observations that the community of the tall *Juncus effusus* was hardly grazed, no elaborate herbage consumption was estimated. In these three communities in October 1981 only the standing crop was measured.

The *Agrostis capillaris* and the *Holcus lanatus* communities showed maximal net accumulation of live herbage in early summer. Net accumulation was quite constant in the *Leontodon autumnalis* community during the growing season (Fig. VIII.6). No more increase in live herbage was found in any plant community after October.

The lowest estimated herbage utilized was in the

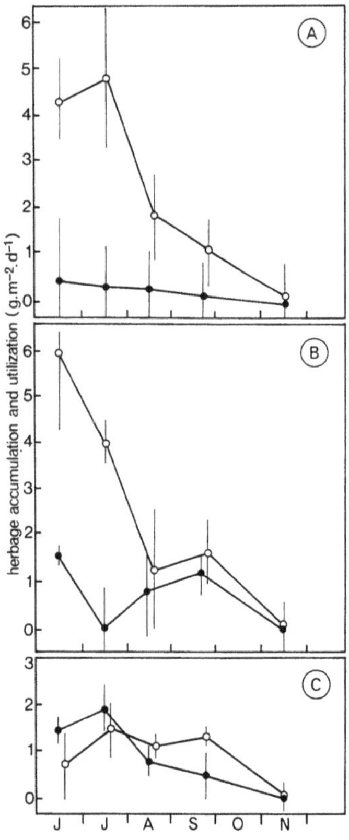

Fig. VIII.6A–C. Net accumulation of live herbage in areas protected from grazing for 4–6 weeks (open circles) and utilization measured as the difference between live herbage under the exclusion cages and in the open plot (filled circles) including S.E. in 1981 in A: *Agrostis capillaris* community, B: *Holcus lanatus* community and C: *Leontodon autumnalis* community (after De Leeuw & Bakker, 1986).

Agrostis capillaris community (Fig. VIII.6). The *Holcus lanatus* community did not appear to be utilized in July, although removal estimates were high in the other months. Utilization was highest in July in the *Leontodon autumnalis* community, but decreased afterwards. No utilization occurred in the three communities from October to January.

The overall utilization factor of each community from June to November was 0.89 in the *Leontodon autumnalis,* 0.30 in the *Holcus lanatus* and only 0.09 in the *Agrostis capillaris* community (Table VIII.2). When the initial standing crop in June, which is also partly utilized, was taken into account, the figures were 0.55, 0.22 and 0.07, respectively. Very little herbage was utilized in the *Juncus effusus* community in early spring. The *Juncus acutiflorus* community was grazed to a larger extent during the spring, until the shoots of the dominating species grew too tall and from September onwards when the above-ground standing crop died back and newly formed shoots were grazed. The relatively small area of the *Agrostis stolonifera* community was quite heavily grazed as long as the stand was not inundated.

VIII.C.3. Utilization factor in salt marsh communities (Schiermonnikoog, grazed by cattle and geese)

Cages were placed in section B of the salt marsh area (see Fig. IV.17) in the *Puccinellia maritima, Juncus gerardii, Festuca rubra, Armeria maritima, Ammophila arenaria* and *Juncus maritimus* communities between May and November 1981 and again between March and May 1982. Five to ten cages were used in each community. Since there were few swards of the *Elymus pycnanthus* community that were large enough to be sampled and as it was already known that no foraging occurred in *Artemisia maritima* swards, no cages were placed in these two communities.

Table VIII.2. Net herbage accumulation, initial standing crop, herbage utilization by sheep and utilization factor from June until November 1981 on the reclaimed grassland of the Westerholt (after De Leeuw & Bakker, 1986).

Community of	Net herbage accumulation gm^{-2}	Initial standing crop gm^{-2}	Herbage utilization by sheep gm^{-2}	Utilization factor decimal	Utilization factor taking into account initial standing crop decimal
Leontodon autumnalis	165	100	145	0.89	0.55
Holcus lanatus	432	165	129	0.30	0.22
Agrostis capillaris	382	118	345	0.09	0.07

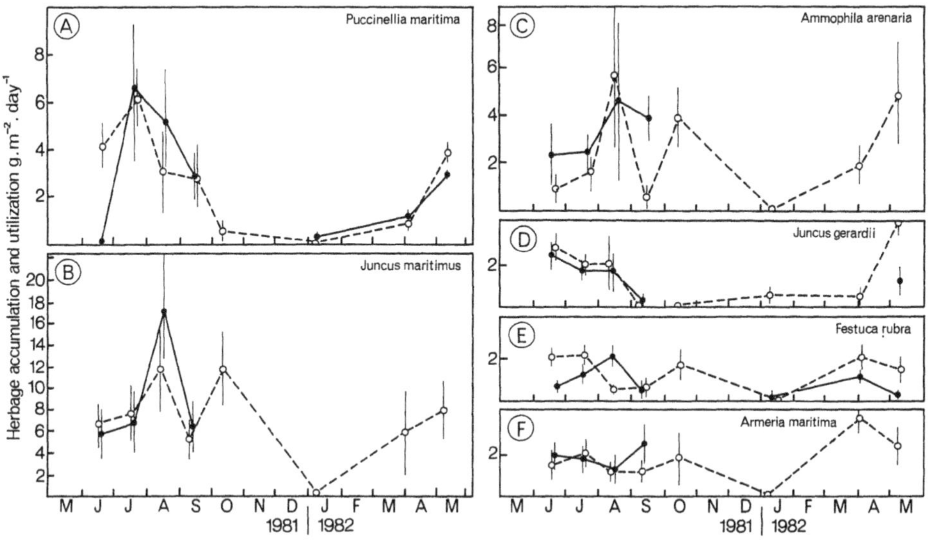

Fig. VIII.7A–F. Net herbage accumulation in areas protected from grazing for 4–6 weeks (open circles) and utilization measured as the difference between live herbage under the exclusion cages and in the open plot (filled circles) including S.E. in the period May 1981–May 1982 in A: *Puccinellia maritima* community, B: *Juncus maritimus* community (note deviating scale), C: *Ammophila arenaria* community, D: *Juncus gerardii* community, E: *Festuca rubra* community, F: *Armeria maritima* community (after Looijen & Bakker, 1987).

In general, net herbage accumulation was high in the spring and early summer and decreased later in summer with a slight increase again in the autumn (Fig. VIII.7). The communities showed different peaks of herbage accumulation in different intervals. The utilization factors (Table VIII.3) indicate that nearly all or possibly more than the net accumulation was utilized by the grazing cattle. This can only be the case if part of the above-ground standing crop at the start of the grazing period is also utilized. The net annual herbage accumulation, taking the initial standing crop, herbage utilized by cattle (June to November 1981) and geese (November 1981 to June 1982) and the resulting utilization factors into account, are given in Table VIII.4.

No accumulation was recorded in the measurements of differences in herbage mass between November 1981 and March 1982 (Fig. VIII.7). Geese, however, start grazing the marsh in winter. Some herbage production must, therefore, take place since only very young protein-rich tillers are consumed (Prins et al., 1980). Herbage utilized by geese only occurred in the *Puccinellia maritima, Juncus gerardii* and *Festuca rubra* communities and took mostly place in the early spring with peaks in May.

The herbage utilized by cattle showed a maximum, at different times, for different communities (Fig. VIII.7). Maximal utilization in the *Juncus gerardii* community took place in June, in the *Puccinellia* community in July, in the *Festuca rubra, Ammophila arenaria* and *Juncus maritimus* communities in August and in the *Armeria maritima* community in September.

The utilization factor was, on average, 0.53 but

Table VIII.3. Net herbage accumulation, herbage utilization and utilization factor during the grazing season of cattle (June to November 1981) in different salt marsh plant communities (Looijen et al., 1984).

Community of	Net herbage accumulation gm^{-2}	Herbage utilization gm^{-2}	Utilization factor decimal
Puccinellia maritima	506	415	0.82
Festuca rubra	177	144	0.82
Juncus gerardii	198	174	0.88
Juncus maritimus	943	1054	1.12
Armeria maritima	180	229	1.27
Ammophila arenaria	268	386	1.44

varied from 0.37 in the *Festuca rubra* to 0.62 in the *Juncus maritimus* community.

The utilization factor for cattle and geese together varied from 0.47 in the *Armeria maritima* to 0.78 in the *Puccinellia maritima* community (Table VIII.4).

VIII.C.4. Decomposition of organic matter

The decomposition rate of organic matter was measured in several plant communities, using the litter bag method, in order to enable an estimate of decomposition of ungrazed biomass. Mainly standing dead plant material, both stems and leaves, was collected in November 1981 on Schiermonnikoog and in January 1982 on the Westerholt and replaced (in litter bags) at each collecting site. The decay rates of cellulose strips was also measured as a standard for comparison.

In the Westerholt reclaimed grassland area (see section V.C.2) the plant communities studied were those of *Agrostis capillaris* and *Holcus lanatus* on moist soil, *Agrostis stolonifera, Juncus acutiflorus* and *J. effusus* on wet soils. The decay rate of plant material is shown in Fig. VIII.8. In order to compare the results of all plant communities, decomposition rates were expressed as the number of days necessary for 40% decomposition to take place. It can be concluded that the decomposition rate of cellulose was higher in wet than in moist places (Table VIII.5). This was, however, not the case with the decay rate of plant material, since the initial C/N ratio tended towards a negative relationship with the decomposition rate.

The decay rates of plant material in six salt marsh communities are shown in Fig. VIII.9. The cumulative percentages weight loss amounted to approximately 40% during the period November to February, but decreased afterwards with the exception of the *Ammophila* and *Armeria* communities. The other communities were situated lower with respect to Ordnance Datum and were covered with some silt mixed with organic matter, as a result of gales in March and April. The organic matter could not be separated properly from the original organic matter and this resulted in an increase in weight relative to that of the preceding sampling date. From April onwards, a low decay rate was found in the *Armeria maritima* and *Ammophila arenaria* communities of the higher salt marsh, a high percentage weight loss in the *Puccinellia maritima, Juncus gerardii* and *Festuca rubra* communities of the lower salt marsh and an intermediate decay rate in the *Juncus maritimus* community (Table VIII.6). These differences are more or less in accordance with the decay rates of cellulose strips, namely after one year 60–100% weight loss in the communities of the lower salt marsh and only 15–25% weight loss in the communities of the higher salt marsh.

Table VIII.4. Net annual herbage accumulation (including initial standing crop), herbage utilization, utilization factor by cattle and geese and dunging intensity in 1981/1982 in different salt marsh plant communities (– = not measured) (after Looijen & Bakker, 1987).

Community of	Net annual herbage accumulation gm^{-2}	Herbage utilization by cattle gm^{-2}	Utilization factor by cattle decimal	Herbage utilization by geese gm^{-2}	Utilization factor by cattle and geese decimal	Number of cattle dung patches ha^{-1} day^{-1}
Puccinellia maritima	767	415	0.54	181	0.78	20 ± 2
Juncus gerardii	429	174	0.41	38	0.49	17 ± 2
Festuca rubra	388	144	0.37	75	0.57	16 ± 2
Armeria maritima	487	229	0.47	–	0.47	11 ± 1
Ammophila arenaria	686	386	0.56	–	0.56	8 ± 1
Juncus maritimus	1707	1054	0.62	–	0.62	15 ± 1
Elymus pycnanthus	–	–	–	–	–	12 ± 2
Artemisia maritima	–	–	–	–	–	17 ± 1

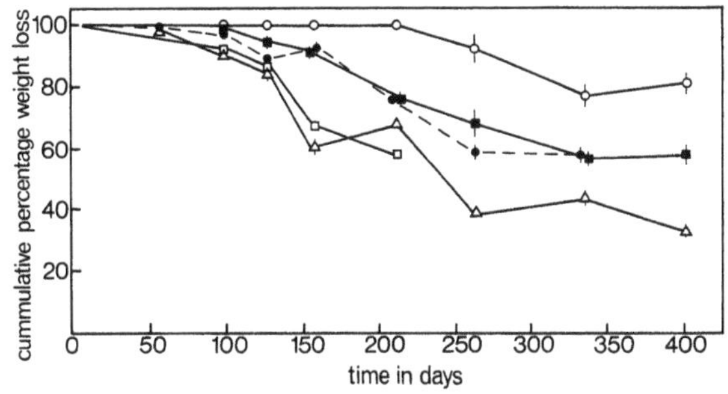

Fig. VIII.8. Cumulative percentage weight loss (± S.E.) of oven dried plant material from 2 mm mesh nylon bags with some 6 mm perforations placed at soil surface level under the litter layer if present in five plant communities: *Juncus acutiflorus* (○), *Juncus effusus* (●), *Agrostis stolonifera* (□), *Agrostis capillaris* (■) and *Holcus lanatus* (△).

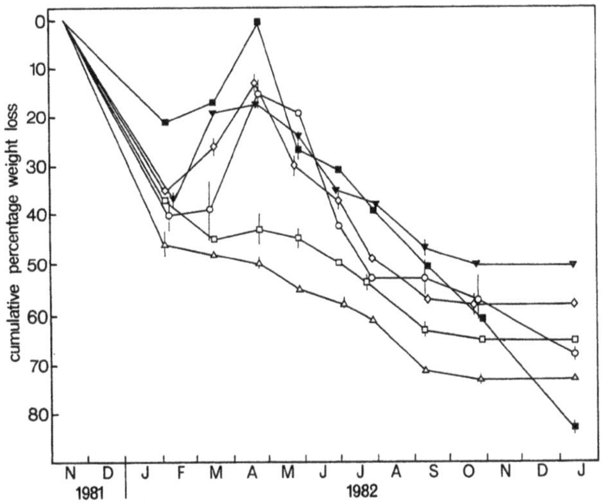

Fig. VIII.9. Cumulative percentage weight loss (± S.E.) of oven dried plant material from 2 mm mesh nylon bags with some 6 mm perforations placed at soil surface level under the litter layer if present in six communities: *Puccinellia maritima* (○), *Juncus gerardii* (■), *Festuca rubra* (◇), *Armeria maritima* (□), *Ammophila arenaria* (△) and *Juncus maritimus* (▼) (after Looijen & Bakker, 1987).

VIII.C.5. Above-ground standing crop and litter accumulation

The combined results of herbage accumulation, herbage utilization and decomposition are reflected in the above-ground standing crop. In the reclaimed grassland of the *Westerholt* study area (Fig. VIII.10) the above-ground standing crop, including standing dead material, but excluding litter, was highest in October and decreased until May. The *Agrostis capillaris* community reached a peak value at 510 gdw · m^{-2}, the *Holcus lanatus* community at 420 gdw · m^{-2}, whereas the *Leontodon autumnalis* community reached only 210 gdw · m^{-2}. The *Juncus effusus, Juncus acutiflorus* and *Agrostis stolonifera* communities reached peak standing crops of 1130, 830 and 730 gdw · m^{-2}, respectively, in October. The utilization factor (Table VIII.2) in the *Leontodon autumnalis* community is apparently high enough to prevent litter accumulation since no litter was found throughout the year. The part of the accumulated herbage which has not been utilized, apparently, suffers from rapid decay. The utilization factor in the *Agrostis capillaris* community is very low and much organic material is left for decomposition. This material shows a weight loss of only 40% after one year and litter accumulation may be expected. Indeed litter amounts ranging

Table VIII.5. Number of days necessary for 40% decomposition of plant material and cellulose, initial C/N ratio of plant material in different plant communities in mixed grassland.

	Agrostis capillaris community	Holcus lanatus community	Juncus effusus community	Juncus acutiflorus community	Agrostis stolonifera community
Number of days necessary for 40% decomposition of plant material	310	220	260	>400	190
Number of days necessary for 40% decomposition of cellulose	260	245	210	145	120
Initial C/N ratio of plant material	54	40	30	50	29

between 110 and 460 gdw · m^{-2} have been found. The *Holcus lanatus* community shows an intermediate utilization factor, but the remaining material shows weight losses of 60–70% after one year, therefore only a little litter accumulation can be expected. An amount of 90gdw · m^{-2} has only been found locally.

In most *salt marsh* communities the living above-ground standing crop outside the exclosures increased to 200 gdw · m^{-2} during the spring and early summer and decreased again to values below 100 gdw · m^{-2} in the late summer and autumn (Fig. VIII.11). The figures of the *Juncus maritimus* community fluctuated between 200 and 400 gdw · m^{-2}. The amount of dead organic matter, including both standing dead material and surface litter, however, increased during the late summer or in the autumn. Only in the *Armeria maritima* community did it decrease in that period. The total above-ground standing crop fluctuated between 200 and 300 gdw · m^{-2} in most communities and was around 400 gdw · m^{-2} in the *Armeria maritima* community in the spring and the summer and between 800 and 1200 gdw · m^{-2} in the *Juncus maritimus* community. The utilization factors ranging between 0.45 and 0.65 (and 0.78 in the *Puccinellia maritima* community) imply that some organic material is left. This is indicated by the increasing amounts of dead material in late summer and the autumn. This material shows weight losses of 35 to 45% in a few winter months, therefore, litter accumulation can be expected. No litter at all was, however, found in all the salt marsh plant communities, probably due to an under-estimation of the decomposition with the litter bag method (Looijen & Bakker, 1987).

Table VIII.6. Percentage weight loss of plant material and cellulose-strips from litter-bags over the period April 22–October 24, 1982 in different salt marsh plant communities.

Location	Community of	Herbage	Cellulose
Lower/	Puccinellia maritima	42	96
upper	Juncus gerardii	61	76
salt marsh	Festuca rubra	46	58
Transition	Juncus maritimus	34	93
Upper salt marsh/	Armeria maritima	22	15
dune	Ammophila arenaria	23	22

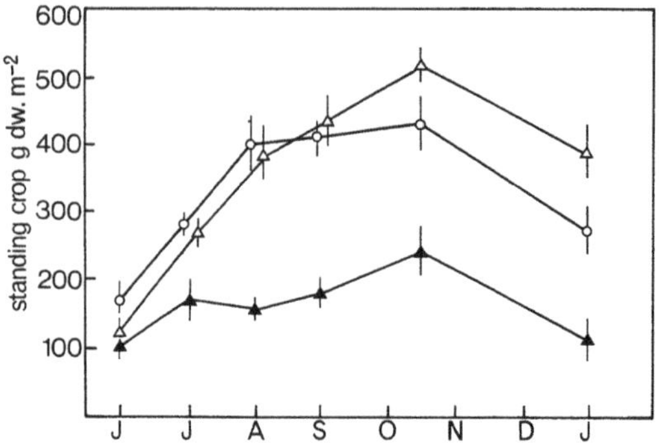

Fig. VIII.10. Above-ground standing crop including S.E. in grazed areas in 1981 in *Agrostis capillaris* community (△), *Holcus lanatus* community (○), *Leontodon autumnalis* community (▲) (after De Leeuw & Bakker, 1986).

The above-ground standing crop, including standing dead and litter material, was measured in the grazed and the abandoned salt marsh from 1973 to 1978 and in 1982. The average values vary between different plant communities and during the grazing season (Fig. VIII.12). The somewhat higher average above-ground standing crop values in the grazed area in the communities of *Juncus maritimus, Juncus gerardii* and *Festuca rubra* compared to the similar communities in 1981 (Fig. VIII.11) can be attributed to the overall low standing crop figures in 1981 (see Fig. IX.1.A–F). A relatively large proportion of the standing crop in the *Juncus gerardii* community was already utilized in June, whereas the *Festuca rubra* communities did not show similar figures earlier than August. The herbage utilization was lowest in the *Juncus maritimus* community. The decreased herbage utilization from June to August in the *Artemisia maritima* community can be attributed to the fully expanded individuals of *Artemisia maritima* in August.

VIII.D. Grazing patterns

VIII.D.1. Grazing patterns on the salt marsh (Schiermonnikoog)

Differences in grazing intensity, whether in occu-

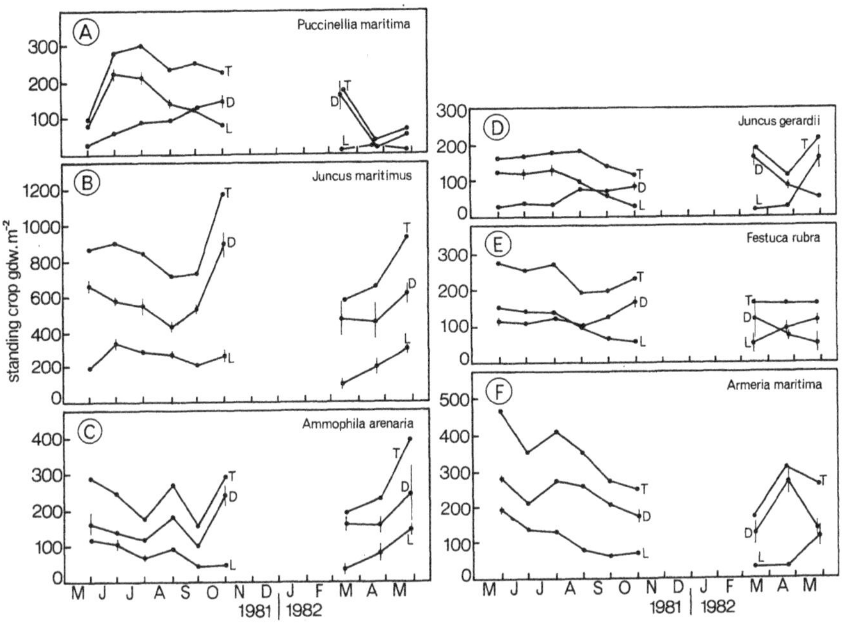

Fig. VIII.11A–F Above-ground standing crop (T) including living (L) and dead (D) material (± S.E.) in the period May 1981 – May 1982 in A: *Puccinellia maritima* community, B: *Juncus maritimus* community (note deviating scale), C: *Ammophila arenaria* community, D: *Juncus gerardii* community, E: *Festuca rubra* community, F: *Armeria maritima* community (after Looijen *et al.*, 1984).

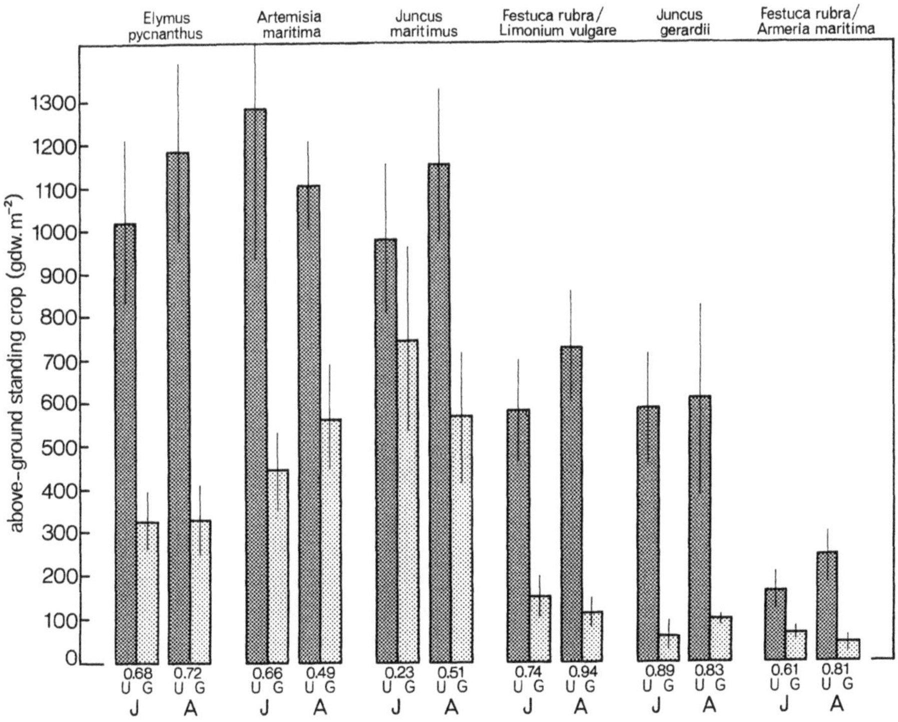

Fig. VIII.12. Average above-ground standing crop including litter (± S.E.) in untouched (first bar) and grazed (second bar) salt marsh plant communities in June and in August. The difference between grazed and untouched i.e. the fraction utilized is indicated below the bars.

pancy (section VIII.B) or in utilization factor (section VIII.C) are particularly interesting for nature conservation if they result in patterns in the structure of the sward. This is only possible if such patterns of closely grazed areas and lightly grazed patches are constantly for several seasons and years since different plant communities adapted to these environmental circumstances cannot emerge with a fast shifting pattern of grazing intensity.

The occupancy of the salt marsh section B (see Fig. IV.17), where grazing by cattle was resumed in 1972, is shown in Table VIII.7. The differences between the communities appear constant, suggesting a constant pattern of occupancy. The differences between the communities in occupancy, however, are in fact not important since the high stocking rate of 1.6 animal · ha^{-1} results in a closely grazed sward all over the salt marsh except for the *Juncus maritimus* community which has a much higher above-ground standing crop (see section VIII.C.5). This difference emerged soon after the resumption of grazing and maintained itself. The position of the *Juncus maritimus* community remained constant as can be seen from repeated vegetation mapping (Fig. VII.13). It can be concluded that a grazing pattern has developed on the salt marsh resulting in extensive closely grazed areas and smaller lightly grazed patches of the *Juncus maritimus* community. These patches range from nearly one ha to small isolated clumps of *Juncus maritimus* plants (Photo VIII.1).

Table VIII.7. Total number of dung patches · ha^{-1} during the grazing season (May-October) in three salt marsh plant communities in 1976 and 1981 (after Looijen et al., 1984).

Plant community	Number of dung patches · ha^{-1}	
	1976	1981
Juncus gerardii	1600	2010
Festuca rubra	1500	1900
Juncus maritimus	1350	1830

VIII.D.2. Macro-patterns in mixed grassland (Westerholt)

The occupancy of the grassland area by sheep was measured from July 1974 to July 1975 and from May 1977 to May 1978 in six of the thirteen sections of the pasture each of these initially containing a uniform sward (see also Fig. VIII.3 and 4). Some communities had such a large area that these were subdivided into several sections. Since the dunging intensity was measured in sections and the initially uniform *Holcus lanatus* dominated sward, including sections 1, 12 and 13, had differentiated from 1972 onwards into the interwoven *Holcus lanatus* and the *Leontodon autumnalis* communities, the dunging intensities for both communities had to be considered together. Table VIII.8 shows the dunging intensities in the sections with different plant communities during the growing season (May to October) and the winter half year (November to April). The occupancies in 1974–75 and 1977–78 were positively correlated ($r = 0.821$, $P<0.05$). This was also the case for the two growing seasons ($r = 0.713$, $P<0.05$) and the winter half years ($r = 0.754$, $P<0.05$). Hence, the occupancy rate within the sections was constant from 1974 to 1978. The sheep occupied the *Holcus lanatus*/*Leontodon autumnalis* communities most and the *Juncus effusus* community least. The *Agrostis* communities took intermediate positions. The lower overall amount of dung in 1977 can be explained by sheep being diverted away because of the local increase of grassland species in the heathland area. *Erica tetralix* and *Calluna vulgaris* dominated that whole area during the first years of the experiment.

A negative correlation ($r = -0.929$, $P<0.02$) between dunging intensity and the height of the canopy was found in the study area in 1977–78. In this area the patterns of occupancy, therefore, also reflect the foraging behaviour. Since a relationship was found between grazing intensity and the proportion of short turf areas, the latter can also be taken as a measure of occupancy. The same sequence of occupancy rate with respect to the canopy height and the proportion of closely grazed areas in 1977 and in 1982 was found for the dunging intensity. It is, therefore, concluded that the occu-

Photo VIII.1. Patterns of closely grazed areas by cattle with a short turf and lighter grazed areas featuring clumps of *Juncus maritimus*.

pancy rate in most of the grassland area remained constant for a long period.

The different occupancy rates of the plant communities in the Westerholt area run parallel with the utilization factors (section VIII.C.2). The latter resulted in closely grazed *Leontodon autumnalis/ Holcus lanatus* communities with sparse litter accumulation, a hardly grazed *Juncus effusus* community with much litter accumulation and an intermediate *Agrostis capillaris* community. Since these communities cover relatively large areas (see Fig. V.13) this pattern is reffered to as a macro-pattern.

VIII.D.3. Micro-patterns in mixed grassland (Westerholt)

On a more detailed scale, a mosaic of taller tufts interspersed with shorter turf emerged in the area with relatively large utilization factors (Photo's VIII.2 and 3). The heavily grazed areas were cov-

Table VIII.8. Dunging intensities during two separate years in several plant communities in the mixed grassland area (after De Leeuw & Bakker, 1986).

Sections	Plant communities	Dunging intensity ($ml \cdot m^{-2}$)					
		1974–75			1977–78		
		May–Oct.	Nov.–April	Year	May–Oct.	Nov.–April	Year
1	Holcus lanatus/	238	46	284	78	89	167
12	Leontodon autumnalis	90	93	183	95	91	186
13		200	69	269	88	57	145
2	Agrostis stolonifera	98	40	138	18	6	24
4	Agrostis capillaris	57	23	80	31	34	65
3	Juncus effusus	6	12	18	1	0	1

Photo VIII.2. The Westerholt study area at the start of the mowing (left) and the grazing (right) regimes.

Photo VIII.3. The outlined area of Photo VIII.2 after eight years of mowing and sheep-grazing featuring a micro-pattern in the grazed area in April. The light patches are hardly grazed and contain standing dead material whereas the interlocking dark areas are heavily grazed.

ered with the *Leontodon autumnalis* community, whereas the lighter grazed patches contained the *Holcus lanatus* community. The diameter of the taller tufts can range from 0.5 m to several metres. This mosaic is referred to as the micro-pattern. The heavily grazed areas had a canopy height ≤5 cm and little litter accumulation, lightly grazed patches had a canopy height ≥10 cm and more litter accumulation. Due to the sharp boundaries no intermediate height class was distinguished.

The micro-pattern was first charted in the autumn of 1979 and thereafter repeated annually until the autumn of 1983 (Fig. VIII.13). In order to determine whether the micro-pattern was fixed, the map patterns were compared using an overlay with 200 randomly placed points. A chi-square based point correlation test was used to determine their correlation. Centres clearly existed in both the lightly grazed taller canopy and in the heavily grazed lower turf. At the fringes of these centres the canopy was either taller or shorter. It is noticeable that the patterns resemble each other more after a short interval of time than after a longer interval (Fig. VIII.14). After two years the resemblance did not diminish any further and remained significant ($P<0.05$). It is, therefore, concluded that the micro-patterns which were found did not change randomly from year to year, but were fixed under the grazing regime. The grazing regime resulted in great differences in soil resistance, measured with a penetrometer, namely heavily grazed areas showed values of 11.5 (± 2.3) $kg \cdot cm^{-2}$ and lightly grazed patches of 5.5 (± 0.7) $kg \cdot cm^{-2}$ in 1981.

Species composition in the heavily grazed areas and lightly grazed patches is shown in Table VIII.9. The presence of species differed only slightly, but their biomasses differed considerably. The amount of *Agrostis capillaris* in the lightly grazed patches and of *Anthoxanthum odoratum, Hypochaeris radicata* and *Trifolium repens* in the heavily grazed areas is striking. The larger amount of dead plant material in the lightly grazed patches indicates litter accumulation. It probably caused the limited occurrence of mosses. The heavily grazed areas resembled the permanent plots with an intermediate grazing intensity (see Table V.7). Thus species

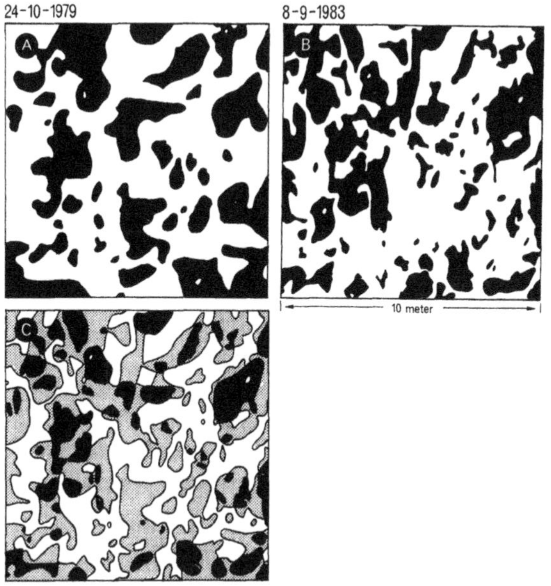

Fig. VIII.13A–C. Micro-pattern in grassland (10 × 10) m^2 as developed after seven years A: October 1979 and after eleven years B: September 1983 of sheep-grazing and C: the difference between 1979 and 1983. White = canopy ≤5 cm height; black = canopy >10 cm height or canopy containing much litter; dotted = canopy high in 1979 and low in 1983 or the reverse (after Brandsma, 1984).

composition and abundancy, as well as litter accumulation, on the scale of the entire grassland area (macro-pattern), can also be found in micro-patterns within the permanent plots.

The boundaries between the centres with a tall canopy and a closely grazed turf apparently shifted within a fringe zone (Fig. VIII.13). The number of grid points with a tall canopy in a micro-pattern increased during the spring and summer and decreased again in the autumn and winter, whereas the number of points with closely grazed canopy showed the opposite trend (Table VIII.10). It appeared from sequential charting that particularly the area with a tall canopy extended between April and July 1983 (Fig. VIII.15D). The diameter of ten lightly grazed patches increased during the summer of 1983 and decreased again during the winter of 1984 (Table VIII.11).

The increase and decrease of lightly grazed patches mentioned above was also quantified by estimations of the utilization factors of the tall and

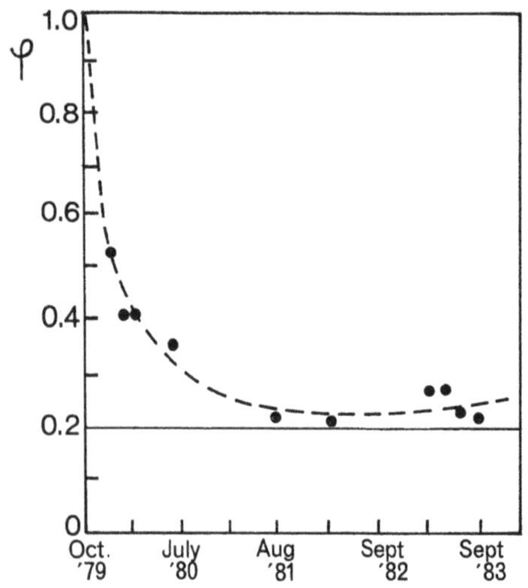

Fig. VIII.14. Correlations between subsequent micro-pattern chartings between October 1979 and September 1983. The level of significance (P<0.05) is indicated (φ = 0.196) (after Brandsma, 1984).

Table VIII.9. Live and dead aerial biomass (gdw · m^{-2})(± S.E.) in heavily grazed areas and lightly grazed patches, measured in sample plots of 78.5 cm^2 (n = 5) in September 1981 (after Bakker et al., 1984).

Number of species	Heavily grazed 11	Lightly grazed 11
Agrostis capillaris	15.4 ± 3.2	306.6 ± 36.7
Holcus lanatus	15.7 ± 5.6	71.7 ± 14.6
Anthoxanthum odoratum	13.6 ± 4.8	5.6 ± 3.6
Poa pratensis	5.6 ± 1.9	7.0 ± 1.8
Cynosurus cristatus	0.1 ± 0.1	4.4 ± 3.8
Lolium perenne	0	1.1 ± 1.1
Festuca rubra	0	0.9 ± 0.8
Total monocots	50.4 ± 7.2	379.3 ± 44.6
Trifolium repens	19.2 ± 4.0	11.1 ± 2.6
Ranunculus acris	3.7 ± 1.1	15.9 ± 8.2
Leontodon autumnalis	27.5 ± 8.0	36.4 ± 18.9
Taraxacum spec.	0.5 ± 0.3	0
Hypochaeris radicata	5.5 ± 4.4	0
Trifolium pratense	0.3 ± 0.3	0
Stellaria media	0	0.5 ± 0.5
Total dicots	56.7 ± 9.8	63.9 ± 24.6
Green aerial biomass	107.1 ± 10.9	461.2 ± 31.8
dead plant material	93.2 ± 6.6	393.1 ± 66.1
mosses	10.6 ± 4.7	0
percentage monocots	47.0 ± 5.4	86.1 ± 6.7

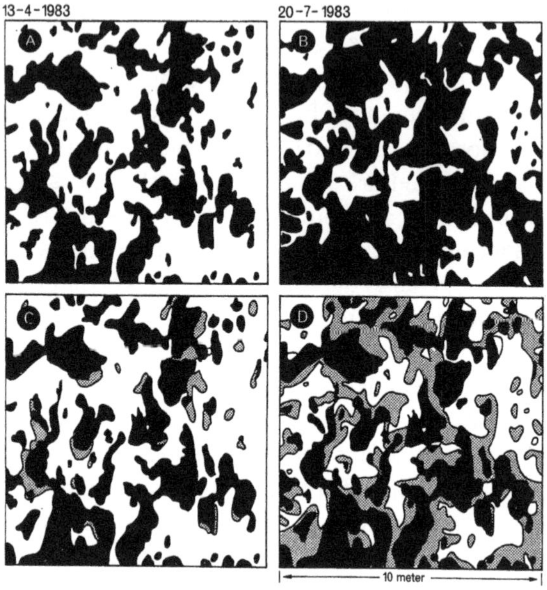

Fig. VIII 15A–D. Seasonal change in micro-pattern in grassland (10 × 10) m^2 after eleven years of sheep-grazing A: April 1983; B: July 1983 (white = canopy ≤ 5 cm height; black = canopy >10 cm height or canopy containing much litter); C: extension low canopy between April and July (dotted area); D: extension of tall canopy between April and July (dotted area) (after Brandsma, 1984).

low sward in three intervals between mid-April and mid-September 1983. The relative herbage accumulation and herbage utilization in each interval in the tall and low sward are shown in Fig. VIII.16 A–B. The relative herbage accumulation and utilization are lower in the tall sward than in the short turf during the growing season except for the interval April to May when herbage utilization is

Table VIII.10. Number of points of a grid (200 points) in two micro-patterns (I and II) (10 × 10) m^2 with tall sward and with short sward in 1983/1984 in a mixed grassland (after Brandsma, 1984).

	13-4-83	8-6-83	20-7-83	8-9-83	13-3-84
I tall sward	62	100	98	73	63
short sward	138	100	102	127	137
II tall sward	72	115	125	105	89
short sward	128	85	75	95	111

minimal in both sward types (the measured negative herbage utilization must be an artefact which can probably be attributed to the wet spring resulting in an avoidance of the very wet places that emerged). The resulting utilization factors are shown in Fig. VIII.16 C and indicate a higher herbage accumulation than utilization in spring. All herbage accumulated in the low sward was removed in June, whereas more herbage was utilized in both the high and low sward than accumulated in August. This process resulted in a high standing crop of the tall sward and small standing crop of the low sward.

If the process which has been described is transferred to the fringes of the micro-pattern it can be concluded that the lightly grazed patches spread during the spring and early summer because of a higher herbage accumulation than utilization and that these fringes are removed again by sheep during the late summer and autumn when herbage utilization exceeds accumulation.

VIII.D.4. Experiments in micro-patterns

The cause of the constancy of the micro-patterns (Photo VIII.4) has been investigated by some experiments in small areas featuring the pattern. Two areas (10 × 7)m^2 were cut once in April and two more areas (10 × 7)m^2 were cut once in April and twice in May 1983 in order to egalize the micro-patterns. Moreover, two areas (6 × 4)m^2 were excluded from grazing in April 1983 (Photo VIII.5). After sequential charting in April and September 1983 the map patterns were compared using an overlay with 57 randomly placed points. The number of points with high and low canopy in the experiments is shown in Table VIII.12. The area with a tall canopy increased in the exclosures, but the micro-patterns remained until September. All points with a low canopy had, however, disappeared in the spring of 1984 and therefore the micro-pattern had too. Though the area with a low canopy increased by cutting three times, the newly emerged pattern did not correspond to the original micro-patterns. The same was the case in one of the areas cut once, but in the other area the newly emerged patterns corresponded to the original.

Table VIII.11. Changes in diameter of ten slightly grazed patches (minimal diameter 20 cm), increasing, constant, decreasing, respectively, in 1983/1984 in a mixed grassland (after Brandsma, 1984).

Period	Increasing	Constant	Decreasing
14-4 to 1-6-1983	4	3	3
1-6 to 13-7-1983	10	0	0
13-7 to 13-9-1983	2	2	6
13-9-1983 to 13-3-1984	0	1	9

It can be concluded that the sheep play an important role in creating and sustaining the observed micro-patterns in the vegetation.

VIII.D.5. Forage quality

VIII.D.5.a. Salt marsh communities

The percentages of crude protein of samples of several plant communities and of the species *Artemisia maritima* and *Elymus pycnanthus* are shown in Fig. VIII.17. They were the very dominant species in their respective communities, during the growing season of 1981. Only the *Puccinellia maritima-* and *Juncus gerardii* community and *Artemisia maritima* showed high values during the season and only *Elymus pycnanthus, Artemisia maritima* and the *Puccinellia maritima* community had a decreasing percentage during the season. It can be concluded that no relationship is found between percentage crude protein and the utilization factor.

The cell wall percentages are given in Fig. VIII.18. Only the *Puccinellia maritima* community and *Artemisia maritima* showed lower values than the mean of 67.6%.

The percentages in-vitro digestion of all organic matter and of the cell wall, respectively, are shown in Fig. VIII.19. Only *Elymus pycnanthus* showed a decreasing digestion during the season. The in-vitro digestion of the cell walls was always lower than that of all organic matter, because the latter included the cell contents, 98% of which are digested (Deinum, 1976). The samples of the *Juncus gerardii,* the *Festuca rubra* and the *Puccinellia maritima* communities and of *Artemisia maritima*

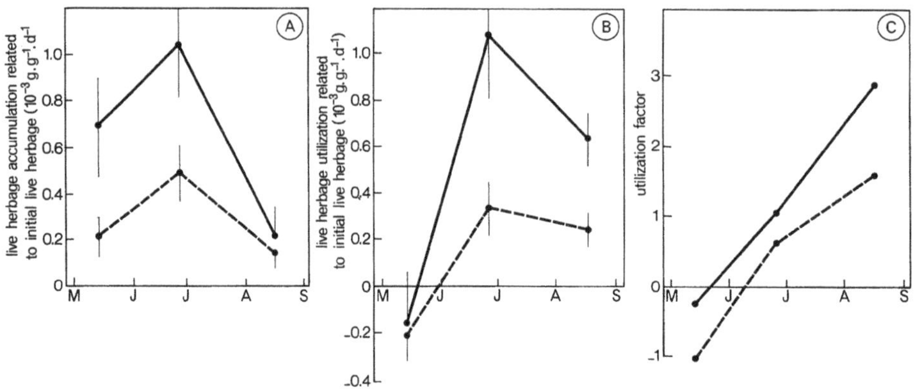

Fig. VIII.16A–C. A: Net live herbage accumulation related to the initial live herbage (± S.E.) in areas protected from grazing for 4–6 weeks, B: utilization of live herbage related to the initial live herbage measured as the difference between live herbage under the exclusion cages and in the open plot (± S.E.) and C: the utilization factor in heavily grazed areas (solid line) and lightly grazed patches (broken line) in 1984 in the micro-pattern of the Westerholt area (after Van Berkel & Schoemaker, 1985).

showed a higher digestion than the mean.

The percentages crude protein, in vitro-digestion of organic matter and of the cell walls in the *Juncus gerardii*, the *Festuca rubra*, the *Puccinellia maritima* communities and in *Artemisia maritima* were significantly higher and the percentages cell wall significantly lower ($P<0.002$, Wilcoxon-Mann-Whitney test) than those of the *Armeria maritima*, the *Ammophila arenaria*, the *Juncus maritimus* and the *Elymus pycnanthus* communities.

Photo VIII.4. Micro-pattern with heavily grazed areas (dark) and lightly grazed patches (light).

Photo VIII.5. The micro-pattern has disappeared within one year after the exclusion of sheep.

There are negative correlations between the mean cell wall percentages and crude protein percentages (r = −0.905, P<0.01) and between the mean cell wall percentages and in-vitro digestion of all organic matter (r = −0.723, P<0.05). The mean percentages crude protein and in-vitro digestion of all organic matter are, however, strongly positively correlated (r = 0.792, P<0.01).

VIII.D.5.b. Mixed grassland communities

Only the forage quality of the closely grazed areas and lightly grazed patches of the micro-patterns

Table VIII.12. Number of points with tall sward and short sward in April and September 1983 in areas cut once in April, cut once in April and twice in May and areas excluded from grazing animals from April onwards. The significance of the correlations between sequential chartings in April and September is indicated, * : P < 0.05 (after Van Berkel & Schoemaker, 1985).

	April		September		Correlation between chartings in April and September
	Number of points in		Number of points in		
	tall sward	short sward	tall sward	short sward	
cut once	35	21	19	37	0.088 n.s.
	19	39	40	18	0.230 *
cut three	35	22	25	32	0.047 n.s.
times	38	12	23	27	0.049 n.s.
exclosure	29	27	46	10	0.483 *
	21	51	51	21	0.412 *

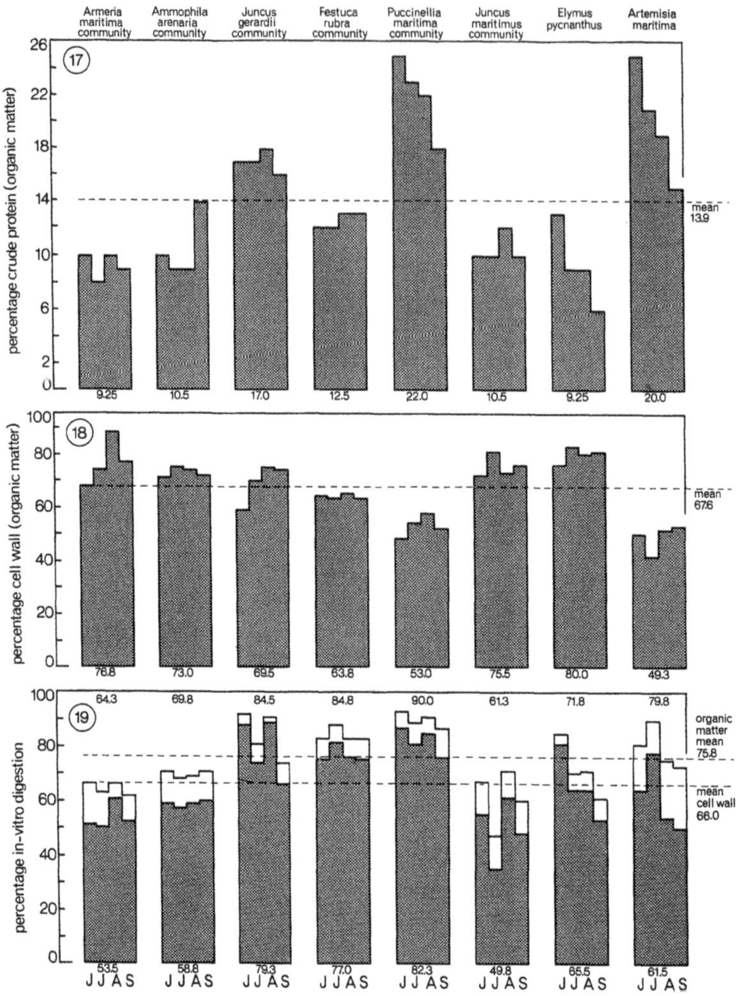

Fig. VIII.17. Percentages crude protein (based on organic matter) of green biomass in different salt marsh plant communities and two dominant species in 1981 (after Looijen et al., 1984).

Fig. VIII.18. Percentages cell wall (based on organic matter) of green biomass in different salt marsh plant communities and two dominant species in 1981 (after Looijen et al., 1984).

Fig. VIII.19. Percentages in-vitro digestion of organic matter (highest bar) and of cell wall (lowest bar) of green biomass in different salt marsh plant communities and two dominant species in 1981 (after Looijen et al., 1984).

have been determined in the mixed grassland Westerholt study area. The percentages of crude protein of green material of samples from both components during the spring and summer of 1983 are shown in Fig. VIII.20. They were high in the early spring and dropped to about 15% in late spring in both the heavily grazed areas and lightly grazed patches, but remained at the same level in the closely grazed area from that period onwards. They did, however, drop to below 10% in the lightly grazed patches.

The cell wall percentages are shown in Fig. VIII.21. Although they increased to somewhat over 50% during the summer in the closely grazed areas, they always remained about 10% lower than those of the lightly grazed patches.

The percentages in-vitro digestion of all organic matter and of the cell wall, respectively, are given

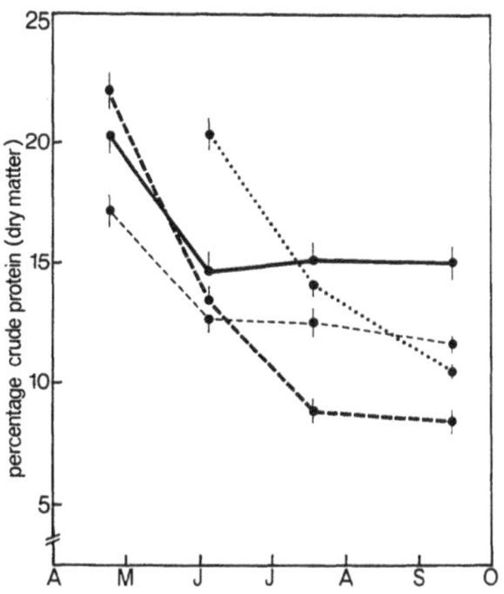

Fig. VIII.20. Percentages crude protein (based on dry matter) of green biomass (± S.E.) in regularly heavily grazed areas (thick solid line), in lightly grazed patches (thick broken line), in excluded areas after heavily grazing (thin broken line) and in grazed patches after three times cutting of the tall sward in spring (dotted line) in mixed grassland in 1984 (after Van Berkel & Schoemaker, 1985).

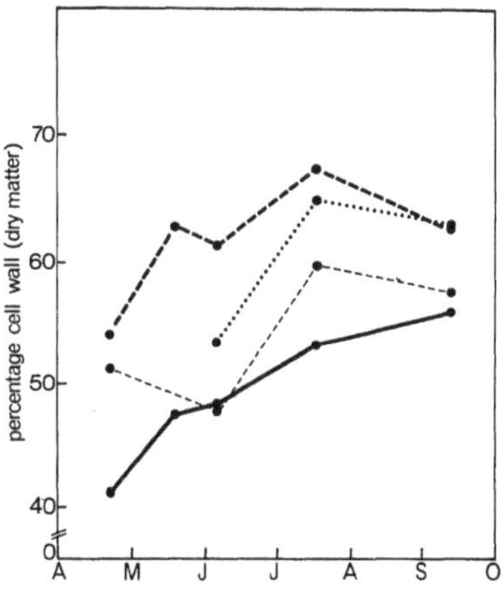

Fig. VIII.21. Percentages cell wall (based on dry matter) of green biomass in regularly heavily grazed areas (thick solid line), in lightly grazed patches (thick broken line), in excluded areas after heavily grazing (thin broken line) and in grazed patches after three times cutting of the tall sward in spring (dotted line) in mixed grassland in 1984 (after Van Berkel & Schoemaker, 1985).

in Fig. VIII.22. The in-vitro digestion of the cell walls was lower than that of all organic matter, whereas both were higher in the heavily grazed areas than in the lightly grazed patches. The correlations between percentages crude protein, cell wall and in-vitro digestibility could not be calculated, because the samples had been pooled. The trends are, however, similar to those of the salt marsh plant communities in the previous section.

Experiments were carried out in order to egalize the micro-patterns (section VIII.D.4) and could influence the forage quality parameters discussed in this section. The crude protein percentages and cell wall percentages were also determined in the experimental plots i.e. in the former heavily grazed areas and in the formerly lightly grazed patches in the areas which were cut three times in 1983. Cutting increased the percentage crude protein (Fig. VIII.20) and decreased the cell wall percentage of former lightly grazed patches (Fig. VIII.21). The exclusion of formerly heavily grazed areas did not result in a lower percentage of crude protein and a higher cell wall percentage than in lightly grazed patches. It perhaps takes more than one growing season to reach the values of the lightly grazed patches. Rosén (1982) did find a lower crude protein content in a three years ungrazed *Avenetum* as compared with the grazed counterpart in limestone grasslands of Oeland, Sweden.

VIII.D.6. Optimal foraging

Optimal foraging can be defined as the natural selection for maximizing the energy intake rate whilst foraging (Pyke et al., 1977; Pyke, 1984). This definition might be translated into the efficiency of the intake of good forage quality i.e. high digestibility, high nitrogen or protein, low fibre, per unit of time. No figures are available on bite size and biting rate of cattle and sheep in the study areas, nor on the quality of the food actually ingested by the herbivores which can be more selective in their feeding than scientists can imitate by clipping (Crawley, 1983). It is, however, known that the proportion of cell contents (a.o. protein) decreases and that of cell wall increases during senescence of plants (Osbourn, 1980) and that stems have a lower digestibility than leaves (Deinum & Van Soest, 1979; Hacker & Minson, 1981). It, therefore, seems likely that the ratio green:dead biomass or leaves:stems in green biomass can be regarded as a measure of the efficiency of intake of good quality forage.

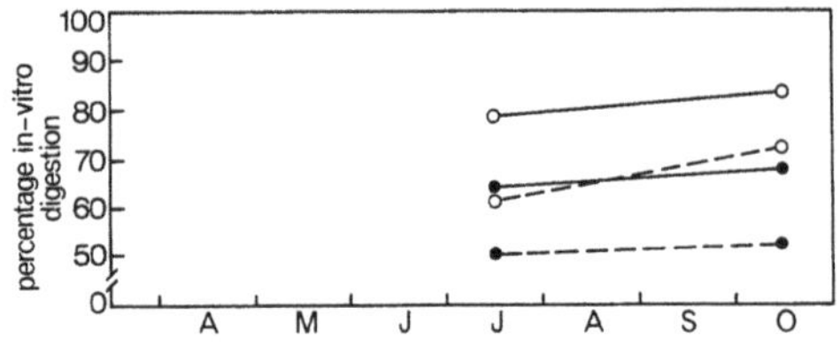

Fig. VIII.22. Percentages in-vitro digestion of organic matter (solid line) and of cell wall (broken line) of green biomass in heavily grazed areas (open circles) and in lightly grazed patches (closed circles) in mixed grassland in 1984 (after Van Berkel & Schoemaker, 1985).

The *Puccinellia maritima, Juncus gerardii* and to a lesser extent *Festuca rubra* communities show a high green:dead biomass ratio (Fig. VIII.23) on the salt marsh.

The heavily grazed areas show both a higher green:dead biomass and leaves:stem ratio of green biomass than the lightly grazed patches in the mixed grassland area (Fig. VIII.24).

Most salt marsh communities and the heavily grazed areas in the mixed grassland have such a high utilization factor that hardly any tiller or flowering stalk escapes from grazing. This results in a very short turf. Are there, however, any species which are preferentially grazed by the cattle or sheep? The frequencies with which individual species are gnawn during the grazing season in several salt marsh plant communities are shown in Fig. VIII.25. Several species, like *Triglochin maritima* and *Plantago maritima,* which occur in a relatively low abundancy were preferentially grazed in some plant communities, whereas *Glaux maritima* and *Juncus maritimus* were avoided.

The frequency of the occurrence of several species as well as the occurrence of grazed leaves in mixed grassland are shown in Fig. VIII.26. *Anthoxanthum odoratum* was only preferentially grazed in the lightly grazed patches, *Trifolium repens* and *Lotus uliginosus* were only avoided in the heavily grazed areas, whereas *Ranunculus* spp. were avoided in both cases. Mean percentages of grazed leaves increased from 22% in May and 39% in June to 57% in September in the heavily grazed area and from 10% and 20% to 22% in the lightly grazed patches.

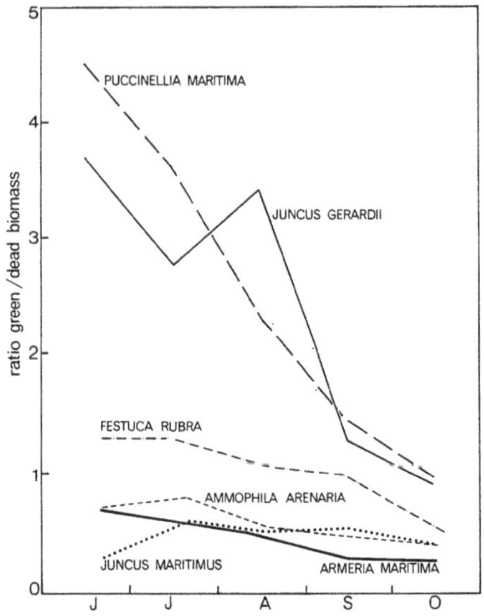

Fig. VIII.23. Mean ratio green to dead biomass in different salt marsh plant communities in 1981 (after Looijen et al., 1984).

The question as to whether the pattern of preferential foraging was initiated by the presence of preferred species or by randomly foraging will be discussed in section VIII.E.4.

VIII.E. Comparison of grazing regimes

VIII.E.1. Net herbage accumulation

Although many problems are encountered when estimating the net herbage accumulation using the difference method (Milner & Hughes, 1968), exclusion cages are the only technique which can be used extensively in grazed swards (Parsons et al., 1983). Most authors refer to whatever is measured with the difference method as above-ground net primary production or herbage production. The net herbage accumulation data, i.e. the result of production and decomposition of the present salt marsh study, fall within the range of herbage production values published for northern European salt marshes (see Table VIII.13). Since most of these authors used the same or similar methods, their data presumably suffer from the same inaccuracies as those encountered in the present

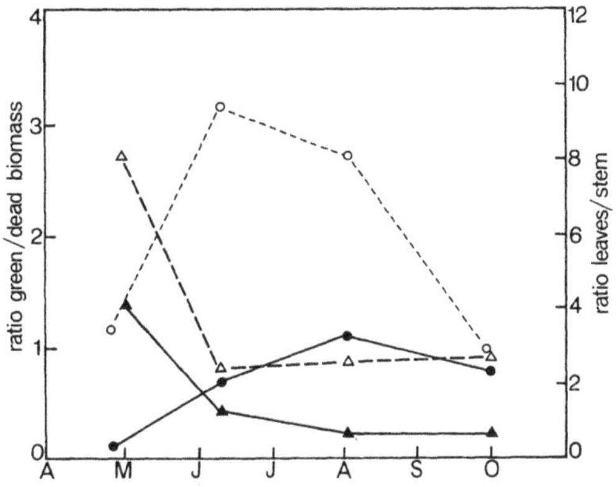

Fig. VIII.24. Mean ratios of green to dead biomass (circles) and of leaves to stems (triangles) in heavily grazed areas (broken line) and in lightly grazed patches (solid line) in mixed grassland in 1984 (after Van Berkel & Schoemaker, 1985).

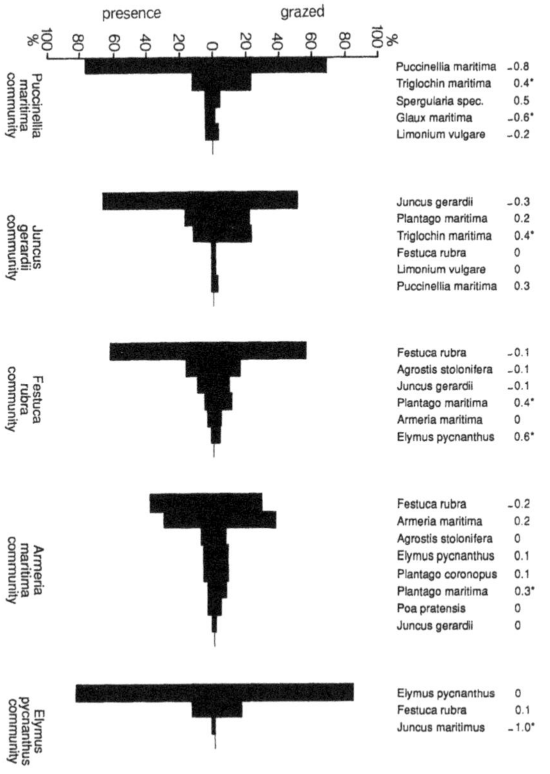

Fig. VIII.25. Percentages of species present and of grazed individuals in different salt marsh communities in the period between May and September 1981. The preference (values between 0 and 1) and avoidance (values between −1 and 0) independant of the relative abundance according to Jacobs (1974) are indicated. The significant ($P < 0.05$) deviations from the expected percentages of grazed individuals are indicated with an asterisk (after Looijen et al., 1984).

study. Herbage accumulation in the tall-growing *Juncus maritimus* community is more comparable to values found in north American salt marshes (Keefe, 1972; Turner, 1976; Kirby & Gosselink, 1976; Hopkinson et al., 1978; Eilers, 1979), although Groenendijk (1984) found comparable values in *Spartina anglica* and *Elymus pycnanthus* communities in the south western part of the Netherlands (Table VIII.13). All these cases concern tall-growing species.

Since Smalley's method ignores any herbage losses due to decomposition, it may result in underestimations when used to measure herbage production. Several authors have reported production values obtained with Smalley's method to be up to 2–3 times lower than values obtained with the paired plot method of Wiegert & Evans (see Table VIII.13). This is more or less confirmed by the results of the litter bag experiment. Weight losses of 22–61% occurred in the communities from April to October 1984. Similar losses (over the same period) were found by Buth et al. (1982) for *Spartina anglica, Elymus pycnanthus* and *Halimione portulacoides* stems and leaves in the Dutch Oosterschelde salt marsh. As herbage utilization was estimated from differences in herbage mass inside and outside the cages at the end of each sampling interval, and it is not known to what extent decomposition proceeds faster outside the cages than inside,

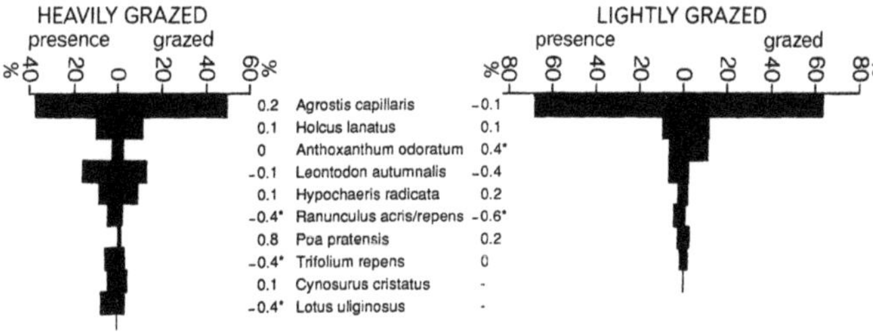

Fig. VIII.26. Percentages of species present and of grazed leaves in the heavily grazed areas and lightly grazed patches in mixed grassland in the period between May and September 1984. The preference (values between 0 and 1) and avoidance (values between −1 and 0) independant of the relative abundance according to Jacobs (1974) are indicated. Significant ($P<0.05$) deviations from expected percentages of grazed leaves are indicated with an asterisk (after Van Berkel & Schoemaker, 1985).

herbage utilization data cannot be adjusted accordingly. The utilization factor is, therefore, not corrected for decomposition losses. The decomposition rate was measured in order to enable some estimation of the disappearance of ungrazed biomass, if any.

It is difficult to find net herbage accumulation data of similar plant communities to those in the present mixed grassland area. Klapp (1965) gave values between 400 and 600 gdw · m^{-2} for grazed dry to moist *Lolio-Cynosuretum* communities with fertilizer application in Germany. Welch & Rawes

Table VIII.13. Production values for grazed salt marsh plant communities in N-Europe expressed as g dry organic matter · m^{-2} · $year^{-1}$; values between brackets are adjusted herbage accumulation data to account for losses due to decomposition (after Looijen & Bakker, 1987.

Community of	Location	Production	Method	Author
Puccinellia maritima	Island of Schiermonnikoog The Netherlands	767 (1089)	Smalley	this study
Puccinellia maritima	Island of Terschelling The Netherlands	278-451	Smalley	Ketner 1972
Puccinellia maritima/ Limonium vulgare/ Triglochin maritima	Oosterschelde estuary The Netherlands	837	Smalley	Lammens & Van Eeden 1977 (cited in Joenje & Wolff 1979)
Juncus gerardii	Island of Schiermonnikoog The Netherlands	429 (690)	Smalley	this study
Juncus gerardii	salt meadows along the Baltic Sea	430	modification of paired-plot	Wallentius 1973
Puccinellia maritima/ Juncus gerardii	Hamburger Hallig N-Germany	380-478	peak standing crop (?)	Hansen 1982
Plantago maritima/ Limonium vulgare (ungrazed)	Island of Terschelling The Netherlands	462-547 553-564	Smalley paired-plot	Ketner 1972
Festuca rubra	Island of Schiermonnikoog The Netherlands	388 (566)	Smalley	this study
Festuca rubra	Island of Schiermonnikoog The Netherlands	400-500	Smalley (?)	Alberda 1974
Festuca rubra*	,,	1013	Smalley (?)	Alberda 1974
Lolium perenne*	Island of Schiermonnikoog The Netherlands	939 (1258)	Smalley	this study
Inland Lolium perenne pastures*	The Netherlands	1500	CO_2 exchange	Lantinga, 1985
Juncus maritimus	Island of Schiermonnikoog The Netherlands	1707 (2287)	Smalley	this study
Spartina anglica	Oosterschelde estuary The Netherlands	1162-1649 2139-2659	Smalley paired-plot	Groendendijk 1984
Elytrigia pungens (=Elymus pycnanthus)	Oosterschelde estuary The Netherlands	474-878 1416-1787	Smalley paired-plot	Groenendijk 1984

* artificial fertilizers applied

(1965) reported 195 gdw · m^{-2} and Perkins et al. (1978) 271 gdw · m^{-2} in grazed but unfertilized *Agrostis capillaris* communities in lowland in the Pennines and upland in Snowdonia, respectively. In our study area, which had not been fertilized since 1971, the values fall within this range. Perkins et al. (1978) calculated that the mean harvestable yield of 271 gdw · m^{-2} was equivalent to 865 gdw · m^{-2}, adjusting for losses due to senescence and decomposition. Results of the litterbag experiment revealed that weight loss of between 40 and 60% occurred in most communities after one year. Perkins et al. (1978) recorded a 70% disappearance in an *Agrostis capillaris* grassland after one year, but this was using a larger mesh (4 mm × 6 mm). The disappearance of litter in an *Agrostis capillaris/Holcus lanatus/Anthoxanthum odoratum* sward using a 7 mm mesh litterbag was remarkably quick, namely c. 50% after 100 days (Dickinson, 1983).

VIII.E.2. Herbage utilization is not the same as consumption

The herbage utilization, which is measured with the difference method, is often called herbage consumed (Milner & Hughes, 1968; Heady, 1975; Stoddart et al., 1975; see also Esselink et al., 1989). Putman (1986) reported offtake as percentages of production of over 100% in the stable system of the New Forest (United Kingdom) and mentioned vegetation senescence as the cause of overestimation of consumption. However, herbage utilization which is measured in this way is not only consumption, but also any herbage spoiled by trampling, fouling and decomposition ('t Mannetje, 1978). Estimates of the herbage consumed as a percentage of the herbage utilized, range from 30–50% for cattle (cf. Esselink et al., 1989). Parsons et al. (1984) reported an almost 80% over-estimation when using the exclusion method as opposed to a measured intake.

A theorethical intake level of 6.5 kgdw per animal per day, or 1100 kgdw · day^{-1} for the entire flock (calculation according to Van Es, 1978) was calculated from the increase in body weight of the cattle (Photo VIII.6) on the salt marsh over the grazing period (0.7 kg · day^{-1} P. Oosterveld, unpublished), their net energy requirements for growth (9190 kJ · day^{-1}), maintencance (25870 kJ · day^{-1}), walking (6215 kJ · day^{-1}), grazing (3650 kJ · day^{-1}) and excretion of excess nitrogen (200 kJ · day^{-1}) and the net energy content of their feed (6990 kJ · $kgdw^{-1}$). Total herbage utilized by the cattle in the communities of the study area was 1040 kgdw · day^{-1}. Herbage utilized in the total grazed area, containing similar plant communities (Fig. IV.17), was then estimated at 4160 kgdw · day^{-1}. Consumption, as a percentage of herbage utilization, would then have been 25%, close to the 30–50% mentioned above.

This percentage would be expected to be higher in the case of grazing by geese, since 'damage' due to trampling is likely to be less. Some estimates of this can be obtained from counts of the droppings of barnacle geese (J. Beekman, unpublished) and digestion measurements. An average of 18 droppings · m^{-2} was counted in the studied communites from 14 February to 17 April 1982. Since the digestion of the dominant species of these communities is about 35% at this time of the year (Ebbinge et al., 1975) and the weight of one dropping is approximately 0.75 gdw (Prins & Ydenberg, 1978), this would amount to an average herbage intake of approximately 21 gdw · m^{-2}. The amount of herbage utilized during the same interval was approximately 39 gdw · m^{-2}. The herbage intake, as a proportion of herbage utilized, would, therefore, seem to have been about 54%. Ebbinge et al. (1975) found a similar difference between herbage intake, as estimated from dropping counts, digestion measurements and 'grazed' herbage as estimated from exclusion cages. These results point to serious over-estimations of herbage intake, when only using the exclosure technique. This depends on the type of grazer.

The percentage of utilized herbage which is consumed by the sheep in the inland mixed grassland communities could not be measured since the grassland area was not representative of the whole fenced-in area, and the herbage accumulation and utilization were not estimated in the remaining woodland and heathland area.

Photo VIII.6. Weighing of the cattle from the Oosterkwelder salt marsh.

VIII.E.3. Grazing intensity: occupancy or utilization factor?

There were considerable differences between the salt marsh communities both with regard to the utilization factor by the cattle (Table VIII.4) and the occupancy as measured by the numbers of dung patches. No correlation was, however, found between these two variables, nor between the number of dung patches and the herbage utilized. This could be expected in the case of the *Artemisia maritima* community as the species is not eaten by the cattle, but the animals spent much time resting in the community, as indicated by the high number of dung patches.

It seems likely that the animals have to spent more time in low-growing swards than in taller stands in order to reach a similar intake level (see also Crawley, 1983). Since the intake per bite or unit of time is lower, the amount of herbage consumed may consequently be higher in tall-growing swards, whilst the number of dung patches is lower than in the short turf. Hence the structure of the sward, in particular its height, might contribute tot he absence of a correlation between the occupancy and the utilization factor.

In mixed grassland communities, which were studied, a positive relationship was found between dunging intensity (Table VIII.8) and the indices of utilization factor (Table VIII.2). In Scotland Welch (1984) found a good correlation between the dung deposition rate of large herbivores (including sheep) and biomass changes of the heather. However, beyond a certain treshold, trends were affected by processes other than defoliation, namely, trampling and large amounts of dung, killing heather plants. Hilder (1966), working in Australia, found that sheep confined in paddocks may contaminate relatively small areas of pasture with high concentrations of dung, chiefly in the areas where they rest at night. Hunter (1962) found up to a fivefold difference in dunging intensity on patches with similar comparative grazing intensities which was thought to result from the sheep's use of special areas for resting. Red deer data from Charles et al. (1977) also suggest a twofold difference in dung return when comparing grasslands with wet heath and bog, whereas comparative grazing intensities showed a fourfold difference.

It is obvious from these data that it would be wrong to conclude that dunging intensities always give a correct impression of foraging activities. Although it is a time-consuming process resulting in rough figures, the 'difference' method, using exclusion cages, gives an estimate of the utilization factor. The easy measurement of dunging intensities can only provide preferences for plant communities with respect to presence or occupancy. The method chosen depends on the problem which has to be solved.

VIII.E.4. Development of grazing patterns

Hunter (1962), Klötzli (1965), Wielgolaski (1976), Charles et al. (1977) and Job & Taylor (1978) found macro-patterns in plant communities which were each dominated by a different species, because preferential grazing resulted in grazing at different intensities. Thus the grazing intensities reflect the preference for or avoidance of dominant species. This is also the case in the Westerholt study area where the improved grassland sections are preferred during the summer and the heathland sections during the winter. In most of these situations grazing probably took place for long periods, so the composition of the vegetation at the beginning of the grazing is not known, nor is the possible previous existence of a macro-pattern. The plant communities mentioned in those studies all occurred on particular soil types. The macro-patterns of plant communities and hence the macro-patterns of grazing intensities, therefore, reflect the affinities between vegetation types and soil types. It can be concluded that the observed patterns of grazing intensity are controlled by the patterns of plant communities.

Little information is available on the development and constancy in grazing patterns. This is in contrast to the Westerholt study area where a fairly constant pattern developed in an initially homogeneous *Holcus lanatus* sward (see section VIII.D.2). Hunter (1962) reported that over a three years observation period the seasonal preference among sward types in Scotland varied consistently. Hunter and Nicholson et al. (1970) found micro-patterns in vegetation stands due to the avoidance of species. The latter described the development of *Nardus stricta* tussocks in a *Nardus stricta*/*Festuca ovina*/*Molinia caerulea*/*Anthoxanthum odoratum* sward under grazing with a stocking density of 2–5 sheep ha^{-1}. Single plants or clumps of *Anthoxanthum odoratum* or young *Molinia caerulea* shoots provided the loci which were initially moderately grazed. While grazing increased around the edges of these patches, material tended to be taken less selectively and then the initial centres expanded outwards producing a patchwork of short-tufted areas. The close-grazed areas eventually joined and grazing pressure over the entire area became relatively uniform, except for isolated areas that were partly rejected – usually *Nardus stricta* tussocks. The constancy of the afore-mentioned micro-pattern seems probable because the avoidance of individual species will remain. Hunter (1962) suggested that the taller tufts consist of one avoided species because the sheep graze on the grass between the *Juncus effusus* clumps. Some micro-patterns as a result of grazing may be originated from the presence of isolated tussocks or clumps of individuals of avoided species in the sward before grazing started.

The taller tufts of lightly grazed patches in the micro-pattern of the Westerholt area do not, however, consist of one single species, but have nearly the same species composition as the heavily grazed areas. Since species which are usually avoided are not found, it is suggested that random grazing in the initially uniform *Holcus lanatus* sward at the beginning of the grazing experiment initiated the development of the micro-patterns by creating patches with young protein-rich tillers which are sustained by the sheep. Since the stocking rate is low in relation to the production, higher stands which are untouched remain untouched. Norman (1957) also reported patchiness with respect to canopy height, within two years, in a chalk pasture trial in the United Kingdom brought about by under-grazing. Arnold (1964) described that, in pastures lightly stocked with sheep, selective grazing is concentrated initially in the better parts of the pasture. The vegetation in the poorer areas, therefore, becomes taller, older and less attractive to the sheep, so that, in time, foraging becomes more and more aggregated into the better areas.

Daines (1980) gave a description of the development of a micro-pattern under cattle-grazing in the species-rich successional grassland of Dohne Sourveld in South Africa. On entering a camp the animals select species of their choice. These species are 'creamed' to obtain the most nutritious food possible, removing less than 50%, of the available leaf of these species. Initially this creaming operation is confined to some most preferred species, but when the available bulk from this source becomes insufficient they extend their foraging to less preferred species. Then the animals do not move to the least preferred species, but return to the most preferred species which are the more heavily grazed, removing more than 50% of the available mass of these species. This cycling of grazing from the most preferred to less preferred species is repeated when forage becomes limiting. Where a low stocking density is applied it results in the uneven utilization of better species and the ignoring of the ungrazed grasses. No information is available on the constancy of the resulting micro-pattern. On the contrary, the author only gives suggestions as to how to remove these patterns of uneven use of grasslands, since like many grazing studies, he focusses on agricultural improvement.

The Serengeti grassland system in Tanzania can be regarded as an old and stable system. The system also contains patterns, which seem to show, however, no equilibrium but a shifting mosaic of successional states. Animal effects, like defoliation, trampling and deposition of nutrients are supposed to oscillate more or less constantly in time and space resulting in a quasi-equilibrium (McNaughton, 1983).

VIII.E.5. Selectivity in grazing

Some plant species are grazed selectively in the salt marsh, but only if they are present in small quantities in plant communities dominated by other species. This selectivity does not seem to be in accordance with the observed high herbage removal in some plant communities in the spring. The latter does, however, correspond with high green: dead biomass ratios. It seems likely, therefore, that cattle grazing salt marsh communities at a high stocking rate select plant communities providing fresh green biomass, probably after an initial selection of individual species. The stocking rate is such that all plant communities are grazed with nearly similar utilization factors. Moreover, the crude protein and in-vitro digestion percentages hardly decrease during the grazing season. Milner & Gwynne (1974) reported similar results from the heavily sheep grazed halophyte swards (*Armeria maritima/ Plantago maritima* community) on St. Kilda. They suggest that selection by the sheep for species palatability is only possible during the late spring and early summer when there is a considerable choice. Selection is limited by availability during the rest of the year. It is noteworthy that the percentages of crude protein and the in-vitro digestion of the halophytic swards of St. Kilda did not decrease during the summer period. Moreover, their mean values were similar to those found in the Schiermonnikoog salt marsh. This is in marked contrast with the findings of Tyler (1971) on an ungrazed *Juncus gerardii* community along the Baltic in Sweden. The spring values were somewhat lower than those found in our study area, but the percentages of crude protein of the seperate species *Juncus gerardii, Agrostis stolonifera, Eleocharis palustris uniglumis, Glaux maritima* and *Plantago maritima* decreased from 13% in May to values between 4 and 7% in late summer and the autumn. *Triglochin maritimum* showed a percentage of crude protein of 25 in May, 13% in the summer and 4% in the autumn. Jensen et al. (1987) also found a decrease of percentages of crude protein in ungrazed *Juncus gerardii* communities along the North Sea coast in Denmark. Percentages of 20 in May decreased to 16% in the summer and in the autumn. They found 18% in the *Puccinellia maritima* community in May, 23% in the summer and 20% in the autumn, which might be due to a prolonged period of growth in contrast with that of the *Juncus gerardii* community. The levels of percentages of crude protein are in agreement with those found in this study. This was also the case for the in vitro digestion of those plant communities.

In the study area which is grazed at a high stocking rate, selection can only be expected if forage quality drops below certain minimal levels for the functioning of the stomach. Those levels are between 8 and 10% crude protein (Van Es, 1979), or more precisely between 6 and 7% for maintenance and between 8 and 9% for growth of 0.5 $kg \cdot day^{-1}$ for cattle (Little, 1982 and Agricultural Research Council, 1984 in Bokdam, 1987). Such levels were not found in the salt marsh communties. The only selection mechanism left then is the avoidance of species. This actually takes place in the case of *Artemisia maritima* and *Juncus maritimus.*

The micro-pattern resulting from the low stocking rate in the mixed grassland area provides some more ideas on selectivity in grazing. A rough ranking of a number of individual plant species which occur in the study area is given in Table VIII.14. This is done according to their preference by sheep as derived from assessments in the literature. The actual grazing frequency, however, deviates from this sequence and, moreover, many species are grazed at the same frequencies. This divergence is mainly due to the fact that the diets of large vertebrate herbivores are composed of many plant species. Apparently 'suboptimal' species may be added to the diet for a number of reasons (Crawley, 1983); (i) the abundance of preferred foods is too low to allow the animals to restrict their feeding to these; (ii) it is not worth being more particular as the costs of discriminating are greater than the rewards; (iii) the foods may not differ sufficiently that the animals need to (or can) discriminate between them.

The present study demonstrates that *Agrostis*

Table VIII.14. Survey of the preference during the summer period of a number of species occurring in the mixed grassland study area according to assessments in the literature (Milton, 1953; Kershaw, 1958; Lodge, 1959; Beddows, 1961; Kruijne et al., 1967; Klapp, 1971; Duffey et al., 1974; Milner & Gwynne, 1974; Wielgolaski, 1975).

1 Trifolium repens	Leontodon autumnalis
	Taraxacum spec.
2 Cynosurus cristatus	Plantago major
2 Lolium perenne	Trifolium dubium
3 Poa pratensis	
3 Poa trivialis	
3 Phleum pratense	
3 Dactylis glomerata	
3 Festuca pratensis	
4 Festuca rubra	
5 Holcus lanatus	
6 Anthoxanthum odoratum	
6 Agrostis capillaris	
7 Festuca ovina	
7 Deschampsia flexuosa	
7 Elymus repens	
8 Nardus stricta	
8 Molinia caerulea	
8 Cyperaceae	
8 Juncaceae	

capillaris is grazed to a small extent in lightly grazed patches, but much more in heavily grazed areas, so its preference might depend on the condition in which sheep keep their forage. Moreover, the range in digestion of a single species is very large because of the differences in digestion of different plant parts and hence can exceed the differences between plant species (Hacker & Minson, 1981). Nicholson et al. (1970) found that the 'unpalatable' species *Nardus stricta* is eaten in a dense mass of interwoven leaves and shoots, but avoided when *Nardus* tufts were fairly isolated. The 'preferred species' concept should, therefore, be treated with caution. Preference should be considered against the background of the choice the foraging animal actually can make (Davies, 1925).

Repeated defoliation in the heavily grazed areas stimulates tillering (Harper, 1977; Witchi & Michalk, 1979). More young leaves and a higher leaf: stem ratio result in a higher preference (Davies, 1925), probably due to a higher percentage of crude protein and a higher digestion of forage (Stobbs, 1973B). The balance in different plant parts is very important since the leaf sheath and stem digestion decrease much quicker than that of the leaf (Hacker & Minson, 1981). Frame & Hunt (1971) found that the frequently defoliated herbage had a higher percentage of crude protein than that of other treatments. The contribution of stems to the total biomass was much higher in the vegetation protected from grazing for 14 years than in the adjacent grazed area (McNaughton, 1983).

Standing dead material discouraged the utilization of green forage by deer and in an *Agropyron spicatum*-dominated sward in the United States both cattle and deer preferred sites where litter was removed by burning or cutting (Willms et al., 1980). In a uniform *Phalaris/Trifolium* pasture in Australia, Langlands and Bennett (1973) found that the ability of sheep to select green material of high digestibility was impaired at low stocking rates. This was due to the dilution of green forage by dead material which had not been consumed in previous years. Grazing and litter accumulation in lightly grazed patches were negatively correlated under sheep-grazing in Wales and in Scotland according to Job & Taylor (1978) and Hodgson & Grant (1981) and under cattle-grazing in Idaho according to Leege et al. (1981). Welch (1986) found that the incidence of grazing for *Nardus stricta* was higer on tussocks grazed in the previous summer than on tussocks ungrazed in the previous summer. He supposed that the presence of old dead shoots deterred the cattle and to a lesser extent the sheep. Van Soest (1982) quoted other authors who found that the intake of tropical forages decreased below a tresshold of 7% crude protein.

Heavy defoliation can result in a reduction of pasture biomass and hence in a decline in food availability. Bite size then reduces and the biting rate and foraging time increase as a compensation for both sheep (Allden & Whittaker, 1970) and cattle with a standing crop between 400 and 700 $gdw \cdot m^{-2}$ (Chacon & Stobbs, 1976) and less than 100 $gdw \cdot m^{-2}$ (Scarnecchia et al., 1985). The intake of cattle may be restricted when bite size is lower than the critical level of about 0.3 g, which cannot be compensated for by an increase of the number of bites and grazing time as suggested by Stobbs (1973A). He reported a bite size of 0.40 g for cattle in a tall sward with stemmy plants and of 0.52 g in a low canopy with a more leafy dense sward of tropical pasture. The bite size was positively correlated with the leaf: stem ratio and hence with crude protein and in vitro digestion of the forage (Stobbs, 1973B). Later Ludlow et al. (1982) indicated that the forage yield per bite is positively correlated to biomass concentration i.e. the mass of foliage per unit of canopy volume. Cow bite size will fall below the critical level of 0.3 g at plant biomass concentration below 0.8 $mg \cdot cc^{-1}$. Most vegetation with a canopy height over 40 cm in the Serengeti system would have a biomass concentration lower than that maintenance level. By keeping the vegetation lower, particularly below 10 cm, grazers in the Serengeti can increase their harvesting rate strongly (McNaughton, 1984). Short (1986) mentioned in a review a range of 41 to 70 bites · min^{-1} and a bite size of 0.02 to 0.40 for sheep, but gave no information on the quality of the ingested forage with different biting rates and bite sizes.

It can be concluded that the heavily grazed areas have a higher forage quality than the lightly grazed patches. There is a preference for specific sites with a high forage quality. Like good farmers the herbivores maintain these sites, including different species, by their feeding strategy (see also McNaughton, 1979; Drent & Prins, 1987). In a review of the interrelationships between the factors influencing the intake and digestibility of forage by ruminants Heard (1980) mentioned that digestibility is negatively correlated with ageing, but ageing is positively correlated with yield so that the most economical growth period before harvest for the farmer is a compromise between digestibility and yield. This suggests that the feeding strategy of cattle and sheep which graze at a low stocking rate and therefore have plenty of available forage, must select on digestibility. It seems, however, unlikely that herbivores can select on the basis of digestibility according to Kenney & Black (1984). They suggested that digestibility is correlated with the ease of chewing and swallowing or the speed with which forage can be eaten, since selection for a forage was more related to the potential rate of intake than the in vitro digestion of the forage (Black & Kenney, 1984). Heavy grazing results in a sward structure without standing dead material or litter and a high leaf : stem ratio and/or biomass concentration with subsequent high protein content and high digestibility of the forage and hence attracts the animals again. Light grazing results in a low protein content and high cell wall percentage of the forage with litter accumulation and subsequent avoidance by the animals.

CHAPTER IX

Cutting regimes

IX.A. Introduction

This chapter deals with the change in the yields with different cutting regimes starting from an abandoned grassland (salt marsh) and from formerly fertilized grasslands (reclaimed heathland and valley grasslands).

The questions to be considered are (i) how did the above-ground standing crop, the root biomass and the shoot to root ratio change during a number of years? Since the amounts of the above-ground phytomass can affect the performance of species, the second question raises, namely, (ii) what is the relationship between above-ground standing crop and species diversity? As the above-ground standing crop is always measured at the time when the cutting regime is carried out (which can vary between June and September and cutting was often carried out twice a year) it does not exactly represent peak standing crop, but gives a measure of the net accumulation since the previous treatment.

The results of the cutting regimes considered in this chapter are compared to those in the literature. The comparison between hay-making and grazing will be made in Chapter X.

IX.B. Above-ground standing crop

IX.B.1. Salt marshes (Schiermonnikoog)

In 1972, thirteen years after its cessation, grazing was resumed in part of the Oosterkwelder salt marsh study area (see section IV.C). The effects of renewed grazing and continued abandoning were compared and a number of cutting regimes (cutting once in June or in August and cutting twice in both June and in August) were imposed in different plant communities (see section VII.C).

The average above-ground 'peak' standing crop including litter varies between different plant communities in the untouched salt marsh, namely, the community of *Artemisia maritima* 1100 ± 100 gdw · m^{-2}, of *Elymus pycnanthus* 1182 ± 206 gdw · m^{-2}, of *Juncus maritimus* 1156 ± 179 gdw · m^{-2}, of *Festuca rubra*/*Limonium vulgare* 727 ± 128 gdw · m^{-2}, of *Juncus gerardii* 604 ± 220 gdw · m^{-2} and of *Festuca rubra*/*Armeria maritima* 243 ± 61 gdw · m^{-2}.

The high standing crop of the untouched communities of *Artemisia maritima, Elymus pycnanthus* and *Juncus maritimus* is heavily influenced by the cutting regimes. The average maximal standing crop during the study period in the annual August cutting regime varies between 25% and 35% and that of the twice cutting regimes only between 10% and 20% of the untouched standing crop (Fig. IX.1 A–F). These figures range from 45% to 55% and from 20% to 30% respectively for the small standing crop of the untouched communities of *Juncus gerardii, Festuca rubra/Limonium vulgare* and *Festuca rubra/Armeria maritima.*

The above-ground standing crop figures reveal clear fluctuations (Fig. IX.1 A–F). The high values in 1974 might be attributed to the mild preceeding winter (Fig. IX.2). The very dry spring and summer of 1976 apparently caused low standing crop values in that year (Fig. IX.3). Sea water inundations in late spring might have been responsible for the low standing crop in 1981. The very high standing crop in 1983 can probably be attributed to the very wet winter and spring of that year.

Only the June hay-making regime in the *Elymus pycnanthus* community resulted in a significant ($P<0.05$) decrease (cf. II.C.3) of the above-ground phytomass during the study period.

IX.B.2. Mixed grasslands (Westerholt)

The study area Westerholt was originally heathland but had been partly ploughed to a depth of 40 cm and then sown to grass in the 1950's. It comprised of heathland, deciduous woodland and pasture (see section V.A). In October 1972 most of the area was fenced-in and sheep were introduced and a cutting regime was adopted for the remainder. Since 1973, part of the area under hay-making has been applied with fertilizers, with only nitrogen (50 kg $N \cdot ha^{-1}$), in order to grow a high crop and hence remove with the cut herbage more phosphorus and potassium. The uppermost 5 cm of the grassland was removed in another area in May 1973. Finally, two small grassland areas were ploughed in September 1978, one area to a depth of 20 cm and the other to 60 cm in order to bring the unchanged C-horizon to the surface. The whole ungrazed grassland area was mown once a year and the cuttings were removed in the period between July and September (see section V.A). Standing crop figures were only estimated in the dry and moist parts of the ungrazed Westerholt study area.

The above-ground standing crop did not change significantly with any of the management practices, only tendancies were detected. In the *Holcus lanatus* community (see section V.C.2) which had received no fertilizer applications since 1971, it remained about 400 $gdw \cdot m^{-2}$ until 1979 and then decreased to 300 $gdw \cdot m^{-2}$ and to 200 $gdw \cdot m^{-2}$ during the last few years (Fig. IX.4A). These figures are low compared to the above-ground phytomass which ranged from 360–810 $gdw \cdot m^{-2}$ for comparable *Lolio-Cynosuretum* stands in the Netherlands measured by Fliervoet (1984). The low values in 1974 and 1976 seem to be due to the large precipitation deficit (Fig. IX.5). The standing crop in the area annually receiving 50 kg $N \cdot ha^{-1}$ increased from c. 400 to c. 800 $gdw \cdot m^{-2}$ until 1977, and dropped to around 600 $gdw \cdot m^{-2}$ after 1979 and to below 500 $gdw \cdot m^{-2}$ during the last few years (Fig. IX.4B). It was significantly ($P<0.05$, sign-test) higher than in the site without fertilizer applications. Standing crop values in the area, after sod removal in 1973, fluctuated around 200 $gdw \cdot m^{-2}$ (Fig. IX.4C). Ploughing in 1978 resulted in an initial increase of the standing crop to c. 500 $gdw \cdot m^{-2}$, probably as a result of decomposition of the ploughed sod, but from 1983 onwards, the values gradually dropped below 200 $gdw \cdot m^{-2}$ (Fig. IX.4D).

The root biomass was 1200–1300 $gdw \cdot m^{-2}$ a few years after the cessation of fertilizer applications (Fig. IX.4 A–B). It dropped dramatically and then gradually increased again to 400 $gdw \cdot m^{-2}$ in both the unfertilized area and the area receiving nitrogen fertilizer. The fraction of living roots amounted to nearly 100%. The root to shoot ratio increased initially, but diminished later on. The root biomass remained very low for a few years after sod removal (Fig. IX.4C), but then gradually increased to 600 $gdw \cdot m^{-2}$, apparently as a result of the building up of a new rootsystem.

IX.B.3. Lower course valley grasslands (Lammertshem)

Several cutting practices were carried out in recent reserve valley grasslands viz. in the upper course study area Loefvledders and in the lower course study area Lammertshem (see Fig. VI.2). Hay-

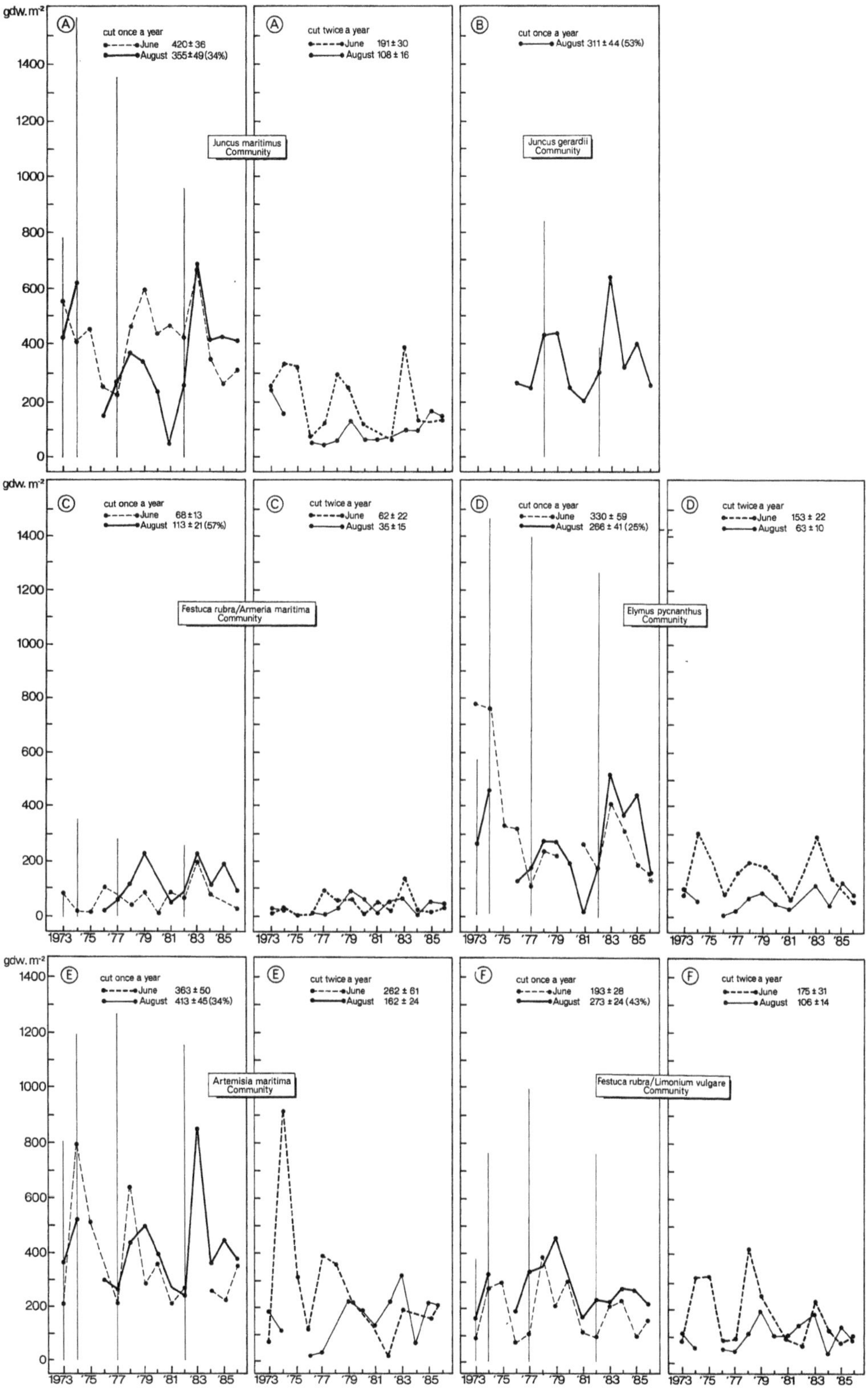

Fig. IX.1A–F. Above-ground standing crop including litter under different cutting regimes in various salt marsh plant communities. * Significant change according to Spearman's rank correlation test ($P<0.05$). The average values ($\pm$ S.E.) over the study period and their percentage of the average peak standing crop (August) in the untouched area are indicated. Bars indicate the peak standing crop in the untouched areas in several years.

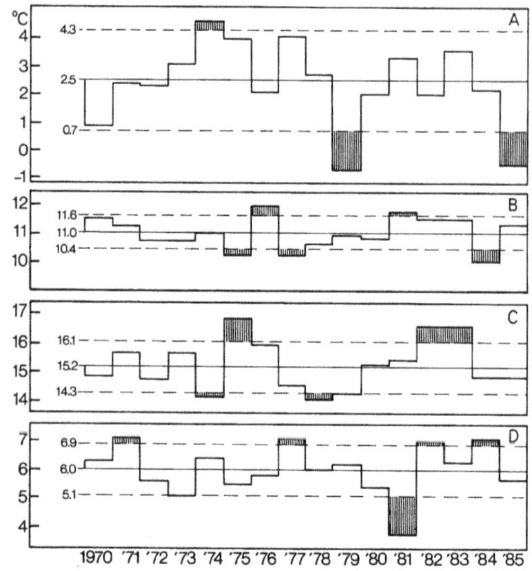

Fig. IX.2A–D. Mean temperature (° C) ± (S.D.) for the period 1970–1985 at Eelde in A: the first, B: the second, C: the third and D: the fourth quarter of a year. Temperature data from Schiermonnikoog are similar (after Royal Netherlands Meteorological Institute, De Bilt).

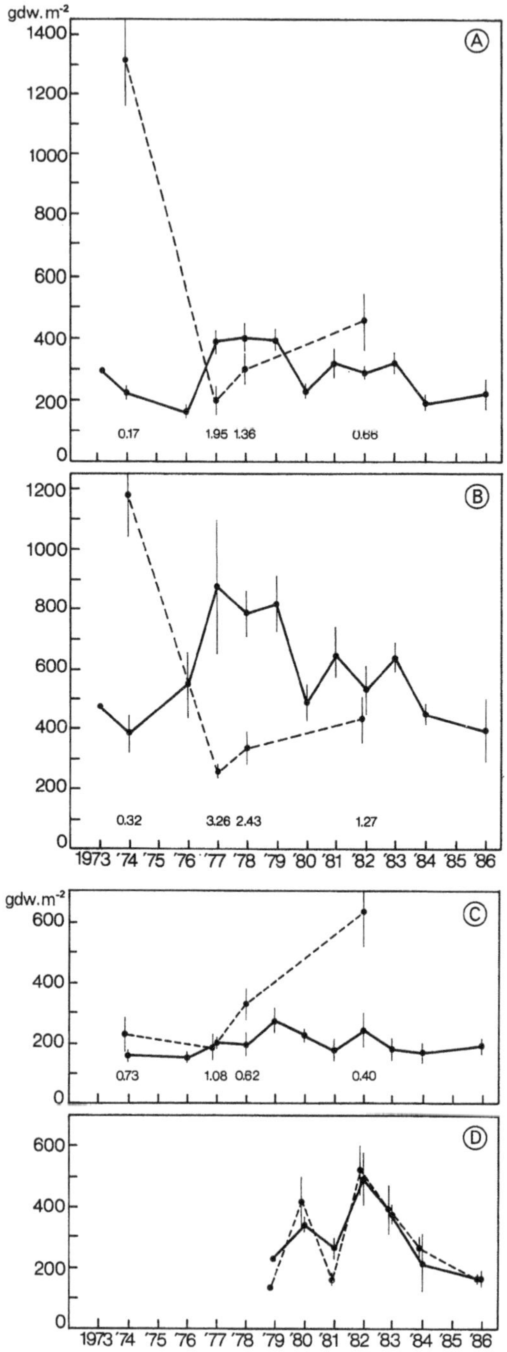

Fig. IX.4A–D. Above-ground standing crop (± S.E.) (solid line) and below-ground biomass (± S.E.) (dotted line) in the *Holcus lanatus* community of the Westerholt area under A: hay-making without fertilizer applications since 1971, B: hay-making with Ammonium Nitrogen Limestone fertilizer (50 kg $N \cdot ha^{-1}$), C: hay-making without fertilizer applications after sod removal in 1973, D: hay-making without fertilizer application after ploughing to 20 cm (solid line) and 60 cm (broken line) in 1978. The shoot to root ratio is indicated below the graphs.

making in July was the management usually practised by the State Forestry Commission and was, therefore, regarded as the standard practice with which other practices could be compared. Other management regimes were compared to hay-making in July in older reserves with species-richer communities (see Fig. VI.3).

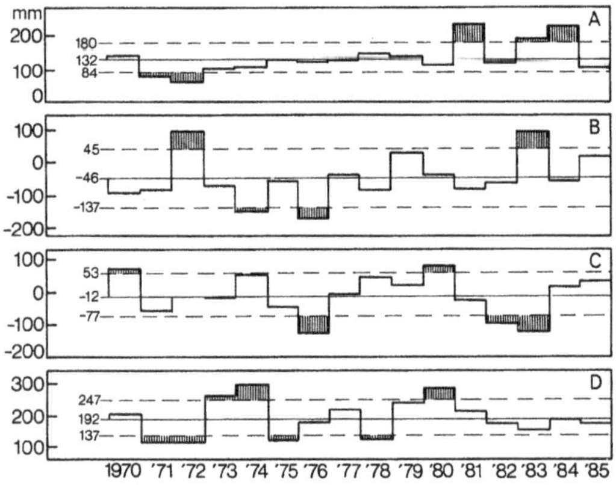

Fig. IX.3A–D. Rainfall surplus (mm) (± S.D.) at Schiermonnikoog in A: the first, B: the second, C: the third and D: the fourth quarter of a year (after Meteorological Station Free University, Amsterdam).

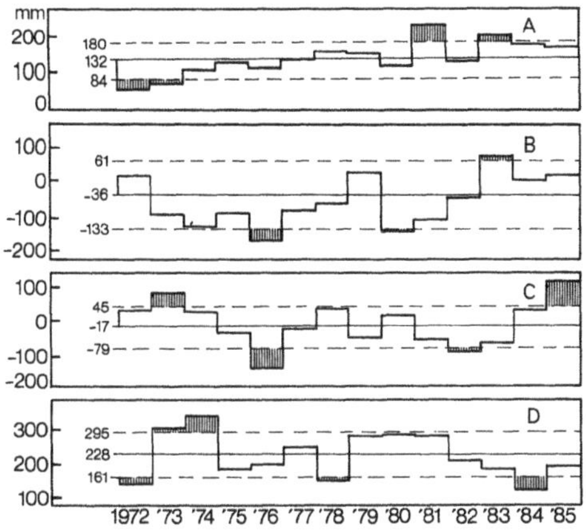

Fig. IX.5A–D. Rainfall surplus (mm) (± S.D.) at Eelde in A: the first, B: the second, C: the third and D: the fourth quarter of a year (after Royal Netherlands Meteorological Institute, De Bilt).

The above-ground standing crop in the *Ranunculus repens/Caltha palustris* community (see section VI.C.1.b) in the lower course area of the river valley fluctuated roughly around 500 to 600 gdw · m^{-2} under different once a year hay-making regimes without fertilizer applications since 1972 (Fig. IX.6 C–F). These figures are in agreement with the measurements by Fliervoet (1984), which ranged from 500 to 600 gdw · m^{-2} in comparable *Calthion palustris* stands in the Netherlands. Hay-making or mulching every second year in September resulted in an above-ground standing crop amounting to c. 700 gdw · m^{-2} (Fig. IX.6 A–B). The standing crop of the twice a year cutting regime decreased to c. 400 gdw · m^{-2} in July and remained c. 300 gdw · m^{-2} in September (Fig. IX.6 G), resulting, therefore, in a total yield of 700 gdw · m^{-2}.

The root biomass diminished rapidly from nearly 1000 gdw · m^{-2} to 200–400 gdw · m^{-2} and only showed a later increase in the cutting regime by scythe (Fig. IX.6 A–G). The fraction of living roots amounted to c. 90%. The shoot to root ratio increased in all management practices.

The above-ground standing crop in the *Valeriana officinalis/Phalaris arundinacea* community (see section VI.C.1.c) fluctuated strongly between 500

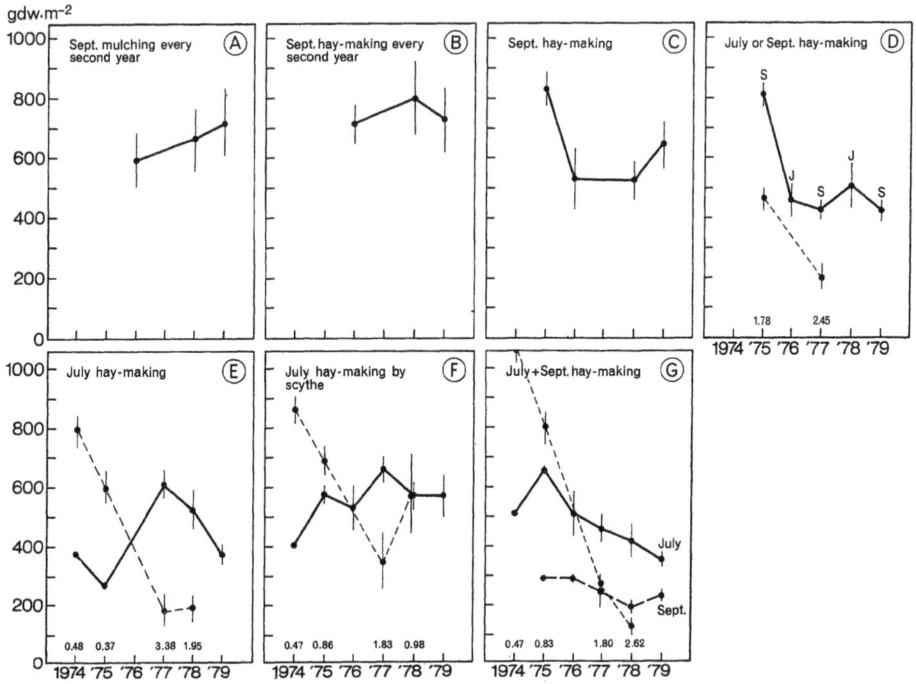

Fig. IX.6A–G. Above-ground standing crop in July and/or September (± S.E.) (solid line) and below-ground biomass in July (± S.E.) (dotted line) in the *Ranunculus repens/Caltha palustris* community of the recent reserve of the Lammertshem area under different management regimes. The shoot to root ratio is indicated below the graphs.

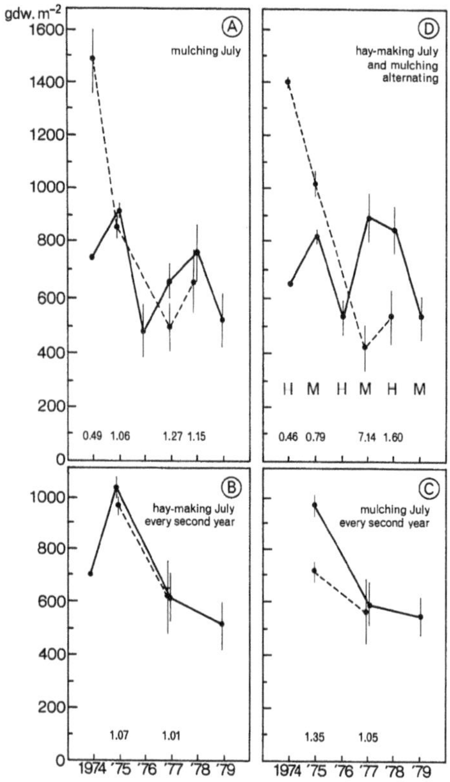

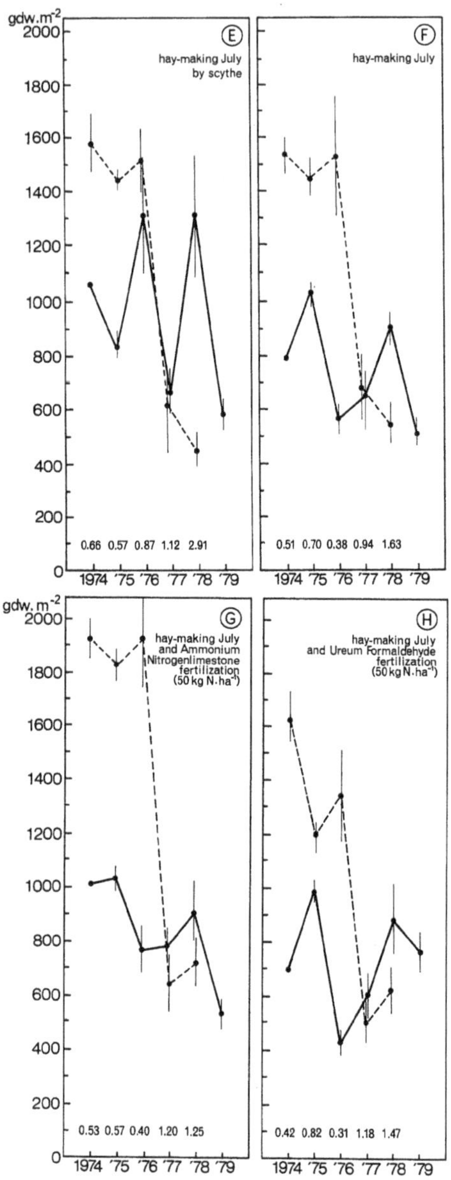

Fig. IX.7A–H. Above-ground standing crop (± S.E.) (solid line) and below-ground biomass (± S.E.) (dotted line) in the *Valeriana officinalis*/*Phalaris arundinacea* community of the older reserve of the Lammertshem area under different management regimes. The shoot to root ratio is indicated below the graphs.

and 1000 gdw · m^{-2} (Fig. IX.7 A–H). Fliervoet (1984) measured higher figures in similar stands in the Netherlands ranging from 1060–1170 gdw · m^{-2}. The low values in 1976 can be attributed to the very dry spring and summer (Fig. IX.5), but the high values in 1978 and the low ones in 1979 are not correlated with climatological data.

In the *Valeriana*/*Phalaris* community with much *Valeriana officinalis* (Fig. IX.7 A–D) the root biomass remained between 1500 and 2000 gdw · m^{-2} until 1976, but then dropped sharply to 500–600 gdw · m^{-2}. The fraction of living roots maintained itself at c. 75%. The root biomass in the *Valeriana*/*Phalaris* community with much *Ranunculus acris* initially amounted to 1500 gdw · m^{-2} and gradually decreased to 500 gdw · m^{-2} (Fig. IX.7 E–H). The fraction of living roots increased from c. 30% in 1975 to c. 80% in 1977 and 1978. The shoot to root ratio increased in all management regimes.

IX.B.4. Upper course valley grasslands (Loefvledder)

The above-ground standing crop in the *Agrostis stolonifera/Poa pratensis* community (see section VI.C.1.b) in the upper course area of the river valley showed distinct fluctuations apart from some trends (Fig. IX.8 A–G). The set-back of 1976 can be attributed to the very dry spring and summer of that year. The decrease in 1979 to 1981 was, however, due to very wet winters particularly those of 1978/1979 and 1980/1981 (Figs. IX.5 and IX.9). Moreover, the winter of 1979 was very severe (Fig. IX.2). The high 1983 values seem to correlate with a wet spring followed by a warm and dry summer. The set-back of 1985 might be attributed to the preceding severe winter.

The initial above-ground standing crop was 700 gdw · m^{-2}, three years after the cessation of fertilizer applications. This is in agreement with figures of *Poo-Lolietum* stands in the Netherlands which ranged from 650–770 gdw · m^{-2} measured by Fliervoet (1984). It maintained itself at that level with the July, July by scythe and both July and September hay-making regimes up to 1977/1978 and then dropped to 200–300 gdw · m^{-2}, except in 1983 (Fig. IX.8 E–G). It fluctuated for a long time around the level of 500 gdw · m^{-2} with the September and July or September alternating hay-making regimes and did not drop to lower levels before 1985 (Fig. IX.8 C–D). Only the hay-making regimes in July, by both machinery and scythe, in September and in both July and September, resulted in a significant decrease in the above-ground standing crop. The September hay-making regime every second year showed a remarkable increase in the standing crop in the last few years of the study period (Fig. IX.8 B). Mulching in September every second year resulted in a standing crop which was always higher than 500 gdw · m^{-2} (Fig. IX.8 A).

The root biomass declined sharply from 1000 gdw · m^{-2} to 200 gdw · m^{-2} and remained low, except with the July hay-making regimes, where it increased again to 400 gdw · m^{-2}. The shoot to root ratio diminished accordingly. The fraction of living roots amounted 90%.

The above-ground standing crop in the *Festuca rubra/Agrostis capillaris* community (see section VI.C.1.b) fluctuated strongly between 200 and 600 gdw · m^{-2} (Fig. IX.10 A–H), without showing high values and set-backs in the same years as in the *Agrostis stolonifera/Poa pratensis* community. Dickinson & Polwart (1982) also presented figures which range from 200 to 600 gdw · m^{-2} above-ground standing crop in an *Agrostis capillaris, Holcus lanatus, Anthoxanthum odoratum* stand in the United Kingdom.

The initial root biomass was extremely high with values between 1300 and 2700 gdw · m^{-2}. It decreased rapidly and reached values of c. 500 gdw · m^{-2} in most management practices. The fraction of living roots varied between 70% and 90%. The shoot to root ratio increased initially and then decreased in most management regimes.

IX.C. Comparison of cutting regimes

IX.C.1. Changes in above-ground standing crop

The above-ground standing crop of the salt marsh fluctuated considerably as a result of fluctuating weather conditions during the study period and did not reveal any trend. Such a trend could not be expected, since no changes, with respect to the application of fertilizers, were introduced.

The application of artificial fertilizers had ceased in the inland study areas and a decreasing standing crop was, therefore, expected, in particular after hay-making. Large year-to-year fluctuations were, however, also measured in the inland plant communities. Changes in above-ground biomass from year to year were also measured in a *Crepido-Juncetum* by Fliervoet (1984). These ranged from 450 to 910 gdw · m^{-2} without remarkable alterations in species composition. A minority of cases did reveal a significantly decreasing trend in the above-ground standing crop as a result of hay-making or mulching without fertilizer applications of formerly agricultural hayfields and/or pastures (Table IX.1). The high initial standing crop decreased suddenly after one or two years and fluctuated strongly in some communities after that. Even where the new management practices had been

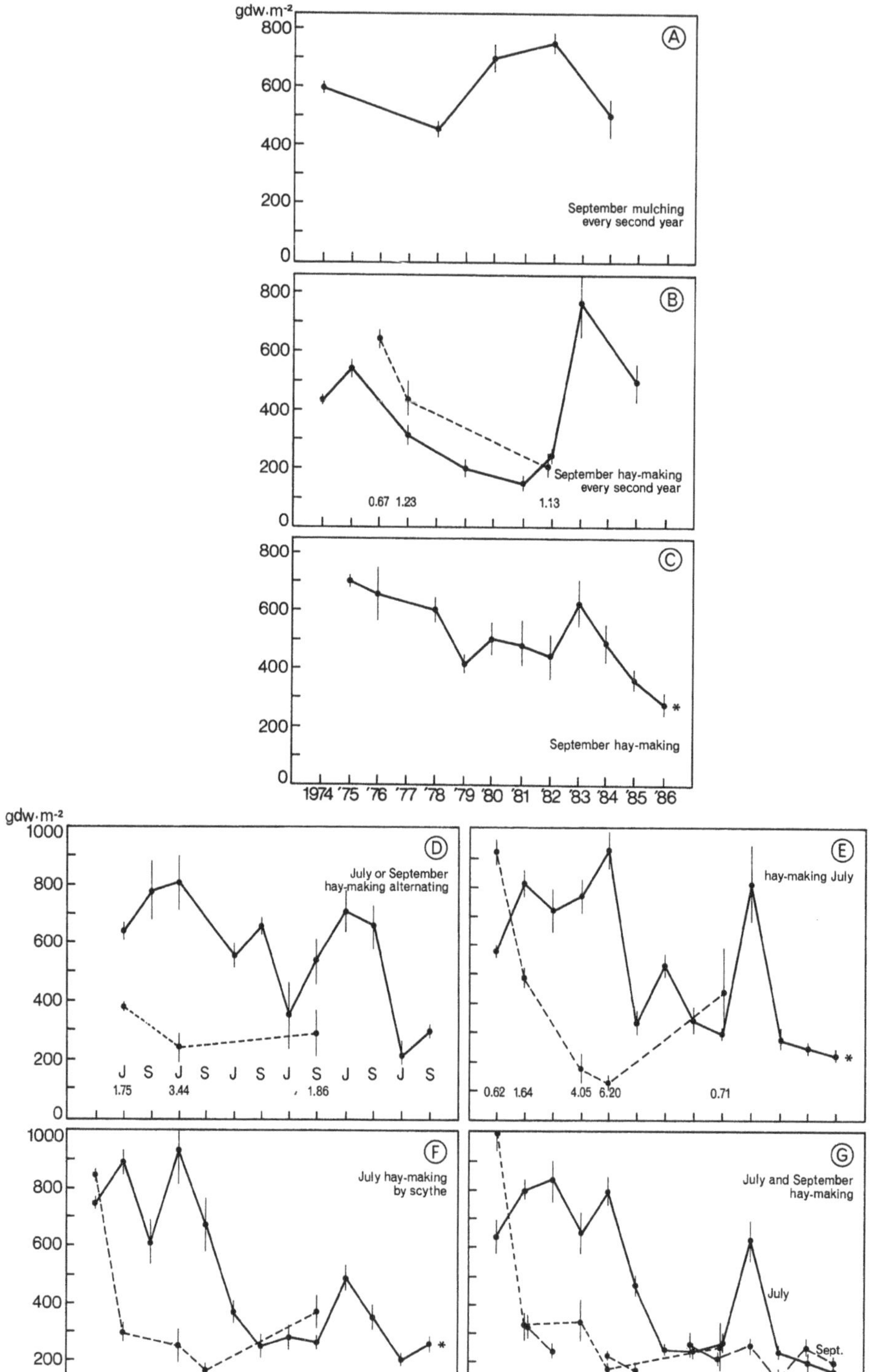

Fig. IX.8A–G. Above-ground standing crop in July and/or September (± S.E.) (solid line) and below-ground biomass in July (± S.E.) (dotted line) in the *Agrostis stolonifera/Poa pratensis* community of the recent reserve of the Loefvledder area under different management regimes. * Significant change according to Spearman's rank correlation test (P<0.05). The shoot to root ratio is indicated below the graphs.

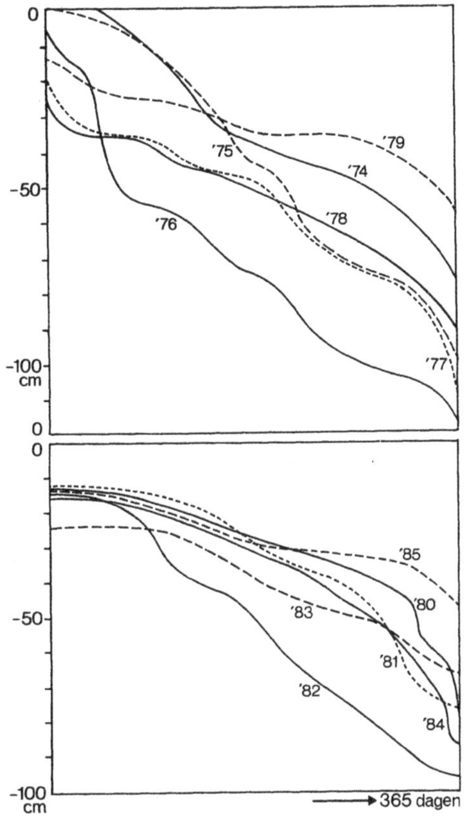

Fig. IX.9. Duration lines in the *Festuca rubra/Agrostis capillaris* community in the older reserve of the Loefvledder area during the period 1974 to 1985.

initiated many years ago most communities still showed fluctuations. In some cases Schiefer (1983) and Schreiber & Schiefer (1985), however, refer to these fluctuations as a decrease, but they might possibly do this too readily.

Schiefer (1983) suggested that the communities which maintained a high standing crop have a natural productive soil in common, optimal soil moisture conditions and optimal temperatures. Most communities with a significant decrease reached values associated with intermediate yields according to Klapp (1965), ranging from 400 to 600 gdw · m^{-2}. The communities in which the effects of earlier fertilizer applications had disappeared quickly and where there was a yield associated with poor grasslands (300 gdw · m^{-2} or less), had dry environmental conditions in common. It cannot, however, be concluded that the above-ground standing crop declined rapidly in all dry conditions under hay-making without fertilizer applications. Other factors such as the intensity of former agricultural practices also exert an influence. This probably explains the fact that standing crop values of between 200 and 300 gdw · m^{-2} are reached in both the *Holcus lanatus* community in the Westerholt area and the moist *Agrostis stolonifera/Poa pratensis* and *Festuca rubra/Agrostis capillaris* communities in the Loefvledder area.

IX.C.2. Root biomass and shoot to root ratio

The amount of below-ground biomass can vary considerably during the year (Table IX.2). Maximal figures were nearly always found during the summer or early autumn. It can, therefore, be assumed that our measurements of below-ground biomass at the end of July (immediately after hay-making) featured approximately maximal values during the season. The sharp decline of the root biomass in most management practices after the cessation of fertilizer applications is striking, particularly when compared with the slow decline, if at all, of the above-ground standing crop (Figs. IX.4, 6, 7, 8, 10). Several authors, cited by Snaydon & Howe (1986), have shown that root competition has more effect than shoot competition. It seems likely that below-ground competition is usually of greater importance than above-ground competition in unfertilized plant communities. The later increase of root biomass in this study has yet to be explained.

The initial decrease and later increase of the root biomass in many cutting regimes in this study seems to contradict with the results of a pot experiment by Dirven and Wind (1980). They found an optimum for the root biomass of both *Lolium perenne* and *Agrostis capillaris* in a series of nitrogen application.

The shoot to root ratios, calculated in the plant communities occurring in the present study immediately after cutting, seem to correspond with those found in other grassland communities (Table IX.3). It is, however, remarkable that this was only the case during the first years of the July hay-making regime. Five years after the start of the

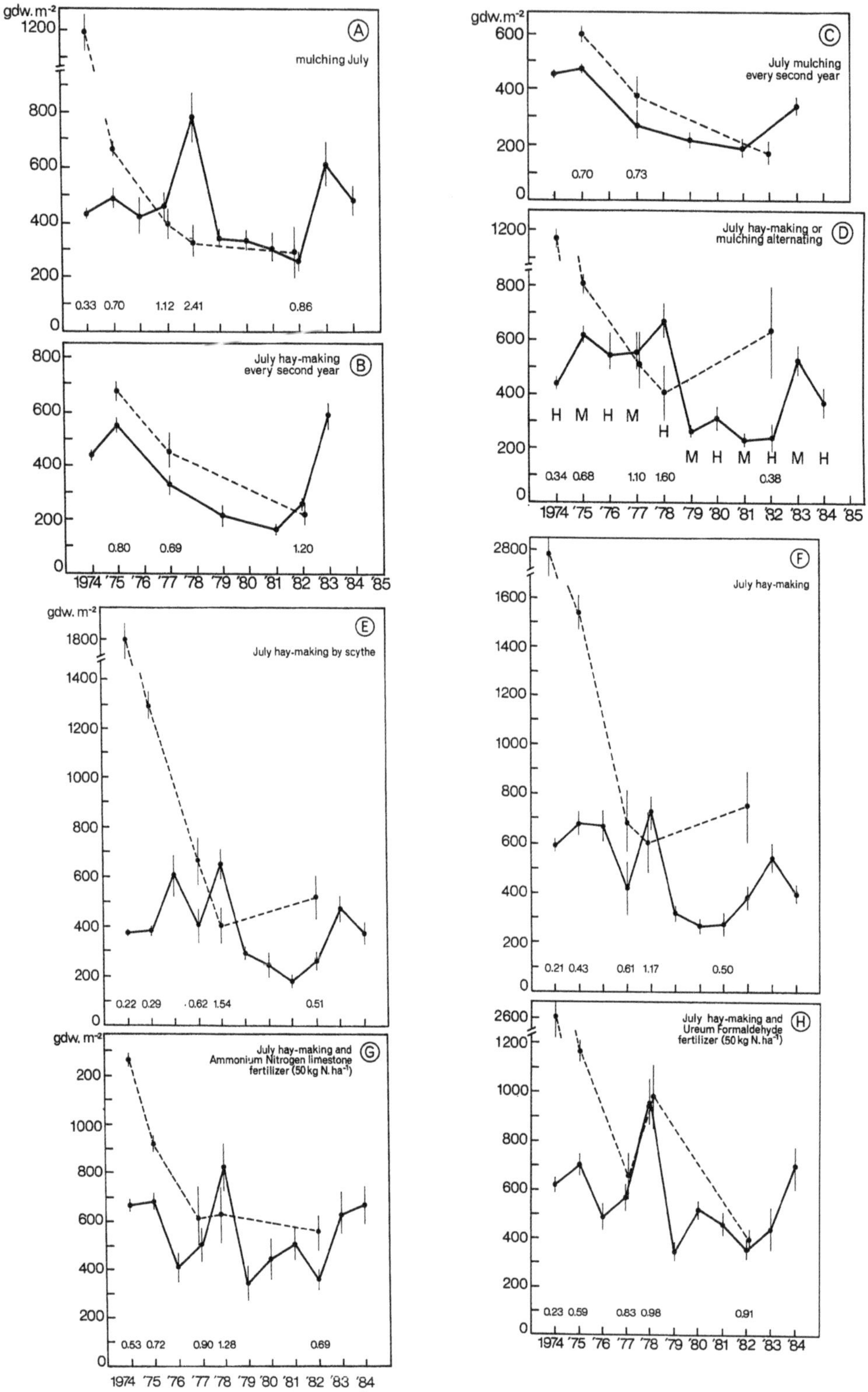

Fig. IX.10A–H. Above-ground standing crop (± S.E.) (solid line) and below-ground biomass (± S.E.) (dotted line) in the *Festuca rubra/Agrostis* capillaris community in the older reserve of the Loefvledder area under different management regimes. The shoot to root ratio is indicated below the graphs.

Table IX.1. Changes in above-ground standing crop (± : fluctuating; -- : sudden decrease after the first few years, but no significant decreasing trend was found over the whole study period; -* significant decrease) with hay-making or mulching without fertilizer applications in formerly fertilized communities. The annual frequency of hay-making (H) or mulching (M) is indicated.

Plant community	Soil	Initial standing crop	Management regime			Changes in standing crop	Assessment
			Number of years	Practice	Period		
Holcus lanatus	sand	330	12	H1	July	±	This study
Ranunculus repens/Caltha palustris	peat	400	5	H1	July	±	,,
,,	peat	840	4	H1	September	--	,,
,,	peat	530	6	H2	July/Sept	±	,,
Agrostis stolonifera/ Poa pratensis	peat	630	13	H1	July	-*	,,
,,	peat	740	11	H1	September	-*	,,
,,	peat	680	13	H2	July/Sept	-*	,,
Arrhenatherion	clay	1100	9	H2	June/Sept	-*	Wind 1980
Lolio-Cynosuretum	sand	1000	5	H2	June/Sept	-*	,,
Arrhenatherion	clay	430	6	H1	June	±	Oomes & Mooi 1981
,,	clay	550	6	H1	September	±	,,
Poo-Lolietum	clay	1100	11	H1	June	--	Oomes & Mooi 1985
,,	clay	1100	11	H2	June/October	-*	,,
Molinio-Arrhenatheretea	clay	900	20	H2	June/October	--	Elberse et al. 1983
,,	peat	520	6	H3	June/July/Aug	--	Korevaar 1986
Arrhenatheretum	loam	310	6	H1	October	±	Willems 1980
Anglico-Cirsietum	peat	760	4	H3	July/Aug/Oct	-*	Kapfer 1987
,,	peat	520	4	H1	October	±	,,
Lolio-Cynosuretum	peat	530	4	H3	July/Aug/Oct	--	,,
,,	peat	390	4	H1	October	±	,,
,,	peat	340	4	H2	July/October	±	,,
,,	peat	250	4	H1	October	±	,,
Arrhenatherion	loam	280	10	M2	June/Augustus	±	Schreiber & Schiefer 1985
Cynosurion	loam	940	10	M2	June/Agustus	±	,,
,,	loam	690	7	M1	August	±	,,
Calthion/Juncion acutiflori	peat	430	10	M2	June/Augustus	±	,,
Mesobrometum	loam	300	10	M2	June/Augustus	±	,,
,,	loam	170	10	M1	August	±	,,
Arrhenatheretum	loam	900	13	H2	?	±	Schiefer 1983
,,	loam	400	16	H2	?	±	,,
Meo-Festucetum	loam	600	15	H2	?	±	,,
Lolio-Cynosuretum	loam	450	14	H2	?	±	,,
Meo-Festucetum	peat	750	1	H2	?	±	,,
,,	loam	600	13	H2	?	±	,,
,,	clay	600	7	H2	?	±	,,
Arrhenatheretum	loam	650	14	H2	?	±	,,
,,	sand	700	11	H2	?	±	,,
Cirsio-Polygonetum	clay	500	7	H2	?	±	,,
Meo-Festucetum	loam	350	14	H2	?	±	,,
,,	loam	300	17	H2	?	±	,,
Arrhenatheretum	loam	350	6	H2	?	±	,,
,,	loam	350	5	H2	?	±	,,
Festuco-Cynosuretum	sand	150	7	H1	?	±	,,
Nardo-Galion	sand	100	8	H2	?	±	,,
Lolio-Cynosuretum	loam	700	7	H5	?	±	,,
Arrhenatheretum	sand	650	7	H2	?	±	,,
,,	clay	600	11	H2	?	±	,,
,,	sand	500	11	H2	?	±	,,
,,	sand	750	7	H2	?	±	,,
Festuco-Cynosuretum	sand	700	8	H2	?	±	,,
Meo-Festucetum	clay	700	7	H2	?	-*	,,
,,	loam	800	10	H2	?	-*	,,
Arrhenatheretum	loam	650	7	H2	?	-*	,,
,,	loam	950	15	H2	?	±	,,
,,	sand	800	7	H2	?	-*	,,
,,	sand	700	14	H2	?	-*	,,
Lolio-Cynosuretum	sand	950	6	H5	?	-*	,,
,,	loam	750	11	H3	?	±	,,

management practices the shoot to root ratio reached high values, which were only found in recently sown leys with little below-ground biomass (Steen, 1985) and in *Spergulo-Corynephoretum* communities (Fliervoet, unpubl.). These high figures can be attributed to a much higher set-back of the below-ground biomass compared to the above-ground biomass. Later a recovery of the below-ground biomass was found and shoot to root ratios of about 1 were calculated (Figs IX. 4, 6, 7, 8, 10), still higher than those assessed in the literature for semi-natural grasslands. Shoot to root ratios over 1 were only mentioned for a ley and a grassland receiving fertilizer applications, whereas most semi-natural grasslands revealed low to very low ratios (Table IX.3). It is, however, clear that those ratios correspond with widely diverging below-ground biomass and that the latter can also show large differences within a single plant community. Stuart Chapin (1980) also mentioned low shoot to root ratios in infertile habitats without, however, giving any details. Goedewagen & Schuurman (1950) and Dirven & Wind (1980) mentioned an increasing shoot to root ratio after fertilizer applications, hence the reverse might be expected after ceasing such applications. Indeed, Gisi & Oertli (1981) found a lower shoot to root ratio in an abandoned than in a fertilized grassland. Clarkson & Hanson (1980) mentioned, from a physiological point of view, that nutrient deficiency can cause some redistribution of growth within the plant, root growth often occurring relatively more rapidly in relation to the shoot, hence decreasing the shoot to root ratio. It seems to be in disagreement with our findings, but it is questionable whether greenhouse experiments lasting a few weeks can be translated into the field situations with longlasting experiments. Application of 50 kg $N \cdot ha^{-1}$ in the *Festuca rubra*/*Agrostis capillaris* community showed somewhat higher shoot to root ratios, in agreement with the afore-mentioned, which can be attributed to a higher above-ground standing crop (Fig. IX.10 G and H). This effect was not measured in the *Valeriana officinalis*/*Phaleris arundinacea* community. Werger (1983) stresses that the shoot to root ratio often increases under 'unfavourable'

Table IX.2. Below-ground biomass in several mostly unfertilized plant communities during a year and the month in which the maximum value is reached. * Only 40% of the below ground biomass is found in the 0–20 cm soil layer.

			Below-ground biomass			
Country	Plant community	Management	Depth (cm)	Range ($gdw.m^{-2}$)	Maximum	Assessment
Central Sweden	Festuca pratensis second year ley	hay-making	0-50	680-950	September	Hansson & Andren 1986
South Sweden	Elymus repens / Holcus lanatus third year ley	hay-making, 50 $kgN.ha^{-1}$	?	230-540	September	Steen 1985
Switzerland	Colchico-Mesobrometum	hay-making fertilized	0-20	500	?	Gisi & Oertli 1981
South Sweden	Filipendula ulmaria	untouched	0-20	780-1170	August	Balsberg 1982
South Sweden	Trifolium medium / Festuca ovina	hay-making	0-30	1100-2200	July	Nilsson 1970
Central England	Agrostis capillaris / Holcus lanatus	hay-making	0-15	400-680	June	Dickinson & Polwart 1982
Wales	Agrostis capillaris / Festuca ovina	grazing	0-30	460-680	August	Perkins et al. 1978
South Poland	Lolio - Cynosuretum	grazing	0-15	600-800	March	Plewczynska-Kuras 1974
South Germany	Cynosurion	mulching	0-30	750-1350	August	Schreiber & Schiefer 1985
Switzerland	Colchico- Brachypodietum	untouched	0-20	1100	?	Gisi & Oertli 1981
The Netherlands	Triglochin maritima	untouched	0-20	500-6800	September	Groenendijk 1987*)
The Netherlands	Spartina anglica	untouched	0-20	3800-5000	September	Groenendijk 1987*)
The Netherlands	Elymus pycnanthus	untouched	0-20	3200-4500	July	Groenendijk 1987*)
The Netherlands	Halimione portulacoides	untouched	0-20	4000-7800	November	Groenendijk 1987*)

Table IX.3. Below-ground biomass and shoot to root ratio in a number of grassland communities. Most communities received no fertilizer applications during various years of hay-making or grazing as indicated. * Only 40% of the below-ground biomass is found in the 0–20 cm soil layer.

Plant community	Management (number of years)	Depth (cm)	Below-ground biomass (gdw.m^{-2})	Shoot to root ratio June/July	Assessment
Agrostis stolonifera /Poa pratensis	hay-making (5)	0-15	130	6.20	This study
Spergulo-Corynephoretum	untouched (20)	0-20	130	5.99	Fliervoet (unpub.)
Elymus repens / Holcus lanatus	hay-making/ fertilized	?	390	2.10	Steen (1985)
Ranunculus repens / Caltha palustris	hay-making (5)	0-15	200	1.95	This study
Valeriana officinalis / Phalaris arundinacea	hay-making (5)	0-15	550	1.65	This study
Holcus lanatus	hay-making (5)	0-15	300	1.35	This study
Festuca rubra / Agrostis capillaris	hay-making (5)	0-15	600	1.20	This study
Colchico - Mesobrometum	hay-making / fertilized	0-20	500	1.02	Gisi & Oertli (1981)
Spergulo - Corynephoretum	grazing (?)	0-20	250	0.95	Fliervoet (unpub.)
Agrostis capillaris / Holcus lanatus	hay-making (?)	0-15	680	0.90	Dickinson & Polwart (1982)
Cynosurion	mulching (5)	0-30	850	0.65	Schreiber & Schiefer (1985)
Lolio-Cynosuretum	grazing (8)	0-20	500	0.63	Fliervoet (unpub.)
Poo - Lolietum	grazing / fertilized	0-20	730	0.60	Fliervoet (unpub.)
Agrostis stolonifera / Poa pratensis	hay-making (1)	0.15	900	0.60	This study
Arrhenatheretum elatioris	hay-making (6)	0-20	1730	0.57	Fliervoet /unpub.)
Arrhenatheretum elatioris	hay-making (6)	0-20	1540	0.55	Fliervoet (unpub.)
Valeriana officinalis / Phalaris arundinacea	hay-making (1)	0-15	1540	0.50	This study
Filipendula ulmaria	untouched (?)	0-20	1170	0.50	Balsberg (1982)
Ranunculus repens / Caltha palustris	hay-making (1)	0-15	780	0.50	This study
Lolio - Cynosuretum	grazing / fertilized	0-20	980	0.46	This study
Colchico - Brachypodietum	untouched (?)	0-20	1100	0.46	Gisi & Oertli (1981)
Calthion palustris	hay-pasture (?)	0-20	1420	0.38	Fliervoet (unpub.)
Agrostis capillaris / Festuca ovina	grazing (?)	0-30	680	0.35	Perkins et al. (1978)
Galio- Koelerion	grazing (?)	0-20	1390	0.27	Fliervoet (unpub.)
Elymus pycnanthus	untouched	0-20	3510	0.26	Groenendijk (1987)*
Lolio - Cynosuretum	grazing (?)	0-15	690	0.25	Plewczynska - Kuras (1974)
Halimione portulacoides	untouched	0-20	5310	0.24	Groenendijk (1987)*
Spartina anglica	untouched	0-20	4400	0.22	Groenendijk (1987)*
Sedo - Thymetum	hay-making (6)	0-20	1780	0.20	Fliervoet (unpub.)
Trifolium medium / Festuca ovina	hay-making ?)	0-30	2200	0.20	Nilsson (1970)
Festuca rubra / Agrostis capillaris	hay-making (1)	0-15	2000	0.20	This study
Holcus lanatus	hay-making (1)	0-15	1300	0.20	This study
Lolio - Cynosuretum	grazing (8)	0-20	2100	0.20	Fliervoet (unpub.)
Valeriano - Filipenduletum	hay-making (20)	0-20	7800	0.16	Fliervoet (unpub.)
Juncus gerardii	untouched	0-35	4200	0.15	Tyler (1971)
Junco - Molinion	hay-making (20)	0-20	6460	0.14	Fliervoet (unpub.)
Crepido-Juncetum	hay-making (20)	0-20	4260	0.11	Fliervoet (unpub.)
Cirsio - Molinietum	hay-making (20)	0-20	500	0.08	Fliervoet (unpub.)
Cirsio - Molinietum	hay-making (10)	0-20	4560	0.08	Fliervoet (unpub.)
Cirsio - Molinietum	hay-making (20)	0-20	490	0.07	Fliervoet (unpub.)
Cirsio - Molinietum	hay-making (20)	0-20	6660	0.07	Fliervoet (unpub.)
Triglochin maritima	untouched	0-20	5200	0.07	Groenendijk (1987)*
Calthion palustris	hay-pasture (12)	0-20	11290	0.06	Fliervoet (unpub.)
Calthion palustris	hay-pasture (12)	0-20	11400	0.06	Fliervoet (unpub.)

conditions which can also imply a shortage of nutrients. He cites, however, short lasting experiments and it is not known whether analogous results would be obtained from longer lasting impoverishing experiments in which the species composition changes and new species with their own characteristics appear.

Only Dickinson & Polwart (1982) measured below-ground biomass in an *Agrostis capillaris/Holcus lanatus* stand over a period of three years. They found that the initial 1000 gdw · m^{-2} decreased to 500 gdw · m^{-2} after two years without fertilizer applications which is in agreement with our results. They give no explanation for this large decrease.

The below-ground biomass was relatively small in lots that had been agriculturally exploited the year before the hay-making without fertilizer application started in our study area. The lots with a longer history of being unfertilized (*Valeriana officinalis/Phalaris arundinacea* and *Festuca rubra/Agrostis capillaris* communities) showed a larger below-ground biomass. This is more or less in agreement with other studies (Table IX.4), although there are several exceptions.

IX.C.3. Above-ground standing crop and species diversity

A negative relationship became apparent between above-ground standing crop, including litter, and phanerogamic species diversity in the present study (Fig. IX.11). It is not possible to fit an optimum curve through the points like Grime (1979) did in his general model. It can only be concluded that the highest species diversity (≥20 species · 4 m^{-2}) was found between 200 and 400 gdw · m^{-2}, which is not in agreement with Grime's optimum between 500 and 600 gdw · m^{-2}.

Because of the small variation in standing crop or/and species diversity very few relationships can be detected in the separate study areas. Very low standing crop can be found at the heavily trampled and hence eroded lower salt marsh and immediately after sod removal in the mixed grassland of the Westerholt.

Hendrikse & Wuring (1981) studied the relationship between peak standing crop and species diversity in several plant communities in the upper course of the valley grasslands (Anlo) (Fig. IX.12). They found the largest species diversity between

Table IX.4. Below-ground biomass in several plant communities from assessments in the literature. The communities are ranked according to the period after sowing or decreasing soil fertility in each assessment.

Country	Plant community	Depth (cm)	Below-ground biomass (gdw.m^{-2})	Assessment
The Netherlands	3 years ley	0–20	890	Goedewagen & Schuurman 1950
The Netherlands	7 years ley	0–20	780	Goedewagen & Schuurman 1950
The Netherlands	12 years ley	0–20	1090	Goedewagen & Schuurman 1950
The Netherlands	15 years ley	0–20	1000	Goedewagen & Schuurman 1950
The Netherlands	> 20 years ley	0–20	980	Goedewagen & Schuurman 1950
Germany	dry Lolio-Plantaginion	0–30	510	Kmoch 1952
Germany	dry Arrhenatherion	0–30	940	Kmoch 1952
Germany	moist Lolio-Plantaginion	0–30	1050	Kmoch 1952
Germany	moist Arrhenatherion	0–30	1430	Kmoch 1952
Germany	Festuco-Cynosuretum	0–30	1530	Kmoch 1952
Germany	Mesobromion	0–30	1630	Kmoch 1952
Germany	Nardo-Galion	0–30	2030	Kmoch 1952
Germany	Caricion curto-nigrae	0–30	3060	Kmoch 1952
The Netherlands	Poo-Lolietum	0–20	560	Dirven & Wind 1982
The Netherlands	Molinio-Arrhenatheretea	0–20	1790	Dirven & Wind 1982
The Netherlands	Festuco-Cynosuretum	0–20	2150	Dirven & Wind 1982
The Netherlands	Nardo-Galion	0–20	1074	Dirven & Wind 1982

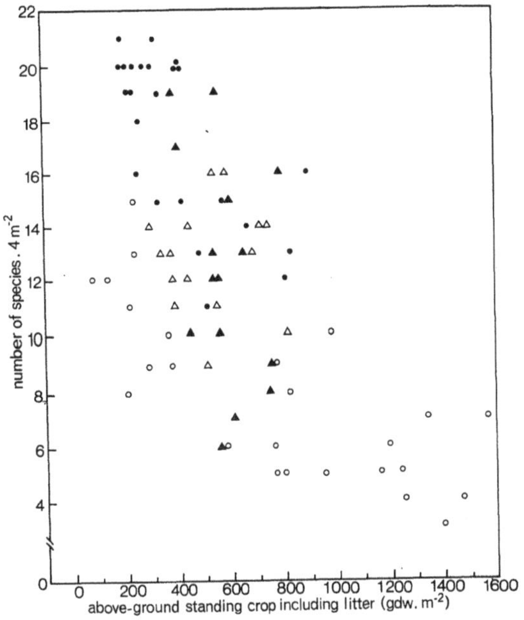

Fig. IX.11. Number of species $\cdot$ 4 m^{-2} as related to the above-ground standing crop including litter in salt marsh Oosterkwelder (open circles), mixed grassland Westerholt (closed circles), upper course grassland Loefvledder (open triangles) and lower course grassland Lammertshem (closed triangles) under different management practices during the study period.

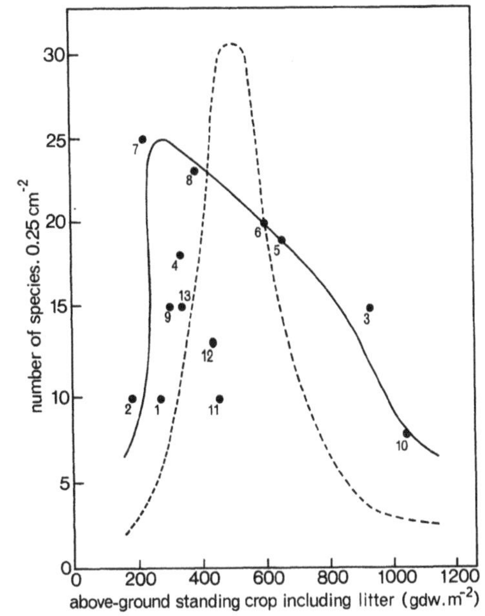

Fig. IX.12. Peak above-ground standing crop as related to the number of species $\cdot$ 0.25 m^{-2} in some plant communities along the upper course of the Anlo area (dashed line is the relation according to Grime's 1979). 1, 2 *Festuco-Cynosuretum;* 3, 4, 5, 6, 8, 9 *Calthion palustris;* 7 *Juncion acutiflori;* 10 *Rumex obtusifolius* stand; 11 *Agrostis capillaris* stand; 12, 13 *Lolio-Cynosuretum* (after Hendrikse & Wuring, 1981).

200 and 400 gdw $\cdot$ m^{-2} above-ground standing crop. Vermeer & Berendse (1983) found an optimum between 400 and 500 gdw $\cdot$ m^{-2} in other grassland and wetland communities and Willems (1980) around 200 gdw $\cdot$ m^{-2} in chalk grassland, which was, therefore, also lower than Grime's (1979) optimum between 500 and 600 gdw $\cdot$ m^{-2}. Wheeler & Giller (1982), however, counted the highest number of species at 1500 gdw $\cdot$ m^{-2} in herbaceous fen stands in the United Kingdom.

Some points could not be fitted (Fig. IX.12) in an ideal optimum curve like Grime's results. The points 11, 12 and 13 have their short period of hay-making without fertilizer applications in common. The explanation can be that their standing crop is fairly low, but the expected increase of the number of species did not (yet) occur. Oomes & Mooi (1981; 1985) started hay-making without fertilizer applications. They found that the species diversity did not always increase despite a large lowering of the standing crop from the second year onwards. This is in agreement with the results in this study. Schiefer (1983) found that new species only emerged in grasslands in Germany when the standing biomass was below 350 gdw $\cdot$ m^{-2}. These findings are, in fact, in agreement with Grime's (1979) general model on the relationship between the annual maximum standing crop including litter and the species density in vegetation. In selecting the sites, Al-Mufti et al. (1977) excluded vegetation of recent origin i.e. they sampled sites in abandoned land which had not been grazed or burned for a period of at least 20 years. Each site was situated in an area of landscape containing a diversity of long-established vegetation types and it is considered that the numbers of species which were attained correspond to an equilibrium between the habitat conditions prevailing at each site and the reservoir of species in its vicinity. Such an equilibrium was apparently not yet reached in our study areas and other afore-mentioned studies and it takes time before the developing vegetation types are 'saturated', if it occurs at all with the present state of increasing fragmentation of source areas.

The reason that the pattern of the relationship between the annual maximum standing crop, including litter, and the species density is lost in this study (Fig. IX.11) could be the comparison of different management practices viz. abandoning, haymaking and mulching with different litter accumulation. Grime (1979) found a similar deviation of the pattern in environments under fluctuating patterns of vegetation management.

The species corridor found in some communities in the present study (Fig. IX.11) and by Hendrikse & Wuring (1981) (Fig. IX.12) was much broader than in Grime's model. The points 3, 5 and 6 were responsible for this phenomenon in Fig. IX.12. Their common factor is that they originate from *Juncus acutiflorus* stands, which explains their high standing crop. Their erect growth form, however, permits a great deal of light to reach the soil. The above-ground phytomass of a stand, but also its canopy structure is apparently important for the species diversity. Wheeler & Giller (1982) also pointed to discrepancies between standing crop and species diversity. Shading in vigorous stands of *Phragmites australis* resulted in small species diversity. Regular mowing and litter removal, however, resulted in a small standing crop but high light attenuation, due to a high shoot density of *Phragmites* and hence a small number of species.

CHAPTER X

Changing soil conditions and stand structure as causes of floristic changes: conclusions and synthesis

X.A. Introduction

The causes of the observed changes in the species composition described in previous chapters, will be discussed in this chapter. Attention will be focused on soil fertility, soil salinity and the structure of the sward.

Restoration management in the inland study areas was aimed at reducing the residual effects of earlier fertilizer applications. It became clear that in many cases the changes in species composition were indeed attended by an increase in the poor to rich species ratio, indicating decreasing soil fertility conditions as a cause of changes (see section VI.D.1.c). The question is whether a decrease in soil fertility can actually be measured. The reverse, i.e. the application of artificial fertilizer, did increase soil fertility, measured by chemical analysis. Elberse et al. (1983) found a gradual P_2O_5 and K_2O increase in the top soil (0–5 cm) layer over a period of 20 years of fertilizer treatment (40 kg $P_2O_5 \cdot ha^{-1} \cdot yr^{-1}$ and 60 kg $K_2O \cdot ha^{-1} \cdot yr^{-1}$) on heavy clay soil in the Netherlands, whereas the control plots showed low P and K-figures. Prins (1983) found an accumulation of mineral N with fertilizer application of more than 400 kg $N \cdot ha^{-1} \cdot yr^{-1}$ on sand and clay soils in the Netherlands. Korevaar (1986) found an increase of P_2O_5 and K_2O over a period of six years of fertilizer treatment (150 kg $P_2O_5 \cdot ha^{-1} \cdot yr^{-1}$ and 300 kg $K_2O \cdot ha^{-1} \cdot yr^{-1}$) on peat soils in the Netherlands. Section B deals with soil impoverishing as a cause of floristic changes.

The soil fertility was measured as soil nutrient content. Only a few parameters were adopted, namely, N_{total}, P_{total}, $K_{exchangeable}$, C/N quotient. Although these variables are of a complex character, they do allow comparison with the results of other authors.

In coastal areas with abandoned salt marshes the restoration management aimed at the removal of accumulated litter. The management practices resulted in an increase of lower salt marsh species in higher salt marsh areas (see section VII.C). The question is whether this process is attended by an increased soil salinity. The reverse, i.e. the cessation of grazing, did decrease soil salinity. Westhoff & Sykora (1979) measured a lower soil salinity in an abandoned sward in an embanked salt marsh, than in control plots where grazing was continued. Section C deals with soil salination as a potential cause of floristic changes.

The new hay-making or grazing regimes might

possibly have changed the sward structure, thus giving access to new species. These concurrent influences of supposed nutrient removal and mechanical factors will be discussed in sections D and E of this chapter.

X.B. Impoverishing

This section considers the impoverishing of the soil as a possible cause of floristic changes. The aspects dealt with are (i) comparison of the poor to rich species ratio on wet and dry soils with July hay-making, (ii) comparison of soil chemical analyses and poor to rich species ratios on the same soil under different hay-making regimes including July hay-making, (iii) comparison of soil chemical analyses and poor to rich species ratios under hay-making and grazing, (iv) the actual nutrient removal related to the nutrient pool in the soil, and (v) some aspects of nutrient cycling.

X.B.1. Changes in poor to rich species ratios on wet and dry soils

The management practice used for the restoration of the valley grasslands in the nature reserve Stroomdallandschap Drentsche A is mainly hay-making in July. The results of this management practice have been shown in chapter VI. The first case, which will be considered is vegetational changes from 1974 onwards on wet peaty soil (see Table VI.12) and on dry sandy soil (see Table VI.10) under hay-making without fertilizer applications since 1972. The subdivision into wet and dry conditions is based on the bundles of duration lines obtained in several places (Fig. VI.4).

The changes in the number of rich and poor species are shown in Fig. X.1. The poor/rich ratio increased with time in both wet and dry areas (Fig. X.2), but it was always higher and increased in an absolute sense more rapidly in the wet area. The relative importance of the new hay-making regime and the cessation of fertilizer applications will be discussed later in this chapter.

The quantitative poor/rich cover ratio shows larger fluctuations than the qualitative quotients.

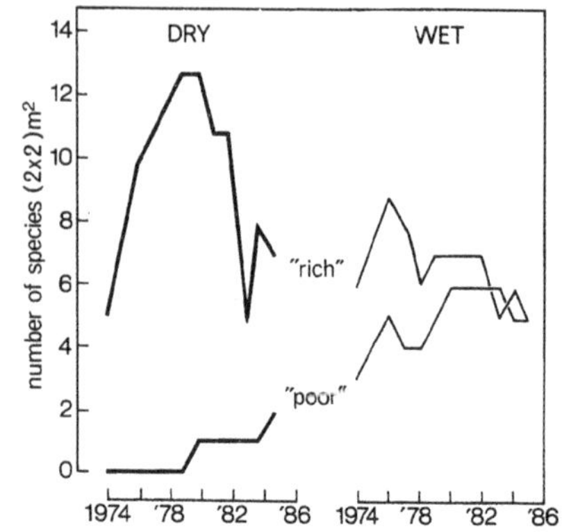

Fig. X.1. The numbers of species indicating nutrient-rich and nutrient-poor soil conditions in wet and dry areas cut for hay in July and receiving no fertilizer applications after 1972 (after Bakker, 1987).

Schmidt (1985) obtained similar results, possibly caused by large year-to-year fluctuations in the degree of cover of some species, though Durwen (1982), Böcker et al. (1983) and Melman et al. (1985) found little differences between the two parameters.

In this study the qualitative poor/rich ratio will be used, because of its greater year-to-year consistency.

X.B.2. Comparison of hay-making regimes

Different cutting regimes were started in 1972, in the Loefvledder area, in order to study the effects of the timing and frequency of hay-making in the absence of fertilizer application. The study area had a homogeneous substrate and fertilizer application had been stopped in 1971. The grassland consisted of a *Agrostis stolonifera*/*Poa pratensis* community.

No species indicating poor conditions appeared in the abandoned site, nor under the management practices of burning, hay-making in July by scythe, mulching in September every second year (Fig. X.3). The poor to rich species ratio became highest with hay-making twice a year (July and September). Poor species appeared quickly with September hay-making and later with July hay-making. Moreover, the above-ground standing crop decreased significantly under the last three manage-

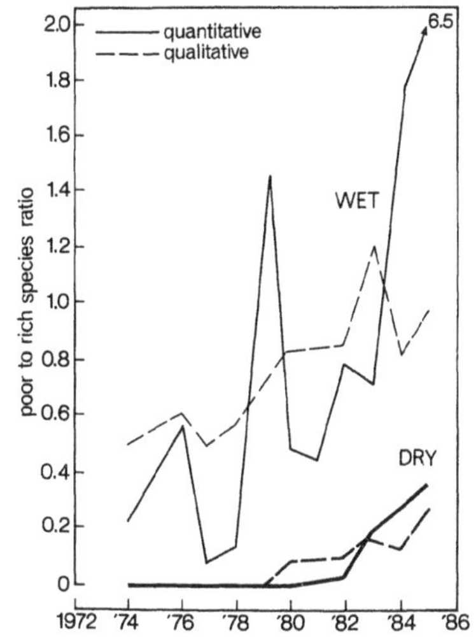

Fig. X.2. Changes in the poor to rich species ratio, based either on the number of species (qualitative) or the cover % of species (quantitative) in wet and dry areas cut for hay in July and receiving no fertilizer applications after 1972 (after Bakker, 1987).

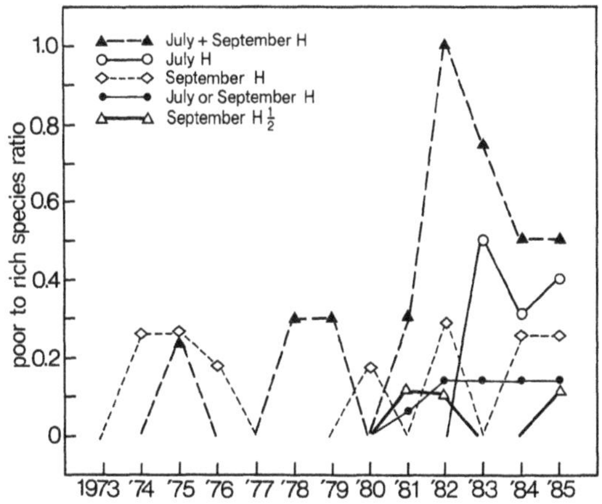

Fig. X.3. Changes in the poor to rich species ratio in different management regimes in the *Angrostis stolonifera/Poa pratensis* community of the recent reserve of the Loefvledder area. The ratio remained zero in the regimes abandoning, burning, hay-making with the scythe in July and mulching every second year in September. H: hay-making, $^1/_2$: every second year. The plots had received no fertilizer applications after 1971.

ment regimes (see Fig. IX.8). The same was the case for July hay-making by scythe, where no poor species appeared at all.

At first the soil chemical analyses showed that the lot in which the hay-making regimes were performed, demonstrated some heterogeneity (Fig. X.4 A–I) two years after the start of the various practices. These initial spatial differences seem relatively unimportant compared to the changes in time measured thereafter. The most striking general trend was the decrease of pH (KCl) in all the regimes. This trend did not, however, continue during the last few years in the abandoned, burned, and September mulched sites, in particular in the 5–20 cm soil layer. According to Schreiber & Schiefer (1985) this might be attributed to the process of accumulation of Ca-compounds in the litter layer and subsequent leaching of those compounds into the deeper soil layers. N_{total} only fluctuated in all management regimes. P_{total} often shows the lowest values in 1978. This phenomenon is even more striking for $K_{exchangeable}$ with the exception of the July hay-making regimes (both with machinery and scythe), and the site which was cut twice, which showed a slight decrease. The changes of P_{total} and $K_{exchangeable}$ in this study are not in agreement with assessments in the literature. Wind (1980) and Elberse et al. (1983) also mentioned some decrease of potassium after a period of hay-making without fertilizer applications. Oomes & Mooi (1985) found a decrease of pH and both P_{total} and $K_{exchangeable}$. Schiefer (1983), however, mentioned both decreases and increases of the P_{total} and $K_{exchangeable}$ and a constant or decreasing pH (Table X.1).

Miles (1985, 1987) stresses that plants not only depend on soil chemical factors, but do themselves induce changes in the soil, which results in complex inter-relationships. Pegtel (1987) reviews examples of the dynamic character of the soil system and the changes in availability of nutrients as a result of release by weathering of parent materials, dissolvation of difficult soluble compounds, decay of organic matter in the soil, absorption by plant roots. Nevertheless, the unexpected increase of P_{total}, P_{water} and $K_{exchangeable}$ in a number of management regimes leads to some hypotheses. Miles (1987) points to the significant changes of the pH of surface soil, as in this study, which can occur within a few years and the considerable effects of a changing

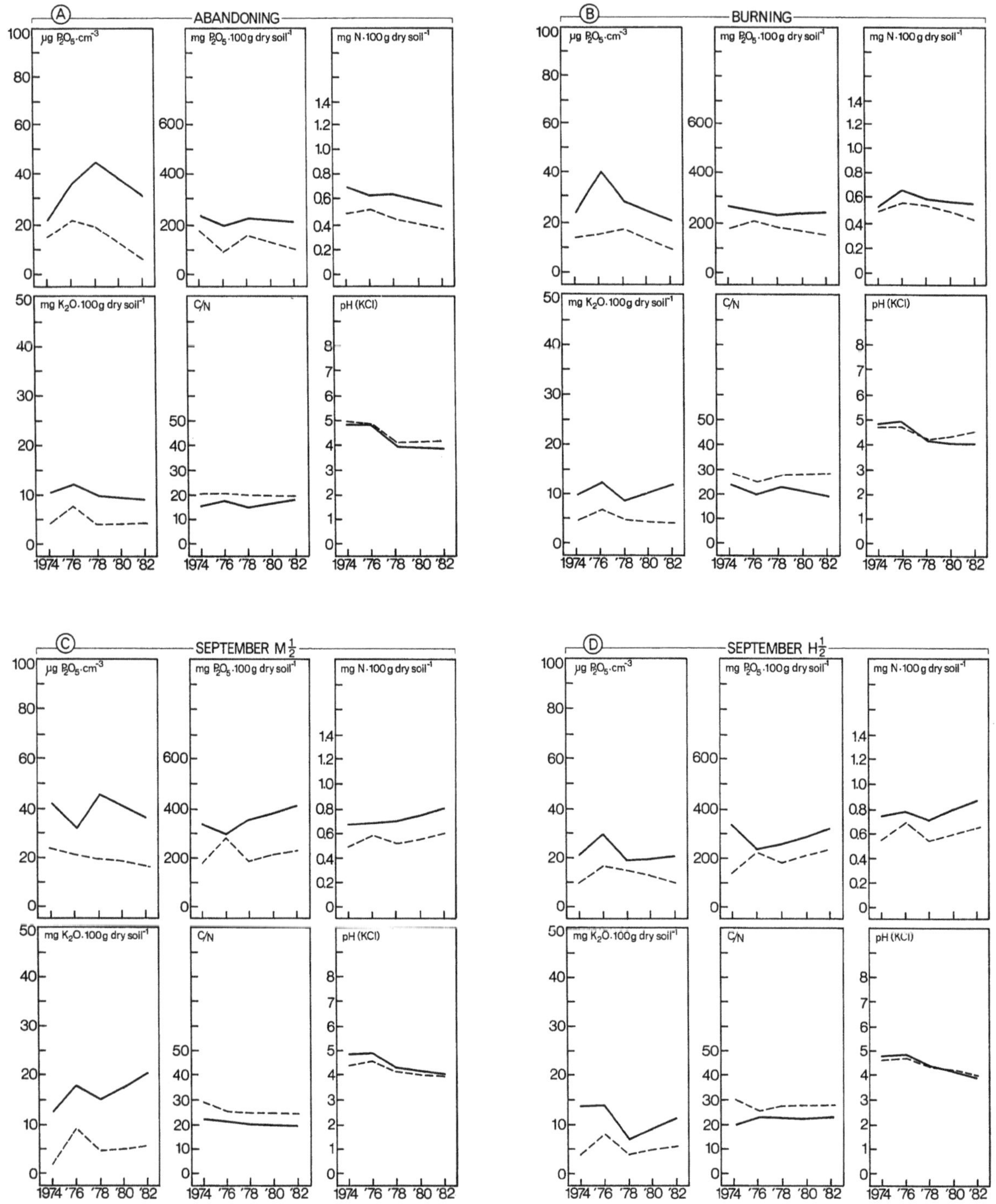

Fig. X.4A–D. Changes in the soil chemical parameters (solid line: 0–5 cm, broken line: 5–20 cm) P_{water}, P_{total}, N_{total}, $K_{exchangeable}$, C/N and pH (KCl) in different management regimes in the *Agrostis stolonifera*/*Poa pratensis* community of the recent reserve of the Loefvledder area. H: hay-making, M: mulching, ½: every second year. The plots had received no fertilizer applications after 1971.

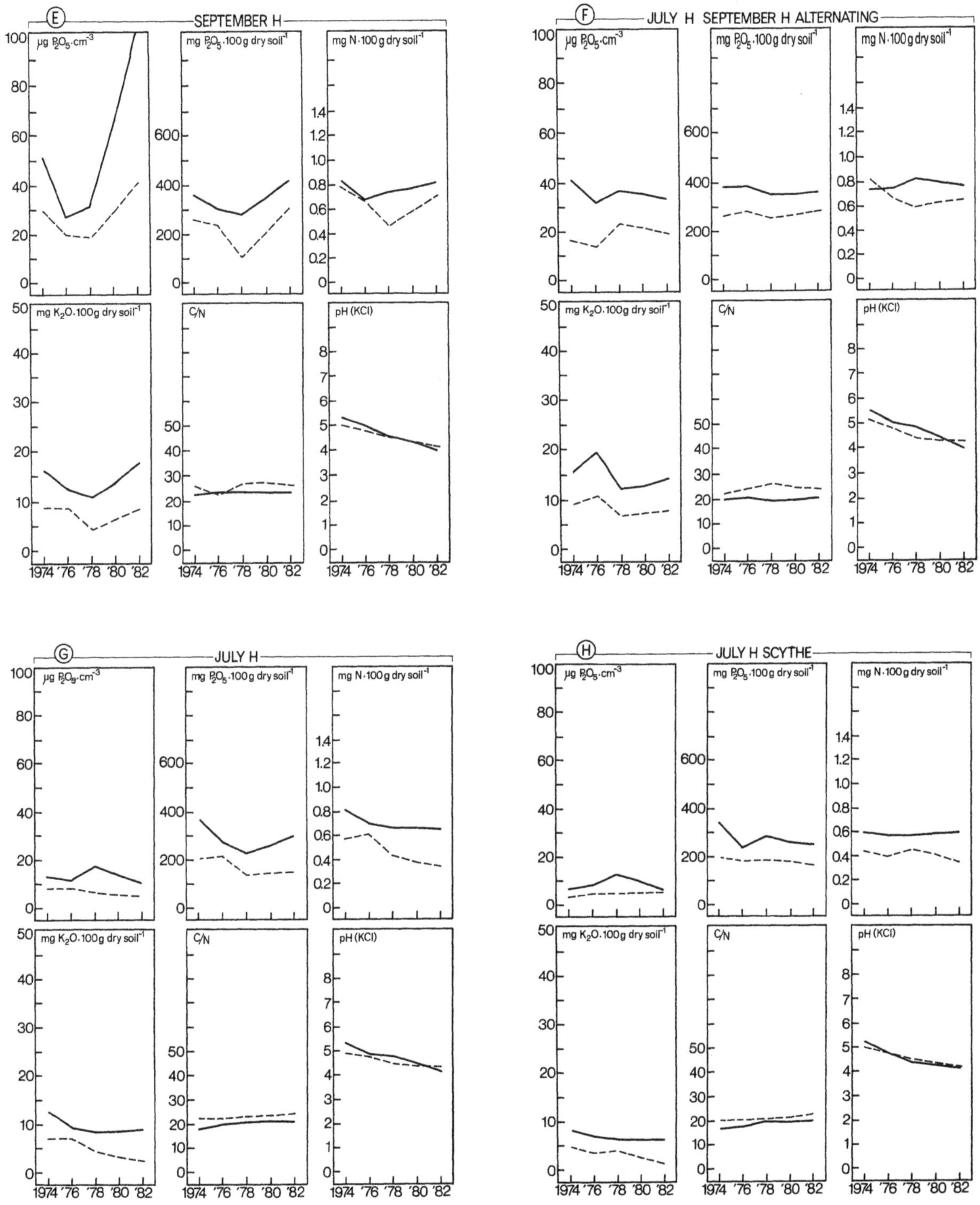

Fig. X.4E–H. Changes in the soil chemical parameters (solid line: 0–5 cm, broken line: 5–20 cm) P_{water}, P_{total}, N_{total}, $K_{exchangeable}$, C/N and pH (KCl) in different management regimes in the *Agrostis stolonifera/Poa pratensis* community of the recent reserve of the Loefvledder area. H: hay-making, M: mulching, $^1/_2$: every second year. The plots had received no fertilizer applications after 1971.

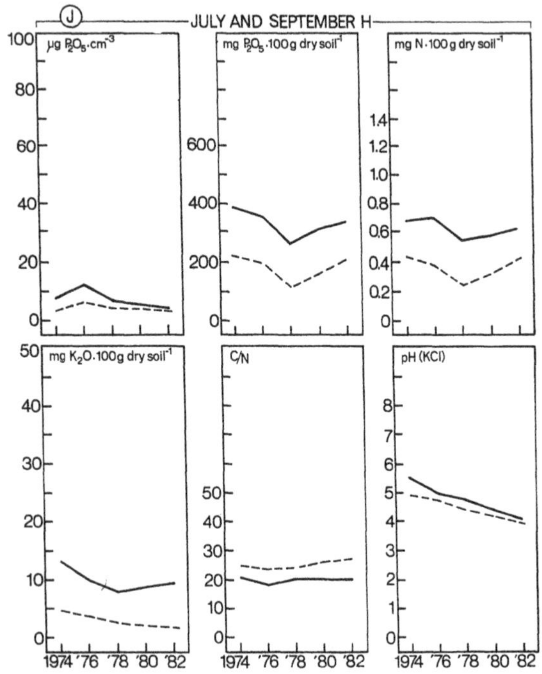

Fig. X.4J. Changes in the soil chemical parameters (solid line: 0–5 cm, broken line: 5–20 cm) P_{water}, P_{total}, N_{total}, $K_{exchangeable}$, C/N and pH (KCl) in different management regimes in the *Agrostis stolonifera/Poa pratensis* community of the recent reserve of the Loefvledder area. H: hay-making, M: mulching, 1/2: every second year. The plots had received no fertilizer applications after 1971.

pH on other soil parameters. The decrease in pH from 5 to 4 in all but the abandoned, burned and September mulched regimes in the *Agrostis stolonifera/Poa pratensis* community might have resulted in a higher solubility of ions from the iron compound (for example occluded phosphate), particularly in wet soil conditions (Kinzel, 1982). Iron is indeed present in high concentrations since even bog ore-digging took place in the past in the river valley (Booy, 1986). The wet years after 1978 (Fig. IX.9) prevented leaching of the ions and, therefore, perhaps facilitated the increase of $K_{exchangeable}$ and P_{water}. The only explanation for the increase of P_{total} can be the exploitation of deeper soil layers by the root system resulting in the pumping of nutrients from soil layers deeper than 20 cm.

In a *Festuca rubra/Agrostis capillaris* community, in another part of the Loefvledder study area, different cutting regimes were also started in 1972. Fertilizer applications had already been ceased in 1967. The poor to rich species ratio reached high figures in the July hay-making regimes (even with the addition of Nitrogen Limestone) with the exception of the practices which involved the addition of Ureum Formaldehyde (Fig. X.5). The annual July mulching regime also revealed very low poor to rich species ratios. There was no relationship with the above-ground standing crop, since this only fluctuated with all management practices (see Fig. IX.10).

Nearly all soil chemical parameters only revealed fluctuations (Fig. X.6 A–H) with minimum values for $K_{exchangeable}$ in all management regimes except for the addition of Nitrogen Limestone. It is remarkable that the soil of the *Festuca/Agrostis capillaris* community with many species indicating poor conditions, showed higher N_{total} and $K_{exchangeable}$ figures in many sites than the soil of the *Agrostis stolonifera/Poa pratensis* community with only a few poor species.

Although some trends can be seen, the increase of P_{water} and $K_{exchangeable}$ in the *Festuca rubra/Agrostis capillaris* community is less pronounced than in the *Agrostis stolonifera/Poa pratensis* community. This is probably due to the fact that the pH was already 4 at the start of the different management practices.

X.B.3. Comparison of grazing and hay-making

July hay-making and sheep-grazing were compared in the Westerholt study area from 1972 onwards. Fertilizer applications had ceased in 1971. The study was carried out in a *Holcus lanatus* community which was locally closely grazed, building mosaics with the *Leontodon autumnalis* community.

The poor to rich species ratio was much higher in the lightly grazed and slightly higher in the heavily grazed area than in the hay-making regime (Fig. X.7), despite the deposition of sheep dung in the grazed area. Three to five poor species were present in both heavily and lightly grazed areas, but the number of rich species dropped after three years from 9 to 1 in the lightly grazed area (overgrown by *Agrostis capillaris*) and remained high in the heavily grazed area until it dropped to five

Table X.1. Changes in the top soil of some chemical parameters during a number of years of hay-making without fertilizer applications in various plant communities. –: not measured. Chemical analyses refer to the top 5 cm in the Netherlands, whereas those data are wanting in Germany.

Country	Plant community	Soil	Number of years of hay-making without fertilizer applications	Change in soil chemical parameters pH KCl	P_2O_5 (mg.100g dry soil^{-1})	K_2O (mg.100g dry soil^{-1})	Assessment
The Netherlands	Lolio-Plantaginion	peat	8	5.4-4.2	360-300	16- 8	this study
The Netherlands	Festuco-Cynosuretum	peat	8	4.1-3.7	330-420	33-47	this study
The Netherlands	Lolio-Cynosuretum	sand	8	4.6-4.6	-	22-14	Wind 1980
The Netherlands	Lolio-Cynosuretum	clay	8	7.0-6.4	-	28-23	Wind 1980
The Netherlands	Arrhenatheretum	clay	17	4.3-3.8	7- 9	28-20	Elberse et al.1983
The Netherlands	Poo-Lolietum	sand	10	5.2-4.5	60- 36	20- 6	Oomes & Mooi 1985
Germany	Festuco-Cynosuretum	sand	7	4.9-4.6	4- 2	20-20	Schiefer 1983
Germany	Arrhenatheretum	loam	10	5.5-5.0	5- 2	7-10	Schiefer 1983
Germany	Arrhenatheretum	loam	7	6.8-6.8	3- 4	29-36	Schiefer 1983
Germany	Arrhenatheretum	sand	14	5.6-5.2	2- 7	12- 8	Schiefer 1983
Germany	Arrhenatheretum	loam	12	6.6-5.9	2- 2	8- 8	Schiefer 1983
Germany	Lolio-Cynosuretum	sand	4	5.1-5.1	9- 4	11- 8	Schiefer 1983
Germany	Lolio-Cynsoretum	loam	14	5.6-5.6	17-11	8-12	Schiefer 1983

during the last three years. The 'rich' species *Lolium perenne* persisted better in the grazed than in the cut area (Fig. X.8 A) in spite of an initial lower degree of cover. The intermediate species, *Anthoxanthum odoratum* (Fig. X.8 B), increased sharply in the cut area, but only slightly in the heavily grazed area. These findings seem to be in accordance with a more fertile soil in the grazed area, because of the return of dung and urine. The abundancy, however, of the 'poor' species *Hypochaeris radicata* initially increased sharply (Fig. X.8 C), especially in the heavily grazed area, but it later declined again.

Before discussing the results of the soil chemical analyses of the Westerholt area, it should be noted that the values of P_{water}, P_{total} and N_{total} are considerably lower than those of the afore-mentioned Loefvledder area Fig. X.9A–F.

Soil chemical composition showed no difference between grazing and hay-making and no decrease during the study period (Fig. X.9 A–B). The only difference is the initially high P_{water} in the mown site, but it soon dropped to levels comparable to those in the grazed site. It can be concluded that no relationship is found between soil chemical composition and changes in the poor/rich ratios in the heavily grazed area and the hay-making practice. Only the degree of cover of *Lolium perenne* (Fig. X.8 A) and P_{water} in the mown site seem to be correlated.

The poor to rich species ratio was similar in both the hay-making regimes without and with 50 kg $N \cdot ha^{-1}$ fertilizer applications (Fig. X.7). The same was the case for most soil chemical criteria. Only P_{total} and $K_{exchangeable}$ were a little higher in the 5–20 cm soil layer in the fertilized than in the unfertilized area. This could be attributed to the higher standing crop which has pumped nutrients from soil layers deeper than 20 cm. Marrs (1985) compared the yield of *Lolium perenne* and its content of N, P and K. He found a higher yield on the fertilized soil and not only a higher content of N, but also of P, whereas no difference was found for K. An extra removal of P was, therefore, possible after the addition of nitrogen. The recovery of N was only 20% and Marrs supposed that the remainder was leached or had incorporated into the soil organic matter. Egloff (1986) also found a recovery of 22% in summer cuts, but only 8% in an autumn cut.

The poor to rich species ratio was much higher in slightly moist grassland areas, in which *Erica tetralix* and *Calluna vulgaris* established from the first year after sod removal, than in drier areas with sod removal (Fig. X.7). However, the latter showed higher ratios than the other hay-making regimes. P_{water}, P_{total} and N_{total} were indeed lowest in the soil

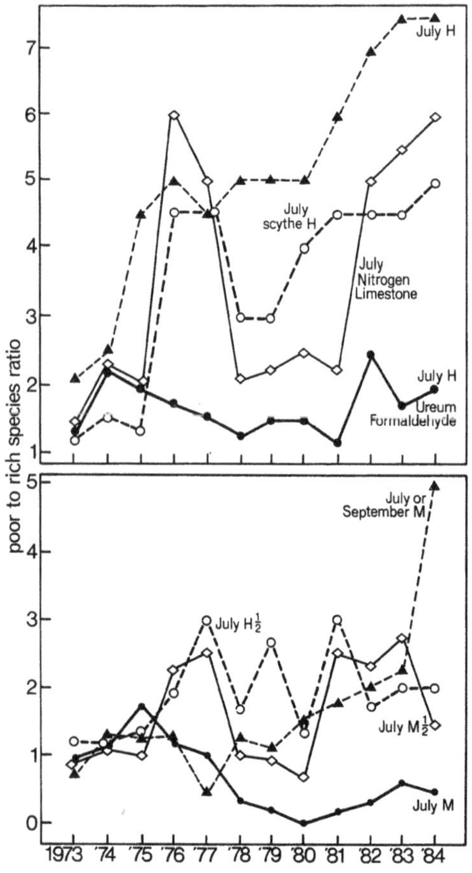

Fig. X.5. Changes in the poor to rich species ratio in different management regimes in the *Festuca rubra/Agrostis capillaris* community in the older reserve of the Loefvledder area. H: hay-making, M: mulching, $^{1}/_{2}$: every second year. The plots had received no fertilizer applications after 1966.

in the area with sod removal, whereas the C–N ratio was higher indicating a less fertile soil (Fig. X.9 C).

Differences in soil chemical attributes in grazed and abandoned heathland were only found in higher P_{water} and $K_{exchangeable}$ values in the grazing treatment (Fig. X.9 E–F).

X.B.4. Nutrient removal

The amount of nitrogen removed in hay (both the July and September hay-making regimes) in the *Agrostis stolonifera/Poa pratensis* community of the Loefvledder study area was similar in 1975 and 1983, but less phosphorus and potassium were removed in 1983 than in 1975 (Table X.2). No difference could be detected between the amounts removed in July or in September from the present study area. This is in agreement with the results of Dickinson & Polwart (1982) measured in a productive (600 gdw · m^{-2}) *Agrostis capillaris* community. It is not, however, common for the nutrient removal to be constant during the season. It is obvious that hay-making in May removed smaller amounts of nutrient than later cuts (Schmidt, 1981; Dickinson & Polwart, 1982; Buytendorp et al., 1983). A smaller amount of nutrient removal was found in late autumn than in summer in low productive (200–300 gdw · m^{-2}) *Mesobromion* (Buytendorp et al., 1983) and *Molinietum* (Kapfer & Pfadenhauer, 1986; Egloff, 1986).

More nutrients were removed in hay in the twice-cut (Table X.3) than with once-cut regime. This could be attributed to a greater amount of biomass removed in 1975. In 1983, however, the twice cut material contained higher nutrient percentages than the once cut material resulting in higher nutrient removal in the first treatment (Table X.4). In general the amount of nutrients removed did not appear to be closely correlated to the amount of biomass removed in the present study. This is not exceptional considering the different assessments in the literature. The amount of nitrogen in removed hay ranges from 15 to 292 kg · ha^{-1}, and 0.75 to 3.69 as percentages of the yield, of phosphorus from 2 to 66 kg · ha^{-1} (0.05 to 1.27%), and of potassium 2 to 230 kg · ha^{-1} (0.10 to 3.57%) respectively (Table X.3). Some trends seem, nevertheless, to occur. The potassium percentage of the hay tends to decrease over the years in plots with hay-making without fertilizer applications (Korevaar, 1986; Schmidt, 1981; Kapfer, 1987), which is in agreement with the results of this study. This seems consistent with the decrease of $K_{exchangeable}$ in the topsoil (Table X.1). Somewhat higher nitrogen and potassium percentages were found in the twice-cut treatment in 1983 than in the once-cut regime in the present study. This is in agreement with Schmidt's (1981) results. He even found increasing nitrogen and phosphorus percentages with four and eight annual cuttings. Special attention needs the idea of

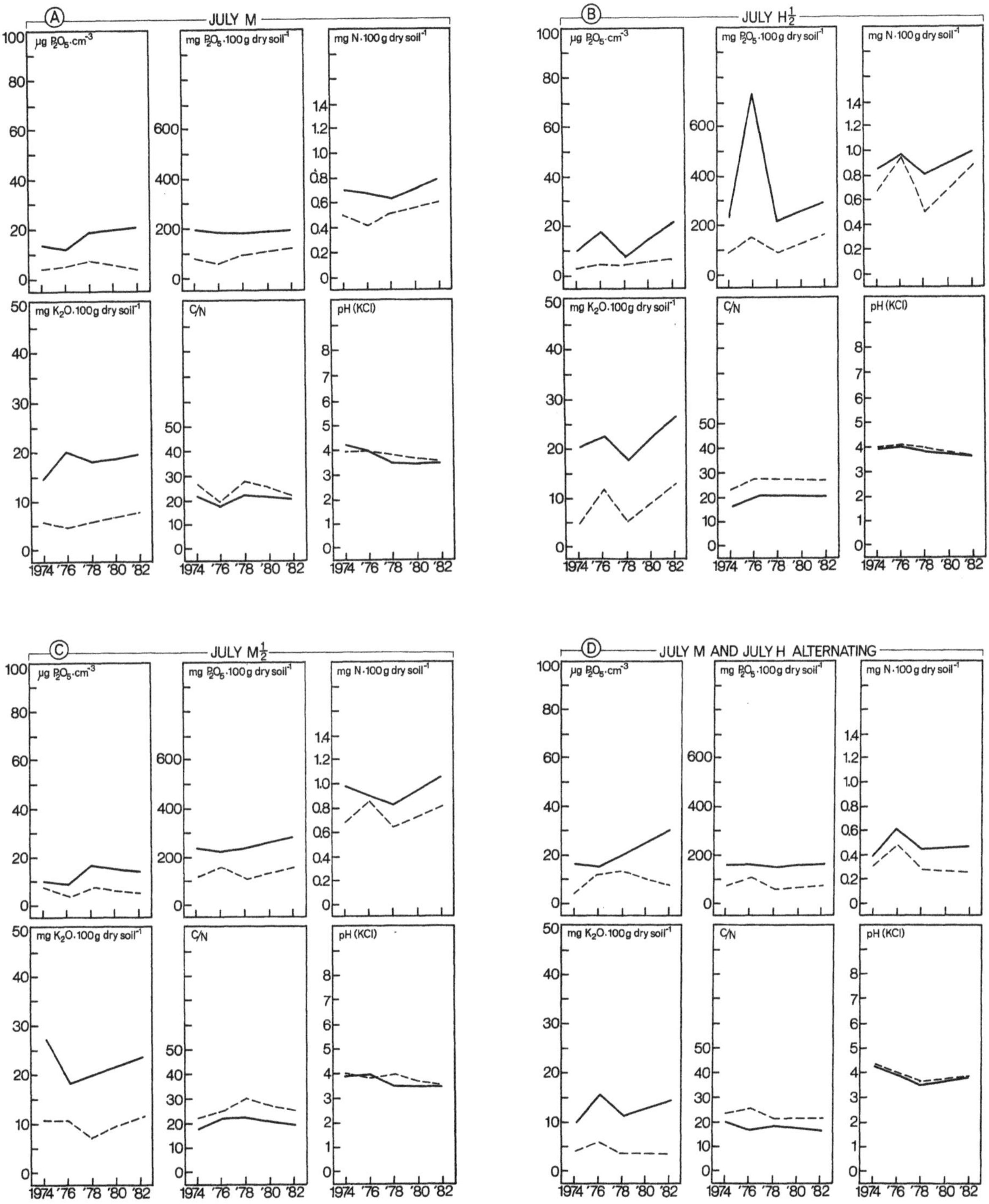

Fig. X.6A–D. Changes in the soil chemical parameters (solid line 0–5 cm, broken line 5–20 cm) P_{water}, P_{total}, N_{total}, $K_{exchangeable}$, C/N and pH (KCl) in different management regimes in the *Festuca rubra/Agrostis capillaris* community in the older reserve of the Loefvledder area. H: hay-making, M: mulching, $^1/_2$: every second year. The plots had received no fertilizer applications after 1966.

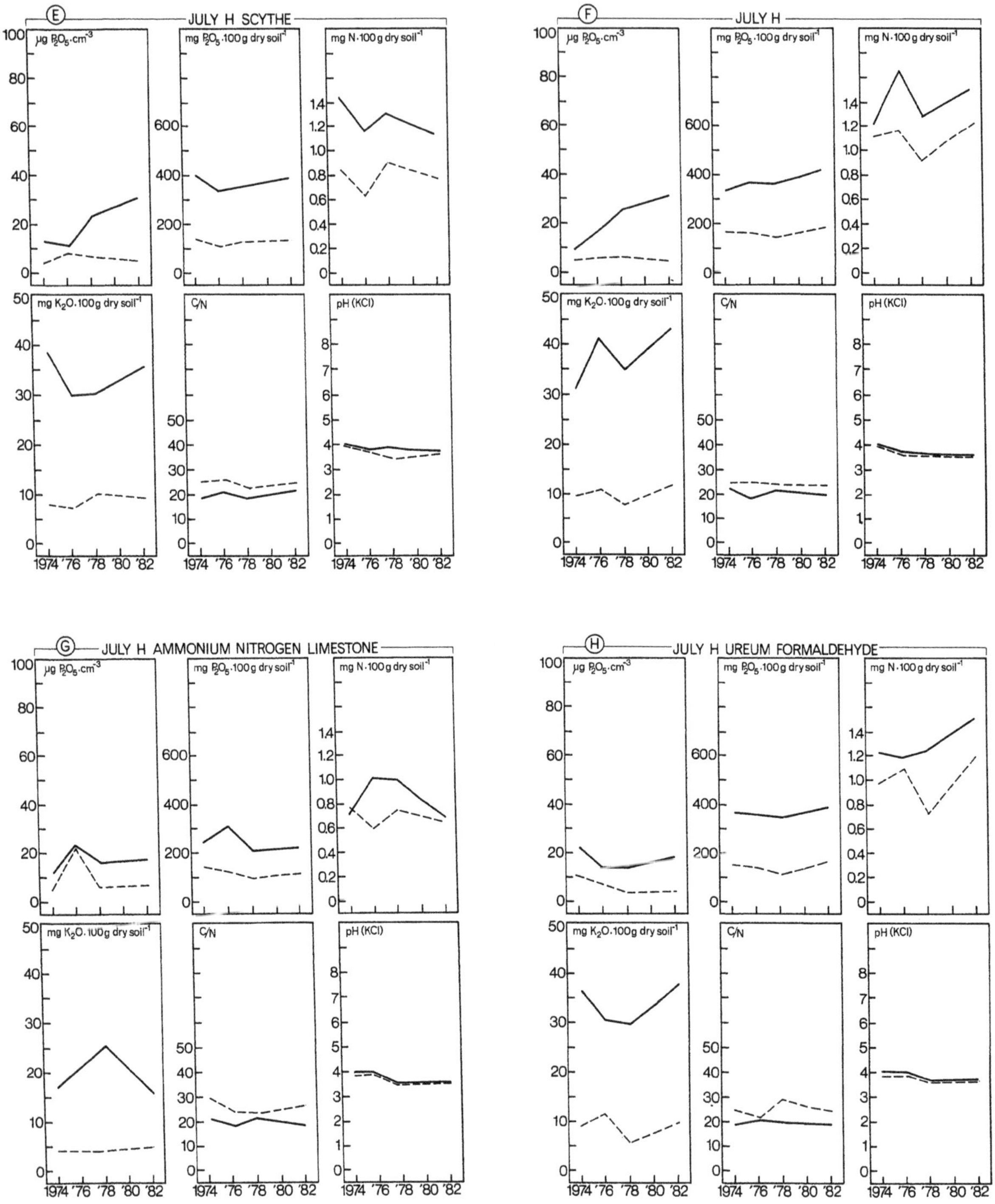

Fig. X.6E–H. Changes in the soil chemical parameters (solid line 0–5 cm, broken line 5–20 cm) P_{water}, P_{total}, N_{total}, $K_{exchangeable}$, C/N and pH (KCl) in different management regimes in the *Festuca rubra/Agrostis capillaris* community in the older reserve of the Loefvledder area. H: hay-making, M: mulching, $^1/_2$: every second year. The plots had received no fertilizer applications after 1966.

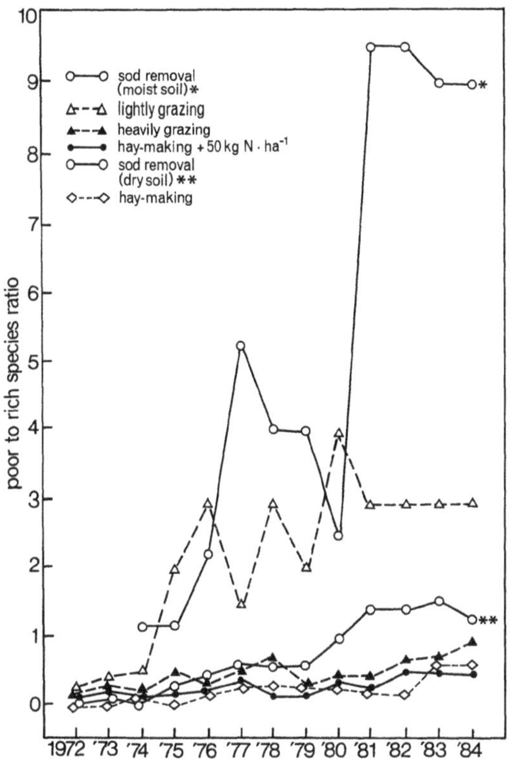

Fig. X.7. Changes in the poor to rich species ratio in different management regimes in the *Holcus lanatus* community in the Westerholt area. The plots had received no fertilizer applications after 1971.

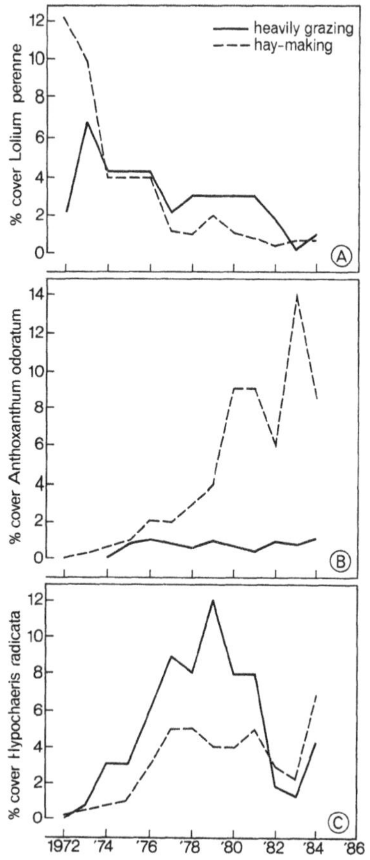

Fig. X.8A–C. Changes in the cover % of A: *Lolium perenne,* B: *Anthoxanthum odoratum,* C: *Hypochaeris radicata* in plots cut for hay in July or heavily grazed. The plots had received no fertilizer applications after 1971 (after Bakker, 1987).

Marrs (1985) that the growth of cereal crops, without the application of fertilizers, might remove more nutrients from a site than the annual inputs from natural sources. He indeed removed 35 kg $N \cdot ha^{-1}$, 10 kg $P \cdot ha^{-1}$ and 61 kg $K \cdot ha^{-1}$ in the United Kingdom, but these figures are lower than most nutrient removals with hay.

Grazing resulted in nutrient removal by former agricultural practices when sheep were allowed to graze on heathland or chalk grassland at daylight and fenced into leys or arable fields at night in order to collect the manure. Present day grazing implies local nutrient removal and enrichtment elsewhere (Wilkinson & Lowry, 1973). A net transport of 1 kg $N \cdot ha^{-1}$ and 0.5 kg $P \cdot ha^{-1}$ with dung from grassland to heathland with resting places was measured from July to October in our study area (Bakker et al., 1983). Rawes & Heal (1978) estimated net transfers of 0.3 kg $N \cdot ha^{-1}$, 0.03 kg $P \cdot ha^{-1}$ and 0.06 kg $K \cdot ha^{-1} \cdot yr^{-1}$ from blanket bog to grassland in Scotland which is probably not significant in view of the errors involved in the estimate. Latrines also receive more nutrients than areas which are closely grazed by ponies (Pratt et al., 1986). This resulted in a higher phosphorus (a factor of 1 · 2) and potassium (a factor of 1 · 7) content of the soil under latrine areas than under grazed areas in the New Forest (Putman, 1986). Hilder (1966) and Czerwinsky et al. (1974) also measured a distinct increase in potassium in the topsoil layer of camping sites. Baradziej (1978) reported considerably increased values of potassium and nitrogen and a slight increase of phosphorus in sheep folds. During the herbage intake dung and urine can, however, be voided, too. In our

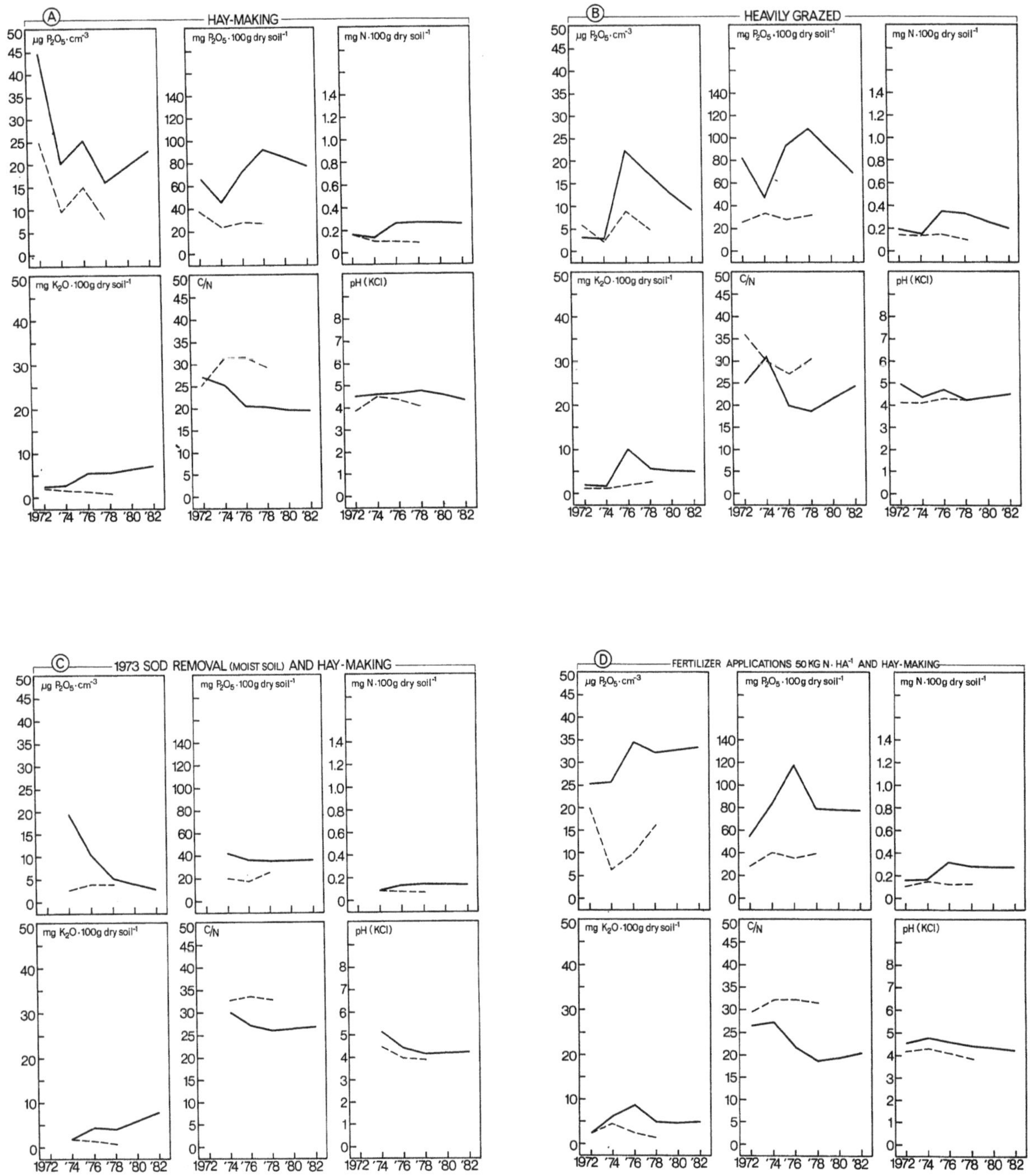

Fig. X.9A–D. Changes in the soil chemical parameters (solid line 0–5 cm, broken line 5–20 cm) P_{water}, P_{total}, N_{total}, $K_{exchangeable}$, C/N, pH (KCl) in different management regimes in the *Holcus lanatus* community (A–D) and in the *Calluna vulgaris* community (E–F) in the Westerholt area. The grassland plots had received no fertilizer applications after 1971.

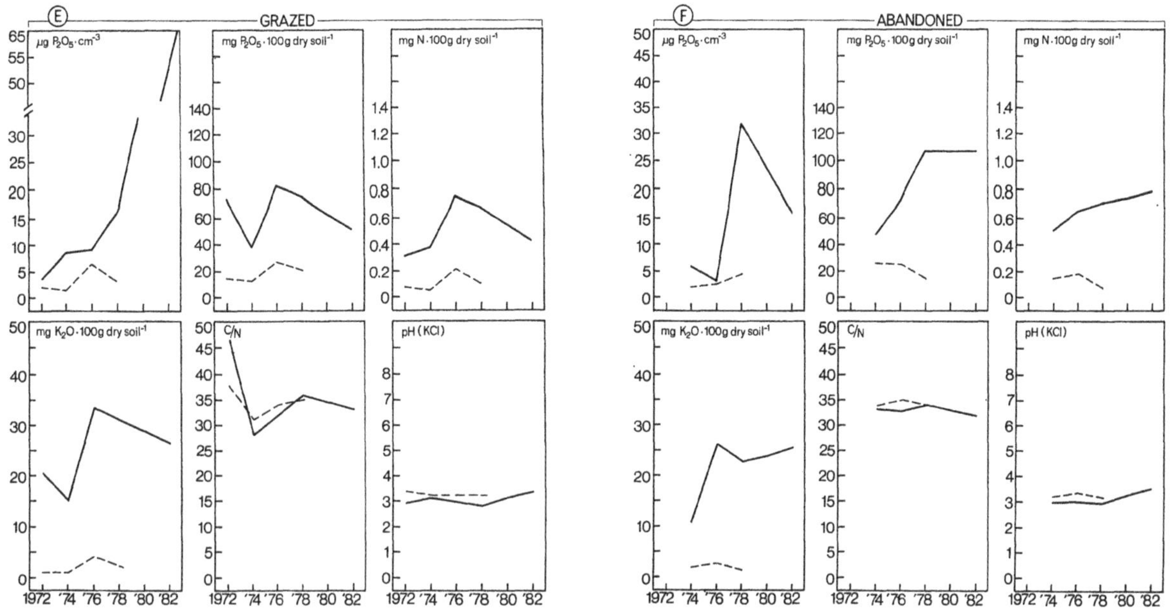

Fig. X.9E–F. Changes in the soil chemical parameters (solid line 0–5 cm, broken line 5–20 cm) P_{water}, P_{total}, N_{total}, $K_{exchangeable}$, C/N, pH (KCl) in different management regimes in the *Holcus lanatus* community (A–D) and in the *Calluna vulgaris* community (E–F) in the Westerholt area. The grassland plots had received no fertilizer applications after 1971.

study area less nutrients were, therefore, removed with grazing than with hay-making (Table X.5).

A net loss of nutrients in a grazed area can only be expected from wool, meat and carcasses. These losses amount a few $kg \cdot ha^{-1} \cdot yr^{-1}$ and are often more than compensated by the input from rainfall (Green, 1972; Perkins, 1978; Bülow-Olsen, 1980A). The removal of nutrients measured locally, i.e. the amount consumed minus the return by dung (and urine), or the amount removed in hay, accounts for a small percentage of the nutrient pool in the topsoil (Table X.6). The nutrient removal related to the nutrient pool in the topsoil is in agreement with literature assessments. Dickinson (1984) suggested even lower figures in estimating that less than 1% of total ecosystem nutrients were removed per annum as cuttings in *Agrostis capillaris* grasslands. In the *Deschampsia flexuosa* community mentioned by Bülow-Olsen (1980A) only the % removal of potassium is very high, due to the low level of the K pool. Again, the compensation by precipitation is considerable and, is in fact underestimated since dry precipitation is not taken into account. The total deposition of nitrogen for instance amounted to $37 kg \cdot ha^{-1} \cdot yr^{-1}$ in 1980 in our study area (Fransen, 1987). Deposition is in particular important in untouched stands in which ammonium is assimilated by green leaves in the spring and summer period ($5.9 kg \cdot ha^{-1}$), whereas under cut stands only $2.5 kg \cdot ha^{-1}$ was deposited (Heil et al., 1988). It can be concluded that nutrient depletion, if any, is apparently a very slow process.

X.B.5. Nutrient cycling

Some conclusions of the previous sections are reviewed with respect to the phenomena involved with impoverishment of the soil. The poor to rich species ratio became highest with hay-making twice a year, but the above-ground standing crop not only decreased significantly in this, but also in other hay-making regimes, hence a relationship can only be found in some cases between the decrease of above-ground standing crop and the increase of the poor to rich species ratio. This is also the case for the relationship of each of these two parameters and soil chemical attributes. The poor to rich species ratio was slightly higher in the heavily grazed than in the hay-making regime, but the changes in cover degree of individual poor and rich species were not always in agreement with this difference. Moreover, soil chemical analyses showed no differences between grazing and hay-making. The absence of trends in measured soil chemical parameters can probably be explained by the very small amount of nutrients which are actually removed as compared to the nutrient pool in the topsoil. Hence, the changes in standing crop are perhaps better related to the cycling of nutrients.

In our study area the yield in grazed communities is considerably higher than in hay-making treatments (Table X.7). This is in agreement with the few comparative studies made in grasslands without fertilizer applications (Table X.7). The nutrient removal related to the nutrient pool, nevertheless, appeared to be very low in both grazed and mown grassland communities. This can be attributed to the conversion of herbage into faeces and urine. A grazing animal returns more than 90% of the nutrients it consumes, as dung and urine (Wilkinson & Lowry, 1973; Perkins, 1978). The nutrients in urine are immediately available to plants, those in dung less so but the decay rate of faeces is considerably higher than standing dead or litter material (Perkins et al., 1978). Grazing, stimulates the decomposition of organic material (Breymeyer, 1978; Perkins et al., 1978), and hence mineral cycling, sometimes up to tenfold, of N (Floate, 1970; Spedding, 1971), P (Floate et al., 1973; Harrison, 1978), Ca, Mg, Mn and P (Bülow-Olsen, 1980A).

Table X.2. Seasonal change of total nutrients (kg · ha^{-1}) in above-ground standing crop and litter in several plant communities. –: not measured. Data refer to the first cut in all assessments.

Plant community	Nutrient		May	June	July	September	January	Assessment
Agrostis stolonifera / Poa pratensis community (The Netherlands)	N (standing crop including litter)	1975	-	-	107	104	-	This study
		1983	-	-	122	108	-	
	P (standing crop including litter)	1975	-	-	41	44	-	
		1983	-	-	19	21	-	
	K (standing crop including litter)	1975	-	-	55	73	-	
		1983	-	-	40	39	-	
Molinio-Arrhenatheretea (Germany)	N (standing crop including litter)	1972	18	-	-	20	-	Schmidt 1981
		1978	5	-	-	26	-	
	P (standing crop including litter)	1972	3.0	-	-	3.9	-	
		1978	0.7	-	-	5.7	-	
	K (standing crop including litter)	1972	14	-	-	27	-	
		1978	2	-	-	33	-	
Molinio-Arrhenatheretea (United Kingdom)	N standing crop		70	-	100	85	78	Dickinson 1982
	litter		25	-	10	60	85	
	total		95	-	110	145	153	
	P standing crop		7	-	8	5	8	
	litter		3	-	2	6	9	
	total		10	-	10	11	17	
	K standing crop		0.1	-	0.1	0.1	0.1	
	litter		1.0	-	0.5	2.5	2.6	
	total		1.1	-	0.6	2.6	2.7	
Molinion (Switzerland)	N (standing crop including litter)		-	36	50	44	-	Egloff 1986
	P (standing crop including litter)		-	2.0	2.6	2.1	-	
	K (standing crop including litter)		-	34	59	40	-	
Mesobromion (The Netherlands)	P (standing crop including litter)		2	-	11	13	2	Buytendorp et al. 1983

Cuttings are returned with mulching, so similar effects to those with grazing might be expected. In our study area, material resulting from mulching in July every year was collected and the decay rate was estimated by means of the litter bag method. Litter bags were brought into contact with the soil in the mulched management regime and were also attached to uncut plants, 5–10 cm above soil surface in the mulching every second year regime. This was in order to estimate the decay rate of standing dead plant material.

The decay rate was relatively high from the first of August 1982 onwards and it dropped to zero during the winter months. The decomposition started again in the early spring but its rate remained low (Fig. X.10). Mulched material disappeared significantly quicker (sign test $P<0.05$) than standing dead material. Bradshaw (1980) also

Table X.3. The amount of biomass and nutrients ($kg \cdot ha^{-1}$) removed by hay in different plant communities. The percentages of N, P and K related to the yield are indicated.

Country	Plant community	Soil	Period of hay-making and numbers of years of hay-making without fertilizer applications		Yield	N	%	P	%	K	%	Assessment
The Netherlands	Lolio-Plantaginion	peat	Jl/Sept	3	11500	156	1.36	66	0.57	91	0.79	this study
				11	9000	176	1.96	30	0.33	93	1.03	
The Netherlands	Poo-Lolietum	sand	Jn/Oct	11	4400	100	2.27	14	0.32	22	0.50	Oomes & Mooi 1985
The Netherlands	Hayfield	?	first cut	0	5000	108	2.16	32	0.64	128	2.56	Rinsema 1979
The Netherlands	Poo-Lolietum	peat	Jn/?	1	10140	258	2.54	41	0.40	230	2.27	Korevaar 1986
				5	6000	154	2.57	22	0.37	49	0.82	
	Poo-Lolietum	clay	Jn/?	1	7250	175	2.41	22	0.30	159	2.19	
				4	8930	222	2.49	26	0.29	134	1.50	
	Mol.-Arrhenatheretea	peat	Jn/?	1	5200	148	2.84	21	0.40	78	1.50	
				6	4260	157	3.69	17	0.40	33	0.77	
Germany	Mol.-Arrhenatheretea	loam	Oct	3	2040	20	0.98	4	0.20	27	1.32	Schmidt 1981
				9	2780	26	0.94	6	0.21	33	1.19	
	Mol.-Arrhenatheretea	loam	Jn/Aug	3	1500	22	1.47	5	0.33	38	2.53	
				9	4820	89	1.85	15	0.30	109	2.26	
	Mol.-Arrhenatheretea	loam	My/Jn/Jl/Sept	3	2900	56	1.93	11	0.38	65	2.24	
				9	5190	138	2.66	18	0.35	96	1.85	
	Mol.-Arrhenatheretea	loam	My/My/Jn/Jn/	3	1930	43	2.23	8	0.41	43	2.23	
			Jl/Aug/Sept/Oct	9	4650	153	3.29	19	0.42	92	1.98	
Germany	Festuco-Cynosuretum	sand	?	7	4400	91	2.07	8	0.18	92	2.09	Schiefer 1983
	Arrhenatheretum	loam	?	12	5100	95	1.86	11	0.22	104	2.04	
	Arrhenatheretum	loam	?	2	6800	153	2.25	17	0.25	212	3.12	
				6	4400	85	1.93	7	0.16	157	3.57	
	Arrhenatheretum	sand	?	3	4000	91	2.28	8	0.20	40	1.00	
	Arrhenatheretum	loam	?	16	6700	129	1.93	49	0.54	145	2.16	
	Lolio-Cynosuretum	sand	?	2	9100	292	3.21	49	0.54	195	2.14	
				5	4800	133	2.77	20	0.42	53	1.10	
	Lolio-Cynosuretum	loam	?	2	7000	202	2.89	39	0.56	108	1.54	
				6	6700	190	2.84	36	0.54	94	1.40	
Germany	Angelico-Cirsietum	peat	Jn/Aug/Oct	4	6000	138	2.30	56	0.93	52	0.87	Kapfer 1987
			Oct	4	4900	89	1.82	35	0.71	47	0.96	
	Lolio-Cynosuretum	peat	Jn/Aug/Oct	4	3000	77	2.57	17	0.57	15	0.50	
			Oct	4	3100	65	2.10	12	0.39	15	0.48	
	Lolio-Cynosuretum	peat	Jn/Aug/Oct	4	2100	55	2.62	11	0.52	8	0.38	
			Oct	4	1800	43	2.39	9	0.50	6	0.33	
Switzerland	Molinion	peat	Sept	always	4000	44	1.10	2	0.05	40	1.00	Egloff 1986
	Molinietum	peat	Oct	always	2100	30	1.43	4	0.19	6	0.29	
	Molinietum	peat	Oct	always	2000	15	0.75	2	0.10	2	0.10	
United Kingdom	Hayfield	?	?	?	3160	55	1.74	6	0.18	48	1.52	Bradshaw 1980
	Agrostis capillaris	?	?	?	1345	39	2.90	5	0.37	26	1.93	
	Festuca rubra	?	?	?	2065	61	2.95	7	0.34	35	1.69	

mentioned that the return of cuttings will increase the rate of nutrient cycling since the decay of cut herbage is more rapid than that of standing dead material in an *Agrostis/Festuca* community.

Buth & Voesenek (1987) studied decomposition of salt marsh plant rates in the sediment surface and in the canopy habitat. They found a significant higher decomposition on the sediment surface in only 50% of the experiments and concluded that the initial composition of plant material was the main factor affecting decomposition rate.

The spread of *Holcus lanatus* and *Rumex acetosa* and the appearance of *Anthriscus sylvestris* seem to be related to enhanced nutrient cycling and an accumulation of humus with the mulching regime in this study area. It is, however, difficult to generalize on this conclusion since Wells (1980) found no difference in floristic composition in chalk grassland plots from which the cut material had been removed for ten years, compared with those to which it was returned. This was possibly because of the lack of humus accumulation due to the high pH in chalk grassland. Schreiber & Schiefer (1985) did find floristic changes with mulching. The change, however, involved the increase of species indicating nutrient poor soil conditions, which is in contrast with our results. This was attributed to the small or decreasing amount of roots at the time of the decomposition of the material which was returned after the second cut in August. Schreiber & Schiefer hypothesized that the nutrients released after the decay of the organic material leached due to insufficient absorption in the root system.

Table X.4. Yield and mineral content of hay in two years in the *Agrostis stolonifera/Poa pratensis* community in the once cut July and September regimes and in the twice cut July + September regime (after Bakker & De Vries, 1985B).

Hay-making regime	1975			1983		
	Yield ($gdw \cdot m^{-2}$)					
July	810			820		
September	700			600		
July	800			650		
September	350			250		
July + September	1150			900		
	Mineral content ($g \cdot kgdw^{-1}$)					
	N	P	K	N	P	K
July	13.1	5.1	6.7	14.9	5.2	10.8
September	14.8	6.3	10.4	17.5	7.8	6.5
July	11.3	4.8	6.3	17.7	6.7	3.6
September	19.2	7.9	11.9	24.7	9.5	6.8

The amounts of nutrients that became available to the vegetation through dung, urine and decomposition of plant material during the growing season in *Festuca/Agrostis* pasture was examined by Bradshaw (1980), using data of Perkins et al. (1978). The amounts of the major nutrients contained in the total production were 1.6 to 2.6 times higher than the amounts in the total biomass. Dickinson (1984) calculated that the nutrients in the total annual production in an *Agrostis capillaris* hayfield varied between 0.5 and 2.2 times the nutrients in biomass. There was, therefore, at least a 50% and up to 220% turnover of all nutrients in the vegetation on an annual basis. In particular a high root turnover was found. This turnover takes place in a very small proportion of the total amount of nutrients. Six percent or less of the major nutrients in the whole ecosystem entered the vegetation (including litter), which means that 94% or more of most nutrients were found in root debris, organic matter and other soil fractions (Dickinson, 1984). Nilsson (1970) calculated that 94% of all organic matter in a hay-field system also consisted of humus. Perkins (1978) indicated that even less than 2% of the major nutrients was found in the vegetation including litter.

It is obvious that the mineralization of organic matter, whereby nutrients are released in a form which the plant can utilize, is a very important process in the ecosystem. The rate of decomposition depends on the origin of the material, temperature and soil moisture (Schreiber & Schiefer, 1985) and is, therefore, very dependent on the weather conditions. This could explain the large year-to-year fluctuations in above-ground standing crop. Moreover, it suggests that circumstances governing the mineralization rate of organic matter are more important than the removal of nutrients in hay. Unfortunately, no data are available of sequential mineralization measurements during a

Table X.5. Nutrients ($kg \cdot ha^{-1}$) removed with grazing and haymaking during the period April to August 1975 in a *Holcus lanatus* community (after Vreugdenhil & van Wieren, 1976).

	N	P
Consumed	32	12
Dung	11	6
Urine	?	?
Removed with grazing	<21	< 6
Removed with hay-making	35	18

Table X.6. Nutrient pool in the topsoil, amount of nutrients removed (their percentages related to the nutrient pool) by hay-making or grazing in different plant communities and the annual nutrient input by rainfall (kg · ha^{-1}).
* Data for Drenthe are derived from Centraal Bureau voor de Statistiek (1983).

Plant community	Management	Nutrient	Pool	Depth (cm)	Removal (%) in the year under study		Rainfall input	Assessment
A.stolonifera/Poa pratensis	hay-making	N	4750	0-20	110	(2.5)	16 *)	this study
Holcus lanatus	hay-making	N	1280	0-5	35	(2.7)	16 *)	this study
Holcus lanatus	grazing	N	1600	0-5	21	(1.3)	16 *)	this study
Agrostis/Festuca	grazing	N	10460	0-30	110	(1.1)	18	Perkins (1978)
Lolio-Plantaginion	agriculture	N	5390	0-20	-		-	De Both et al.(1983)
Lolio-Plantaginion	agriculture	N	10810	0-20	-		-	De Both et al.(1983)
A.stolonifera/Poa pratensis	hay-making	P	2000	0-20	20	(1.0)	0.07 *)	this study
Holcus lanatus	hay-making	P	370	0-5	18	(4.9)	0.07 *)	this study
Holcus lanatus	grazing	P	420	0-5	6	(1.4)	0.07 *)	this study
Agrostis/Festuca	grazing	P	3020	0-30	9	(0.3)	2	Perkins (1978)
Deschampsia flexuosa	grazing	P	40	0-10	-0.1	(0.3)	1	Bülow-Olsen (1980A)
Angelico-Cirsietum	hay-making	P	1590	0-20	50	(3.1)	-	Kapfer (1987)
Lolio-Cynosuretum	hay-making	P	730	0-20	20	(2.7)	-	Kapfer (1987)
Lolio-Cynosuretum	hay-making	P	480	0-20	15	(3.1)	-	Kapfer (1987)
Lolio-Plantaginion	agriculture	P	1510	0-20	-		-	De Both et al.(1983)
Lolio-Plantaginion	agriculture	P	960	0-20	-		-	De Both et al.(1983)
A.stolonifera/Poa pratensis	hay-making	K	5300	0-20	60	(1.1)	-	this study
Agrostis/Festuca	grazing	K	13610	0-30	5	(0.03)	3	Perkins (1978)
Deschampsia flexuosa	grazing	K	46	0-10	10	(22)	15	Bülow-Olsen (1980A)
Angelico-Cirsietum	hay-making	K	5190	0-20	70	(1.4)	-	Kapfer (1987)
Lolio-Cynosuretum	hay-making	K	720	0-20	25	(3.5)	-	Kapfer (1987)
Lolio-Cynosuretum	hay-making	K	1080	0-20	15	(1.4)	-	Kapfer (1987)
Lolio-Plantaginion	agriculture	K	940	0-20	-		-	De Both et al. (1983)
Lolio-Plantaginion	agriculture	K	930	0-20	-		-	De Both et al. (1983)

long period of hay-making or grazing without fertilizer applications. If it was available such data could, however, only give more insight into the soil chemical aspects of impoverishing, but could not explain the observed floristic changes. As will be discussed in section X.D, repeated defoliation by grazing animals renders an open canopy despite the high decomposition of dung voided. In grasslands with low productivity, suppression of the dominant grass by defoliation is of far greater importance than the return, or non-return of cut material (Duffey et al., 1974).

Table X.7. Yield (kg · ha^{-1}) with unfertilized grazing and hay-making treatments in some plant communities.

Plant community	Grazing	Hay-making	Assessment
Holcus lanatus	4500	3200	this study
Lolio-Cynosuretum	3065	1300	Wind 1980
Arrhenatheretum	7150	4775	Elberse et al. 1983

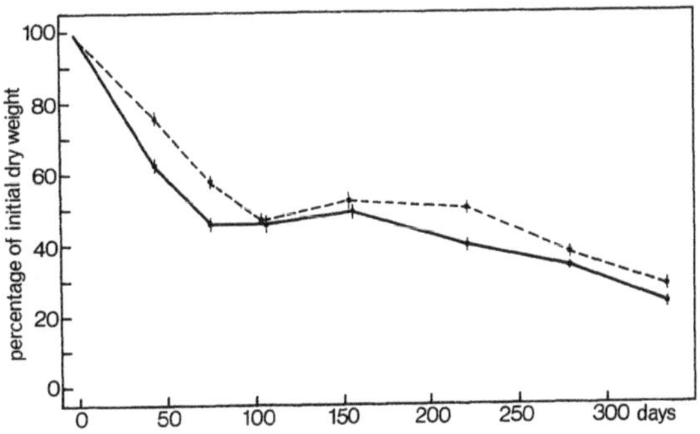

Fig. X.10. Decay rate (± S.E.) of green plant material after mulching in July 1982 (solid line) and 5–10 cm above soil surface in standing material (dotted line) in an *Agrostis stolonifera/Poa pratensis* community.

X.C. Salination

In this section the soil (0–5 cm) salinities in various plant communities under the management regimes of grazing, hay-making and abandoning are considered. Moreover, a gradient from a low dune towards the salt marsh is considered in detail in comparing soil (0–1 cm) salinities in grazed and abandoned sites. Finally, the results will be compared to those found in the literature.

X.C.1. Different plant communities

The course of the top-soil (0–5 cm) salinity in some salt marsh plant communities showed strong seasonal fluctuations and year-to-year differences (Fig. X.11 A–B). The salinity trends were similar in all plant communities studied. Winter levels were lower than those in the summer. The high salinity period was much longer in 1980 than in 1981. The exceptionally low summer values in August 1980 can be attributed to previously high precipitation. Great fluctuations in the salinity in the uppermost layer were also discovered by Tyler (1971). The soil water is diluted by rainwater in the spring, but salinity increases in the dry summer period, followed by a dilution in late autumn. Beeftink (1965) preferred relating salinity to soil solution, because plants come in contact with salts in solution. During dry periods, however, salts crystallize and disappear from solution in the soil. This could be the reason for the poor ($P>0.05$) correlation between salinity related to soil solution and dry soil in June 1980 (May was very dry), and in April 1981 (March was very wet). These extreme weather conditions in particular affect soil salinities differently in the plant communities due to their different soil characteristics. Salinity related to dry soil can be regarded as a more or less absolute figure through which one can possibly make a better distinction between the various communities (Fig. X.11 A) than salinity related to soil solution (Fig. X.11 B).

The ranking of salinities in the different plant communities (according to Spearman's correlation coefficient) differed slightly for salinities related to dry soil and to soil solution, respectively. Communities with clay soil (*Juncus gerardii, Juncus maritimus,* and *Festuca rubra*/*Limonium vulgare*) showed highest salinities, possibly due to a smaller pore volume and subsequent higher capillary rise (Ijjas, 1969).

The sequences of soil salinities with abandoning, mowing and grazing are given in Table X.8. These sequences differed for both the various communities and for the two salinity parameters. It can be concluded from these data that the soil salinity is not influenced by the management practice. Higher salinities due to evaporation with reduced canopy (Schmeisky, 1974; Hansen, 1982, citing several authors) were not found in the topsoil (0–5 cm) in the present study. This does not agree with assessments in the literature (Table X.9).

X.C.2. Gradient from dune to salt marsh

Further investigations were started. It was remarkable that, in contrast with other studies, the topsoil (0–5 cm) salinity was not higher in the grazed saltmarsh than in the abandoned salt marsh. Moreover, Kinzel (1982) suggested that germination might be reduced by higher salt concentrations in the top 1 cm layer of soil.

Figure X.12 shows the analysis of the vegetation along the gradient from a low dune to the salt marsh in the grazed and abandoned area in 1983. *Elymus repens* dominated the dunes in the abandoned area with different dominating species lower down the gradient namely *Festuca rubra, Elymus pycnanthus* and *Artemisia maritima,* respectively. These species were hardly found in the grazed area where there were more lower salt marsh species.

The soil profile in the gradient was similar in the whole gradient except for a thinner clay layer in the highest site and showed a rooted layer, a humus-rich clay layer and sand subsoil. Samples were taken in the uppermost (top 1 cm) layer of soil, in the lowermost 5 cm of the clay layer and in the uppermost 5 cm of the underlying sand layer. Pore volume in the 0 to 1 cm layer of soil was highest in the abandoned area of all the sites (Table X.10), but not in the deeper layers of the soil. The percentage of organic matter was highest at the two lowest sites (3 + 4). Grazed areas had a higher organic matter content at sites 1 + 2 (but not at 3 + 4). The

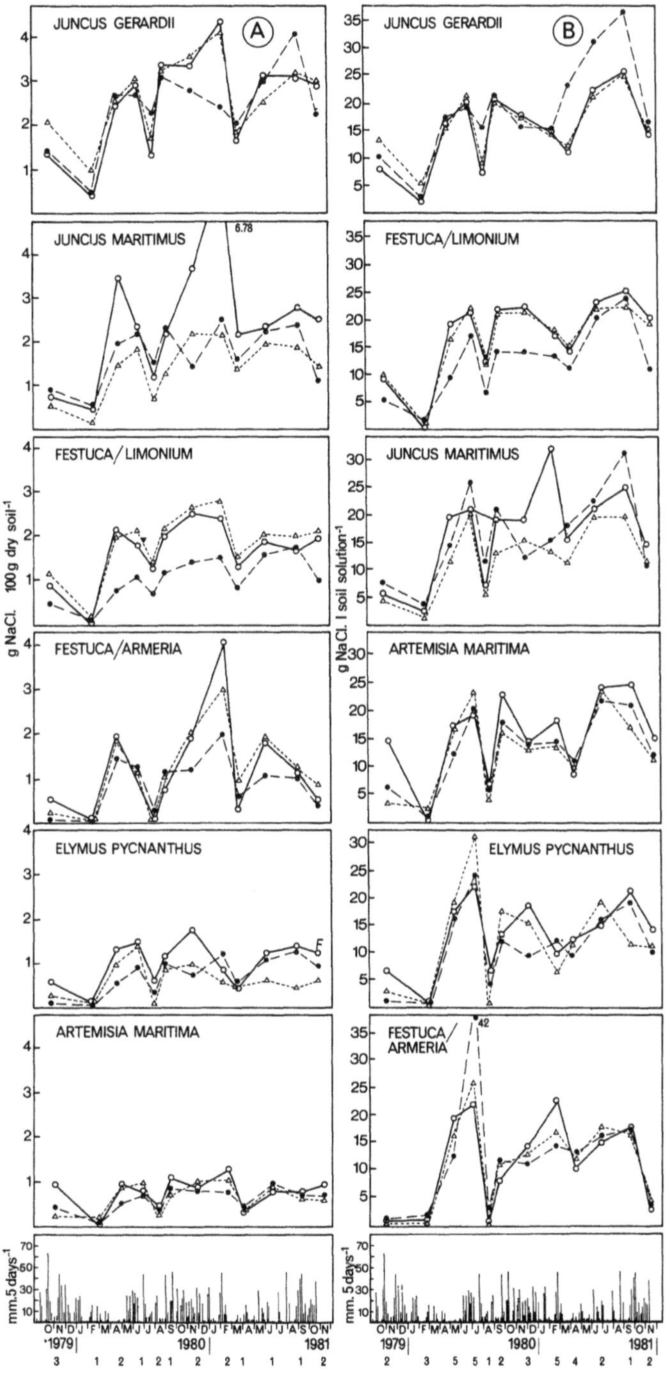

Fig. X.11A–B. Top-soil salinity (0–5 cm) in mixed samples from 1 m^2 from October 1979 until November 1981 in various salt marsh plant communities A: salinity in g NaCl · 100 g dry soil^{-1}, B: salinity in g NaCl · l soil solution^{-1}. Solid lines: abandoning, broken lines: grazing, dotted lines: mowing. Precipitation is given in mm per five-days period at Schiermonnikoog. Correspondence of the actual sequences in soil salinity with the given sequence from top to bottom is given for each sample date. 1: $P < 0.005$, 2: $P < 0.01$, 3: $P<0.002$, 4: $P<0.05$, 5: not significant in Spearman's rank correlation test (after Bakker, 1985).

Table X.8. Sequence of top-soil salinities (0–5 cm) with different management practices (Abandoning, Mowing, Grazing) per community according to the sign test ($P<0.05$), different symbols indicate significant differences. 1: salinity as gram NaCl per 100 gram dry soil, 2: salinity as gram NaCl per liter soil solution.

Juncus gerardii		Festuca/Limonium		Artemisia maritima		Juncus maritimus		Elymus pycnanthus		Festuca/Armeria	
1	2	1	2	1	2	1	2	1	2	1	2
M a	G a	M a	M a	M a	A a	A a	A a	A a	A a	M[2] a	G a
G[1] b	A b	A b	A a	G b	G b	G b	G b	G b	M b	A b	M a
A b	M b	G c	G b	A ab	M c	M c	M c	M b	G c	G c	A b

[1] Only in autumn/winter, [2] only in summer/autumn (absolute figures in Fig. X.11) (after Bakker, 1985).

moisture content at all four sites was much lower in the dry summer period than in the wet spring. The more organic matter found in the soil, the greater the quantity of water retained in the dry summer months.

The ground water table was clearly related to elevation along the gradient and salinity was increasingly higher from the dune to the salt marsh in all layers (Fig. X.13). The uppermost cm showed the greatest fluctuations (high values after the dry summer of 1982, lower ones after rainy periods in October, an increase after inundation following gales in December, a decrease after the very wet spring in 1983 and an increase after the dry summer). The salinity in the clay layer, found at lower soil levels, showed similar but smaller fluctuations. Little fluctuation was found in the lowest layers which were studied. There was no constant difference in salt concentration between grazed and abandoned salt marsh except for the uppermost cm of soil in the lowest part of the gradient, in which salinity was greater in the grazed than in the aban-

Table X.9. Top-soil salinities in grazed salt marshes and embanked areas compared to abandoning (+: higher, −: lower, =: equal). [1] No chemical analysis, but observation by eye: white salt crust in grazed areas attributed to capillary rise. [2] Mowing experiment.

Plant community / dominant species	Area S: salt marsh E: embankment	Soil layer cm	Period	Soil salinity in grazed areas compared to abandoning	Assessment
Juncus gerardii	S. The Netherlands/	0–5	two years	−	this study
Juncus maritimus	S. Waddensea	0–5	two years	−	this study
Festuca rubra/Limonium	S. Waddensea	0–5	two years	−	this study
Festuca/Armeria maritima	S. Waddensea	0–5	two years	−	this study
Elymus pycnanthus	S. Waddensea	0–5	two years	−	this study
Artemisia maritima	S. Waddensea	0–5	two years	−	this study
Ammophila arenaria	S. Waddensea	0–1	one year	=	this study
Festuca rubra/Armeria	S. Waddensea	0–1	one year	=	this study
Elymus pycnanthus	S. Waddensea	0–1	one year	=	this study
Artemisia maritima	S. Waddensea	0–1	one year	+	this study
Puccinellia maritima	S. Germany/Waddensea	0–10	April–December	+	Hansen 1982
Puccinellia maritima	S. Germany/Baltic	0–5	three years	+	Schmeisky 1974
Juncus gerardii/F. rubra	S. Germany/Baltic	0–5	three years	+	Schmeisky 1974
Juncus gerardii	S. Finland/Baltic	[1]	one moment	+	Siira 1970
Juncus gerardii	S. Finland/Baltic	[1]	one moment	+	Kauppi 1967
Festuca rubra	E. The Netherlands	?	one moment	+	Westhoff & Sykora 1979
Salicornia ssp.	E. The Netherlands	0–5	July–October	+[2]	Joenje 1978

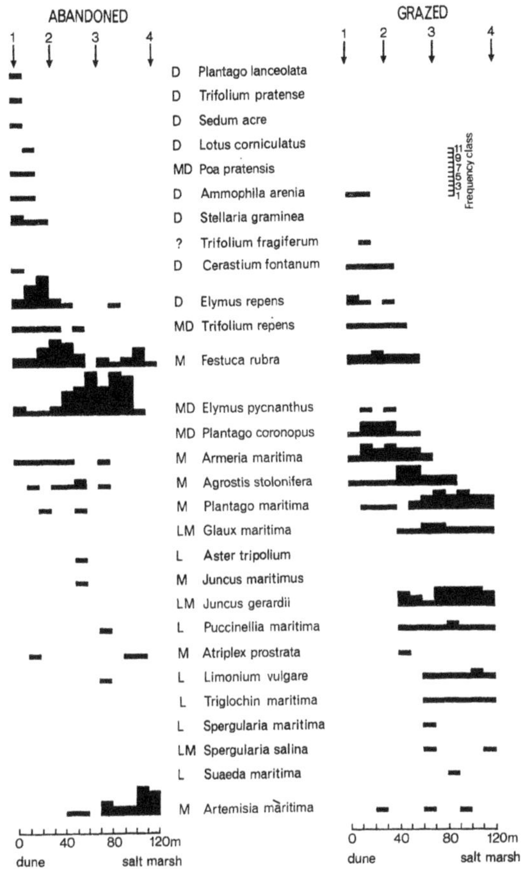

Fig. X.12. Occurrence of species at the grazed and abandoned area in 10 m sections along a gradient from a low dune (1) to the lower salt marsh (4). Frequency classes were derived from point quadrat analysis: class 1 = 1 to 9 hits, class 2: 10 to 19 hits etc. D: dune species, M: mid- and upper salt marsh species, L: lower salt marsh species, ?: distribution unknown (after Bakker et al., 1985).

doned area (sign-test; $P<0.05$). The differences in salinity found along the gradient were in agreement with Rozema's (1976) results. He also found topsoil (0.5 cm) salinity gradients in summer and autumn which corresponded with vegetation zoning from a beach plain towards a 2 m high dune at Schiermonnikoog. The water content of the dune was low compared to the sites lower down the gradient. Rozema found that the higher soil surface was too high above groundwater level for capillary rise during the summer period. A capillary rise of 65 to 100 cm could be expected at an evaporation of $1\,mm \cdot d^{-1}$ in moderately fine sand (Schlichting & Blume, 1966). It is, therefore, probable that water from the water lens beneath the gradient could not reach the topsoil at site 1. This could explain the very low moisture content and soil salinities at the top of the gradient studied during the summer period. There is evidence that soil compaction due to trampling enhances capillary rise (Ijjas, 1969) and this effect probably explains the high moisture content in the upper part of the study area during the spring. Lower down the gradient Rozema (1976) found capillary rise of saline groundwater and evaporation and, therefore, salt accumulation in the uppermost soil and during the summer period, which is in agreement with the results of the present study.

X.C.3. Higher salinity in grazed salt marshes?

Several authors mentioned higher topsoil salinities in grazed or mown salt marshes or embanked sand flats than in abandoned areas (Table X.9). This is not in agreement with the present study, which over a two year period showed that topsoil salinity (in bulk soil samples 0 to 5 cm) was not higher in grazed than in abandoned areas. Only the soil salinity of the uppermost 1 cm was higher temporarily and locally in grazed than in abandoned areas.

Gray & Scott (1977) showed that above the main zone of distribution *Puccinellia maritima* was restricted to local hollows (hoofmarks?) or areas opened up by turf cutting. Driftline material can cover the established vegetation which then dies off and leaves gaps. Species which normally occur lower down the salt marsh can temporarily be found higher on the salt marsh in these gaps (Bertness & Ellison, 1987). This phenomenon seems of general importance. The larger species diversity, which is often established and higher percentage of lower salt marsh species with grazing than with abandoning, can probably be attributed to the short and open canopy with patches of bare soil throughout the year, provided that accessibility is no problem. The role of the canopy structure will be discussed later in this chapter.

X.D. Structure of the sward

The light interception by the canopy under different management practices will first be considered in this section. Secondly, the influence of the canopy structure on the emergence and survival of seedlings from sown seeds is discussed.

X.D.1. Light interception

The structure of the sward is quantified by measurements of light interception, amount of bare soil and litter accumulated. The figures presented in section IX.B are estimates of the above-ground standing crop present at the time of cutting in different regimes in the Loefvledder study area. They might be regarded as an impression of the net herbage accumulation of the site including both production and decomposition. The net herbage accumulation in the *Agrostis stolonifera/Poa pratensis* community appears to be similar with July hay-making, September hay-making and both July and September hay-making (Fig. IX.8 A, C, G).

The structure or the density of the canopy differed considerably between management practices in 1982 (Fig. X.14). The canopy density, expressed as a percentage non-visible points at different heights, increased during the spring and early summer at all management sites. It had, however, already reached its maximal value at the abandoned site in June. The vertical view through the vegetation was higher in the September and in both the July and September hay-making regimes than in the other re-

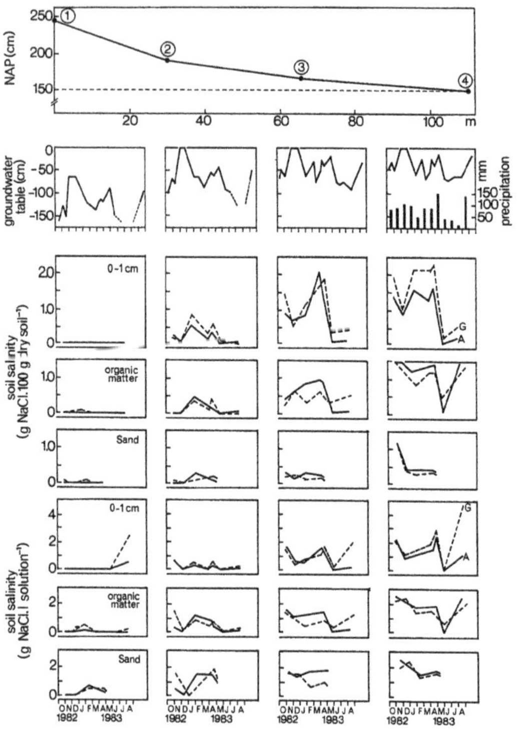

Fig. X.13. Lay-out of the gradient from a low dune (1) to the lower salt marsh (4), groundwater table (including monthly precipitation), soil salinity (g NaCl · 100 g dry soil^{-1} and g NaCl · l solution^{-1}) in the topsoil (0–1 cm), the humus-rich clay layer and the underlying sand layer in grazed (G) and abandoned (A) areas (after Bakker et al., 1985).

Table X.10. Pore volume (%) and organic matter (%) in the 0–1 cm layer of soil and in the upper part of the sand layer, and moisture content in the upper cm of soil and the clay layer rich in organic matter in April and July on the four study sites (Fig. X.13) in abandoned (A) and grazed (G) areas (after Bakker et al., 1985).

	Site 1		Site 2		Site 3		Site 4	
	A	G	A	G	A	G	A	G
Pore volume (%)								
0–1 cm	47	42	71	60	62	45	62	41
sand	43	42	42	41	41	40	39	40
Organic matter (%)								
0–1 cm	1	11	6	9	22	21	20	20
sand	0.9	0.7	0.6	0.7	1.4	1.0	1.6	3.4
Moisture content (g water · 100 g dry soil^{-1})								
24.4.83 0–1 cm	13	75	85	135	150	130	95	95
organic matter	10	26	21	43	105	85	110	85
15.7.83 0–1 cm	0.1	0.1	29	15	32	20	13	11
organic matter	0.1	0.5	11	11	30	45	55	60

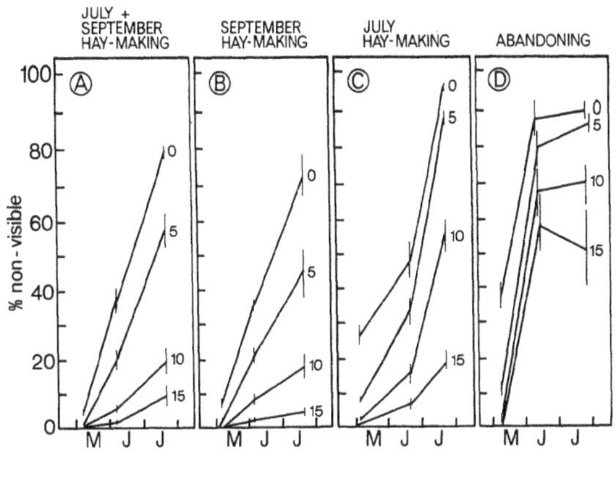

Fig. X.14A–D. Average density (± S.E.) (percentage non-visible points out of 100 dots on a calibrated stick of 1 m) at ten randomly chosen sites at various levels above soil surface in different management regimes in the *Agrostis stolonifera/Poa pratensis* community in the recent reserve of the Loefvledder area (after Bakker & de Vries, 1985B).

gimes in May. Moreover, the percentage of litter coverage differed considerably among the management regimes (Fig. X.15). A relatively low coverage of litter was found in winter and early spring on plots cut in both July and September (0–40%) and in September only (20–50%), as opposed to July cutting (30–80%), and uncut plots (70–95%) (Photo X.1).

A continual decrease in light during the growing season was also found by Campino Johnson (1978) in an *Arrhenatheretum* in Germany. He, however, mentioned a higher light interception in a site where the hay was cut twice than in an abandoned site which is not in agreement with the present results. This discrepancy might be attributed to artificial fertilizer applications in the cut area with probably subsequent rapid growth and the fact that the site was only left abandoned for two years which was little time to accumulate litter which enhances light interception.

The heavily grazed areas in the *Holcus lanatus* community in the Westerholt study area showed less litter and more bare soil than the lightly grazed patches but more living plants (Table X.11). Moreover, the soil in lightly grazed patches received c. 20% full daylight, whereas the soil in the heavily grazed areas received nearly 100% (Fig. X.16).

The grazed areas along the salt marsh gradient discussed in section X.C.2, featured less litter, more bare soil and more light reaching the soil than the abandoned areas (Table X.12).

X.D.2. Plant responses

The structure of the sward, as discussed in the previous section, is expected to affect the ability of species to spread or to invade. The fate of artificially sown seeds of several species, sown in the *Agrostis stolonifera/Poa pratensis* community in the Loefvledder study area in November 1977, was for example investigated with different management practices. Although seedlings emerged in all regimes, the emergence and survival to juvenile and flowering individuals was highest in both the September and twice cut plots and lowest in the uncut plots (Fig. X.17). Since a 'poor' species (*Rhinanthus angustifolius*) and a 'rich' species (*Bromus*

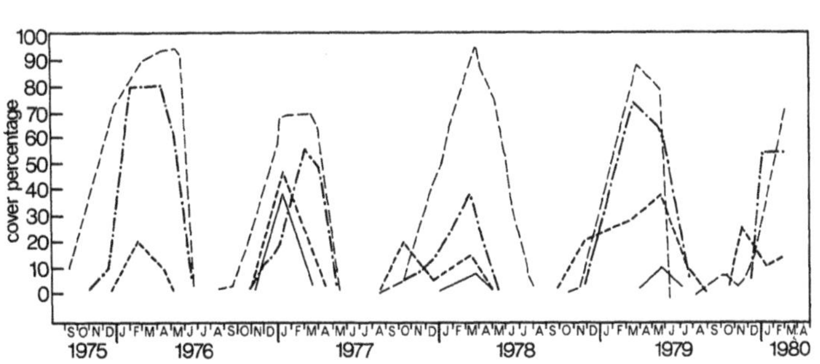

Fig. X.15. Cover percentage of litter in the period 1975–1980 in different management regimes in the *Agrostis stolonifera/Poa pratensis* community in the recent reserve of the Loefvledder area. Solid line: July and September hay-making; thick broken line: September hay-making; thin broken line: abandoned; broken line with dots: July hay-making (after Bakker et al., 1980).

Photo X.1. The recent reserve of the Loefvledder study area in May 1984. Dead standing crop is lightly coloured and decreases in the sequence abandoned (left), cut in July (third site from right), cut in September (middle) and cut in both July and September (right).

Table X.11. Structural characteristics (± S.E.) of heavily and lightly grazed patches in a *Holcus lanatus* community in May and October 1983: percent cover estimated by point quadrat analysis (* P<0.05 according to Student test) (after Hendriks, 1984).

	Heavily grazed	Lightly grazed
May 1983		
living plants	78.1 ± 1.2	68.1 ± 1.9*
mosses	13.4 ± 1.7	6.9 ± 1.9*
litter	11.4 ± 1.1	92.4 ± 0.8*
bare soil	5.2 ± 1.3	0.3 ± 0.2*
October 1983		
living plants	63.2 ± 2.1	15.4 ± 0.9*
mosses	6.1 ± 0.7	5.5 ± 0.4
litter	43.2 ± 2.7	97.4 ± 0.7*
bare soil	4.7 ± 0.8	1.4 ± 0.4*

hordaeceus) reacted in a similar way to cutting regimes it is likely that the structure of the sward plays a more important role than the nutrient status of the soil in this experiment.

Grazing intensity also influences the emergence and the fate of the seedlings. When seed was sown in the *Holcus lanatus* community in the Westerholt study area in January 1983, more seedlings emerged and survived until October in the heavily grazed areas with a short turf than in the lightly grazed patches (Table X.13). Seedlings of some species, however, also disappeared in the heavily grazed area.

On 15th March 1983 seeds of two salt marsh species were sown along a gradient from a small dune down to the salt marsh. The lower salt marsh species hardly emerged in the dune area, but more seedlings were found lower down the gradient (Fig.

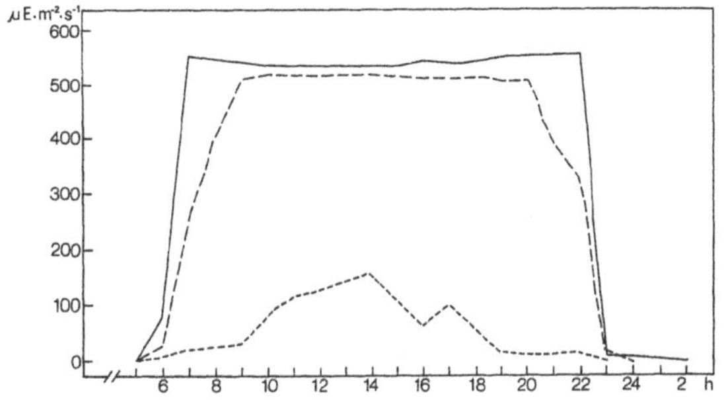

Fig. X.16. Light reaching the soil in heavily grazed areas (broken line) and lightly grazed patches (dotted line) in the Westerholt area in July 1983 during an unclouded 24 h period, compared to the light intensity above the canopy (solid line) (after Hendriks, 1984).

X.18). More seedlings emerged in the grazed than in the abandoned area, but most of those which emerged, died.

In all three study areas the afore-mentioned relationship of seedling emergence and survival with the differences in the structure of the sward is obvious. Litter accumulation can prevent seedling establishment (Silvertown, 1980; Willems, 1983A), but can also increase microbial attack of emerged seedlings (Grime, 1979) and cause desiccation before their roots can reach the soil (Grubb, 1977). In this study rosette plant species such as *Hypochaeris radicata* as well as *Taraxacum* sp., *Leontodon autumnalis, Bellis perennis* and *Plantago major,* in particular, establish better in heavily grazed areas. These species have a higher frequency percentage in a grazed area than in a hay-field (Elberse et al., 1983). It can be assumed that the short turf in the closely grazed sward provided more safe sites for these rosette plants than the hay-making regime with its taller canopy. Indeed Blom (1977), Harper (1977), Mølgaard (1977), Fenner (1978) and Schenkeveld & Verkaar (1984) also demonstrated a better development of *Plantago major, Bellis perennis, Taraxacum* sp., *Hypochaeris radicata* and chalk grassland species, respectively, in a short turf than in a dense sward. The experimental creation of open patches in a dense sward or litter layer also favoured seedling establishment of *Rhinanthus angustifolius* (De Hullu, 1985), *Solidago* spp. (Goldberg & Werner, 1983) and several biennial species (Gross & Werner, 1982).

The emergence and survival of seedlings is positively correlated to the percentages of full sunlight reaching the soil surface (Mølgaard, 1977; Goldberg & Werner, 1983; Schenkeveld & Verkaar, 1984). The leaf canopy not only changes the quantity of light but also its quality at soil level. The lower red: far red ratio under a plant canopy can reduce the emergence and seedling establishment of various species (Silvertown, 1980).

X.E. Synthesis

The nutrient status of the soil is considered to be very important in determining the composition of inland plant communities. The nutrient status of the soil is expected to decline after the application of a fertilizer is ceased. The process is speeded up when nutrients are removed in the hay crop. This would lead to a lower above-ground standing crop and hence a more open structure of the sward, facilitating the establishment of weak competitors

Table X.12. Area of bare soil, thickness of litter layer and the amount of light reaching the soil on grazed (G) and abandoned (A) areas along the gradient (Fig. X.13) (* $P < 0.002$ according to Fischer's test) (after Bakker et al., 1985).

	Site 1		Site 2		Site 3		Site 4	
	G	A	G	A	G	A	G	A
Bare soil (number of 306 hits)	27	25	86	8*	178	16*	223	84*
Litter layer (cm) n = 5	1.8	4.6*	1.0	6.6*	0.6	13.2*	0.3	1.8*
Light reaching soil (April)	100	28*	100	34*	100	21*	100	60*
(% of grazed) n = 5 (June)	100	40*	100	23*	100	21*	100	15*

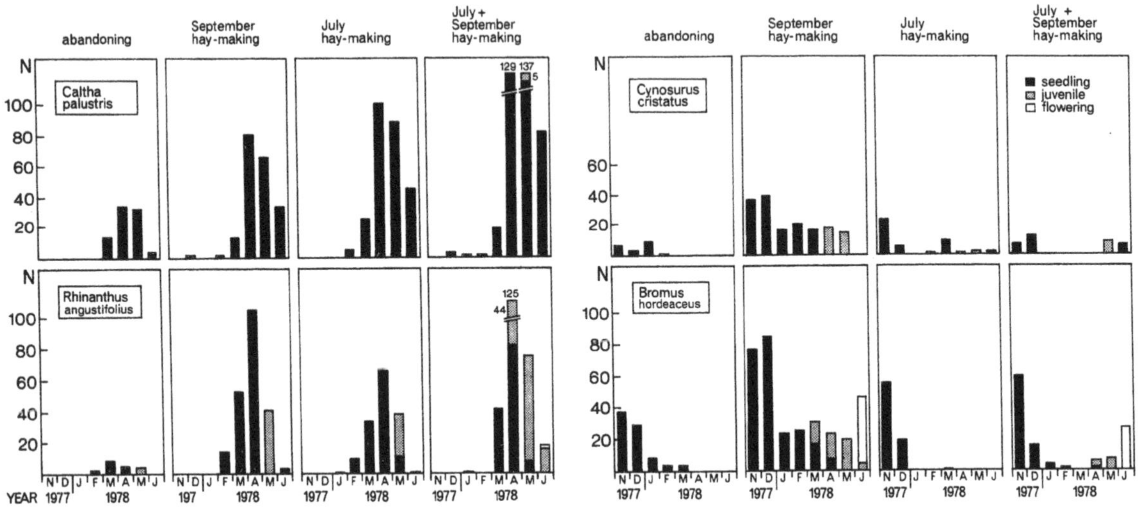

Fig. X.17. The effect of cutting regimes on the total number of seedlings in 1978, when sown in November 1977. The hay-making was first applied in 1973 and the plots had received no fertilizer applications since then (after Bakker et al., 1980).

(slowly growing species) which are often species adapted to nutrient-poor soil conditions. In this, and several other published studies, the above-ground standing crop usually decreased, but the yields were strongly influenced by the weather conditions. There was little relationship between current mineral removal and the yield of standing crop. In general, the removal of minerals was very low, compared to the mineral pool in the soil and mineral inputs by precipitation were usually comparable to mineral removal. Moreover, the strong fluctuations in an above-ground standing crop suggest the influences of weather conditions upon mineralization rates of dying plant material and litter.

The mineralization rate is likely to be a good measure of nutrient availability in fen communities (Vermeer & Berendse, 1983) and is positively correlated to the above-ground production in heathland communities (Berendse, 1986). It is, however, difficult to measure and likely to be affected by other factors such as drainage of peaty soils. The vegetation itself plays an important role in these processes. Grootjans et al. (1985) pointed out that a lowering of the groundwater table in a poorly fertilized *Calthion palustris* hay-field resulted in high N-mineralization of the fen peat soil. Most of

Table X.13. The seedlings emerged and survived until October 1983 after sowing 150 seeds of each species in January of the same year in heavily grazed areas and lightly grazed patches (after Hendriks, 1984).

	Heavily grazed		Lightly grazed	
	Emergence	Early establishment	Emergence	Early establishment
Nardus stricta	17	13	4	0
Narthecium ossifragum	10	0	2	0
Danthonia decumbens	16	12	0	0
Gentiana pneumonanthe	14	0	0	0
Potentilla erecta	0	0	1	1
Succisa pratensis	29	24	34	12

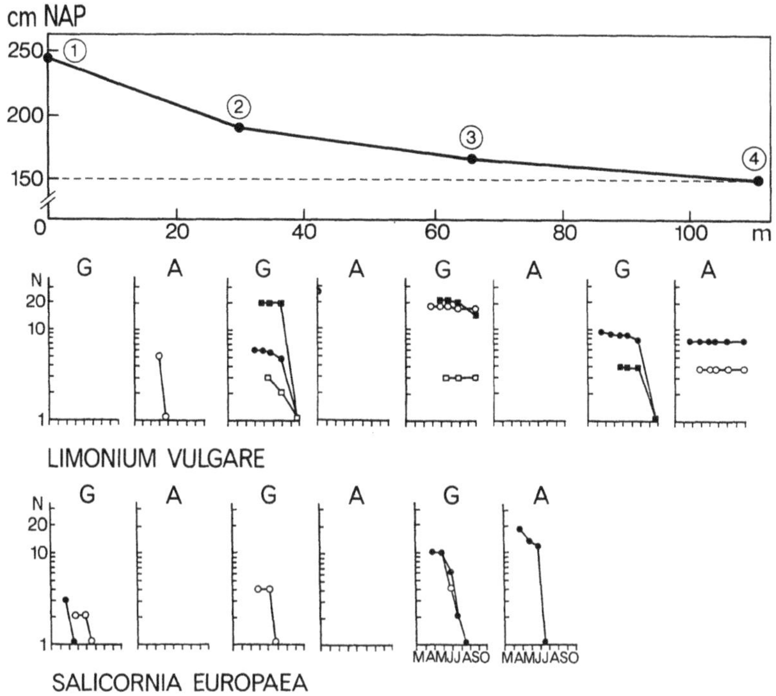

Fig. X.18. Survivorship curves, log (N+ 1), of cohorts of seedlings of *Limonium vulgare* and *Salicornia europaea* in grazed (G) and abandoned (A) areas along a transect from a low dune (1) to the lower salt marsh (4) (after Bakker et al., 1985).

the mineralized NO_3^- was absorbed by plant roots with a subsequently increased yield. Lowering of the groundwater table in unfertilized *Cirsio-Molinietum,* also resulted in high N-mineralization of the fen peat, but as the NO_3^- which was produced was not absorbed by the plant roots, no vegetational response could be recorded (Grootjans et al., 1986). Mineralization of the top soil with a subsequent decrease of the poor to rich species ratio might have been related to the superficially drainage in some lots of the Anlo study area (section VI.D.1.c).

Changes of the hydrological conditions with a resulting higher groundwater table can also change the vegetation. If the influence of inundation water increases and that of seepage water decreases eutrophication can take place, despite a higher groundwater table and hay-making without fertilizer applications (section VI.D.3.a). Similar effects were recorded in the Stroomdallandschap Drentsche A 3 km downstream (Bakker et al., 1987).

The effects of soil on plant species and vegetation types is usually discussed. Miles (1985, 1987) stressed the fact that vegetation influences soil properties and hence soil productivity. In the course of a few years changing patterns of vegetation will cause correspondingly changed patterns of labile soil properties, such as pH. The soil pH largely controls the C/N ratio of litter and humus and thus the rates at which these decompose, and, for example, at which nitrogen is mineralized. Because of selective grazing at low stocking rates, the structure of grassland swards tends in time to become progressively heterogeneous. This results in heavily grazed areas and lightly grazed patches. Such relatively ungrazed patches tend to develop a mat of decaying surface organic matter. Kirkwood (1964) found considerably contrasting surface soil properties which had developed over a period of 15 years under areas with and without such mats.

Pegtel (1987) discussed the measurement of soil chemical properties in order to describe changes in soil fertility as a result of changes in management. He elucidated the complexity of soil chemistry dealing with seasonal changes, spatial differences, interactions between roots and soil, inter- en intra-specific interaction of plants and concluded that soil chemical analyses, particularly in mesotrophic and oligotrophic semi-natural communities, can hardly be interpreted for ecological purposes. He argued that a better method of assessing changing soil fertility is to be found in using undisturbed turfs or seedling stands of individual plants for bio-assays in the greenhouse. It was found that the productivity of soil samples from twice cut sites was slightly, but consistently, lower than that of abandoned sites from which no nutrients were removed in hay and soil fertility had, therefore, decreased slightly.

There is no immediate set-back in the above-ground yield after the cessation of fertilizer applications. This is in accordance with the fact that the nutrient pool in the soil seems fairly comparable to that of valley grasslands where fertilization was continued (Table X.6). De Hullu (1985) did not find higher N, P and K_{total} values for fertilized soil than for soils in which *Rhinanthus angustifolius* had invaded after the cessation of fertilizer applications. This can also be seen from Table II.2. The amounts of nutrients removed with hay are such a small fraction of the pools present that the latter can render enough nutrients for high yields for a number of years. It is not known why the root biomass shows a marked decline a few years after the cessation of fertilizer applications. It is now clear that both nutrient removal and artificial fertilizer additions (c. 250 kg $N \cdot ha^{-1}$, c. 100 kg $P \cdot ha^{-1}$ and c. 100 kg $K \cdot ha^{-1}$ in agricultural practices) are marginal with respect to the nutrient pool in the soil with the probable exception of very nutrient-poor soils. An eventual decrease of standing crop after the cessation of fertilizer applications and often large year-to-year fluctuations can probably be attributed to changes in the mineralization rates of organic material, which were not, however, measured in the present study. It can be concluded that the nutrient pool can only be effectively depleted by sod removal.

Mineral removal by grazing or hay-making without fertilizer applications seems to be a very slow process. Wells (1980) suggested this on the basis of results from chalk grassland experiments. Nevertheless the species-diversity increased a little in this and several other studies and the proportion of species indicating nutrient-poor soil conditions increased relatively quickly. The performance of these nutrient-poor species depended on safe sites created in the sward, in which the timing and frequency of the removal of the above-ground standing crop was very important, provided the seeds of species expected to establish were present.

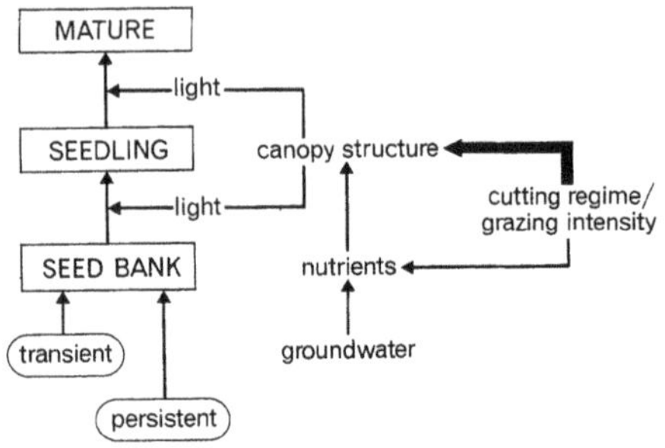

Fig. X.19. Diagrammatic representation of the environmental factors influencing the establishment of new species. The thickness of the arrow indicates the importance of the effect (after Bakker, 1987).

It can, finally, be concluded that the changes caused by the start of hay-making or grazing without fertilizer application can be measured by a slow overall increase in species-diversity and a quicker increase of species indicating nutrient-poor soil conditions. These responses are mainly due to changes in the structure in the canopy as a result of the timing and frequency of above-ground standing crop removal and hardly to impoverishment of the soil by nutrient removal which seems to be a long-term process (Fig. X.19). The same conclusion was reached by Oomes & Mooi (1985). The temporarily increase of species indicating nutrient-rich soil conditions, despite hay-making without fertilizer application, might also be attributed to changes in the structure of the sward and the emergence of bare soil. Hülbusch (1986) also suggested that open space on the soil might be more important than its nutrient status. After 15 years the imposition of a new management has not yet resulted in other constant plant communities, since the vegetational succession is still going on, indicating that an equilibrium has not as yet been established with the new management practices.

The nutrient status of the soil can strongly be influenced by lowering of the groundwater table and subsequent mineralization of the top-soil. The species composition also depends on the quality of the groundwater, i.e. the proportion of groundwater with a short period in the soil, deep groundwater with a long period in the soil and inundation water from the brook. It can, therefore, be concluded that management practices inside a nature reserve can seriously be countered by hydrological changes outside a reserve even with slightly higher groundwater tables.

Soil salinity is a very important factor in determining the composition of coastal plant communities and the occurrence of halophytes. Evidence suggests, that, although capable of germinating on the lower salt marsh until $10\,gNaCl \cdot l^{-1}$, glycophyte seedlings are not able to compete with the surrounding halophytic species. Despite the germination of seedlings of halophytes in glycophytic vegetation, halophytes are poor competitors on the abandoned salt marsh. The fate of the seedlings of any particular species, therefore, corresponds with the place of that taxon in the vegetation along the gradient of halophytes and glycophytes, provided the seeds of that particular species were present.

Summarizing the results, it appeared that seed dispersal and germination responses to temperature and chilling cannot be the limiting factors in the establishment of salt marsh and dune species in the study area. Grazing in the abandoned area enhances the number of species by removing the standing crop and accumulated litter. Upper salt marsh and dune species cannot spread down towards the lower salt marsh because of their low salt tolerance. Lower salt marsh species can, however, and do invade the upper salt marsh and dunes since they grow very well in less saline environments provided the light is not intercepted by the canopy. A similar picture emerges from other studies, e.g. an experimental field study of plant zonation in an Alaskan salt marsh, where many species could potentially survive, revealed that species occurring in zones along a physical gradient are limited by physiological tolerance towards one end of the gradient, and by competitive ability towards the other (Snow & Vince, 1984). Bertness & Ellison (1987) also carried out transplantation experiments and reached a similar conclusion. They found that the potential plant success can be strongly interfered by consumers which can reduce the seed set of some species in the stand and hence their interference ability.

PART FIVE

Population dynamics

CHAPTER XI

Seed dispersal and seed bank

XI.A. Introduction

In part three, on vegetation dynamics, it appeared that many species have become newly established in permanent plots in the study areas. Newly established species in a lot can have been present in the seed bank in the soil before emergence. This is only the case for species with a persistent seed bank sensu Thompson & Grime (1979). Some of the new species were certainly present in the borders or along ditches, where they had found a refugium as a result of agricultural practices or after abandoning. These individuals can act as a source population from which species can spread again after changes in the management regimes aimed at the restoration of earlier plant communities. Many newly established species were, however, not present in or near the actual stands in the study areas at the start of the management practices.

The dispersal of diaspores of chalk grassland species barely exceeds 2 m (Verkaar et al. 1983). Even the dispersal of seeds from anemochorous species does not always cover large distances and most of the seeds of some of those species land within the neighbourhood of the parent plants (Harper, 1977; Ter Borg, 1979). Ninety percent of the seeds in untouched plants of non-anemochorous *Rhinanthus angustifolius* are dispersed within a radius of less than 25 cm. Mowing during seed ripening resulted in a dispersal of over 2 m and the hay-making process added another 6 to 7 m sometimes in the same direction. The greatest dispersal distances which were observed were, therefore, 8 to 9 m (Ter Borg, 1985). Birds and grazing animals are another means by which dispersal distances might be enlarged, within lots, through their faeces and wool. Agnew & Flux (1970) found that the pelt of a hare contained seeds of over 100 species. It is evident that it takes a long time for many species to invade a new lot. This was already indicated in section IX.C.3 where it was noted that the developing vegetation types contained less species than was expected from the low above-ground standing crop. Hay-making machinery may play a role in this process.

This chapter considers the questions (i) to what extent are seeds of different species dispersed by the faeces and wool of grazing animals and by hay-making machinery and (ii) what is the composition of the seed bank beneath the plant communities in the study areas and to what extent can persistent and transient seed banks be discerned?

XI.B. Seed dispersal

XI.B.1. Dispersal in the cattle- and geese-grazed salt marsh (Schiermonnikoog)

The presence and number of viable seeds in cattle dung patches and geese pellets, which were collected near the top of a small dune on the salt marsh, were investigated in November 1982, using samples each of which were 20 cm^3. The same procedure was repeated in March 1984 for driftline material, which was deposited near the top of a small dune covered with *Festuca rubra, Armeria maritima, Plantago coronopus* and *Ammophila arenaria* (see site 1 in Fig. X.12).

The species of which viable seeds were found in driftline material and in geese pellets and dung patches are listed in Table XI.1. All the lower and some mid salt marsh species which were in the driftline seed bank were not found in the established vegetation of the site. Waisel (1972) reported seeds of *Halimione portulacoides, Limonium vulgare, Puccinellia maritima, Salicornia* spp., *Suaeda maritima, Triglochin maritima, Elymus pycnanthus* and *Atriplex prostrata,* in driftline material in the United Kingdom but their viability was not tested. Koutstaal et al. (1987) found that the floating capacity in basins of several halophytic species amounted to weeks and that most species did not lose their germination capacity. Geese pellets contained many seeds of *Salicornia europaea* and *Juncus gerardii.* Dung patches contained many seeds of *Juncus gerardii, Spergularia salina* and *Agrostis stolonifera.*

Table XI.1. Mean number of germinable seeds ± S.E. in the seed bank of 20 cm^3 driftline material (n = 10), geese pellets (n = 5) and dung patches (n = 5). D: dune species; M: mid- and upper salt-marsh species, L: lower salt-marsh species (according to Bakker & Ruyter, 1981) (after Bakker et al., 1985).

		Driftline material	Geese pellets	Dung patches
D	Plantago lanceolata	1.0 ± 0.5		
D	Rumex crispus	0.5 ± 0.3		
D	Stellaria graminea	0.3 ± 0.2		
D	Bromus hordeaceus hordeaceus	0.5 ± 0.3		
MD	Elymus pycnanthus	74.0 ± 22.0		
MD	Poa pratensis	1.6 ± 0.6		0.6 ± 0.4
MD	Plantago coronopus	5.4 ± 1.2		
?	Polygonum aviculare	1.1 ± 0.5		
M	Atriplex prostrata	1.6 ± 0.9		
M	Agrostis stolonifera	5.4 ± 1.4		2.6 ± 0.6
M	Festuca rubra	132.2 ± 23.4		0.6 ± 0.4
M	Artemisia maritima	5.7 ± 1.5	0.2 ± 0.2	
M	Potentilla anserina	4.3 ± 1.6		
M	Plantago maritima	2.5 ± 0.8		
M	Armeria maritima	0.8 ± 0.4		
LM	Juncus gerardii	22.5 ± 5.5	0.6 ± 0.3	9.4 ± 2.1
LM	Spergularia salina	0.3 ± 0.2		3.4 ± 0.8
L	Limonium vulgare	4.7 ± 0.1		
L	Suaeda maritima	1.2 ± 0.4		
L	Triglochin maritima	0.4 ± 0.2		
L	Salicornia europaea	0.5 ± 0.3	2.2 ± 0.4	

Addenda: species occurring in the seed bank in one or two subsamples in the driftline- Rumex acetosella (D), Cerastium fontanum (D), Carex distans (M), Glaux maritima (LM), Spergularia maritima (L), Mentha aquatica (?), Poa annua (?), Poa trivialis (?), Plantago major (?), Lolium perenne (?).

Seeds of a number of lower salt marsh species could have been transported to the higher salt marsh and dunes by tidal waters. Dijkstra (1984) stated that only 5% of deposited seeds of a number of species remained in an open *Puccinellia maritima* vegetation after two tidal floodings. It is not known whether or not seeds are removed by tidal action from within a closed canopy. Koutstaal (pers. comm.) established that considerably more seeds are transported by the flood than by the ebb stream. A net transport towards the upper salt marsh and dunes can, therefore, be expected. The occurrence of viable seeds of mid and lower salt marsh species in driftline material deposited on the dunes also suggests that seeds of this group of species are more easily transported up the gradient. One of the ecological roles of driftlines appears to be the concentration of masses of seeds of many species in restricted areas.

Geese pellets and dung patches were also found in driftline material, but most of these remained after being deposited in the grazed area. They may, in part, account for the spread of *Spergularia salina, Juncus gerardii, Agrostis stolonifera* and *Salicornia europaea*. De Vries (1961) found viable seeds of *Salicornia europaea, Glaux maritima* and *Trifolium fragiferum* in duck pellets in the Netherlands. Siira (1970) reported that the seeds of *Juncus gerardii, Agrostis stolonifera* and *Plantago maritima* retained their capacity to germinate after passing through the alimentary canal of the mallard duck in Finland. Cattle dung in Switzerland contained *Agrostis capillaris, Poa annua, P. pratensis, Lolium perenne, Festuca rubra, Trifolium pratense, T. repens, Prunella vulgaris, Plantago lanceolata, P. major* and *Luzula campestris* (Müller, 1955) which also occur on the higher salt marsh and small dunes of the study area. Boeker (1959) recorded in Germany *Trifolium repens, T. pratense, Poa annua, P. pratensis, Agrostis capillaris* and *Festuca rubra* in cattle dung.

XI.B.2. Dispersal in the sheep-grazed mixed grassland (Westerholt)

The presence of viable seeds in sheep pellets and wool fragments was detected in July and November 1975. The species of which viable seeds were found in the grassland and heathland area are shown in Table XI.2. It is clear that viable seeds of species which occur in the grassland area may have been transported to the heathland area via dung pellets and/or wool fragments. Seeds of those species were also dispersed within the grassland area. Only the seeds of *Calluna vulgaris* were transported from the heathland area into the grassland area. Welch (1985) recorded that many seeds of *Agrostis capillaris, Cerastium fontanum, Poa pratensis, Rumex acetosella, Sagina procumbens,* but also of *Calluna vulgaris* had been dispersed in dung of catlle, sheep and non-domestic herbivores in heather moorland in Scotland. He also found that non-heathland species invaded and spread into the moorland.

Viable seeds of *Plantago lanceolata, Rumex acetosa* and *R. acetosella* were found in sheep droppings by Müller (1955). In Canada Dore & Raymond (1942) recorded viable seeds of 49 species in cattle manure, of which *Agrostis stolonifera, Phleum pratense, Poa pratensis, Trifolium repens, Chenopodium album, Poa compressa, Cerastium fontanum, Plantago major, Veronica serpyllifolia, Ranunculus acris, Plantago lanceolata, Rumex acetosella* and Festuca rubra were most abundant. Seed dispersal by grazing domestic animals is also mentioned by Klapp (1965) and Ellenberg (1978). They suggested that the dispersal of seeds of clover and grasses with agricultural value was facilitated by alternatively grazing of already agriculturally

Table XI.2. Occurrence of germinable seeds in sheep dung pellets and wood fragments collected in the grassland and heathland area. *Calluna vulgaris* was found in the heathland area and all other species in the grassland area. (after Vreugdenhil & Van Wieren, 1976).

	Dung		Wool whole area
	Grassland	Heathland	
Cerastium fontanum	+	·	·
Calluna vulgaris	+	·	·
Sagina procumbens	+	·	·
Poa annua	+	·	·
Elymus repens	+	·	·
Lycopus europaeus	·	+	·
Plantago major	+	·	+
Taraxacum spec.	·	+	+
Poa trivialis	+	+	+
Lolium perenne	+	+	+
Bellis perennis	+	+	+
Holcus lanatus	·	·	+
Leontodon autumnalis	·	·	+

improved land and rough grazing areas. The dispersal takes place either by passing through the gut and being deposited in faeces or by being attached to the exterior of the animal namely on the hoofs and pelt.

Bülow-Olsen (1980B) thought that *Pimpinella saxifraga, Campanula rotundifolia, Galium verum, Festuca ovina* and *Agrostis capillaris* had established in a grazed former heathland overgrown with *Deschampsia flexuosa* in Denmark after the cattle had grazed in a species-rich grassland which contained those species. Hillegers (1985) found seeds of the genera *Agrimonia, Carduus, Cirsium, Galium, Lappa* and *Cynoglossum* attached to the long fleece of Mergelland-sheep in the Netherlands. Moreover, he recorded an extension of the area of *Cynoglossum officinale* after three years in a grazed reserve, whereas this was not found in a similar ungrazed reserve. Both Bülow-Olsen (1980B) and Hillegers (1985) suggested to bring large herbivores to species-rich grasslands and afterwards to areas under restoration management in order to facilitate the dispersal of seeds of species with nature conservation interest.

XI.B.3. Dispersal by hay-making machinery in valley grasslands (Anlo)

Ter Borg (1985) found that seeds of *Rhinanthus angustifolius* can cover distances of 8 to 9 m by means of hay-making machinery. Klapp (1965) and Ellenberg (1978) suggested that hay can play a role in the dispersal of seeds, but they gave no details.

The presence and number of viable seeds attached to hay-making machinery in the valley grasslands were investigated in August 1984, using samples of dust, each of which was 20 cm^3. The viable seeds are listed in Table XI.3. On top of the hay-making machinery many seeds of species were found which occur in the initial stage of succession after the cessation of fertilizer applications, namely, *Lolium perenne, Poa pratensis, P. trivialis, Holcus lanatus* and *Dactylis glomerata.* Also seeds of species which do not occur immediately after the cessation of fertilizer applications (see section VI.D) were found, namely, large numbers of *Rhinanthus angustifolius, Anthoxanthum odoratum, Cynosurus cristatus, Plantago lanceolata* and small numbers of *Festuca rubra, Agrostis capillaris, Lychnis flos-cuculi* and *Prunella vulgaris.* It is clear that the process of hay-making can contribute to the dispersal of seeds by machinery which moves from one lot to another. Whether the dispersed

Table XI.3. Mean number of germinable seeds ± S.E. in a sample of 20 cm^3 attached to hay-making machinery (n = 10). Species of the Rhinanthus-group are not present in stands immediately after the cessation of fertilizer applications.

Species	Mean number of germinable seeds ± S.E.
Lolium perenne	50.7 ± 3.4
Poa pratensis	33.6 ± 6.1
Poa trivialis	57.9 ± 5.5
Holcus lanatus	161.4 ± 8.4
Dactylis glomerata	93.3 ± 4.8
Juncus effusus	11.4 ± 2.6
Rumex obtusifolius	3.9 ± 0.8
Urtica dioica	6.3 ± 1.0
Cerastium fontanum	2.7 ± 0.4
Rumex acetosa	12.3 ± 9.3
Festuca pratensis	3.7 ± 2.0
Alopecurus geniculatus	1.7 ± 0.5
Poa annua	0.4 ± 0.3
Bromus hordeaceus	1.5 ± 0.6
Stellaria media	0.4 ± 0.2
Glyceria fluitans	0.7 ± 0.4
Sagina procumbens	0.3 ± 0.2
Ranunculus repens	2.1 ± 0.6
Agrostis stolonifera	1.4 ± 0.6
Rumex crispus	0.6 ± 0.4
Phleum pratense	0.5 ± 0.3
Elymus repens	0.2 ± 0.2
Epilobium tetragonum	0.5 ± 0.2
Taraxacum spec.	0.1 ± 0.1
Cirsium arvense	0.4 ± 0.2
Leontodon autumnalis	0.1 ± 0.1
Rhinanthus angustifolius	54.5 ± 9.4
Anthoxanthum odoratum	7.2 ± 1.2
Cynosurus cristatus	9.4 ± 2.8
Plantago lanceolata	5.7 ± 1.4
Anthriscus sylvestris	19.7 ± 2.9
Festuca rubra	3.6 ± 0.5
Agrostis capillaris	3.7 ± 1.4
Lychnis flos-cuculi	1.2 ± 0.4
Prunella vulgaris	0.4 ± 0.2
Cirsium palustre	0.4 ± 0.2
Hypochaeris radicata	0.1 ± 0.1
Juncus acutiflorus	0.1 ± 0.1
Agrostis canina	0.2 ± 0.1
Carex ovalis	0.2 ± 0.2

species establish or not depends, among other factors, on the density of the sward (see also chapter XII).

XI.C. The seed bank

XI.C.1. The seed bank in a salt marsh (Schiermonnikoog)

The presence of germinable seeds was recorded in the upper 5 cm of the soil of five replicates (see section II.E.2) in October 1977 and in March 1978. In July 1982 the seeds of each species were recorded in six replicates in the plant communities in which the different management practices were carried out. The species were classified as occurring regularly or irregularly (see Table XI.4). The results from 1977/1978 and 1982 have been taken together since they were similar. The spring samples contained more germinable seeds of the summer annuals *Atriplex prostrata* and *Salicornia europaea* and the perennials *Plantago maritima* and *Glaux maritima.*

Most species which were found in the stands were also represented in the seed bank. Some species, for example, *Juncus maritimus, Trifolium repens, Poa annua* and *Sagina procumbens* were represented in the seed bank, but did not always occur in the established vegetation. These species have affinities with the *Lolio-Potentillion anserinae; Centaurium littorale* and *C. pulchellum* with the *Nanocyperion flavescentis; Plantago coronopus* and *Sagina maritima* with the *Saginion maritimae* (Westhoff & Den Held, 1969). Although the absence of these species in the established vegetation and their occurrence in the seed bank could be attributed to dispersal by inundations, their common ecology of unstable environments and the absence from the seed bank of many other species possibly points to a persistent seed bank (Thompson & Grime, 1979). Species of pioneer or unstable environments from inland communities are also considered to be persistent seeds bank species (see section XI.C.4). Milton (1939) also found *Juncus maritimus* in the seed bank although it was not amongst the established vegetation. In a Baltic sea-shore meadow Jerling (1982) found that *Triglochin maritima,* and to a lesser extent *Potentilla anserina* and *Agrostis stolonifera,* occurred in the seed bank but not in the established vegetation. *Juncus gerardii* and *Glaux maritima* were always found in both the seed bank and the established vegetation, but were over-represented in the seed bank. *Juncus gerardii, Glaux maritima* and *Triglochin maritima* were considered to have a persistent seed bank in this study area, since they occurred no longer in 1981 in the established vegetation, but did occur in the seed bank. These species need a bare soil for germination and establishment and this is not always available on the mid- and upper salt marsh. In this unstable habitat a persistent seed bank might offer a good opportunity for recovery in openings in the turf (Jerling, 1982). Watkinson & Davy (1985) reported that the annual lower salt marsh species *Salicornia europaea* has a transient seed bank in N.W. Europe, where the salt marsh has a constant environment, whereas it has a persistent seed bank in Canada where the length of the growing season is very irregular.

Many species e.g. *Limonium vulgare, Artemisia maritima, Agrostis stolonifera, Festuca rubra, Atriplex prostrata, Potentilla anserina* and *Elymus pycnanthus* only occurred in the seed bank when they were also represented in the established vegetation. This suggests that their seed bank might be transient or persistent. It can also be concluded that their seeds are not as easily dispersed as inundations of salt marshes would suggest. The results in section XI.B.1 showed that most of these species were actually found in driftline material, but it was also suggested that the strong concentration in driftline material prevented an even distribution all over the salt marsh.

XI.C.2. The seed bank in a mixed grassland (Westerholt)

In February 1983 and at the end of July of the same year the presence of germinable seeds was recorded in the upper 2 cm, the layer of 2 to 4 cm and of 4 to 6 cm (Photo XI.1) in the soil of ten replicates (see section II.E.2). The samples were collected from the area which was cut for hay and from the heavily

Table XI.4. Changes in percentage of cover in permanent plots (4 m^2) from 1971 (beginning situation (B); until 1981 under continued abandoning (A), mowing (M) in August and grazing (G), in various plant communities. Classification of species to lower salt marsh (L) <1.30 m N.A.P., mid- and upper salt marsh (M) between 1.30 and 1.80 m N.A.P. and dunes (D) >1.80 m N.A.P. Frequency of species found as readily germinable seeds (S) in the seed bank (0–5 cm) in the abandoned area in 1977/78 and 1982: ●: regularly occurring; •: not regularly occurring (after Bakker, 1985).

Community of		Festuca/Limonium					Artemisia maritima					Juncus gerardii					Juncus maritimus					Festuca/Armeria					Elymus pycnanthus				
Regime		B	A	M	G	S	B	A	M	G	S	B	A	M	G	S	B	A	M	G	S	B	A	M	G	S	B	A	M	G	S
Year (19..)		71	81	81	81		71	81	81	81		71	81	81	81		71	81	81	81		71	81	81	81		71	81	81	81	
Number of species		10	5	11	11	15	8	7	10	12	15	8	10	12	10	7	8	8	8	13	14	13	7	7	13	15	5	3	11	13	19
Limonium vulgare	L	3	4	8	2	•	1		1	1																					
Artemisia maritima	M	6	2	1		•	55	8	1	4	●	1		1					1	1		1		1	1				1	2	•
Puccinellia maritima	L	1			1					8	•	6	4	1	8					2	•									1	
Triglochin maritima	L	2			1	•	1			1	•	8	4	2	12		3	1	1	1	•				1						
Glaux maritima	L/M	3			1	•					•	6	1	1	2	●	10	1		2	●	1			2	•					
Juncus gerardii	L/M	6		1	30	●	1			12	●	35	4	70	4	●	3	1	1	2	●	6	1	1	20	●					●
Agrostis stolonifera	M	3	2	1	20	●	3	1	20	2	●	2	1	1		•	5	2	4		●	10	1	30	20	●	1		1	12	●
Festuca rubra	M	30	50	80	30	●	10	60	80	2	●	3	2		1		20	4	30	1	•	30	20	40	8	•	1	2	70	20	•
Atriplex prostrata	M	1	1	1		•	4	12	2	8	•	6	40	2	1	•	8	50	2	1	●	2		1	1		3	1	1	1	•
Juncus maritimus	M					●					•						20	12	1	10	●					●			1		●
Plantago maritima	M	4		1	4	•	1			1	•		2	4	1					1		10	12	4	4	•			1		
Armeria maritima	M			1	4				1										1			10	2	4	2				1	1	
Poa pratensis	M/D															•					•	2			1						•
Trifolium repens	M/D					•																3	1		1	•			1		•
Potentilla anserina	M																					10	2		1	•					
Elymus pycnanthus	M/D			1				8	1					1												•	30	70	8	8	●
Carex distans	M																					1								1	•
Odontites verna	?																					1									
Salicornia europaea	L			1	1				1	1			1	1	1					1	•										
Spergularia maritima	L				1					2	•					•				1	•					•				2	•
Suaeda maritima	L			1				1	1	2				1	1		1	1		1									1		
Atriplex littoralis	?								1		•																				
Spergularia salina	L/M												1	1	1					2					1					1	
Halimione portulacoides	L							1																							
Rumex crispus	D																										1				
Cerastium fontanum	D																												1		•
Sagina maritima	?					•										•														1	•
Bupleurum tenuissimum	?																													1	
Paraphalis strigosa	?																													1	
Centaurium littorale	?																									•					•
Centaurium pulchellum	?																				•					•					•
Bellis perennis	?																				•										•
Sagina procumbens	?					●																				•					●
Plantago coronopus	M/D					•																									
Poa trivialis	?										•																				
Rumex acetosella	D					•																				•					•
Stellaria media	D										•																				
Poa annua	?										•										•					•					•

Photo XI.1. Different numbers of seedlings emerge from the 0 to 2 cm (left), the 2 to 4 cm and the 4 to 6 cm soil layer (right) from the Westerholt area.

grazed area and lightly grazed patches of the Westerholt area (see chapter V). Seeds of species that were recorded in July, i.e. immediately after the seed rain, were always also found in February. *Leontodon autumnalis, Prunella vulgaris* and *Rhinanthus angustifolius* only emerged from the samples collected in February and apparently needed stratification. The cover percentages of species in the established vegetation and minimal and maximal numbers of germinable seeds of many species in the sites under hay-making and different grazing regimes are shown in Fig. XI.1A–C.

Some species were found as germinable seeds in the soil, but were not found in the established vegetation. Other species, namely, *Juncus effusus, Cerastium fontanum, Holcus lanatus* and *Poa trivialis* featured a small degree of cover, but large numbers of germinable seeds. Most of these species might be considered to be species with a persistent seed bank sensu Grime (1979).

Most species were found in both the seed bank and the established vegetation. In many species a positive relationship is found between the number of germinable seeds and the degree of cover. Some species, namely, *Anthoxanthum odoratum, Hypochaeris radicata, Rhinanthus angustifolius, Leontodon autumnalis* and *Lotus uliginosus* feature a large cover percentage, but small numbers of viable seeds, which suggests that no large numbers of seeds from previous years have been concentrated. Most of these species might be considered to be species with a transient seed bank sensu Grime (1979).

The question as to whether species can be classified as having a transient or a persistent seed bank is discussed in detail in the next sections since the Westerholt and the Anlo study area have many species in common.

XI.C.3. The seed bank in the valley grasslands (Anlo)

The seed bank was sampled roughly in the salt marsh area, namely, the soil layer from 0 to 5 cm in the autumn and the

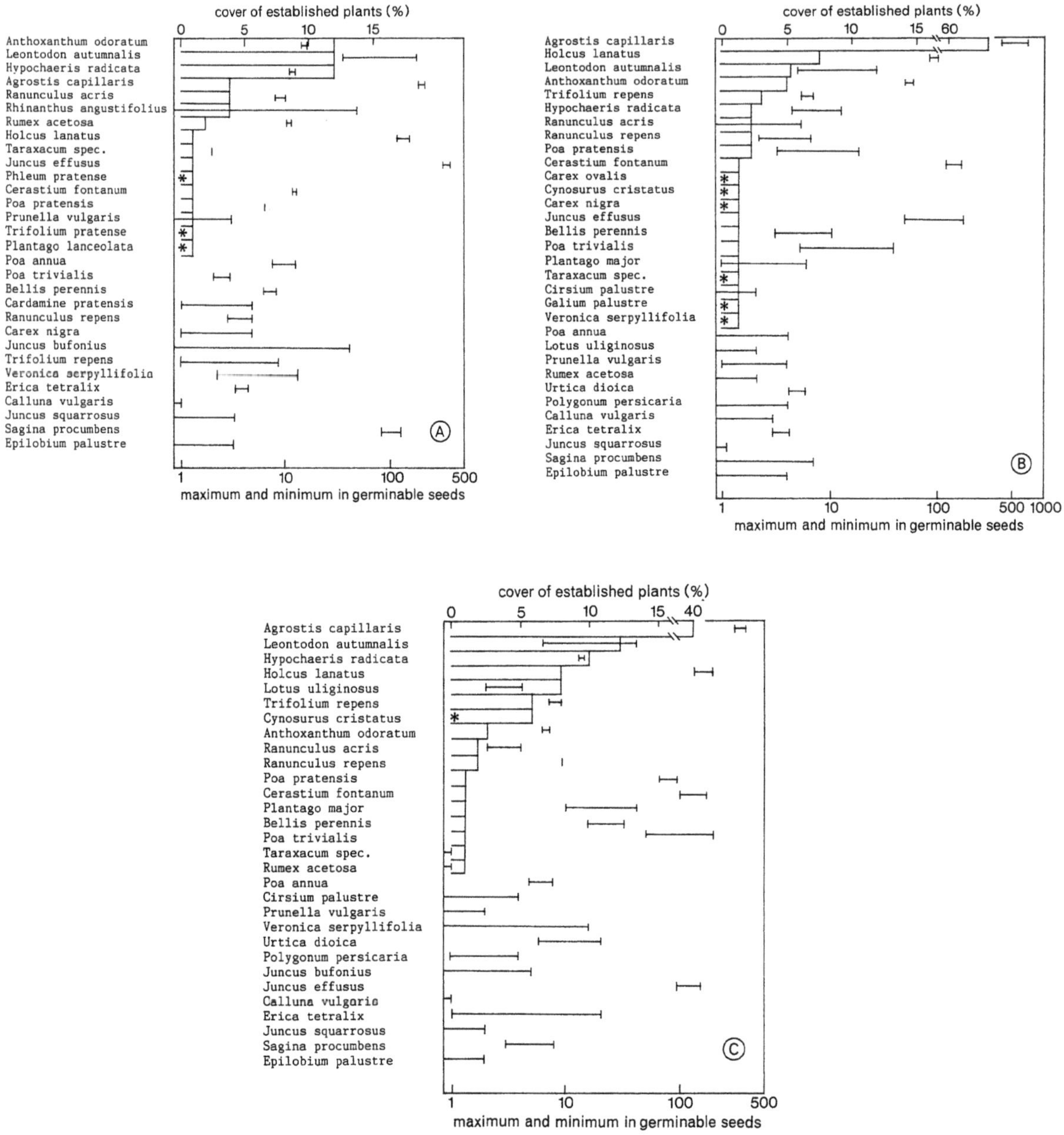

Fig. XI.1A–C. Relation between percentage of cover of established species and germinable seeds recovered from the top 6 cm of the soil profile in A: hay-field, B: lightly grazed and C: heavily grazed pasture in the Westerholt area. The bars at the upper and lower extremities of the vertical lines indicate the maximum and minimum numbers of seeds detected by a standardized procedure of sampling applied in February and in July. The histograms indicate the percentages of cover of established plants. An asterisk means that no seeds were detected (after Brandsma & Bakker, 1984).

spring. In the Westerholt study area sampling was carried out in more detail, namely, the soil layers from 0 to 2, 2 to 4 and 4 to 6 cm in the spring and in late summer. The investigations were most sophisticated in the valley grasslands of the Anlo study area where the soil layers of 0 to 2, 2 to 4 and 4 to 6 cm were sampled from December 1982 until November 1983, according to the following scheme. The seed bank in the Anlo study area was studied in quadrats of 25 m^2. The 0 to 2 cm soil layer was sampled nine times (in the same area) throughout the year with intervals of 6 to 8 weeks, whereas the 2 to 4 and 4 to 6 cm soil layers were sampled three times, with intervals of 5 months. Ten replicates were collected on each sampling date (see section II.E.2).

The total numbers of germinable seeds from the 0 to 6 cm soil layer sampled during a year are given in Table XI.5. The cover percentages of species in the established vegetation near the sample plots are also shown. The apparent disparity between the composition, both qualitatively and quantitatively, of the seed bank and that of the established vegetation has also been found by many other authors (Champness & Morris, 1948; Van Altena & Minderhoud, 1972; review by Roberts, 1981). Some of the causes of the disparity are summarized here.

Whether or not germinable seeds are detected in the seed bank depends on the management of the established vegetation, namely, the grazing intensity and the frequency and timing of cutting, which influence the seed rain. The primary dispersal of the local seed rain can be patchy (Schenkeveld & Verkaar, 1984), seed input from outside the study area is possible (see section XI.B), seeds can be dispersed after the seed rain (secundary dispersal) (Van Breemen & Van Leeuwen, 1983), the numbers of germinable seeds can vary strongly in the course of the year (Thompson & Grime, 1979), the seed bank can contain species which no longer occur in the established vegetation (Thompson & Grime, 1979).

Despite these problems, some conclusions may be drawn from the relationship between germinable seeds in the seed bank and the composition of the established vegetation. If information is available on the dynamics of the established vegetation, 'decreasers', 'increasers' and 'constant' species can be recognised. If the series (Table XI.5) from the *Alopecurus geniculatus* community to the *Juncus acutiflorus*/*Ajuga reptans* community is considered as a successional series (see section VI.D.1), four groups of species can be discerned with respect to the 'changes' in the degree of cover. The group with *Lolium perenne* consists of decreasers, the group with *Ranunculus repens* was present in all lots or revealed no clear pattern, the group with *Ranunculus acris* consists of increasers.

The group of species with *Chamerion angustifolium* was found as germinable seeds in the soil, but was almost totally absent in the established vegetation. Other species within the first two groups, namely, *Poa trivialis, Cerastium fontanum, Juncus effusus, Stellaria uliginosa* and *Juncus bufonius* featured a small degree of cover and, at least in some lots, large numbers of germinable seeds. Except for the anemochorous species *Chamerion angustifolium* (transient) and *Epilobium hirsutum* (persistent according to Thompson & Grime, 1979) most of the species of the *Chamerion* group and the five last mentioned species can be classified as species with a persistent seed bank sensu Grime (1979).

All other species were found in both the established vegetation and as germinable seeds. In the case of *Lolium perenne, Alopecurus geniculatus* and *Trifolium repens* a positive relationship seems to exist between the two parameters, namely a decrease in coverage coincides with a decrease in the number of seeds. These species can be considered as species with a transient seed bank sensu Grime (1979). Although few germinable seeds were found in lots in which the species did not occur in the established vegetation, it seems that most of the species which are considered as increasers do not draw from a persistent seed bank. As the aforementioned shows, a knowledge of which species are not represented in the seed bank can be just as important as knowing which species are represented (Fenner, 1985). In the case of many species it is not always possible to say into which type of seedbank they can be classified simply from the relationship between degree of cover in the established vegetation and the numbers of germinable seeds. The changes in the number of germinable seeds in the course of a year might give a better insight into the type of seed bank concerned.

Table XI.5. The percentage of cover of species in the established vegetation (4 m²) (–: not present, ·: 1–2%, •: 3–5%, ○: 6–10%, ●: 11–20%, ■>20%) and the total numbers of germinable seeds in the 0 to 6 cm soil layer sampled during a year (1983) in a total volume of 1155 cm³) (–: not present, ·: 1–10 seeds, •: 11–50 seeds, ○: 51–200 seeds, ●: 201–1000 seeds, ■>1000 seeds) in different plant communities with different period of hay-making without fertilizer application (after Hoogveld & Muller, 1985).

Community of	Alopecurus geniculatus		Agrostis stolonifera		Anthoxanthum odoratum		Anthoxanthum/ Juncus acutiflorus		Juncus acutiflorus/ Ajuga reptans	
Years of management	1		4		10		15		30	
	cover %	seed bank	cover %	seed bank	cover %	seed bank	cover %	seed bank	cover %	seed bank
Number of species	19	31	21	34	27	39	31	39	25	37
Lolium perenne	•	·	·	·	–	·	–	–	–	–
Bellis perennis	○	○	·	•	–	•	–	·	–	·
Alopecurus geniculatus	●	●	○	■	·	○	–	–	–	–
Poa trivialis	●	●	○	●	·	●	·	●	•	○
Taraxacum spec.	•	○	·	•	·	•	·	•	–	–
Agrostis stolonifera	○	·	■	•	•	•	·	–	–	·
Trifolium repens	•	·	·	–	·	–	–	–	–	–
Festuca pratensis	·	·	•	○	•	•	·	·	–	·
Ranunculus repens	■	•	•	■	■	●	•	○	•	●
Glyceria fluitans	·	·	·	·	–	•	·	·	–	–
Holcus lanatus	■	•	■	■	•	●	●	●	○	●
Rumex acetosa	·	•	•	○	·	●	•	●	·	•
Cardamine pratensis	·	·	·	○	·	○	·	○	·	○
Cerastium fontanum	·	·	·	○	·	○	·	●	·	○
Cynosurus cristatus	·	–	–	–	·	–	·	·	–	–
Juncus effusus	·	○	·	●	·	■	•	■	·	■
Stellaria uliginosa	·	•	–	·	·	·	·	○	–	•
Juncus bufonius	–	•	·	•	·	○	–	•	–	·
Deschampsia cespitosa	–	–	·	–	·	·	–	–	–	–
Ranunculus acris	·	–	·	•	·	●	•	○	·	•
Rhinanthus angustifolius	–	–	–	•	·	•	·	•	•	•
Equisetum palustre	–	–	·	–	·	–	·	–	•	–
Anthoxanthum odoratum	–	–	·	·	■	■	●	●	•	
Plantago lanceolata	·	–	–	–	·	●	•	○	·	•
Myosotis palustris	–	–	·	·	–	–	·	·	–	·
Eleocharis palustris	–	–	–	–	·	–	–	–	–	–
Leontodon autumnalis	–	–	–	–	·	·	·	–	–	–
Galium palustre	–	–	–	–	·	–	·	·	–	·
Carex nigra	–	–	–	–	·	·	·	·	·	–
Festuca rubra	–	–	–	·	·	·	·	•	•	·
Equisetum fluviatile	–	–	–	–	–	–	–	–	·	–
Caltha palustris	–	–	–	–	–	·	·	•	·	·
Ranunculus flammula	–	–	–	·	–	·	·	•	–	·
Juncus acutiflorus	–	–	–	–	·	○	■	■	■	■
Dactylorhiza majalis	–	–	–	–	–	–	·	–	·	–
Cirsium palustre	–	·	–	–	–	–	·	·	·	•
Filipendula ulmaria	–	–	–	–	–	–	·	–	○	–
Lychnis flos-cuculi	·	·	–	·	–	·	·	●	·	○
Rumex obtusifolius	–	–	–	–	–	–	–	–	·	•
Lythrum salicaria	–	·	–	–	–	–	–	·	·	·
Valeriana officinalis	–	–	–	–	–	–	–	–	•	•
Ajuga reptans	–	–	–	–	–	–	–	–	•	·
Chamerion angustifolium	–	·	–	·	–	·	–	–	–	·
Epilobium hirsutum	–	·	–	·	–	–	–	·	–	·
Callitriche spec.	–	·	–	–	–	–	–	·	–	–
Gnaphalium uliginosum	–	–	–	·	–	–	–	–	–	–
Hypericum tetrapterum	–	–	–	–	–	·	–	·	–	·
Juncus articulatus	–	•	–	•	–	●	–	○	–	•
Juncus bulbosus	–	·	–	·	–	•	–	·	–	–
Carex ovalis	–	·	–	·	–	•	·	·	–	–
Mentha aquatica	–	·	–	–	–	·	–	–	–	·
Poa annua	–	○	–	○	–	·	–	–	–	–
Sagina procumbens	–	○	–	·	–	○	–	○	–	·
Stellaria media	–	○	–	•	·	·	–	•	–	–
Urtica dioica	–	·	–	·	–	·	–	·	–	•
Plantago major	–	–	–	·	–	·	–	–	–	–
Polygonum mite	–	–	–	–	–	·	–	·	–	·
Scirpus setaceus	–	–	–	–	–	•	–	·	–	·

The intensive sampling programme of the valley grasslands was carried out in order to enable a subdivision of types of seed banks which match with Grime's (1979) four types of seed banks (see also Thompson & Grime, 1979). The four types are (I) transient seed bank with seeds germinating in the autumn immediately after the seed rain, (II) transient seed bank with seeds germinating in the spring after stratification, (III) persistent seed bank with seeds germinating in both the autumn and spring and containing a small seed pool during the year and (IV) persistent seed bank with seeds germinating in both the autumn and spring and containing a large seed pool which doesn't change much with the seasons and is large in relation to the annual production of seeds. The situation is, in fact, more complicated as can be seen from population biological research in different species. Van der Vegte (1978) and Ter Borg (1985) stated that *Stellaria media* and *Rhinanthus angustifolius*, respectively, feature both a transient and a persistent seed bank.

In this study the seed bank type I is only represented by *Festuca pratensis* (Fig. XI.2A). The seed bank type II is represented by *Rhinanthus angustifolius* and *Caltha palustris* (Fig. XI.2B).

All other species revealed germinable seeds throughout the year and, according to Thompson & Grime (1979) are considered as having a persistent seed bank. Some of these species shown irregular fluctuations or reveal no distinct peak in the number of germinable seeds (Fig. XI.2C), for example, *Alopecurus geniculatus, Rumex acetosa, Juncus acutiflorus, Sagina procumbens* and*Ranunculus flammula* (*Bellis perennis, Juncus bufonius, Glyceria fluitans, Carex ovalis, Stellaria media, Urtica dioica* and *Agrostis stolonifera* are comparable to *Ranunculus flammula*). Other species, mainly grasses, show a clear peak in the number of germinable seeds in late summer immediately after seed dispersal (Fig. XI.2D), namely, *Anthoxanthum odoratum, Holcus lanatus* and *Cerastium fontanum* (*Poa trivialis* and *Stellaria uliginosa* show a picture similar to *Cerastium fontanum*). The last group of species consists mainly of herbs and shows a peak in the winter and early spring, apparently after stratification (Fig. XI.2E), namely, *Ranunculus repens, Plantago lanceolata* and *Lychnis flos-cuculi* (*Ranunculus acris, Cardamine pratensis, Juncus effusus* and *J. articulatus* show a picture similar to *Lychnis flos-cuculi*). It is not always possible to make a rigid distinction between the seed bank types III and IV (Thompson & Grime, 1979). If a low peak in the number of germinable seeds in relation to the numbers of seeds throughout the year is taken as a criterion *Juncus effusus, J. acutiflorus, Cardamine pratensis, Ranunculus flammula, Sagina procumbens, Rumex acetosa, Glyceria fluitans, Carex ovalis* and *Stellaria media* can then be classified as being seed bank type IV, and all the other persistent seed bank species referred to as as being type III.

A discrepancy now seems to exist between the different criteria for the classification of the species into transient or persistent seed bank types. Many species considered as increasers have no seed bank at all in lots in which the species are not (yet) found in the established vegetation, whereas they do have a seed bank throughout the year in lots in which the species occur in the established vegetation. This problem is considered in section XI.D.1.

XI.C.4. The seed bank in different management regimes (Loefvledder)

In November 1981, the presence of germinable seeds was recorded in the upper 2 cm, in the layer of 2 to 4 cm and of 4 to 6 cm in the soil of five replicates (see section II.E.2). The samples were collected from the Loefvledder area in which different management regimes had been practiced since 1973. These were, namely, hay-making in July, hay-making in both July and September, haymaking in September and an uncut site (see chapter VI).

The total numbers of germinable seeds from the 0 to 6 cm soil layer are shown in Table XI.6 as well as the cover percentages of species in the established vegetation in 1973 (this was when the management regimes started) and in 1981 (the summer preceding the sampling of the seed bank). The disparity between the composition of the seed bank and that of the established vegetation has already been discussed in section XI.C.3. The group of species with *Alopecurus geniculatus* consists of decreasers, the group with *Holcus lanatus* was present for the whole study period, the group with *Ranunculus repens* consists of increasers.

The group of species with *Glyceria fluitans* was only found as germinable seeds and not in the established vegetation. These species can be classi-

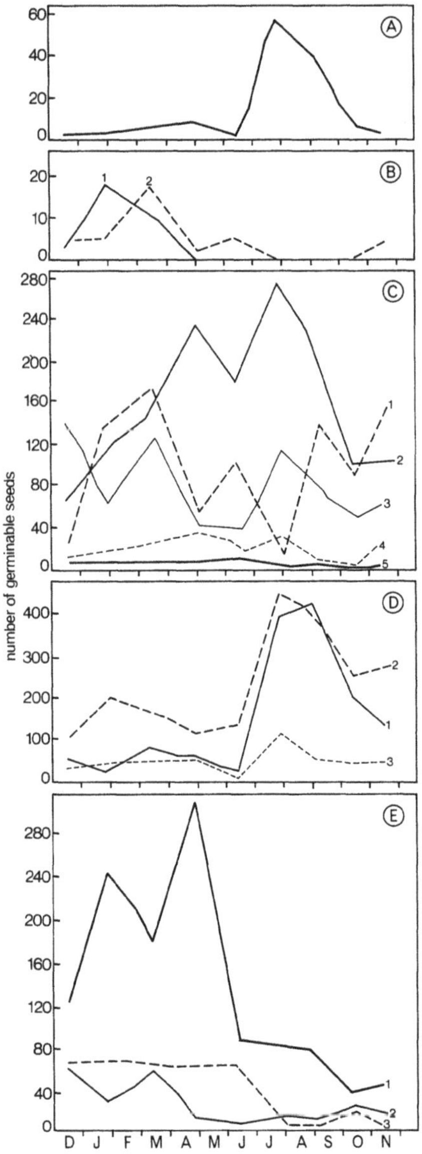

Fig. XI.2A–E. Seasonal variation in the number of germinable seeds in the soil (0–6 cm) on 195 cm², A: example of species in which the occurrence of germinable seeds is restricted to a well-defined peak in the summer (*Festuca pratensis*), B: Examples of species in which the occurrence of germinable seeds is restricted to a well-defined peak in late winter and spring (1: *Rhinanthus angustifolius*, 2: *Caltha palustris*), C: Examples of species in which germinable seeds are found throughout the year with no well-defined peak (1: *Juncus acutiflorus*, 2: *Alopecurus geniculatus*, 3: *Rumex acetosa*, 4: *Sagina procumbens*, 5: *Ranunculus flammula*), D: Examples of species in which germinable seeds are found throughout the year with a peak in the summer (1: *Anthoxanthum odoratum*, 2: *Holcus lanatus*, 3: *Cerastium fontanum)*, E: Examples of species in which germinable seeds are found throughout the year with a peak in late winter and spring (1: *Ranunculus repens*, 2: *Plantage lanceolata*, 3: *Lychnis flos-cuculi*) (after Hoogveld & Muller, 1985).

fied as species with a persistent seed bank sensu Grime (1979), from which no species have as yet emerged. Species which disappeared from the established vegetation, especially in the abandoned site, but still maintained a seed bank, namely, *Alopecurus geniculatus, Poa trivialis* and *Cardamine pratensis* might also be considered as species with a persistent seed bank.

Although the management practices resulted in different compositions of established vegetation, in particular in the abandoned site, the composition of the seed banks did not diverge in the first eight years. The species of the first three groups with the exception of *Alopecurus geniculatus, Poa trivialis* and *Cardamine pratensis* could be considered species with a transient seed bank sensu Grime (1979).

XI.D. The role of the seed bank in restoration management

XI.D.1. How permanent is a persistent seed bank?

Seeds can be incorporated into the soil underneath grasslands after dispersal, whether by natural means, for example, water and small animals or by human interference such as agricultural activities including hay-making or cattle grazing (Harper, 1977). It can be assumed that the longer the viability of the seeds the larger the proportion that becomes buried. Chippindale & Milton (1934) split up the soil profile into horizons of different depths and Wesson & Wareing (1967) excluded surface soil. They demonstrated that buried seeds make up a very high proportion of persistent seed banks and that the seeds of those species which lack persistent seed banks tend not to become buried.

In this study the ratio of the total numbers of germinable seeds in the surface soil layer (0–2 cm) and those in the deeper (2–6 cm) horizons, taking similar soil volumes into account, is used as a criterion for the subdivision of species into transient or persistent seed bank type. The species were classified in five groups $A \leq 0.75 < B \leq 1.25 < C \leq 2 < D \leq 5 < E$, hence the higher the ratio the higher the proportion of seeds in the topsoil and, therefore, the more the species can be considered as

Table XI.6. Changes in percentage of cover of established plants in permanent plots in 1973, beginning situation and 1981, after 8 years of different management regimes and the total numbers of germinable seeds in the seed bank in November 1981 in 575 cm^3 (0–6 cm soil horizon), –: no seeds, ·: 1–10 seeds, •: 11–50 seeds, O: 51–200 seeds, ●: 201–1000 seeds.

	Hay-making in July			Hay-making in both July and September			Hay-making in September			Abandoned		
	Percentage of cover		Seed bank	Percentage of cover		Seed bank	Percentage of cover		Seed bank	Percentage of cover		Seed bank
	1973	81		1973	81		1973	81		1973	81	
Alopecurus geniculatus	4	1	•	1	1	O	1	1	•	3	o	•
Poa trivialis	13	1	•	20	3	O	3	6	O	3	o	·
Agrostis stolonifera	40	30	·	28	10	·	27	7	•	15	1	·
Poa pratensis	3	16	•	12	1	•	18	10	·	20	9	·
Festuca pratensis	12	1	—	9	4	—	10	1	—	6	o	—
Lolium perenne	9	o	—	2	1	—	3	o	—	1	o	—
Cirsium arvense	1	o	—	1	o	—	1	1	—	1	1	—
Trifolium repens	1	o	—	1	o	—	1	1	—	1	o	—
Taraxacum spec.	1	o	—	1	1	—	1	1	—	1	o	—
Holcus lanatus	10	14	O	10	20	O	15	15	•	16	20	•
Cerastium fontanum	1	1	·	1	1	O	1	1	•	1	o	·
Cardamine pratensis	1	1	·	1	1	•	1	o	·	1	o	·
Festuca rubra	3	6	—	1	1	—	4	1	—	2	1	—
Ranunculus repens	3	20	●	5	45	●	3	25	●	11	45	●
Rumex acetosa	1	2	•	1	3	•	1	11	O	1	1	•
Rumex crispus	o	o	—	o	o	—	o	1	—	o	o	·
Potentilla anserina	o	o	—	o	o	—	o	1	·	o	o	·
Rumex obtusifolius	1	1	—	1	o	—	o	o	—	o	15	—
Urtica dioica	o	o	—	o	o	—	o	1	—	o	o	—
Rhinanthus angustifolius	o	o	—	o	1	—	o	1	—	o	o	—
Elymus repens	o	1	· —	o	o	—	o	o	—	o	o	—
Glyceria fluitans	o	o	·	o	o	•	o	o	·	o	o	·
Plantago major	o	o	·	o	o	·	o	o	·	o	o	·
Stellaria uliginosa	o	o	·	o	o	·	o	o	·	o	o	·
Poa annua	o	o	·	o	o	·	o	o	·	o	o	·
Viola palustris	o	o	·	o	o	•	o	o	•	o	o	•
Lythrum portula	o	o	·	o	o	·	o	o	•	o	o	·
Sagina procumbens	o	o	·	o	o	·	o	o	·	o	o	·
Juncus bulbosus	o	o	•	o	o	·	o	o	·	o	o	·
Callitriche spec.	o	o	·	o	o	·	o	o	·	o	o	·
Juncus bufonius	o	o	·	o	o	·	o	o	·	o	o	•
Gnaphalium uliginosum	o	o	—	o	o	·	o	o	·	o	o	·
Juncus effusus	o	o	•	o	o	O	o	o	•	o	o	•

having a transient seed bank. Most species with a low ratio have seeds without appendices, teeths or hairs according to data given by Grime et al. (1981), so their seeds can easily be incorporated into the soil. The results for species occurring in the Westerholt and Anlo study areas are shown in Table XI.7. The results for the salt marsh are not shown because the soil samples were not subdivided into different horizons. The group of species with *Sagina procumbens* has similar numbers or less seeds in the topsoil compared to the deeper horizons, whereas the group of species with *Bellis perennis* has at least twice as many seeds in the topsoil. The group of species with *Epilobium palustre* is some-

Table XI.7. Ratio of the numbers of germinable seeds in the top soil (0–2 cm) and deeper soil horizons (2–6 cm) as classified in five groups: $0<A\leq 0.75<B\leq 1.25<C\leq 2<D\leq 5<E$ in different areas in this study and derived from other studies in the Netherlands. Assessments from the literature show the classification persistent (P) or transient (T) seed bank after criteria of the authors, 1: Thompson & Grime (1979), 2: Grime (1979), 3: Roberts (1981), 4: Pfadenhauer & Maas (1987), 5: Van Gool & Van Der Hoog (1986). Letters in parentheses: classification not quite sure.

	This study						
	Westerholt	Loefvledder	Anlo Brands & Hoekstra (1980)	Anlo Hoogveld & Muller (1985)	Van Zeist (1978)	Hofman & Griffioen (1980)	
Sagina procumbens	A	A	B	A	.	.	P (5)
Poa annua	A	A	.	A	.	.	P (1,4,5)
Gnaphalium uliginosum	.	A	A	A	.	.	
Galium palustre	.	.	A	A	.	.	P (1)
Erica tetralix	A	.	.	.	.	.	
Juncus squarrosus	(A)	.	.	.	.	.	
Juncus bulbosus	.	A	.	.	.	.	
Callitriche spec.	.	A	.	.	.	.	
Polygonum aviculare	.	.	A	.	.	.	
Lythrum portula	.	A	A	.	.	.	
Polygonum hydropiper	.	.	A	.	.	.	
Scirpus setaceus	.	.	.	A	.	.	
Ranunculus flammula	.	.	A	A	.	.	
Rumex obtusifolius	.	.	A	A	.	.	
Glyceria fluitans	.	B	A	A	.	.	
Carex ovalis	.	.	B	A	.	.	
Juncus bufonius	B	A	.	A	.	.	T (5)
Juncus effusus	B	A	A	B	A	.	P (1,5)
Mentha aquatica	.	.	B	A	.	.	
Polygonum persicaria	B	.	A	.	.	.	P (4)
Veronica serpyllifolia	B	.	.	.	.	.	
Polygonum mite	.	.	B	B	.	.	
Lychnis flos-cuculi	.	.	B	B	.	.	P (4)
Viola palustris	.	B	.	.	.	.	
Alopecurus geniculatus	A	A	C	A	.	.	P (5)
Juncus articulatus	C	.	B	A	.	.	P (1,2,4)
Carex nigra	C	.	.	A	.	.	
Myosotis palustris	.	.	C	A	.	.	
Epilobium palustre	C	.	.	.	.	.	
Juncus acutiflorus	.	.	.	C	.	.	
Stellaria uliginosa	.	A	A	D	.	.	
Trifolium repens	B	.	.	D	.	.	P (3,5)
Agrostis capillaris	D	.	.	A	.	.	P (1,5)
Ranunculus repens	D	D	A	C	(A)	E	P (2,5)
Cirsium palustre	D	.	.	.	.	.	P (1)
Leontodon autumnalis	E	.	.	A	.	.	
Cardamine pratensis	E	D	C	A	.	.	T (5)
Urtica dioica	D	(A)	B	C	.	C	P (1,5)
Bellis perennis	C	.	.	D	.	E	P (5)
Plantago major	C	A	.	D	C	.	P (5)
Poa pratensis	D	C	.	.	B	D	
Agrostis stolonifera	.	C	D	.	.	.	P (5)
Holcus lanatus	D	B	C	E	E	.	P (1,2,5)
Cerastium fontanum	D	D	E	C	C	D	P (1,4)
Poa trivialis	E	B	C	C	B	C	P (1,5)/T(4)
Ranunculus acris	E	.	D	C	C	(E)	T (5)
Stellaria media	.	.	E	C	.	.	P (2,5)
Calluna vulgaris	D	.	.	.	.	.	P (1)
Caltha palustris	.	.	D	D	.	.	
Montia fontana	.	.	D	.	.	.	
Rumex acetosa	E	E	E	D	E	.	
Anthoxanthum odoratum	E	.	E	D	E	.	P(1,2)/T(4,5)
Taraxacum spec.	.	.	E	D	.	.	
Festuca pratensis	.	.	(E)	E	.	.	T(1,3,4,5)
Plantago lanceolata	.	.	.	E	.	(E)	P(1,2)/T(4,5)
Hypochaeris radicata	E	.	.	.	.	.	
Rhinanthus angustifolius	E	.	.	E	.	.	T (5)
Lolium perenne	(E)	.	(E)	E	E	(E)	T (1,2,3)
Prunella vulgaris	E	.	.	.	.	.	
Festuca rubra	.	.	.	(E)	E	(E)	T (1,2)

where in between or shows contradictory data. *Poa trivialis* and *P. pratensis* should be classified in the intermediate group when taking the findings of Van Zeist (1978) and Hofman & Griffioen (1980) into account. These authors are in agreement with this study with regard to the other ten species they mentioned.

Only *Caltha palustris* and *Rhinanthus angustifolius* can be classified as being species with a transient seed bank type after applying Thompson & Grime's (1979) criteria (see section XI.C.3). Most of the species of the group with *Bellis perennis* should, therefore, be considered as being species with a persistent seed bank. This is in agreement with the classifications of different authors. Many of the species in this group were, however, not found in the seed bank of lots in which they are absent in the established vegetation, but in which they certainly occurred before the fertilizer applications. This implies that, although many species are considered as having a persistent seed bank according to Thompson & Grime (1979), the longevity of those seeds is too short to generate new populations after the cessation of fertilizer applications. Many species of the group with *Sagina procumbens* (Table XI.7) do feature a seed bank although the species are not present in the established vegetation, and are apparently real longeval. It is, therefore, necessary to distinguish between *persistent* seed banks according to Thompson & Grime (1979) which can play a role in the maintenance of populations after bad seed setting or cutting too early and *permanent* seed banks which can play a role in the regeneration of populations after their disappearance from the established vegetation several years earlier. It is obvious that permanent seed banks are very important in restoration management.

The species with few seeds in the upper soil layer (the group with *Sagina procumbens* in Table XI.7) are mainly pioneers, weeds and heathland species, whereas the species with a great number of seeds in the upper soil layer (the *Bellis perennis* group) consists mainly of species characteristic of grassland which is receiving some fertilizer application (Fig. XI.3). This means that precisely many of the hay field species which were expected to regenerate by the restoration management do not have a permanent seed bank.

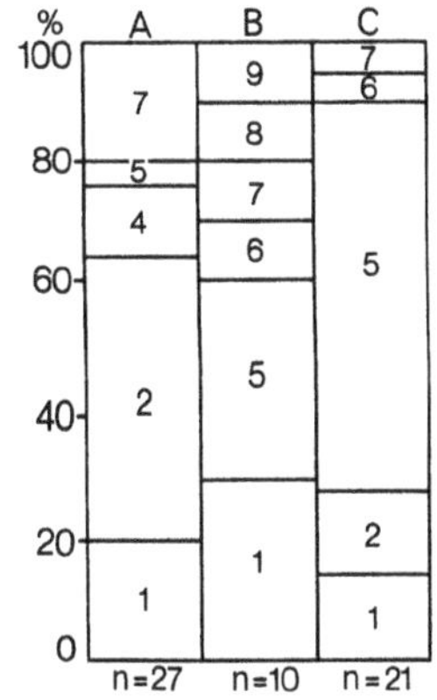

Fig. XI.3. Percentage occurrence of species characteristic of different ecological groups in A: species with less or similar numbers of seeds in the top soil (0–2 cm) compared to deeper horizons (2–6 cm), B: intermediate group, C: species with twice or more as much seeds in the top soil (see Table XI.7). Ecological groups according to van der Meijden et al. (1983B), 1: weeds, 2: pioneers, 4: water and bank plants, 5: grasslands receiving some fertilizer application, 6: dry grasslands, 7: heathland and peat. 8: woodland fringes, 9: woodland.

XI.D.2. Possibilities of regeneration from a seed bank

The restoration of species-rich chalk grassland communities in the Netherlands can be hampered by the absence of characteristic chalk grassland species in the seed bank (Hoffman & Griffioen, 1980; Willems, 1983B). In Belgium Vyvey (1980) only found a few species in the seed bank of a former litter fen twenty years after abandoning a.o. *Juncus* and *Carex* spp. In Germany Pfadenhauer & Maas (1987) reported that typical litter fen species had disappeared from the seed bank about fifteen years after the commencement of fertilizer applications, and only some *Juncus* spp. and *Carex* spp. proved to have a permanent seed bank. Similar conclusions were made by Howe & Chancellor (1983) and Williams (1984) in the United Kingdom. The latter also mentioned *Ranunculus repens.* Chippindale & Milton (1934) had already suggested that *Juncus* spp., *Carex* spp., *Ranunculus repens, Trifolium repens, Bellis perennis, Poa trivialis* and *P. annua* might have permanent seed banks beneath British pastures. In the Netherlands Van Gool & Van Der Hoog (1986) did not answer the question as to whether a permanent seed bank could generate hay-field species under restoration management, since in one lot, the espected species did not emerge eight years after the cessation of fertilizer application, but they did emerge in another lot after sixteen years of restoration management. They did not suggest that those species could have invaded with hay-making machinery (see section XI.B.3), but left open the possibility that a permanent seed bank was present but was too small to detect.

In the Netherlands Van Tol (in Van Gool & Van Der Hoog, 1986) showed that a number of hay-field species were absent in the seed bank of lots receiving fertilizer application, but were still present in unfertilized lots. Since this was the case for both the seed bank and the established vegetation it is suggested that once species have disappeared from the sward their seeds are also no longer found within a period of a few years.

This hypothesis can be tested if the seed input can be prevented. Five sods were transplanted into the experimental garden in June 1980 and clipped in mid July when the annual hay cutting was carried out in the field. The sods were covered by glass in order to prevent seed rain from elsewhere and they were clipped frequently in order to prevent seed input from the plants growing in the sods. The

Table XI.8. Comparison of the numbers of germinable seeds in 575 cm³ (0–6 cm soil horizon) in three successive years in the end of June on frequently cut sods in the experimental garden, where seed input was prevented, and in field plots annually cut in July. E: species containing many seeds in the top soil, A: species containing many seeds in deeper horizons, C: intermediate species (see Table XI.7).

		1980		1981		1982	
		Seed input		Seed input		Seed input	
		Prevented	Not prevented	Prevented	Not prevented	Prevented	Not prevented
E	Holcus lanatus	91	95	34	240	11	44
E	Poa trivialis	10	50	6	70	3	11
E	Poa pratensis	6	4	3	2	1	46
A	Alopecurus geniculatus	13	43	5	40	4	46
A	Rumex obtusifolius	26	0	17	4	5	2
A	Lythrum portula	37	3	21	4	12	5
C	Ranunculus repens	23	19	12	51	17	172
E	Cerastium fontanum	5	0	6	1	2	3
A	Juncus effusus	38	14	23	23	37	0
A	Juncus bufonius	0	1	0	0	6	1
A	Sagina procumbens	18	5	9	2	14	3
A	Gnaphalium uliginosum	7	1	4	4	6	3
C	Stellaria uliginosa	8	1	2	2	6	1
C	Rumex acetosa	3	1	0	16	2	11
E	Agrostis stolonifera	4	10	3	8	3	2

number of germinable seeds which were detected did not decrease during the three years in the group of species with *Juncus effusus* (Table XI.8), which consisted mainly of species with many seeds in deeper soil horizons. The group of species with *Holcus lanatus* tended to show a decreased number of germinable seeds and contained both grassland species with many seeds in the topsoil and in deeper soil horizons. For a number of species this finding is in agreement with the results of the prevention of seeds input on frequently cut plots over a three year period in an *Agrostis capillaris/Festuca rubra* pasture in the United Kingdom (Williams, 1984). *Anthoxanthum odoratum, Cynosurus cristatus* and *Lolium perenne* which were present in small amounts at the start of the experiment, did not feature in any of the subsequent samples. *Poa trivialis* and *Holcus lanatus* declined, but maintained at a lower level. In general, seeds of dicotyledonous species declined less rapidly than those of grasses. *Cerastium fontanum, Ranuncuclus repens* and *Rumex acetosa* maintained their contribution, but that of *Trifolium repens, Plantago major* and *Ranunculus acris* appeared to decrease.

The results of both the present study and those of Williams (1984) are in agreement with the aforementioned suggestion on the disappearance of seeds from the seed bank and results in the maintenance of a few species in a permanent seed bank.

If can be concluded that many species that have disappeared from the established vegetation will also vanish from the seed bank because many of the species occurring in grasslands feature a transient or a persistent seed bank and only a minority a permanent seed bank. Since the permanent seed bank contains mainly pioneers and weeds and not species characteristic of grasslands receiving light fertilizer applications, the latter group of species has to become re-established from populations from elsewhere whether by wind dispersal, grazing animals or hay-making machinery. Another possibility is the willing introduction of species characteristic of unfertilized grassland communities. Pfadenhauer et al. (1987) also suggest the possibility of introduction. Chapter XII considers the possibilities of (re-)introduction.

CHAPTER XII

Re-introduction of seeds

XII.A. Why re-introduce seeds?

Restoration management aims at the increase of species diversity and/or the re-establishment of plant species that have disappeared as a result of the intensification of agricultural practices or abandoning. It was shown in chapter XI that only a limited number of species has a permanent seed bank from which plants can re-establish. Moreover, most of these species appeared to be weeds or pioneers. This means that many of the hay-field species expected to re-establish as a result of restoration management cannot do so from a permanent seed bank.

The dispersal of almost 100% of the seeds which are shed even from many anemochorous species, only covers a few metres. Larger distances can be covered by grazing animals or hay-making machinery moving from one lot to another. In section IX.C.3 it was already mentioned that the developing vegetation types contained less species than would be expected from the already low above-ground standing crop. This fact seems to be in agreement with Ter Borg's statement (1979) that the arrival of many plant species is a matter of chance and may take a long time. Apparently some limiting factors exist in the re-establishment of species during restoration management. These factors can be the dispersal conditions for germination and the density of the sward.

This chapter considers the questions (i) which species can be used for re-introduction without interfering with spontaneously emerged seedlings?, (ii) what are the required conditions for germination with respect to temperature and stratification?, (iii) how successful are the seedlings which emerge?, (iv) to what extent can the fate of the seedlings be influenced by manipulating the density of the sward?

Another question which should be answered here, is namely, on the permissibility of the re-introduction of seeds. In this study seeds were introduced as an experiment and all the individuals which emerged were removed before seed set in order not to influence the successional series.

Some definitions should be given for the sake of clarity. Introduction is the deliberate release of plant species into an area in which it has not occurred for a long time, subjectively defined as the twentieth century. If the species was indigenous in the past the release is called re-introduction (Green, 1981). Taking the whole salt marsh reserve or the valley grassland reserve into account, all the released species do still occur. It is, however, not known whether the released species were found earlier in the different study areas although it is very likely that they were.

There are two views on the permissibility of (re-) introduction. The first one is that the introduction of seeds or other diaspores or even complete individuals (as long as it concerns 'wild gardening') into natural or semi-natural landscapes, in particular nature reserves, should not be allowed since it may imply falsification of the established flora (Westhoff, 1949, 1983). Re-introduction may, however, be justified for the sake of safeguarding a virtually entirely or locally extinct species and only if serious precautions are made (Westhoff, 1949, 1983 and several authors quoted in the latter). The second view is that the (re-)introduction cannot be a problem if it is assumed that all introductions which do not fit into the local environment will not result in the establishment of the species (or ecotypes) concerned. The non-establishment of species in a certain place can then be regarded as an expression of the unsuitability of the habitat for that species (Londo, 1984). This is only the case when seeds are re-introduced, since if mature plants or even complete sods are re-introduced several stages in the life-cycle are skipped. For this reason only the deliberate re-introduction of seeds can be considered. At the present time habitat creation in nature reserves does not include the sowing of species. In the (near) future, however, when the seed bank of the former original vegetation will have disappeared and the isolation of plant populations will have increased to a high degree, many species will have difficulties in establishing because of a lack of nearby seed sources. In such cases sowing experiments could be carried out. In the mean time, experience should be gained on the problems involved with re-introduction as has been done in the present study.

Van Groenendael (1985) puts the question into a broader framework and suggests that the possibilities for the re-introduction of plant species are limited when no attempt is made to stop habitat destruction elsewhere at the same time. It can be argued that the re-introduction of seeds is permissible in order to study the limiting factors in the re-establishment of species during restoration management. Re-introduction is, in fact, only successful if the newly emerged plants can reproduce themselves. Only the first stages of re-introduction were recorded in this study since seed setting was prevented by the pulling out of flowering plants.

The choice of re-introduced species depends on the species present in the established vegetation and/or in the seed bank. Most of the re-introduced species were expected on the basis of the species composition of the established vegetation, although they were not (yet) present. Most information on the limitations of dispersal was obtained from the re-introduction of species which were not found in either the established vegetation or the seed bank (section XII.B).

The establishment of (re-)introduced species depends first of all on the suitability of the habitat for germination. It is important to know about the needs of the re-introduced species with respect to temperature and/or stratification (section XII.C).

Once emerged, the fate of the seedlings of re-introduced species depends for a great deal on the density of the sward. This effect was studied by comparing the fate of seedlings in different plant communities and in stands with different management practices (section XII.D).

XII.B. The re-introduced species

XII.B.1. The salt marsh (Schiermonnikoog)

Lower salt marsh species emerged in many vegetation types after the resumption of grazing and extended their areas to the mid- and upper salt marsh and dunes, wheres dune species did not spread lower down the salt marsh (see chapter VII). The seed bank, however, contained more lower salt marsh species and dune species in the abandoned than in the grazed sites along the gradient (Table XII.1) (total number of species in the seed bank from dune to lower salt marsh in the abandoned area: 20-14-14-9 as opposed to the grazed area: 12-12-11-9). This discrepancy can be put down to the deposition of seeds with driftline material (see section XI.B.1) which is easily 'captured' by the tall sward of the abandoned area, and was studied by the re-introduction of both lower salt marsh and dune species (see Table XII.1).

The re-introduction took place on four sites along a gradient from the lower salt marsh to a low dune (see section X.C.2) in the grazed and abandoned areas on both sides of the fence. On 15 March 1983 batches of 100 seeds of each species were sown in the canopy in each of three plots (10 × 10) cm^2. The species which were known from other experiments to require stratification (section XII.C.1), were stratified in 100% sea water for one month at 5° C before being sown.

Several of the species in the sowing experiment were found in the seed bank and occurred as seedlings. It, therefore, seemed possible that seedlings emerging in the sown plots originated from seeds already present in the soil. In order to correct for this bias, naturally-emerging seedlings were recorded in five sods of (10 × 10) cm^2 at each site during the spring and summer of 1983. The number of seeds sown greatly exceeded the number of seeds in the seed bank and/or emerging seedlings

Table XII.1. Differentiation of the salt marsh vegetation at four sites along a gradient after the resumption of grazing in 1972 illustrated by the percentage of cover in 1981 and the change in the period 1972 to 1981 in permanent plots (= cover percentage unchanged provided fluctuations, >increase, ≫large increase, <decrease, ≪large decrease, Λ irregular appearance and disappearance during the study period and point quadrat analysis in 1981 at site 1 (+ : present, ++ very abundant) at the abandoned and at the grazed area, since this plot is no permanent plot. Seed bank (•≤5 seeds · 100 cm^{-2}, ●>5 seeds · 100 cm^{-2}, ■>20 seeds · 100 cm^{-2}); seedlings (•<1 seedling · 100 cm^{-2}, ●≥1 seedling · 100 cm^{-2}, ■>5 seedlings · 100 cm^{-2}). D = dune species, M : mid- and upper salt marsh species, L: lower salt marsh species. Besides species indicated with an asterisk were sown Aster tripolium, Trifolium fragiferum, Galium mollugo, Cochlearia danica, Rumex crispus (after Bakker et al., 1985).

		site 1						site 2								site 3								site 4							
		abandoned			grazed			abandoned				grazed				abandoned				grazed				abandoned				grazed			
		presence	seed bank	seedlings	presence	seed bank	seedlings	coverage	change	seed bank	seedlings	coverage	change	seed bank	seedlings	coverage	change	seed bank	seedlings	coverage	change	seed bank	seedlings	coverage	change	seed bank	seedlings	coverage	change	seed bank	seedlings
M	Festuca rubra	++	•	•	++	■	●	50	=	●		20	<	•		60	>	•		20	=	•	●	50	=	●		30	>	●	
M	Agrostis stolonifera		•		+	●	•	2	=	•		20	>	●		1	=	■		30	≫	•		2	<	•		20	≫		
D	Ammophila arenaria	+			+																										
D	Sedum acre	+	•																												
D	Trifolium pratense	+					•																								
D	Stellaria media		•	●			•																								
D	Elymus repens	++			++			2	>			1	Λ			1	Λ												Λ		
D	Stellaria graminea	++	■	•		●	•			●																					
D	Holcus lanatus		•	•			•																								
D	Rumex acetosella		■			•																									
D	Bromus hordeaceus hordeaceus*									•																					
M D	Elymus pycnanthus	++	●					2	>	●		1	=	•		20	≫	■	■	1	≪	●			Λ	●			Λ	•	
M	Armeria maritima	+			+		●	2	>	•		20	≫	●		1	<			8	>	•						1	>		
M D	Poa pratensis	+	•			●			<	●			<			1	=	•		1	=										
M D	Trifolium repens	+	•	•	+	●	•	1	=	•		1	=	•	•	1	=	•		2	>	•	•						>		
D	Cerastium fontanum	+	•		+	●	●		>	●			Λ	•	•		Λ	•			Λ										
M D	Plantago coronopus		•		+	●	●		Λ	•		2	>	●				•		1	>	●									
D	Plantago lanceolata*	+	●	•		●	•								•					1	>										
D	Lotus corniculatus								<				<							1	>										
M	Carex distans								<												>										
M	Juncus maritimus		•															•													
M	Plantago maritima*							1	<			4	>	•	•	1	=			1	=	•			<	•		4	=		
M	Atriplex prostrata		●	•		●	•	2	>	●			<	•	•	1	=	■	●		<	•		1	=	●	●		Λ	•	●
?	Lolium perenne												Λ							1	>										
M	Artemisia maritima		•	•				1	>		•	1	>		•			•	●	1	>		•	2	=	●	●		<	●	●
L M	Juncus gerardii		●			●			<	•				•				●			>	•			<	•		30	≫	●	
L M	Spergularia salina*		•			●				•			Λ	■				•			Λ	●				•			Λ	■	●
L	Suaeda maritima												Λ								Λ								Λ		
L	Salicornia europaea*		•							•	•		Λ		•			●									■	1	Λ	•	•
L	Limonium vulgare*		•											•				•						4	=	•		2	=	•	•
L	Triglochin maritima*																								<			1	=		
L M	Glaux maritima																					•			<			1	=	•	
L	Puccinellia maritima																											1	>		
L	Spergularia maritima																											1	Λ		

Addenda: species occurring in the vegetation during one or two years at site 2 abandoned – Odontites verna (?), Erophila verna (?); at site 2 grazed – Leontodon autumnalis (D), Trifolium dubium (?), T. arvense (D), Cochlearia anglica (?), Halimione portulacoides (L); at site 3 grazed – Sagina maritima (?), Potentilla anserina (M) and Cochlearia danica (?) as seedling.

outside the sown plots (Table XII.1). It is, therefore, assumed that, with the possible exception of *Salicornia europaea, Spergularia salina* and *Atriplex prostrata,* the number of seedlings emerging in the sown plots was hardly influenced by naturally occurring seedlings of the same species.

XII.B.2. Mixed grassland (Westerholt)

The restoration management of grazing or haymaking without fertilizer application aimed at the development of a heathland vegetation (*Nardo-Callunetea*) or grassland communities (*Violion caninae*). However, species characteristic for those communities hardly became established in the previously fertilized grassland (see chapter V), whereas only small numbers of a few species were found in the seed bank beneath the pasture (see section XI.C.2). Many species of these communities were re-introduced experimentally, viz. *Festuca rubra, F. ovina, Juncus squarrosus, J. conglomeratus, Molinia caerulea, Nardus stricta, Narthecium ossifragum, Danthonia decumbens, Calluna vulgaris, Erica tetralix, Gentiana pneumonanthe, Hypochaeris radicata, Potentilla erecta, Rumex acetosella* and *Succisa pratensis.* The seeds were sown into the pasture area community in the heavily grazed area, lightly grazed patches and in the hay-field. Since thick carpets of mosses occurred locally in the hay-field, seeds were sown in both the 'bare' hay-field and in 'moss' hay-field. On the 7th January 1983 batches of 50 seeds of each species were sown in the sward in each of three plots $(10 \times 10)\,cm^2$ in the four sites.

The occurrence of naturally-emerging seedlings was recorded in five sods of $(10 \times 10)\,cm^2$ at each site in June. Of the sown species, only *Hypochaeris radicata* was present in the established vegetation and revealed some naturally emerging seedlings in the 'bare' hay-field (Table XII.2). The numbers of emerged seedlings of *Hypochaeris radicata* in the sown plots were corrected by substracting one individual in each plot.

XII.B.3. The valley grasslands (Loefvledder)

Hay-making without fertilizer applications aimed at the development of species-rich grassland communities (*Lolio-Cynosurion/Calthion palustris*). However, species characteristic of those communities hardly established in the previously fertilized grassland (see section VI.D.2). This was probably due to their absence in the seed bank (see section XI.C.4) beneath the recent reserve.

A number of species were re-introduced into the *Agrostis stolonifera/Poa pratensis* community in the hay-making regimes in July, September, both July and September and in the abandoned site. Since a tall forb community can be expected in the latter regime *Bromus hordeaceus, Dactylis glomerata, Epilobium hirsutum, Heracleum sphondylium, Cirsium vulgare, Filipendula ulmaria, Lysimachia vulgaris* and *Lythrum salicaria* were sown. *Anthoxanthum odoratum, Senecio aquaticus, Veronica arvensis, Cynosurus cristatus, Plantago lanceolata, Trifolium dubium, Lychnis floscuculi, Caltha palustris, Rhinanthus angustifolius, Cirsium palustre, Crepis paludosa, Myosotis palustris, Prunella vulgaris, Lotus uliginosus* and *Carex ovalis* were re-introduced since those species would be expected in the sites in which hay-making was carried out. On the 1st November 1977 batches of 100 seeds of each species were sown in the sward in each of three plots $(10 \times 10)\,cm^2$ in the four management practices.

The occurrence of naturally-emerging seedlings was recorded monthly in field plots adjacent to the

Table XII.2. Number of seedlings in five sods $(10 \times 10)\,cm^2$ in June at the Westerholt study area in different sites (after Hendriks, 1984).

	Hay-making area		Grazed area	
	without mosses	with mosses	heavily	lightly
Monocots	21	1	20	–
Leontodon autumnalis	82	13	7	2
Hypochaeris radicata	7	–	2	–
Cerastium fontanum	6	1	10	1
Bellis perennis	–	–	1	–
Rumex spec.	4	–	–	–
Ranunculus spec.	1	–	–	–
Cirsium spec.	1	–	–	–

sowing plots from September 1977 to May 1978. Only seedlings of *Holcus lanatus, Poa* sp., *Rumex acetosa, Ranunculus* sp., *Cerastium fontanum, Urtica dioica* and *Cardamine pratensis* were found, but none of the sown species, which were also absent in the established vegetation.

XII.C. Germination ecology of re-introduced species

XII.C.1. The salt marsh (Schiermonnikoog)

Two batches of 50 seeds of each sown species were tested in the laboratory for their germination response (see section II.E.3). Dry-stored seeds of the dune species *Plantago lanceolata, P. coronopus* (Fig. XII.1 A), *Rumex crispus, Cochlearia danica, Bromus hordeaceus* and *Galium mollugo* (Fig. XII.1 B) showed decreasing germination responses at either high constant or high fluctuating temperatures. Salt marsh species reached higher germination percentages at higher temperatures.

As expected, winter annuals, germinating in the autumn, had optimal germination rates at low temperatures. In contrast, species with plants at the seedling stage in the summer and autumn (Dijkstra, 1984) needed no chilling to reach high germination percentages (Grime et al., 1981). Furthermore, perennials and summer annuals, with germination in early spring, did require chilling.

The increasing germination response of salt marsh species with increasing temperatures contrasted with the optimum curve of the dune species. Germination during warm periods is not a problem for salt-marsh species with relatively high local moisture content, even in the summer period. Dune species seedlings would, however, die from drought after germination in hot periods.

Fresh water stratification enhanced the germination of dune species (Fig. XII.1 A), but stratification in sea water resulted in a reduction in germination (Fig. XII.1 B). Both fresh and sea water stratification (square and triangular symbols) increased the germination of salt marsh species.

The germination percentages at different NaCl concentrations after one month are shown in Table XII.3. The highest germination rates of most of the species were reached under fresh water conditions. In contrast to the performance of halophytic species most dune species showed little germination at concentrations>10 g $NaCl \cdot l\ solution^{-1}$.

Table XII.3 also shows the relative dry weights of different species grown at different salinities for two months. *Atriplex prostrata* and *Salicornia europaea* produced most dry matter at 10 g $NaCl \cdot l\ solution^{-1}$ ($P<0.05$, Student's test). The dry matter of the dune species decreased sharply at concentrations >5 g $NaCl \cdot l\ solution^{-1}$ and that of the salt marsh species at concentrations >10 g $NaCl \cdot l\ solution^{-1}$.

The highest germination for both halophytes and glycophytes occurred under fresh water conditions, a finding which has been reported by several authors in Waisel (1972). In these experiments, germination and dry matter production of glycophytes decreased more than those of halophytes with increasing salinities. This is in agreement with earlier results (e.g. Kinzel, 1982).

XII.C.2. The inland grasslands (Westerholt/Loefvledder)

Two batches of 50 seeds of each species sown in the mixed grassland (Westerholt) were tested in the laboratory for their germination response (see section II.E.3). The seeds which were stratified for a period of one month showed higher germination responses than unstratified seeds for *Juncus squarrosus, Nardus stricta, Narthecium ossifragum, Danthonia decumbens, Erica tetralix, Gentiana pneumonanthe* and *Succisa pratensis,* whereas the reverse was true for *Rumex acetosella* (Fig. XII.2). Moreover, the lower temperature limit is extended by exposure to diurnal fluctuations in temperature for most species, which is in agreement with Grime's et al. (1981) results. Only the germination response of *Juncus conglomeratus* was very low. Higher responses were reached after seven weeks, namely, 20% for stratified seeds and 40% for unstratified seeds.

The germination response of species sown in the valley grassland (Loefvledder) was studied in a more simple way than that for the other study areas

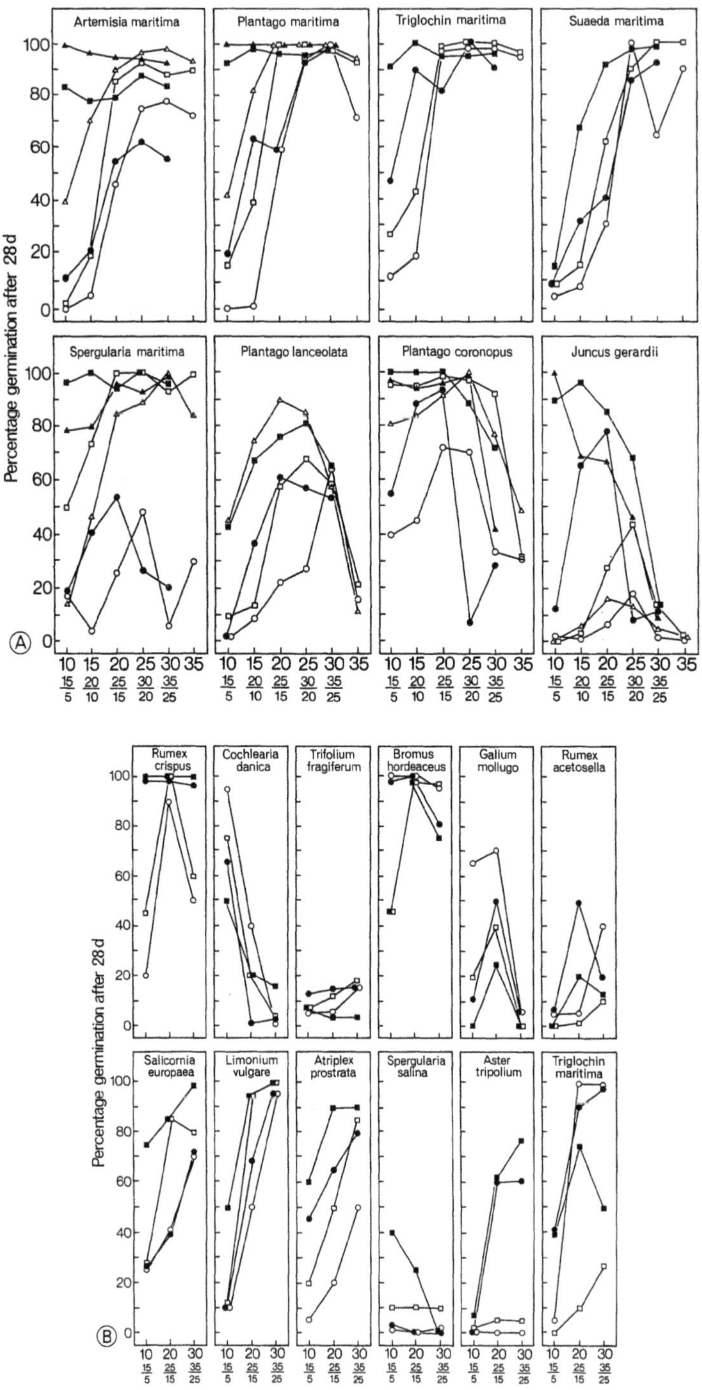

Fig. XII.1A–B. Germination percentages of salt marsh species after 28 days under constant (open symbols) and fluctuating temperatures (closed symbols) of dry stored (circles), one month stratified (squares) and three months stratified seeds (triangles). Seeds had been stratified in A: fresh water and B: in sea water (after Bakker et al., 1985).

because of the large number of species which were involved. From the 21st of October 1977 onwards two batches of 50 fresh seeds of each sown species were tested in sterilized soil in the garden for their germination response. Stratification could, therefore, take place spontaneously since the recording of seedlings went on until July of 1978.

The grass species which were sown and *Veronica arvensis* featured high germination rates and germination responses in the autumn (Fig. XII.3 A). Seeds of *Caltha palustris* and *Senecio aquaticus* showed some response in the autumn, but most seedlings emerged in March, whereas other species of this group (for instance *Cirsium palustre*) showed no seedlings at all in the autumn. All species of this group reached germination responses of over 80% (Fig. XII.3 B). The group which contained species like *Rhinanthus angustifolius* and *Lychnis flos-cuculi* also revealed most seedlings in March, but reached smaller germination responses (Fig. XII.3 C). Finally, a group could be discerned which contained species which emerged mainly in May like *Myosotis palustris* and *Prunella vulgaris* (Fig. XII.3 D). It is interesting to compare the germination responses in the different management regimes (Fig. XII.9 A–D).

A great many of the species sown in the garden were tested for their germination response by D.M. Pegtel (unpublished). His results showed that five different germination response patterns could be distinguished. Fresh seeds of *Holcus lanatus* revealed a germination response of ≥50% (Fig. XII.4 A). Fresh seeds of *Festuca pratensis* hardly showed any response, but the germination response increased strongly after stratification or with diurnal fluctuating temperatures (Fig. XII.4 B). Fresh seeds of *Caltha palustris* only showed a response at higher temperatures, whereas stratification and diurnal fluctuations of the temperatures markedly enhanced the response (Fig. XII.4 C). Fresh seeds of *Cardamine pratensis* did show a response at low temperatures, whereas stratification and diurnal fluctuations of the temperature increased the response (Fig. XII.4 D). Other species, like *Festuca ovina, Juncus conglomeratus* and *Potentilla erecta* from the Westerholt area (see Fig. XII.2), hardly showed a germination response at different regimes.

All species from the Westerholt (Hendriks, 1984) and Loefvledder (unpublished data D.M. Pegtel) study area which were tested for their germination response were classified according to their response pattern (see Figs. XII.2 and XII.4) in Table XII.4. The group with *Holcus lanatus* only contained grasses of which only three showed a higher germination response with diurnal temper-

Table XII.3. Germination percentage after one month and relative dry weight (%) of several species grown at increasing NaCl concentration after two months. Dry weight grown at fresh water is put at 100%. – = not measured (after Bakker et al., 1985).

	Germination (%)					Relative dry weight ± S.E.			
Gram NaCl · liter^{-1}	0	5	10	25	35	5	10	25	35
Salicornia europaea	54	23	22	24	25	234 ± 59	396 ± 42	177 ± 22	156 ± 29
Atriplex prostrata	100	57	37	18	14	132 ± 17	163 ± 24	12 ± 4	6 ± 3
Limonium vulgare	75	68	85	64	42	61 ± 10	88 ± 11	25 ± 4	29 ± 5
Spergularia salina	8	22	5	5	0	103 ± 14	84 ± 17	3	0
Plantago maritima	94	98	99	76	34	87 ± 17	63 ± 12	9 ± 2	2
Aster tripolium	58	47	29	36	0	95 ± 8	76 ± 11	3	1
Triglochin maritima	39	34	37	6	5	64 ± 10	54 ± 12	8	2
Rumex crispus	89	75	74	35	4	83 ± 10	25 ± 7	1	1
Cochlearia danica	94	82	68	21	4	31 ± 6	14 ± 5	0	0
Juncus gerardii	–	–	–	–	–	88 ± 15	29 ± 7	0	0
Plantago lanceolata	78	32	22	0	0	78 ± 12	33 ± 7	0	0
Bromus hordeaceus	100	94	86	5	1	101 ± 11	3 ± 9	0	0
Galium mollugo	93	13	15	0	0	14 ± 4	2 ± 3	0	0

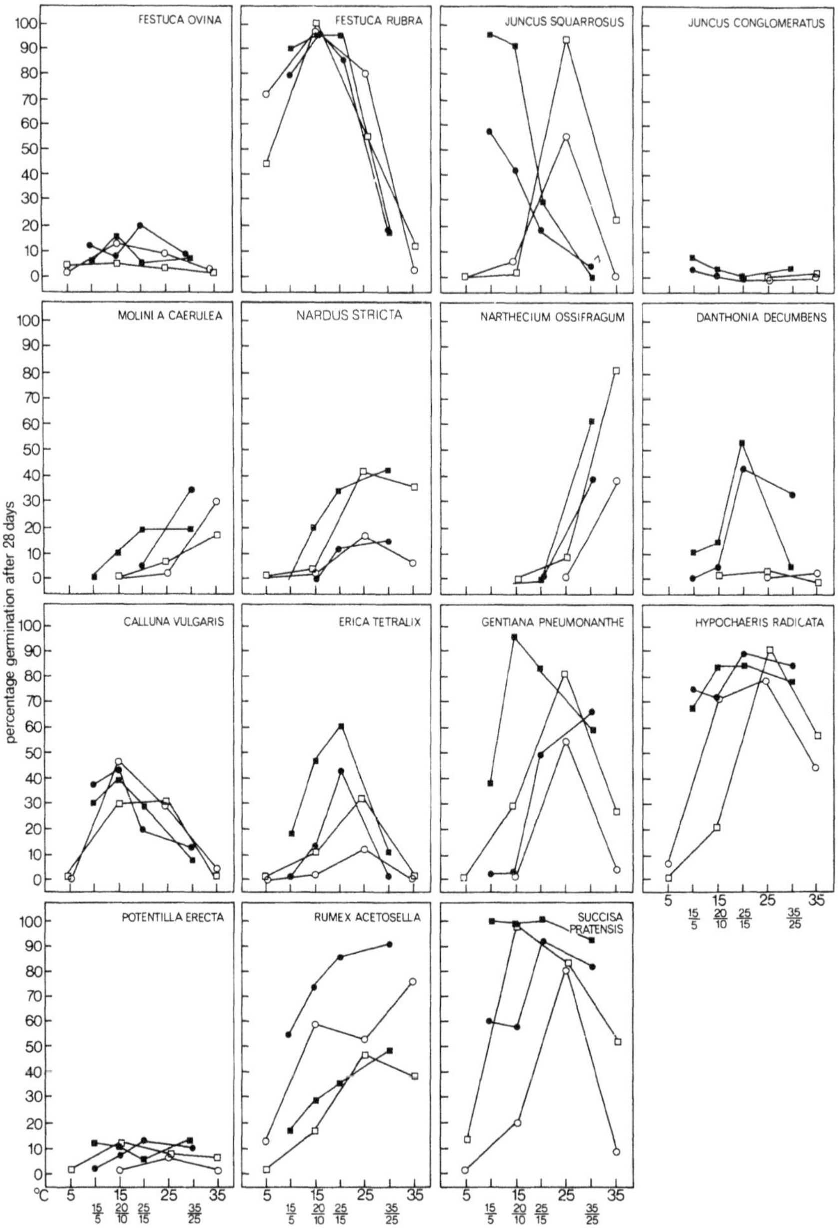

Fig. XII.2. Germination percentages of inland grassland species (Westerholt) after 28 days under constant (open symbols) and fluctuating temperatures (closed symbols) of dry stored (circles) and one month stratified seeds (squares) (after Hendriks, 1984).

ature fluctuations than with constant temperatures. The group with *Festuca pratensis* also only contained grasses, but a larger share of these species showed an enhanced response at fluctuating temperatures. The group with *Caltha palustris* consisted mainly of dicotyledons, whereas diurnal temperature fluctuations enhanced the germination response of most species. The groups with *Cardamine pratensis* and with *Lotus uliginosus* mainly contained dicotyledons.

The period of the emergence of seedlings in the garden (Fig. XII.3) corresponds with the germina-

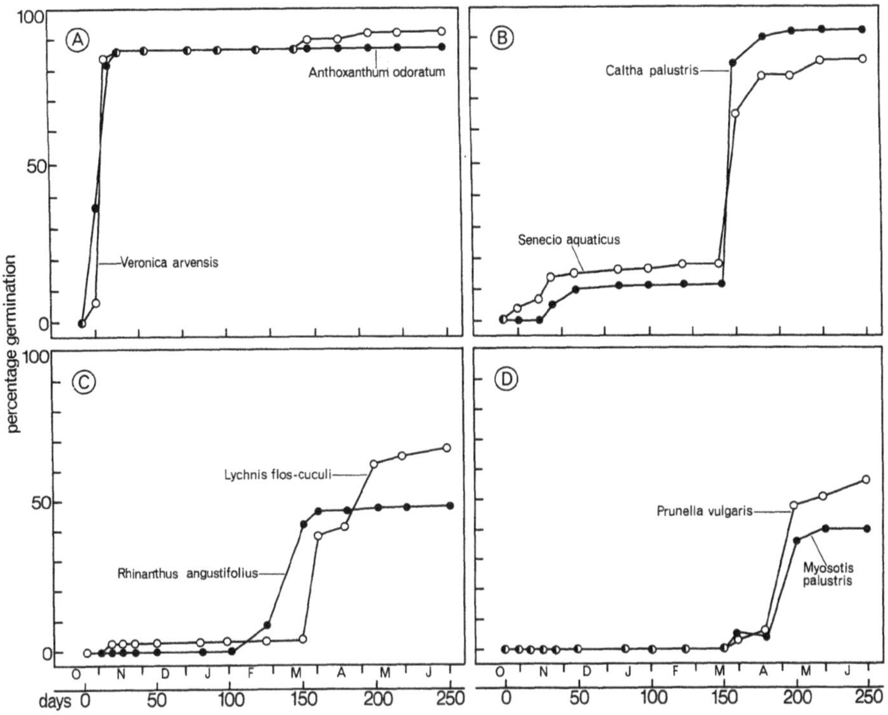

Fig. XII.3A–D. Germination percentages of dry stored inland grassland species (Loefvledder) sown in sterilized soil in the garden at the 21th of October 1977 with A: germination response in the autumn (also *Bromus hordeaceus, Cynosurus cristatus, Dactylis glomerata*), B: high germination response in early spring (also *Plantago lanceolata, Cirsium palustre, Heracleum sphondylium*), C: lower germination response in early spring (also *Epilobium hirsutum, Cirsium vulgare*), D: lower germination response in late spring (also *Trifolium dubium, Crepis paludosa, Sagina procumbens, Filipendula ulmaria, Lotus uliginosus, Carex ovalis, Lysimachia vulgaris*).

tion response in the laboratory (Table XII.4). *Molinia caerulea* and *Nardus stricta* are known to germinate at relatively high temperatures (Grime et al., 1981), which is in agreement with these findings. In the literature a number of species with higher germination responses after stratification in the laboratory, namely *Molinia caerulea* and *Nardus stricta,* or those which featured seedlings in the spring in the garden, namely, *Caltha palustris, Heracleum sphondylium* and *Lotus uliginosus* also showed higher germination percentages after stratification (Grime et al., 1981).

In an *Agrostis* spec./*Holcus lanatus*/*Rumex acetosa*/*Ranunculus acris* community in both the undisturbed plots and in bared patches in the field Van Zeist (1978) found more seedlings of grass species emerged in the autumn and more seedlings of dicotyledons emerged in the spring, which is in agreement with the present study.

XII.D. Re-introduction into the field

XII.D.1. The salt marsh (Schiermonnikoog)

The germination and early establishment of sown species at the different sites along the gradient in grazed and abandoned areas is shown in Fig. XII.5A for some salt marsh species, and in Fig. XII.5B for some dune species. The lower salt-marsh species hardly germinated in the dune area, but more seedlings were found lower down the gradient. Many more seedlings emerged in the grazed area. Most seedlings which emerged died, but in seedlings seen in October some species persisted in the grazed area at the lower part of the gradient, and one species also survived at the abandoned area.

Seedlings of both dune and salt marsh species were found in the grazed and the abandoned area. Most seedlings were, however, found in the grazed

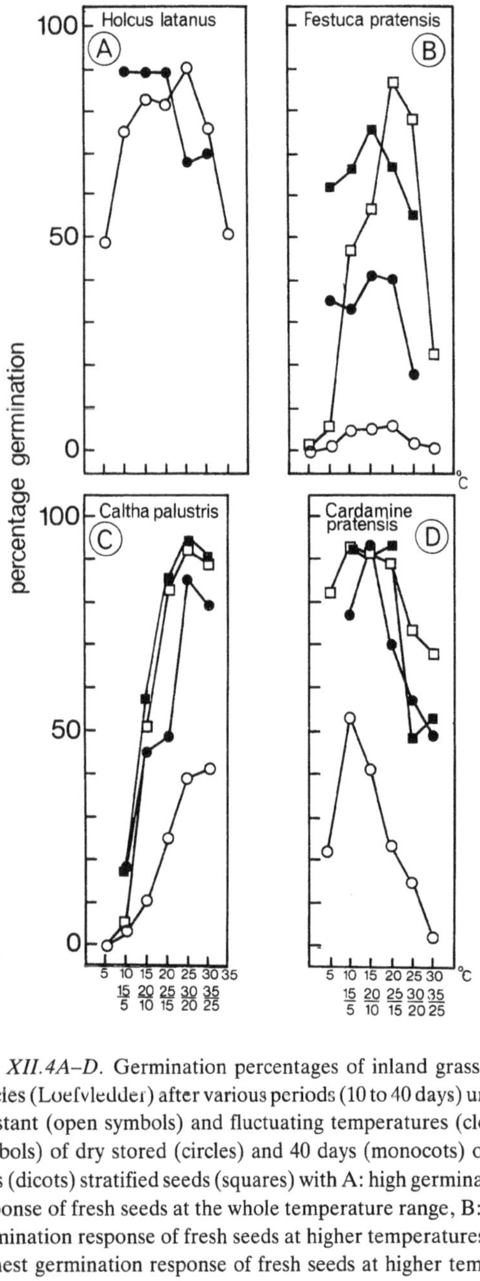

Fig. XII.4A–D. Germination percentages of inland grassland species (Loefvledder) after various periods (10 to 40 days) under constant (open symbols) and fluctuating temperatures (closed symbols) of dry stored (circles) and 40 days (monocots) or 75 days (dicots) stratified seeds (squares) with A: high germination response of fresh seeds at the whole temperature range, B: low germination response of fresh seeds at higher temperatures, C: highest germination response of fresh seeds at higher temperatures, D: highest germination response of fresh seeds at lower temperatures (unpublished data from D.M. Pegtel).

area with the exception of seedlings of salt marsh species which were found in large numbers in the abandoned area on the lowest site of the salt marsh. This can be attributed to the dominance of *Artemisia maritima* which has died off in the winter period and subsequently leaves a great deal of bare soil in the spring. Most seedlings died. In October only seedlings of four species had survived, mainly on the grazed area.

Not only the grazed or ungrazed status of the plots was related to the number of emerged seedlings. In the abandoned area the better seedling emergence of sown glycophytes as compared with the halophytes could be explained by their heavier seeds (Fig. XII.5). Only the small *Cochlearia danica* seeds gave rise to larger numbers of seedlings in the abandoned area.

By March 1984, no seedlings of either salt marsh or dune species were found to have survived. In all the sites, establishment was clearly influenced by soil movement as, with the exception of site 1, the marking pins were found to have been buried (as a consequence of winter gales) under a layer of clay up to 5 mm thick at the lowest site along the gradient.

The fact that no seedlings of the sown species survived during the study period is in agreement with the infrequent establishment of new individuals of lower salt marsh species during the 12 years of observation in the study area.

XII.D.2. Mixed grassland (Westerholt)

The average numbers of seedlings which emerged in May 1983, four months after the sowing of the seeds, in the heavily grazed, lightly grazed, 'bare' hay-field and 'moss' hay-field areas are shown in Fig. XII.6. In the case of 7 of the 14 species which had been sown seedling emergence was significantly highest in the 'bare' hay-field, whereas *Festuca ovina, Juncus conglomeratus* and *Molinia caerulea* only emerged in the 'bare' hay-field area. *Rumex acetosella* did not emerge at all. The numbers of emerged seedlings of *Succisa pratensis* did not differ in any management practice. In the case of some species the heavily grazed plots usually revealed more emerged seedlings than the other regimes, except for the 'bare' hay-field.

The percentages bare soil differed significantly in the various management regimes according to the series 'bare' hay-field > heavily grazing > 'moss' hay-field, lightly grazing (see Table X.11). The numbers of seedlings which emerged were, however, not highest in the 'bare' hay-field in the case of all species. The exceptions were *Succisa pratensis, Danthonia decumbens* and *Gentiana pneumonanthe,* and this could be explained by their heavy seeds.

The survival of seedlings in the different management regimes is shown in Fig. XII.7. The few seedlings which emerged in the 'moss' hay-field and the lightly grazed patches nearly all disappeared between May and October 1983. In the heavily grazed area over 75% of the seedlings of *Nardus stricta, Danthonia decumbens* and *Succisa pratensis* survived and 30% of *Festuca rubra* and *Hypochaeris radicata.* Only in the 'bare' hay-field did seedlings of most species survive, whereas 9 species were still found in June 1984. *Succisa pratensis,* with its heavy seeds, was the only species of which seedlings maintained themselves in all management regimes. The survival percentages in October of seedlings which emerged in May are shown in Fig. XII.8. This survival is correlated with the seed weight of the sown species ($r = 0.607$, $P<0.05$).

XII.D.3. Valley grasslands (Loefvledder)

The total number of seedlings which emerged in the different management practices, old and new, since no cohorts were discerned, is shown in Fig. XII.9 A–D. It is obvious from the comparison with the germination response of the same species in sterilized soil in the garden (Fig. XII.3 A–D), that a similar pattern of germination was found in the field. Only the difference between the higher and lower germination response in early spring in the garden proved not to be consistent with the field data. Some species like *Anthoxanthum odoratum* already showed a germination response in the autumn (Fig. XII.9 A). The other three groups needed stratification since seedlings emerged in the spring, early like *Caltha palustris* (Fig. XII.9 B) and *Rhinanthus angustifolius* (Fig. XII.9 C), or some-

Table XII.4. Germination response patterns of inland grassland species as examplified for some individual species in Fig. XII.2 (Westerholt) and Fig. XII.4 (Loefvledder). An asterisk means that the germination response was enhanced by diurnal fluctuating temperatures, in particular after stratification.

High germination response of fresh seeds at all temperatures (Fig. XII.4A).		
Holcus lanatus	Alopecurus geniculatus	Lolium perenne
Phleum pratense	* Deschampsia cespitosa	Agrostis capillaris
Poa annua	Anthoxanthum odoratum	* Dactylis glomerata
* Poa pratensis	Cynosurus cristatus	
Low germination response of fresh seeds (Fig. XII.4B)		
* Festuca pratensis	* Festuca rubra	Agrostis stolonifera
Arrhenatherum elatius	* Glyceria fluitans	* Danthonia decumbens
High germination response of fresh seeds at high temperatures (Fig. XII.4C)		
* Caltha palustris	* Cirsium palustre	Nardus stricta
* Myosotis palustris	* Lychnis flos-cuculi	* Juncus squarrosus
* Ranunculus flammula	* Achillea ptarmica	* Narthecium ossifragum
Plantago lanceolata	Filipendula ulmaria	Gentiana pneumonanthe
* Lythrum salicaria	Hypochaeris radicata	* Succisa pratensis
* Ranunculus acris	* Agrostis canina	* Erica tetralix
* Mentha aquatica	Juncus effusus	Molinia caerulea
* Prunella vulgaris		
High germination response of fresh seeds at low temperatures (Fig. XII.4D)		
* Cardamine pratensis	* Rumex acetosella	Calluna vulgaris
Cerastium fontanum	Rumex acetosa	
Low germination response of both fresh and stratified seeds		
Lotus uliginosus	Galium palustre	Juncus conglomeratus
Trifolium pratense	Festuca ovina	Potentilla erecta
Trifolium repens		

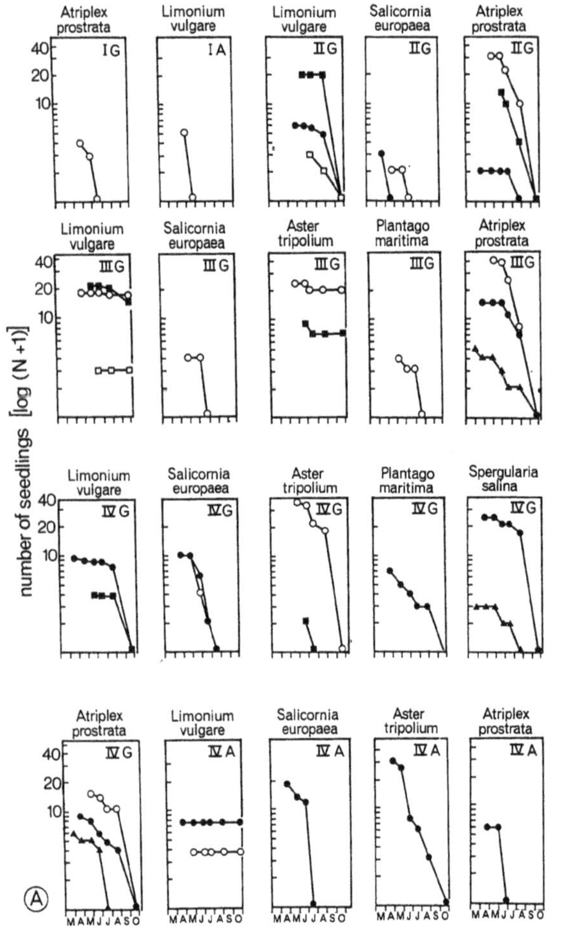

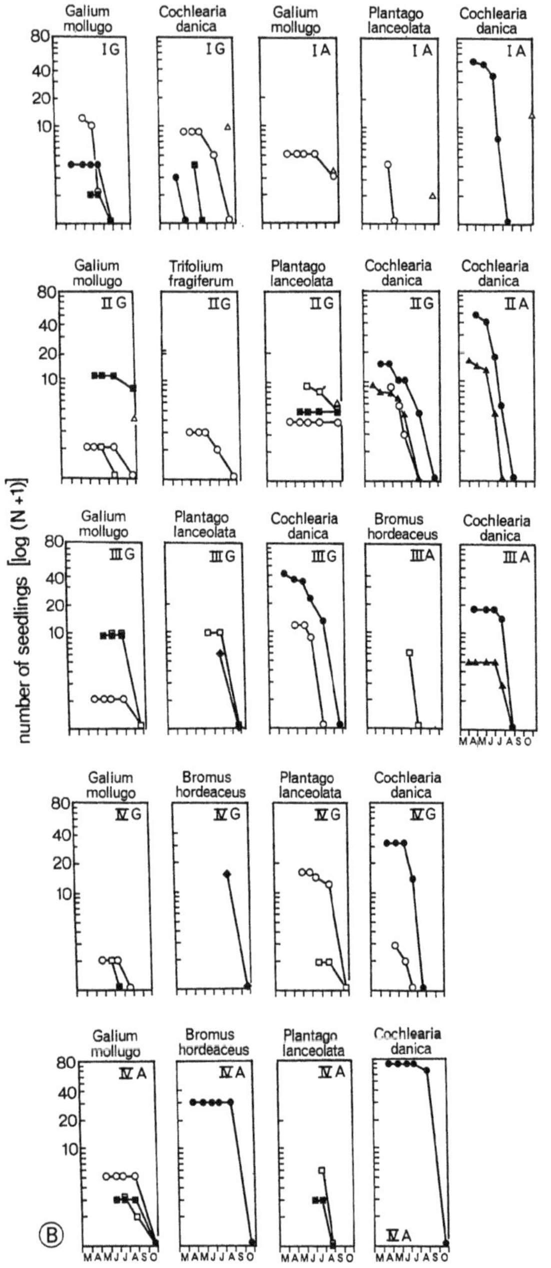

Fig. XII.5A–B. Survivorship curves, log (N + 1), of cohorts of seedlings of species introduced in March 1983 at grazed (G) and abandoned (A) areas of the study sites I (dune) to IV (lower salt marsh) along the gradient of A: lower/mid salt marsh species, B: mid salt marsh/dune species. Seed weight (mg) of species is: *Salicornia europaea* 2.7, *Aster tripolium* 9.4, *Plantago maritima* 5.3, *Spergularia salina* 0.6, *Atriplex prostrata* 17.3, *Bromus hordeaceus* 12.3, *Plantago lanceolata* 14.9, *Cochlearia danica* 2.1 (after Bakker et al., 1985).

what later like *Filipendula ulmaria* (Fig. XII.9 D).

The group with *Anthoxanthum odoratum* (Fig. XII.9 A) already showed a germination response in the autumn. *Dactylis glomerata,* moreover, germinated in all regimes, with the highest response in the July and September hayed plots. The seedlings, however, disappeared in all the regimes during the winter. *Veronica arvensis* did not reveal any seedlings in the plot which was mown in both July and September, but most seedlings were found in the plot which was mown in September.

Other species which showed seedlings in the early spring (Fig. XII.9 B–C) were *Cirsium palustre,* with only a few individuals in all regimes, *Senecio aquaticus* which germinated in all regimes, but by far the most seedlings remained in the plot which was mown in both July and September.

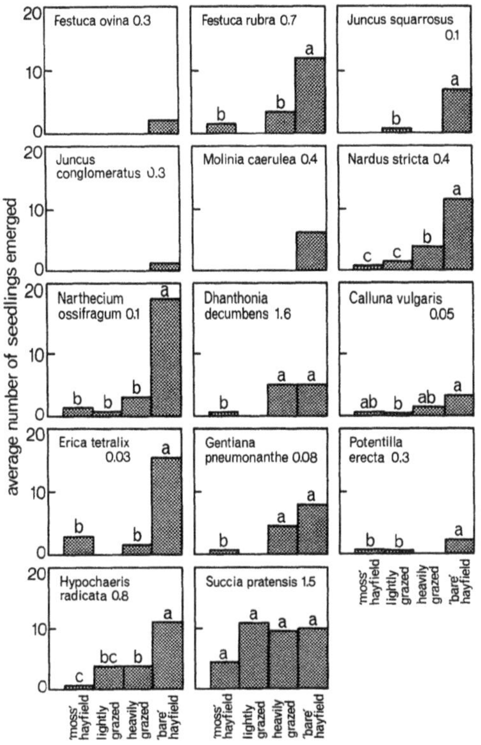

Fig. XII.6. Average number of seedlings emerged (± S.E.) of species introduced in January 1983 in 'moss' hay-field, lightly grazed patches, heavily grazed area and 'bare' hay-field in May. Different letters indicate significant differences, P<0.05 according to the Student test. Seed weight (mg) is indicated (after Hendriks, 1984).

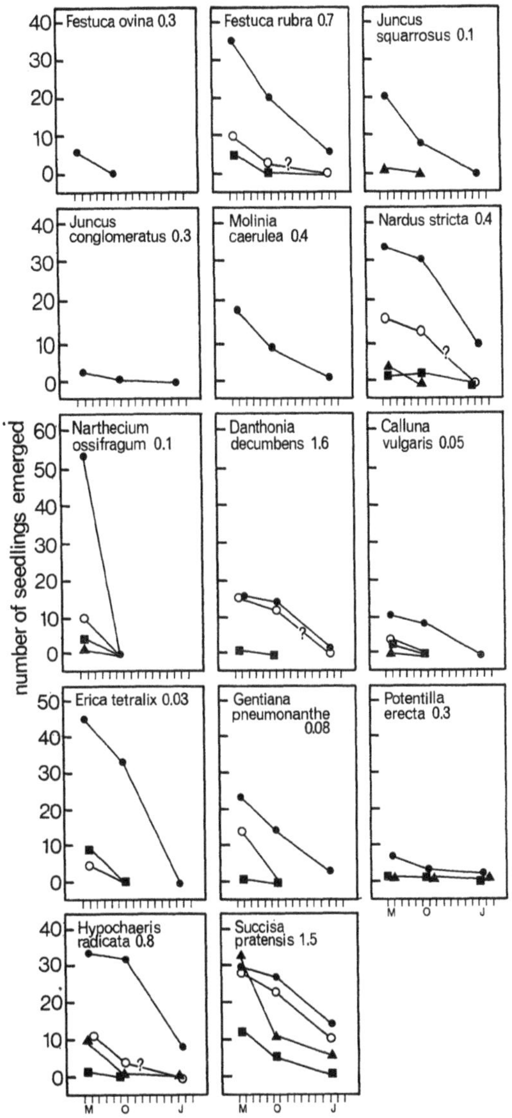

Fig. XII.7. Survivorship curves of seedlings of species introduced in January 1983 in 'moss' hayfield (squares), in lightly grazed patches (triangles), in heavily grazed areas (open circles) and in 'bare' hayfield (closed circles). Question marks in the heavily grazed area indicate that coloured pins had disappeared by trampling of the sheep. Seed weight (mg) is indicated (after Hendriks, 1984).

Seedlings of *Epilobium hirsutum* also appeared in all regimes but only remained in the plots which were cut in July and in both July and September.

Seedlings of species which emerged somewhat later in the spring (Fig. XII.9 D) were also found e.g. *Lotus uliginosus,* but only in the plot which was mown in July, whereas those of *Myosotis palustris* only remained in the plot which was mown in September. Seedlings of *Crepis paludosa* only appeared in the abandoned plot and in the plot which was cut in both July and September.

The numbers of seedlings found in June 1978, eight months after the sowing of the seeds, differed between the management practices. The sequence of numbers of species of which seedlings were found was July + September hay-making (12 species) > September hay-making (11) > July hay-making (11) > abandoned (3). If only species were taken into account of which more than 10 seedlings were found in June, the sequence was July + September hay-making (8 species) > September hay-making (6) > July hay-making and abandoned (2).

There was a trend towards the best emergence and/or maintenance of seedlings of seeds sown in October in plots in which the standing crop was

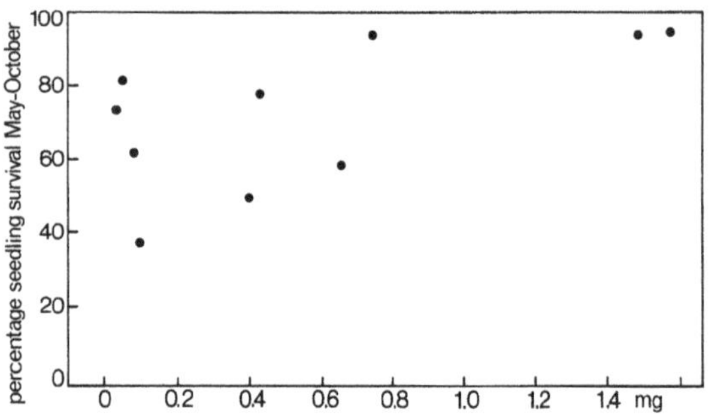

Fig. XII.8. Percentage of seedlings survived between May and September after sowing in 'bare' hayfield in January 1983 related to different seed weights (after Hendriks, 1984).

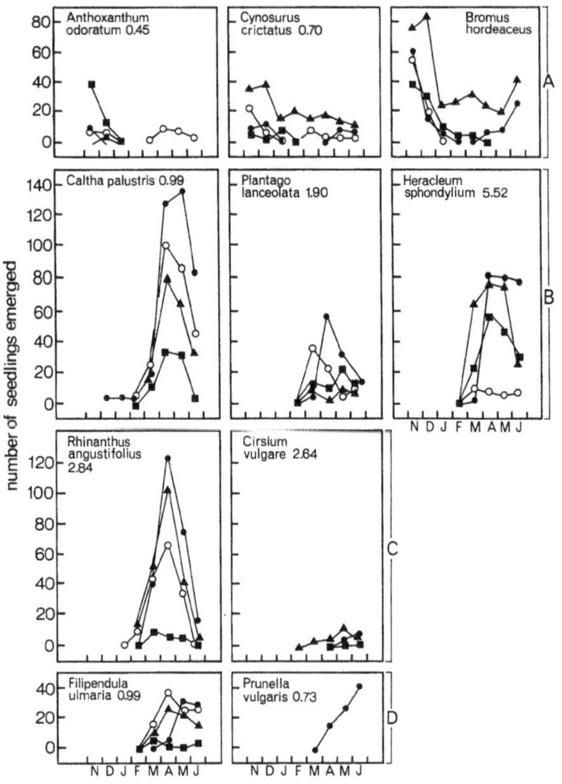

Fig. XII.9A–D. Number of seedlings emerged (old and new) of species introduced in November 1977 in abandoned area (squares), September hay-making (triangles), July hay-making (open circles) and both July and September hay-making regimes (closed circles). Seed weight (mg) is indicated. See Fig. XII.3 for subdivision A–D.

removed in September. The species which had considerable numbers of seedlings in June in the abandoned plot, namely, *Heracleum sphondylium* and *Plantago lanceolata,* have heavy seeds.

Large dynamics in the observed numbers of seedlings can be seen as derived from Fig. XII.9. These dynamics are, in fact even larger since both seedlings and juveniles (second (pair of) leave(s) appeared), and in some cases even flowering adults, were concerned as can be seen in some examples in Fig. X.17. Juveniles of *Anthoxanthum odoratum, Veronica arvensis, Cynosurus cristatus, Bromus hordeaceus, Caltha palustris, Plantago lanceolata, Heracleum sphondylium, Rhinanthus angustifolius* and *Filipendula ulmaria* were recorded. In June 1978, the last month of observation before hay-making in July, a large number of these juveniles had disappeared again. Many seedlings of *Bromus hordeaceus* turned into flowering individuals and only a few of *Veronica arvensis* and *Rhinanthus angustifolius.* All individuals which emerged were removed from the experimental plots in all management regimes. The long-term fate of these seedlings is, therefore, not known.

XII.D.4. Long-term observations on seedlings

The long-term fate of the seedlings of re-introduced species was not recorded in the different management regimes of the Loefvledder area, since all the individuals which emerged were removed before the hay-making in July in order not to influence the successional series. The observation period, therefore, only covered eight months after the sowing in November of the preceding year.

Another re-introduction experiment took place in a lot in the Anlo study area. Stratified seeds of species not occurring in the established vegetation nor in the seed bank were sown at the end of March 1983. The vegetation was dominated by *Holcus lanatus.* Seeds were sown into (i) the sward (ii) plots clipped to 5 cm (which was repeated every two months), and (iii) plots with bare sterilized soil from which invading tillers were removed every two months. Batches of 80 seeds each were sown into five plots $(10 \times 10)\,cm^2$ in each of the treatments. The seedlings which emerged were marked by coloured pins. Only in July 1983 were seedlings

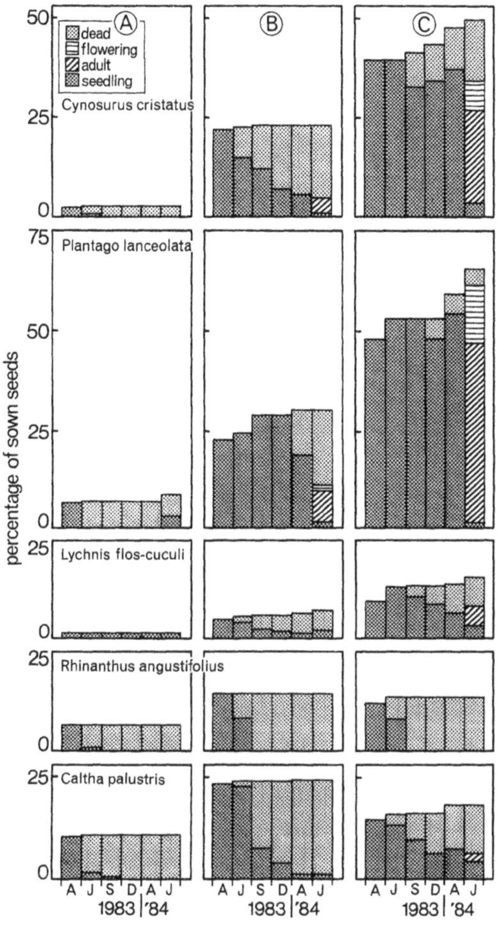

Fig. XII.10A–C. Percentages of different stages in the life cycle related to the number of viable seeds sown in March 1983 in A: sward mown in July 1983, in B: sward repeatedly clipped to 5 cm height and in C: bare plot with sterilized soil and repeatedly removed tillers (after Hoogveld & Muller, 1985).

recorded when hay-making was performed (clipping by hand in the experimental plot in order not to remove the marking pins close to the seedlings) and the observations continued until July 1984 when the adult individuals were removed. The observation period covered, therefore, a period of 15 months.

It is obvious from Fig. XII.10 A–C that the emergence and later development of seeds of *Cynosurus cristatus* and *Plantago lanceolata* was better in the clipped plots (B) and much better in the bare plots (C) than in the untouched vegetation (A). This was also the case, but to a lesser extent, for seeds of *Lychnis flos-cuculi.* There was no difference in the clipped and bare plots for seeds of *Rhinanthus angustifolius.* The seedling emergence of *Caltha palustris* was even lower in the bare plots than in the clipped plots. Hardly any seedling of the two latter species survived, whereas some of *Cynosurus cristatus* and *Plantago lanceolata* reached the flowering stage in the bare plots.

XII.E. Limiting factors in restoration succession

The conclusion from the previous section is that the re-introduction of seeds of species absent from both the established vegetation and in the seed bank can result in the emergence of such species. Real establishment of new species could not be studied, because the emerged plants were removed in order not to interfere with the succession which occurred.

A great number of the experiences mentioned in the literature concerning the (re-)introduction of species in nature reserves involve the transplantation of adult individuals. Sometimes direct transplantation took place, but seeds were often collected and grown to plants in the greenhouse or experimental garden followed by transplantations of those individuals. Such transplantations were successful for many species in Germany (Schwaar, 1980, 1985; Montag, 1980; Glitz, 1980; Ebel & Rauschert, 1982) and in the United Kingdom (Wells, 1983). Although some species did not maintain themselves, a large number of species transplanted in sods had survived and had spread 15 years after the transplantation (Rawes & Welch, 1972) or 5 years after transplantation (Wathern & Gilbert, 1978) in experiments in the U.K. Londo (1984) gives examples of transplantations of some rare species in the Netherlands but these failed.

The (re-)introduction of seeds in nature reserves was only mentioned by Glitz (1980) in Germany and involved *Gentiana pneumonanthe, Dactylorhiza majalis, D. maculata, Narthecium ossifragum* and *Drosera rotundifolia.* The results turned out to be negative. Londo (1984) assumes that the (re-)introduction of seeds was certainly carried out in reserves in the Netherlands, but no results are available due to the poor registration of those (re-)introductions.

More information is available on the introduction of seeds in newly reinstated grasslands e.g.

Table XII.5. Survival of species after a certain number of months of sowing in different management regimes according to assessments in the literature and this study. + : species present at the moment of observation; – : species not present at the moment of observation.

Assessment	This study		This study		Altena (1983)		Wathern & Gilbert (1978)		Wathern & Gilbert (1978)			Londo (1982B)
Months of observation	6	6	10(15)	10(15)	18	18	24(60)	24(60)	26(63)	26(63)	26(63)	24 (80)
Management	Uncut	Twice hay-making	Once hay-making	Lightly grazing	Twice hay-making	Mulching	Bare soil	Uncut	Uncut	Hay-making	Paraquat	Twice hay-making
Heracleum sphondylium	+	+	.	.	+	-	.	.	.	.	.	.
Plantago lanceolata	+	+	.	.	+	+	+ (-)	+ (+)	+ (+)	+ (+)	+ (+)	.
Filipendula ulmaria	+	+	.	.	.	.	.	.	.	.	.	.
Crepis paludosa	+	+	.	.	.	.	.	.	.	.	.	.
Senecio aquaticus	+	+	.	.	.	.	.	.	.	.	.	.
Bromus hordeaceus	-	+	.	.	.	.	.	.	.	.	.	.
Cynosurus cristatus	-	+	.	.	+	+	.	.	.	.	.	+ (+)
Rhinanthus angustifolius	-	+	.	.	+	+	.	.	.	.	.	+ (-)
Caltha palustris	-	+	.	.	.	.	.	.	.	.	.	.
Cirsium vulgare	-	+	.	.	.	.	.	.	.	.	.	.
Prunella vulgaris	-	+	.	.	+	+	+ (+)	+ (+)	+ (+)	+ (-)	+ (+)	.
Festuca rubra	.	.	+ (+)	- (-)	.	.	.	.	.	.	.	.
Juncus squarrosus	.	.	+ (-)	- (-)	.	.	.	.	.	.	.	.
Molinia caerulea	.	.	+ (-)	- (-)	.	.	.	.	.	.	.	.
Nardus stricta	.	.	+ (+)	- (-)	.	.	.	.	.	.	.	.
Danthonia decumbens	.	.	+ (+)	- (-)	.	.	.	.	.	.	.	.
Calluna vulgaris	.	.	+ (-)	- (-)	.	.	.	.	.	.	.	.
Erica tetralix	.	.	+ (-)	- (-)	.	.	.	.	.	.	.	.
Gentiana pneumonanthe	.	.	+ (+)	- (-)	.	.	.	.	.	.	.	.
Hypochaeris radicata	.	.	+ (+)	+ (-)	.	.	.	.	.	.	.	.
Potentilla erecta	.	.	+ (+)	+ (-)	.	.	.	.	.	.	.	.
Succisa pratensis	.	.	+ (+)	+ (+)	.	.	.	.	.	.	.	.
Achillea millefolium	.	.	.	.	+	-	+ (-)	+ (+)	- (+)	+ (+)	+ (+)	.
Leucanthemum vulgare	.	.	.	.	+	+	+ (+)	+ (+)	.	.	.	.
Galium verum	.	.	.	.	+	-	+ (+)	+ (+)	.	.	.	.
Achillea ptarmica	.	.	.	.	+	-	.	.	.	.	.	.
Centaurea jacea	.	.	.	.	+	+	+ (+)	+(+)	+(+)	+(+)	+(+)	.
Trifolium pratense	.	.	.	.	+	+	.	.	.	.	.	.
Crepis biennis	.	.	.	.	+	+	.	.	.	.	.	+ (+)
Pimpinella saxifraga	.	.	.	.	+	+	.	.	.	.	.	+ (+)
Rumex acetosa	.	.	.	.	.	.	+ (+)	+ (+)	+ (+)	+ (-)	+ (-)	.
Knautia arvensis	.	.	.	.	.	.	+ (+)	+ (+)	- (+)	- (-)	+ (+)	.
Lathyrus pratensis	.	.	.	.	.	.	+ (+)	+ (+)	+ (+)	+ (+)	+ (+)	.
Hypericum perforatum	.	.	.	.	.	.	- (-)	+ (+)	+ (+)	+ (+)	+ (+)	.
Primula veris	.	.	.	.	.	.	+ (+)	+ (+)	.	.	.	.
Anthyllis vulneraria	.	.	.	.	.	.	+ (+)	+ (+)	.	.	.	.
Lotus corniculatus	.	.	.	.	.	.	+ (+)	+ (+)	.	.	.	+ (+)
Sanguisorba minor	.	.	.	.	.	.	+ (+)	+ (+)	.	.	.	.
Anthriscus sylvestris	.	.	.	.	.	.	+ (-)	+ (+)	.	.	.	.
Ranunculus acris	.	.	.	.	.	.	+ (+)	+ (-)	.	.	.	.
Medicago lupulina	.	.	.	.	.	.	+ (+)	+ (+)	.	.	.	.
Vicia angustifolia	.	.	.	.	.	.	+ (-)	+ (+)	.	.	.	.
Rhinanthus minor	.	.	.	.	.	.	- (-)	+ (-)	.	.	.	.
Tragopogon pratensis	.	.	.	.	.	.	- (-)	+ (-)	.	.	.	+ (+)
Geranium pratense	.	.	.	.	.	.	.	.	- (+)	+ (+)	+ (+)	+ (-)
Arrhenatherum elatius	.	.	.	.	.	.	.	.	.	.	.	+ (+)
Daucus carota	.	.	.	.	.	.	.	.	.	.	.	+ (?)
Dianthus deltoides	.	.	.	.	.	.	.	.	.	.	.	+ (+)
Helictotrichon pubescens	.	.	.	.	.	.	.	.	.	.	.	+ (+)
Luzula campestris	.	.	.	.	.	.	.	.	.	.	.	+ (+)
Pastinaca sativa	.	.	.	.	.	.	.	.	.	.	.	+ (+)
Ranunculus bulbosus	.	.	.	.	.	.	.	.	.	.	.	+ (?)
Trisetum flavescens	.	.	.	.	.	.	.	.	.	.	.	+ (?)
Briza media	.	.	.	.	.	.	.	.	.	.	.	- (+)
Hieracium pilosella	.	.	.	.	.	.	.	.	.	.	.	- (+)
Scabiosa columbaria	.	.	.	.	.	.	.	.	.	.	.	- (+)
Veronica chamaedrys	.	.	.	.	.	.	.	.	.	.	.	- (+)
Campanula rapunculus	.	.	.	.	.	.	.	.	.	.	.	+ (-)
Euphrasia stricta	.	.	.	.	.	.	.	.	.	.	.	+ (-)

road verges or in experimental gardens. In the U.K. a sowing experiment was carried out by Wathern & Gilbert (1978) in a dense sward dominated by *Lolium perenne* and *Festuca rubra.* Strips were prepared for seeding by cutting to a height of 12 cm, a treatment with paraquat which was in turn followed by oversowing with *Festuca ovina* to prevent erosion, and there was an untouched control (Table XII.5). After sowing in the late summer the experimental area was left unmown for the next five years. A great deal of the sown species flowered in the first or the second year and seeded back into the plots. The seedling establishment of most species was highest in the paraquated strip after the first six months. The results after five years indicated that the crucial period for many herbs is the initial establishment. Although the rapid growth of grasses in the cut and control plots prevented many seedlings from converting to mature plants, those that did survive spread vegetatively.

Another experiment was performed on slightly sloping raw calcareous subsoil (Wathern & Gilbert, 1978). Two areas were sown with commercial grass mixtures, a *Lolium perenne* mixture and a *Festuca* mixture, whereas a third area was left unsown as a control (Table XII.5). All the plots were then oversown with a wildflower mix containing 26 species and left untouched. After two years, 22 of the herbaceous species established in the intermediate *Festuca* sward, while slightly fewer were found in the denser *Lolium* sward and the bare control area contained 17 species. After five years the abundancy of herbs had strongly increased in the *Festuca* sward by both seeding and vegetative spread.

Londo (1982B) studied the succession in an experimental garden in the Netherlands starting from bare soil, with different soil types, on which the vegetation was cut twice a year. The vegetation developed into a species-rich dry grassland. He carried out sowing experiments with 79 species, for the first time three years after the start of the experiment and it was repeated several times. Sixteen species had established after two years, but *Campanula rapunculus, Geranium pratense, Euphrasia stricta* and *Rhinanthus angustifolius* disappeared later (Table XII.5). It took three years after the first sowing for *Hieracium pilosella* and *Veronica chamaedrys* to establish and six years for *Briza media* and *Scabiosa columbaria.* Fifty-nine species did not establish at all.

In contrast to the afore-mentioned authors, Wells et al. (1981) stressed the management of sown areas, since it is crucial to both the establishment and the final composition of the grassland. If the mixture is sown without a nurse crop there may be considerable problems from weeds. These problems are, however, less with a nurse crop but the growth of the nurse crop must itself be controlled if the slower-germinating herbs are to survive. The precise timing of the cuttings depends on the soil fertility and stage of growth of the species.

Van Ast et al. (1987) carried out sowing experiments in November with seeds of *Hypochaeris radicata* in road verges in the Netherlands. More seedlings emerged in the site in which litter was removed than in an untouched site. Moreover, the species maintained itself better in sites mown in May or July with a second cut in November providing gaps for new seedling emergence than in untouched sites or sites mown in the wrong period.

Altena (1983) carried out sowing experiments in grasslands under restoration management in the Netherlands the results of which have previously been mentioned by Oomes & Mooi (1981). The difference with the re-introduction experiments in this study area is that Altena created gaps in most experiments before sowing. Seedling emergence and survival after 18 months in a closed canopy were higher if the seeds were sown in the autumn rather than in the spring, but it was highest if sown in gaps in the spring. Another experiment showed that the width of the gap (10 cm, 5 cm or 2 cm) revealed no difference, but the seedling emergence and survival was always higher in gaps than in a closed sward. Finally, the period of hay-making proved important for the survival of seedlings during the first 18 months. They maintained themselves better with an early cut in May or June followed by a second cut in September than with the first cut in July or in August. A poor survival was found in the mulched site (Table XII.5).

The light interception by the canopy under different management practices and the room available on the soil surface for the emergence and survival of seedlings from sown seeds has already been discussed in section X.D, although some further information is provided below.

Van Tooren et al. (1987) tested the germination of 12 chalk grassland species on soil under a bryophyte layer and on bare soil in the glasshouse. The germination of 9 species was significantly reduced by the bryophyte layer, which is in accordance with the small seedling emergence in 'moss' hay-field in this study. Fenner (1978) demonstrated the importance of the structure of the sward on the establishment of sown species. Seeds of six different closed-turf species were sown into 2.5 cm-wide gaps in short turf (1 cm tall) and tall turf (8 cm) of *Festuca rubra.* The germination and shoot weight were much reduced after 13 weeks in the tall turf as compared to the short turf. A significant correlation was found between the seed weight and dry weight of the three-month-old seedlings in the tall turf. Three species produced flowers after eight months in the short turf and only one species flowered in the tall turf. Gross & Werner (1982) sowed seeds of four biennial species of different seed sizes into swards of different densities with 67, 11 and 0.8% bare ground, respectively. Seedling emergence, survival and growth were highest in the bare ground patches in the case of all four species. The small seeded species only established in the stand with a lot of bare soil, whereas the large seeded species survived in all the stands. In another experiment Gross (1984) again stressed the importance of seed size for seedling emergence and growth. During et al. (1985) found a general tendency of small-seeded species to show higher seedling mortality in chalk grasslands.

Miles (1974) considered the effect of the width of a gap on seedling emergence and establishment. He created bare patches of 25, 250 and 2500 cm^2 in heathland in Scotland. The density of self-sown seedlings which had emerged after one year decreased with increasing patch size, but the survival of seedlings after four years showed the reverse trend. The shelter and humid conditions provided by the surrounding plants probably favoured germination in the small gaps, but this effect can later be countered by competition from the quicker encroaching vegetation in the smaller gaps than in the larger ones. De Hullu (1985) found that *Rhinanthus angustifolius* performed better in early successional stages in 7 cm^2 rather than in 36 cm^2 gaps, but no differences were found in late successional stages as other factors apparently then played a role.

The main cause of death in seedlings of species of herbaceous perennials in chalk grassland appeared to be desiccation (Silvertown & Dickie, 1981). Biotic factors, such as predation, disease and competition also play their role in maintaining a high level of mortality.

The high germination rate of many species in bare patches, as compared to a closed turf, is likely to be attributed to temperature characteristics of those patches. Thompson & Grime (1983) demonstrated that a large number of plant species showed an increased germination response under conditions with diurnal temperature changes. Such regimes simulate conditions of an exposed soil surface, which is the case in a bare patch.

Pre-emption of space by early germination and establishment has been demonstrated in experiments by Ross & Harper (1972) on *Dactylis glomerata* and Weaver & Cavers (1979) on *Rumex crispus.* A difference of only a few days in the time of emergence can result in a marked reduction in survival of the later cohorts. Van Ast et al. (1987) sowed *Hypochaeris radicata* into a dense *Festuca rubra* turf. The shoot weight of *Hypochaeris* was compared with that found after sowing into bare soil and amounted to 62% if sown at the same time as *Festuca,* 11% if sown two weeks after *Festuca* and 4% if sown four weeks after *Festuca.*

Although the experimental (re-)introduction layouts of the present study and those described in the literature are not always comparable, the results all point to similar conclusions. The emergence of individuals which reach the adult and flowering stage after the (re-)introduction of seeds is indeed possible for a number of grassland species (Table XII.5). It is, however, not known whether those individuals can maintain themselves in the long run by vegetative spread or self-sowing.

The naturally occurring temperature regimes do not seem to be a serious limiting factor for the establishment of new species during the process of restoration management. Three limiting factors can be identified for such an establishment. The first one is met if the characteristics of the habitat do not meet the physiological response of the species. The second one is the poor dispersal of many species and the third the lack of gaps once the seeds of a species have reached the spot.

The poor dispersal of diaspores was already discussed in chapter XI. Londo (1982B) tried to make a species-rich dry grassland community in an experimental garden in the Netherlands. He studied the succession during 10 years from bare soil with different soil types supplied from elsewhere. In that period 124 species of plants were recorded, of which most must have emerged from the seed bank. Only 21 out of the 238 species present in the adjacent garden established in the experimental garden, namely, 15 after deliberately sowing and 6 spontaneously. Schmidt (1981) started different cutting regimes on a former arable field in Germany. The km^2 in which the experiment was carried out contained 362 plant species. He studied the succession during 10 years from bare soil. In that period 267 species were recorded out of which 195 occurred in the adjacent 1000 m. The zoochorous species were, in particular, under-represented as compared to the surroundings. These two observations, again, stress the poor dispersal of many species.

Species with large seeds can interfere with the established surrounding vegetation, but small seeded species need gaps, either artificial or natural, for their emergence. Moreover, the germination response of many species can increase due to the conditions of diurnal temperature changes. Although gaps enhance the emergence of seedlings, they must not be too large since the eventual result of establishment can be smaller than in little gaps. Since the seeds of many grasses germinate in the autumn, those species have an advantage over dicotyledons, the species of which most emerge in the spring. Gaps, therefore, seem of particular importance for the spread of dicots. The occurrence of gaps is mainly determined by the management practices, i.e. timing of hay-making or grazing intensity.

PART SIX

Management

CHAPTER XIII

Management

XIII.A. Introduction

The nature management practices of grazing and cutting can be implemented for several purposes. Historical, cultural and aesthetic ones are not discussed here. One of the aims of nature conservation, is the maintenance of or the creation of conditions in natural and semi-natural landscapes for as many species as possible and, in particular, for those which are rare or endangered. The different practices are maintenance management and restoration management (see section I.C). The restoration management discussed in this study, namely that of grazing and cutting did not result in the re-establishment of rare or endangered species. Some species emerging in (fragments of) the *Saginion maritimae* in the salt marsh, in the *Violion caninae* in the reclaimed grassland and in the *Calthion palustris* and *Juncion acutiflori* in the valley grasslands do, however, represent a nature conservation interest.

In this final chapter, the effects of grazing (section XIII.B) and cutting (section XIII.C) are evaluated by comparing them with the aims of restoration management of this study and with other examples mainly in the Netherlands and some from abroad. There also appears to be a shortage of information on the effects of both practices. Although this study only deals with the effects of management practices on plants, the effects on animals should also be taken into account for management plans. Some results taken from the literature are, therefore, mentioned in section XIII.D. The effects of grazing and cutting management and the problems involved are discussed in section XIII.E. Finally, the question to what extent regular farming can be involved in restoration management is dealt with. The last section of this chapter XIII.F summarizes the gaps in the available knowledge on management regimes aiming at the restoration of species-rich grasslands.

XIII.B. Grazing

XIII.B.1. Aims and results of restoration management

The aims of restoration management of species-rich grasslands in the Netherlands by grazing (and hay-making) can be summarized as follows: (i) the prevention of succession to monotonous tall forb communities and woodland, (ii) the restoration of plant communities with rare and endangered species, (iii) the creation of patterns through different grazing intensities locally and (iv) the integration of former agricultural areas and nature reserves in order to remove the distinct boundaries (Oosterveld, 1975; Thalen, 1987; Van Wieren, 1987). These aims are followed by the removal of litter and impoverishment of the soil.

The removal of litter present in an abandoned (Photos XIII.1 and 2) area and the prevention of litter accumulation has been fulfilled in this study. On the salt marsh cattle and geese used so much of the herbage accumulated that only some dead standing material was found in the clumps of *Juncus maritimus*. All the litter present at the start of the grazing experiment had disappeared under a seasonal grazing regime within a few years. In the inland grassland, however, sheep used so little of the herbage accumulated that only a minor part of the area remained free of litter under a seasonal grazing regime, whereas standing dead material and litter was found on the remaining area. The area from which litter is removed or accumulation is prevented by grazing and trampling animals depends on the intensity of grazing. Hillegers (1984) reported the disappearance of litter a few years after the re-introduction of 'Mergelland' sheep in an abandoned chalk grassland. Van Wieren (1985) found that thick carpets of accumulated *Deschampsia flexuosa* litter had disappeared a few years after the introduction of Scottish Highland cattle on a former heathland. Bokdam et al. (1986) found similar results with dairy heifers. These last three experiments involved permanent grazing regimes.

Changes in the chemical composition of the soil on the regularly flooded salt marsh have not been studied. Soil chemical analyses from the reclaimed grassland area of the Westerholt could not reveal changes in the chemical composition of the soil. The appearance of species indicating poorer soil conditions made, however, clear that changes did take place. Van De Laar & Slim (1981) reported similar results. They found a decline of weeds of arable fields, namely, *Stellaria media* and *Apera spica-venti,* whereas species indicating poorer soil conditions, namely, *Agrostis capillaris* and *A. canina* appeared a few years after the introduction of Icelandic ponies on arable fields which had previously been heavily manured.

Although some species indicating more nutrient poor-soil conditions appeared, no rare or endangered species typical of heathlands appeared in the reclaimed grassland of the Westerholt or in the former arable field described by Van De Laar & Slim (1981).Under the environmental conditions prevailing in these study areas it apparently takes more than 15 years, starting from reclaimed, fertilized pastures or arable fields, to restore heathland or heathland-like communities occurring on nutrient poor soils using a grazing regime. In fact, only the initial stages can be reached within 15 years and more intensive practices like several cuts a year or cutting sods are necessary in order to remove sufficient nutrients. The restoration of former characteristic plant communities, including rare and endangered species, seems more successful in chalk grasslands and in areas without tillage of the soil and fertilizer applications, for example in abandoned heathland (Van Wieren, 1985; Bokdam et al., 1986), abandoned salt marsh (this study; P. Oosterveld pers. comm.), abandoned chalk grassland (Hillegers, 1984) and even fertilized chalk grassland (Willems, 1983A). Gibson et al. (1987) reported that after two years of a sheep grazing regime an abandoned arable field in the United Kingdom turned into a vegetation containing many chalk-grassland species after two years of a sheep grazing regime.

The number of species nearly always increased under grazing with low stocking rates. Such a regime resulted in the development of patterns consisting of a mosaic of heavily utilized areas and lightly utilized patches. Species indicating both nutrient-poor soil and nutrient-rich soil could establish and contribute to a relatively high number of species. The patterns or mosaics consisted of short turf and patches with tall grasses or *Juncus* spp. (this study) and accumulated litter on intermediate soils or tall forbs, for example, *Urtica dioica* or *Cirsium* spp. (Van De Laar & Slim, 1981) on heavily fertilized soils. They are still a long way from being like the patterns of, for example, the Borkener Paradies in northern Germany (Burrichter et al., 1980), which is considered to be a good example of the spatial relationship between grassland and woodland including a fringe community. This area has been grazed for centuries, but bushes and trees were cut regularly by farmers in order to maintain the

Photo XIII.1. A layer of litter and dead standing material is present in the abandoned salt marsh area.

Photo XIII.2. Litter and dead standing material are absent in the grazed salt marsh area.

open character of the area (see also Thalen et al., 1987). The fringe community in the Borkener Paradies is an *Alliario-Chaerophylletum,* but the species diversity is limited and, due to the large quantities of dung voided, the vegetation is dominated by *Urtica dioica* (Burrichter et al., 1980). In southern Sweden the 'ängen' do indeed show well developed, stable transitions from woodland to grassland via fringe communities. Their survival is due to an annual hay management in the pasture after which only the aftermath is grazed (Ekstam & Sjögren, 1973; see also Westhoff et al., 1983).

It appeared from an evaluation of 15 years of grazing management throughout the Netherlands carried out by the biggest private organization for nature conservation 'Vereniging tot Behoud van Natuurmonumenten', that the fringe communities between grassland and woodland are the most difficult to manage (H. Piek, pers. comm.). Scrub already present in reserves at the introduction of grazing tended to change into either woodland or grassland. Maybe the disturbance caused by a new management regime plays a role. Grazing at a low stocking density in areas with nutrient-rich soils did result in the appearance of patches with tall forbs, which can be considered characteristic of fringe communities, but they also included such non-characteristic species like *Urtica dioica* and *Elymus repens.* In such cases the authority in charge of the management can either wait a very long time for the development of scrub or it can try to create more nutrient-poor soil conditions by a few years of hay-making or by sod removal.

The boundaries between the heathland (2 ha) and grassland area (6 ha) in the Westerholt study area did become less sharp. This was, however, due to the fact that the heathland became grassy and not because of the change of the grassland into a heathland. Weber (1986) mentioned that farmers in northern Germany include small adjacent woodlands into their pasture areas. These woodlands are grazed and browsed and used as resting place. This process results in groups of trees with a pasture-like undergrowth. The same was observed in the Netherlands with regard to small oak and birch woods and small juniper scrubs (pers. comm. J.J. Barkman). In an area in the southern part of the Netherlands consisting of former arable fields (50 ha) and heathland and woodland (80 ha) some heathland species established in the agricultural area, whereas no arable weeds emerged in the heathland and woodland (Van De Laar & Slim, 1981). This apparent discrepancy supports the theory that nutrient-poor types of landscape must dominate with respect to relief and/or area over nutrient-rich types of landscape in order to obtain floristically valuable gradients (Van Leeuwen, 1966; Oosterveld, 1975).

XIII.B.2. Grazing in perspective

The short-term changes in the standing crop of five species under different grazing regimes in a simple ecosystem, namely, a young embanked sandflat was successfully modeled by Fresco et al. (1987). Three main processes were incorporated in the model viz. grazing, growth and death and variation in abiotic conditions (nitrogen, salinity and moisture). Lippe et al. (1985) and Prentice et al. (1987) simulated the vegetation dynamics of a heathland consisting of three species and found a good match with the cover data recorded over a period of 19 years. They used establishment rates, diameters and heights of individuals and growth constants as parameters. If such models are to be used for long-term prediction the changing ecosystem must be observed for longer periods of time and more parameters should also be involved. The authors could have made some validation of their model by using the changes in the heathland which were observed during the first 10 years, then simulating the vegetation dynamics of the next 9 years and comparing them with the field data of the same period. No such validation was, however, made. Moreover, if the models are to be of interest to nature conservation, they have to be extended to species-rich ecosystems. Berendse & Kwant (1985) built a model which predicts the development of a species-poor heathland vegetation over a period of 50–100 years. Nutrient availability, water conditions, interference of plants, mineralization, decomposition and sheep-grazing were incorporated in the model. Two main problems exist with regard to long-term predictive modeling. The first is of course the validation. Another is how to predict the disappearance of existing species and the invasion of species from outside the observed ecosystem.

It becomes evident from the afore-mentioned that long-term changes in species composition in complex ecosystems under an adopted grazing regime cannot be predicted by using a modeling approach. The species composition can only be estimated by 'best professional judgement' if major abiotic conditions e.g. the groundwater table, groundwater quality, percentage of lime, salinity, nutrient status of the soil, are known. Though species composition and vegetation dynamics may be characteristic of the different substrates, in most cases succession follows the same pattern of structural sequence, namely, (i) short herbaceous vegetation, (ii) tall forb vegetation, (iii) low shrub vegetation and (iv) tall woodland in the absence of grazing. It seems that this macro-structure of the vegetation can, to a certain extent, be predicted. If the major abiotic conditions and the accessibility of the area for immigrating species are known, the species composition depends on the stage of succession and the structure of the vegetation. The various activities performed by animals such as grazing/browsing, trampling and manuring, lead to a change in the structure of the vegetation.

Thalen et al. (1987) found that a modeling approach to simulating changes in vegetation macro-structure under grazing would be very useful. The reasons for such a modeling approach are a.o. (i) to gain further insight into the structure and functioning of the system by recognizing the apparently essential processes and (ii) to develop methods for the prediction of the long-term effects of different grazing regimes. The authors distinguished in the absence of grazing, four main structural classes in the succession from grassland to woodland, namely, (i) short herbaceous vegetation of grasses and forbs, height < 25 cm, (ii) tall herbaceous vegetation of grasses and forbs, height > 25 cm, (iii) low woody vegetation dominated by shrubs and small trees, height < 2 m and (iv) tall woodland, height > 2 m (Thalen et al., 1987). The herbaceous undergrowth in shrubland and herbaceous and shrub undergrowth in woodland were not taken into account in this model. The vegetation texture of any terrain can be described by the percentage area covered by each of these structural classes. Simulation can only be achieved if the relative preference of different herbivores for the macro-structural classes are known. The areas influenced by different intensities at various stocking densities should also be quantified. It is generally assumed that the increase of the impact on the vegetation is proportional to increased stocking density. This is a simplification, since in east Africa Laws (1970) found that larger groups of elephants have a greater effect on certain parts of the habitat, especially the woody area, than the actual density increase for the whole area might suggest.

Few comparative studies have been carried out into the relative preference of different herbivores for the above mentioned macro-structural classes and different types of forage. Putman (1986) compared the use of community types in the New Forest by cattle and ponies under a permanent grazing regime. The cattle spent 60 to 70% of their time on grassland communities, especially re-seeded grasslands, throughout the year and improved grasslands in winter, whereas streamside grasslands and road side verges were used at far lower intensities. They spent 10 to 20% of their time in woodland and the remainder on heathland. The woodland was preferred for shelter at night. This pattern was fairly constant throughout the year. Ponies spent 50% of their time on grassland communities, mostly improved grasslands and made more consistent use of acid grasslands, streamside grasslands and road side verges. They spent 20 to 30% of their time on heathland, mainly the wetter parts and 10 to 20% in woodland and gorse shrubs. Ponies showed some more variation between the seasons than cattle.

The diet of cattle in the New Forest remained remarkably constant throughout the year. It consisted of 70% grasses, 20% heather and 10% others, but no shrubs or trees (Putman, 1986). The pony diet showed marked seasonal changes with 85% grasses of which 20% was *Molinia caerulea* in the spring and summer and 15% others. It consisted of 50% grasses, 20% *Ilex aquifolium,* 10% heather, 10% *Ulex europaeus* and 10% others in winter and early spring. An interaction between the herbivores was found, namely, the cattle preferred to graze the tall grass in the latrine areas left by the ponies.

In Denmark Buttenschøn & Buttenschøn (1982B) compared the selectivity of cattle and sheep under a seasonal grazing regime from June to November. The cattle chose the more lush swards and preferred grasses to other species. The uptake of *Calluna vulgaris* was restricted to the summer period. *Betula pubescens* was also browsed seasonally and other shrubs were avoided. The sheep also chose the grasslands, but preferred dicotyledonous herbs and readily browsed shrubs during the whole season (Buttenschøn & Buttenschøn, 1978; 1982B). The sheep preferred forbs, legumes and some woody species to grasses and sedges and were quite selective in their choice of grasses. Their results are in agreement with a comparison of the diets of cattle and sheep by Cook et al. (1967) in the United States. They found a higher selectivity by sheep, including higher proportions of forbs and browsing.

Grant et al. (1985) carried out comparative studies on the diet selection of sheep and cattle in hill grasslands in Scotland under seasonal grazing regimes from May to October. Sheep, and not cattle, were able to select forbs or certain grasses even when these grew low in the profile or in a fine admixture with other species. Moreover, sheep, but not cattle, tended to avoid dead material and flowerstems. Cattle did have increased proportions of some components in their diets compared to the proportions in the sward, but this could be attributed to non-selective grazing of the upper layers of the sward or the preferential grazing of small areas where the component was more abundant. Grant et al. (1987) came to similar conclusions with regard to grazing on blanket bogs and heathland in Scotland. *Molinia caerulea* was selected and all heather species were avoided by both cattle and sheep. Sheep, however, selected *Molinia* to a greater extent and cattle *Eriophorum vaginatum.* Cattle diets contained more dead material than those of the sheep. The differences between sheep and cattle with respect to grazing height were much smaller than in the grasslands (Grant et al., 1985). Selection in the horizontal plane operated more strongly on bog and heather than on grassland.

In the Netherlands Oosterveld & Slim (manuscript) compared the diet selection of goats and sheep in a stand which was dominated by *Molinia caerulea* with some *Calluna vulgaris* and *Erica tetralix* and *Salix* and *Betula* shrubs. The most striking difference was the goats' very marked preference for *Salix* and *Betula* leaves and bark as compared with the sheep.

Fig. XIII.1 shows the differences summarized by Van Dyne et al. (198C) in the diet of cattle, horses sheep and goats, with respect to the percentages grasses, forbs and shrubs consumed over a period of a year. There is, however, considerable seasonal variability which is not expressed in the figure.

Some relative preferences of cows, sheep, horses and goats for macro-structural classes can be derived from the afore-mentioned, but they cannot easily be quantified. Van Dyne et al. (1980) did quantify the diet on a yearlong basis for grass, forbs and shrubs (Fig. XIII.1), which can more or less be transformed into the four macro-structural classes. They are aware of the considerable seasonal variability as well as location effects. Thalen et al. (1987) quantified the relative preference of some herbivores for the macro-structural classes for their

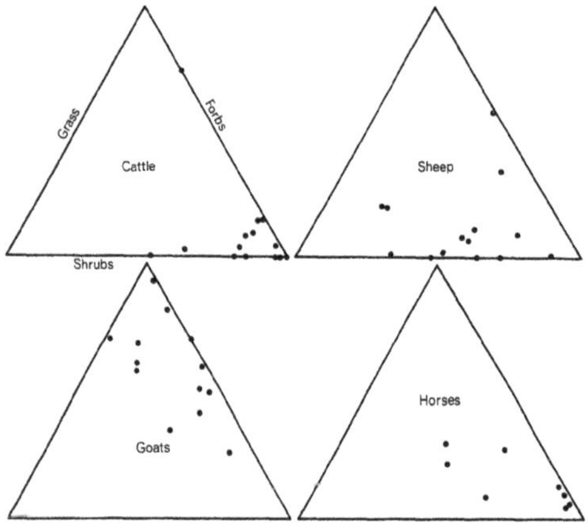

Fig. XIII.1. A graphic summary of the botanical composition of the diet on a yearlong basis for cattle, sheep, goats and horses. The values for each of the components grasses, forbs and shrubs should be read perpendicular to the three axes, respectively ranging from 0 to 100% (after Van Dyne et al., 1980).

modeling approach (Table XIII.1) based on a large number of observations of animal distribution (P. Oosterveld and others, unpublished). Their estimated preference of goats for woody vegetation is lower than the figure presented by Van Dyne et al. (1980), whereas the latter indicate no difference for cattle and horses. Thalen et al. (1987) do not, however, discuss these discrepancies, nor do they incorporate the findings of Grant et al. (1985; 1987) that cattle diets contain more dead material than sheep diets. They quantified the forage preference for different types of forage also on the basis of a large number of observations (Table XIII.2).

In the New Forest Putman (1986) found a concentration of feeding activities for both cattle (54% on a year basis and 75% in the spring and summer) and ponies (34% on a year basis and 55% in the spring and summer) on the small grassland area (7% of the total area), which resulted in local heavy grazing. Thalen et al. (1987) and Bokdam (1987) also mentioned very intensive grazing on only 3 to 6% of the total area of large nature reserves in the Netherlands. Large numbers of observations were summarized into a scheme representing the areas (%) which are influenced by various utilization factors, for different grazing intensities (see Oosterveld, 1979; Thalen et al., 1987) (Fig. XIII.2).

In order to estimate the impact of large herbivores on the successional series of the above mentioned macro-structural classes one must quantify the proportion consumed in relation to the aboveground production. For each of the afore-mentioned (Table XIII.2) types of forage, the amount available was calculated on estimations of (i) the amount of live material already present, (ii) the growth rate, (iii) the time of the year and (iv) the amount of dead material. The food demands of the herbivores were calculated for (i) different herbivore species, (ii) the stocking density and (iii) the time of the year. Estimated percentage values of biomass disappearance as a result of trampling and decomposition were also incorporated (Thalen et al., 1987).

The application of the modeling approach for macro-structural classes could be used for predic-

Table XIII.1. Assumptions of the relative preference of four large herbivore species for four macro-structural vegetation classes. The figures indicate presence in a structural class expressed as percentage of time spent in all classes together (after Thalen et al., 1987).

Herbivore species	Macro-structural classes of vegetation			
	Short herbaceous vegetation	Tall herbaceous vegetation	Low woody vegetation	Tall woodland
Cow	80	10	5	5
Sheep	60	20	15	5
Horse	60	15	15	10
Goat	40	15	30	15

Table XIII.2. Assumed use factor (%) of four large herbivore species for different types of forage by Thalen et al. (1987).

Type of forage	Herbivore species			
	Cow	Sheep	Horse	Goat
Herbaceous, young	100	100	100	100
Herbaceous, old	10	10	10	5
Herbaceous, dead	1	1	1	1
Twigs, young	10	25	25	100
Twigs, old	5	5	5	5
Fruits	100	100	100	100

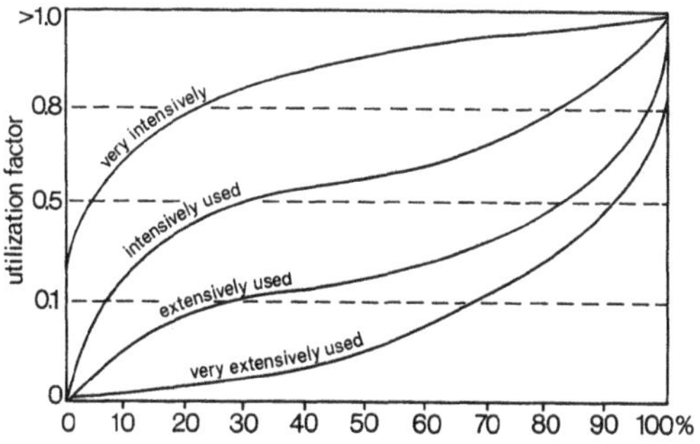

Fig. XIII.2. Areas (%) of a terrain, which are influenced by different utilization factors (after Oosterveld, 1979 and Thalen et al., 1987).

tions for any number of years (Thalen et al., 1987). The authors carried out a sensitivity analysis and found the model to be relatively stable. They also indicate that it is very difficult to validate such models because of a lack of quantitative data on the vegetational structural development for areas that have been or still are grazed/browsed by known stocking densities for a long period. In the Netherlands records are only available for a period of less than two decades. The only validity test for their model was the prediction of an increase in the woody vegetation in the afore-mentioned Borkener Paradies. Here it is indeed known that bushes were regularly cleared by farmers, in the past, in order to maintain the grasslands.

XIII.C. Cutting

XIII.C.1. Aims and results of restoration management

The litter accumulated in the abandoned salt marsh area was removed immediately under a cutting regime which was just above soil level and which included hay-making. The prevention of litter accumulation has been achieved in this study as well as in others (Oomes & Mooi, 1981; Willems, 1983A).

The impoverishing of the soil could not be demonstrated from either soil chemical analyses from the reclaimed grassland area of the Westerholt or from the valley grasslands of Loefvledder. The increase in the ratio between species indicating nutrient-poor and nutrient-rich soil conditions did, however, clearly indicate that some changes took place. This was confirmed by the decline of the aboveground standing crop. Oomes & Mooi (1981), Elberse et al. (1983) and Willems (1983A) recorded similar results.

Although some species indicating more nutrient-poor soil conditions established themselves, no rare or endangered species typical of heathlands appeared in the reclaimed grassland of the Westerholt area. It apparently takes more than 15 years of cutting to restore reclaimed fertilized pastures to heathland or heathland-like communities characteristic of nutrient-poor soils. Such communities did, however, appear in the area from which sods were cut. Van Der Maarel (1980) started hay-making without fertilizer application in a former apple orchard in the Netherlands in which the trees were removed and which had been heavily manured. Removal of the top soil and subsequent cutting resulted much more rapidly in an *Arrhenatherion elatioris* after 8 years than did just mowing.

Not only the species diversity increased in the valley grasslands starting from fertilized pastures, but after 15 years the initial stages of *Calthion palustris* communities developed in seepage areas. The succession from *Calthion palustris* to *Juncion acutiflori* was much slower. The development to the initial stages of *Festuco-Cynosuretum* was also recorded within 15 years on the drier parts, but hardly any change was found during the next 15 years.

XIII.C.2. Cutting in perspective

A successional scheme of the main grassland types in the Drentsche A area in the Netherlands has been derived from many simultaneous recordings of a series of hay-making regimes without fertilizer applications dating back to different times (varying from 0 to 35 years) (Everts et al., 1984). On dry to moist substrates without seepage or inundation the sere runs from *Poo-Lolietum* via *Holcus lanatus* communities and *Lolio-Cynosuretum* to *Festuco-Cynosuretum* in dry sites and to *Juncion acutiflori* in moister ones. Sod removal in the *Festuco-Cyno-*

suretum is supposed to result in *Violion caninae* in dry sites and *Juncion acutiflori* in moister sites. On wet to very wet substrates the sere runs from *Rumici-Alopecuretum* via *Holcus lanatus* communities to *Calthion palustris* communities. A rise of the groundwater table and an increase of the rainwater component in the groundwater by a diminished discharge of seepage water is supposed to result in *Cirsio-Molinietum* in the moist to wet sites and *Caricion curto-nigrae* in the very wet sites, whereas an increase in the inundation time could result in *Magnocaricion* communities.

Sincc this successional scheme takes the main abiotic conditions into account it seems a sound basis for the prediction of species composition. Some successional steps of the scheme were demonstrated in this study, but most of the suggested changes have yet to be demonstrated. So far, such successional changes have not yet been studied by a modeling approach.

XIII.D. Some effects on the fauna

The effects of restoration management regimes have mostly been evaluated on the basis of botanical criteria. Relatively little work has been done on changes in the fauna under grazing and cutting regimes. Research on the avifauna in the Netherlands (Beintema & Müskens, 1987) showed that a delay in the date of the first cutting increased the hatching success of meadow birds. Large areas of the embanked sand flats of the Lauwerszeepolder in the Netherlands became overgrown by *Agrostis stolonifera* and litter and dead standing material accumulated. This process was detrimental for the main function of the area, namely, being a place for wintering and transmigrating water fowl. Grazing with cattle and horses enabled the use as a feeding area for the birds (Fresco et al., 1987). The salt marsh described in this study was not visited by Barnacle geese (*Branta leucopsis*) before grazing was re-introduced. However, one or two years after the re-introduction the birds grazed that part of the salt marsh. Later 31 species of breeding birds with 850 to 1000 pairs on 83 ha of grazed salt marsh and 25 species with 550 to 600 pairs on 77 ha of ungrazed salt marsh were recorded (Van Dijk & Bakker, 1980). This once again suggested that grazing with large herbivores can create favourable conditions for smaller animals. This was also the case with rabbits in a former arable field in the Netherlands which has now been grazed for 15 years (Oosterveld, 1983). Not only the total number, but also the number of species of mice increased in the same area (Thalen et al., 1987).

Schultz (1987) compared grazed and ungrazed salt marshes in the German Wadden area with respect to their function as an area for foraging, resting with high tides and breeding. Foraging Barnacle geese and Brent geese (*Branta bernicla*) preferred grazed salt marshes. It is suggested that the increase of breeding Oystercatchers (*Haemotopus ostralegus*) might be related to the increased exploitation of the salt marshes for grazing purposes, whereas the number of breeding Redshanks (*Tringa totanus*) decreased in the same period. Schultz concluded that the limiting factors for the avifauna were the frequency with which the birds were flushed and the trampling of nests and pulli with increasing stocking density. The composition and the structure of the vegetation was thought to be of less importance.

Recently there has been a growing interest in the evaluation of management regimes on invertebrates, mostly the entomofauna. Some insects are strictly dependend on a single plant species or even parts of it. Other species of insects depend on certain micro-climatological characteristics, which are largely determined by the structure of the sward (Barkman & Stoutjesdijk, 1987). It would be beyond the scope of this study to give a complete survey of the relationship between management regimes and the entomofauna. Only some cases are, therefore, presented without entering into problems of causal relationships.

Siepel et al. (1987) sampled grassland stands on dry sandy soils in the Netherlands. The stands included:

(i) unfertilized stands of *Deschampsia flexuosa* with a production $< 2t \cdot ha^{-1} \cdot yr^{-1}$ which were
  (a) untouched,
  (b) cut in the late summer and
  (c) permanently grazed with 0.2 Scottish Highland cattle $\cdot ha^{-1}$,

(ii) stands with *Holcus lanatus*/*Lolium perenne* with a production of $8t \cdot ha^{-1} \cdot yr^{-1}$ and fertilized with $50kgN \cdot ha^{-1} \cdot yr^{-1}$ which were
  (a) cut in June,
  (b) cut in both June and September and
  (c) grazed during the growing season with 1.8 cattle $\cdot ha^{-1}$ and

(iii) stands with *Lolium perenne*/*Poa annua* with a production of $15t \cdot ha^{-1} \cdot yr^{-1}$ and fertilized with $380kgN \cdot ha^{-1} \cdot yr^{-1}$ which were
  (a) cut monthly from May until October and
  (b) grazed rotationally with 3.8 cattle $\cdot ha^{-1}$.

The management regimes had been carried out for at least three years at the different sites. The invertebrates were subdivided into mesofauna of the soil, macrofauna of the soil, grasshoppers and the remainder of the macrofauna of the vegetation. The numbers of individuals and species of the mesofauna of the soil decreased with higher fertilizer applications, whereas grazing and hay-making produced far fewer effects. The macrofauna of the soil and the vegetation, however, had the highest numbers of individuals and species with light fertilizer applications, which is not remarkable if the oligotrophic soil is taken into account. Hay-making did not influence the macrofauna of the soil, whereas the numbers of individuals and species of the macrofauna of the vegetation were higher than in the grazed or untouched sites. The lowest numbers of individuals and species of both the macrofauna of the soil and the vegetation were found in the grazed sites. The stocking density was, however, high for nature conservation purposes. The numbers of individuals and species of grasshoppers were lowest in the sites with heavy fertilizer applications and high stocking densities.

In southwest Germany Opperman (1987) sampled butterflies and grasshoppers in *Molinion* communities which were partly abandoned and partly cut, in *Calthion* communities with fertilizer application and cut once a year and *Lolio-Cynosuretum* communities cut three times a year. The highest numbers of individuals and species were found in the unmanured *Molinion* communities with a cut late in the season (October or later) or recently abandoned. These values were lowest in the *Lolio-Cynosuretum* communities and intermediate in the *Calthion* communities.

Rahmann et al. (1987) sampled most types of the larger invertebrates in the *Puccinellion maritimae* and the *Armerion maritimae* of salt marshes in northern Germany. The number of individuals decreased in the series abandoned for 10 years > abandoned for 1 year > cut in July > nearly 2 cattle $\cdot$ ha^{-1} > 5 sheep $\cdot$ ha^{-1}. The numbers of species and faunal groups were highest in the abandoned area with a dominance of cicades (*Cicadina*), bugs (*Heteroptera*) and snails (*Gasteropoda*) and decreased to only beetles (*Coleoptera*) and bugs in the grazed areas. Irmler & Heydemann (1985; 1986) compared the effects of different stocking rates (0.5, 1, 1.5 cattle $\cdot$ ha^{-1}) with an abandoned site three years after the withdrawal of the usual stocking density of 2 cattle $\cdot$ ha^{-1}. The number of species did not change, but the species which are sensitive to trampling increased in the abandoned area, namely, springtails (*Collembola*), spiders (*Araneae*) and species feeding on tall plants or parts of them, namely, butterflies (*Lepidoptera*), aphids (*Aphinidae*), bees (*Apidae*) and hover-flies (*Syrphidae*), whereas the coprophagous species of those feeding on algae of the bare soil decreased. It was remarkable that the faunal community of the lower parts of the salt marsh gradient maintained itself best in the heavily grazed site of the higher salt marsh. A similar phenomenon of lower salt marsh plant species was found in this study (see section X.C.).

The results of the afore-mentioned studies show that the application of fertilizers proved, in most cases, to have a negative effect on the invertebrate fauna. Hay-making can be detrimental depending on the period of cutting, namely, an early cut (May) is better than a later one (July) for some groups (Morris, 1979; 1981). No cut at all, cutting every second year or an early cut facilitates the accumulation of litter which is of great importance to a number of invertebrates (Siepel et al., 1987; Decleer, 1987). Grazing proved to have more of a negative effect on the invertebrate fauna than hay-making. This is not so remarkable since the grazing intensity was high in the examples presented. A mosaic of grassland, scrub and woodland which is associated with low grazing intensities feature an important fauna of butterflies (Bink & Van Der Made, 1986). One can conclude, as did Morris (1978), that although an increasing amount of information exists on the effects of management on invertebrate animals, there are still many gaps in our knowledge. There is, in particular, a need for comparative studies to be carried out into the effects of different management regimes or of the same treatment on different faunas. More detailed investigations into the effects of different large herbivores and different grazing intensities are also required (Photo's XIII.3 and 4).

(Members of) different families of invertebrates can require different management regimes. Also the demands for the avifauna, for example transmigrating geese which require a heavily grazed salt marsh, can be contradictory to those for most invertebrates. Finally, the demands of many members of the invertebrate fauna can have an adverse effect on most of the flora and vegetation, for example an early cut or the accumulation of litter. A solution, which is indeed recommended by several authors (Zoller et al., 1984; Handke & Schreiber, 1985; Opperman, 1987; Decleer, 1987; Siepel et al., 1987) can be a mosaic of different cutting regimes and abandoned sites or a rotational grazing system (Irmler & Heydemann, 1986; Rahmann et al., 1987). Grazing with low intensity should also be considered since it creates and maintains a mosaic of short turf and taller forbs with some litter being accumulated and locally abandoned areas.

XIII.E. Aspects of the wider implementation of management practices

XIII.E.1. Hay-making

Hay-making regimes, often in combination with grazing of the aftermath, are applied in large areas in the Netherlands, in particular the moist and wet grasslands. The work is nowadays often carried out with heavy machinery which, in some cases, forces the authorities in charge of the management into superficial drainage resulting in repercussions for the groundwater composition, groundwater table and mineralization of the top soil. There is, moreover, a discussion as to whether grazing of the aftermath can be allowed in wet years. Since the cutting is done by light machines but the removal of the grass with heavy machinery, the question also arises as to whether the swath should be removed or no cutting should be practiced at all in wet years.

The area in which hay-making regimes are carried out by the largest private organization for nature conservation, the 'Vereniging tot Behoud van Natuurmonumenten' included 3431 ha in 1987, of which 1805 ha was cut in the summer period and 1626 ha, namely the reed stands, in the autumn and winter period (Table XIII.3).

Photo XIII.3. Dominance of *Elymus pycnanthus* at the Schiermonnikoog salt marsh, abandoned for 30 years.

The increasing costs of hay-making regimes are an important problem. The costs of both grazing management and hay-making have been calculated for a few nature reserves. In salt marsh areas grazing is Dfl. 125.—.$ha^{-1} \cdot yr^{-1}$ cheaper than hay-making. In inland grassland/heathland areas these figures are Dfl. 30,— to 120.— · $ha^{-1} \cdot yr^{-1}$ if the practices are carried out by the authorities in charge of the management (Thalen, 1981). Kottmann et al. (1985), too, showed that heathland conservation in Germany is less expensive using sheep grazing than mechanical cutting.

XIII.E.2. Grazing

In the 1970's grazing was widely promoted in the Netherlands as a management practice because of the expected high nature conservation interest of the resulting vegetation and lower costs than hay-making regimes (Oosterveld, 1975). In the northern part of the Netherlands 5060 ha of grassland

Photo XIII.4. Short turf dominated by *Festuca rubra* at the Schiermonnikoog salt marsh grazed by 1.6 heifer · ha^{-1} for 15 years.

area in nature reserves were under a grazing regime in 1982, namely, 1900 ha in private property and 3160 ha owned by the State (Van De Waerdt, 1982). Recently 'Natuurmonumenten' carried out a survey of various Dutch landscapes under grazing, which included over 5700 ha in 1987 (Table XIII.3). If the reed stands (which are cut in the winter period) are disregarded the grazed grassland area (1660 ha) was nearly as large as the cut area (1805 ha).

A large variety of herbivores are used in grazing management, namely, ordinary livestock and animals which are more hardened against winterconditions and a lesser forage quality (Oosterveld 1975; Thalen, 1984A). Examples of these old Dutch landbreeds are Dutch landgoats, Lakenvelder cattle, Veluwe sheep, Kempen sheep, Drenthe sheep, Schoonebeek sheep and Mergelland sheep. Animals from abroad are also used, for example Swedish Fjell cattle, Scottish Highland cattle, French Limousin and Charolais cattle and English Galloway cattle. Moreover, animals have been introduced which resemble their extinct wild ancestors like konik horses from Poland and Heck cattle from Germany. The introduction of wisent and elk has been proposed, since they are supposed to have been indigenous and to have created and maintained glades in the primeval forests before the interference of man started (Van De Veen, 1985).

In the Netherlands a discussion has developed concerning the 'naturalness' of grazing systems. Management by grazing started with the view that nature conservation is the aim and the grazing animal and cutting machinery are the means (Thalen, 1984A). In other words the nature conservation interest is restricted to the development of plant communities and does not involve the grazing animals since they are domestic, whether modern or more primitive. The production of meat and hence too intensive grazing as well as the concern for the well-being of these animals tends to exceed the original purpose of the grazing management, namely the manipulation of the vegetation (Kuiper et al., 1981). The introduction of populations of koniks (a primitive breed of horses) and Heck cattle (cross-bred animal resembling the aurochs) implies the development of more complete ecosystems in which these grazers represent a nature conservation interest of their own (Van De Veen, 1985; Vera, 1986). Clason (1986) is, however, of the opinion that from this point of view the introduction of koniks and Heck cattle makes no sense, since they are not more primitive than the indigenous domestic animals of old Dutch landbreeds. Only research on the specific impact on the vegetation of koniks and Heck cattle can solve this dispute. Another point which Clason (1986) makes is that there is no evidence from fossil findings that the wisent ever lived in the Dutch territory. Bottema (1987) presented palynological evidence that hardly any pollen of herbs and grasses have been found in deposits from the Atlantic Period (5000–3000 B.C.) and he, therefore, doubts the occurrence of glades in the primeval forests and, therefore, the presence of large herbivores. It seems that elk and wisent might have played an important role in the establishment of forests which had a more natural structure than the forests which are found in the Netherlands nowadays, but that the new type of forest which would emerge would differ from the 'original' primeval forests. It is, therefore, doubtful whether they can play a role in the restoration of species-rich grasslands which represent semi-natural landscapes.

Table XIII.3. Different landscapes of the biggest private organization for nature conservation in the Netherlands 'Vereniging tot Behoud van Natuurmonumenten' and the area under cutting and grazing regimes in 1987 (after pers. comm. H. Piek).

		Cutting in the summer	Cutting in the autumn/ winter	Cutting in the summer/ grazing the aftermath	Area cut	(%)	Area grazed	(%)	Seasonal grazing	Permanent grazing
Heathland and peat bog	7378	0	0	0	0	0	3099	42	221	2878
Grassland	3465	839	0	966	1805	52	1660	48	1660	0
Reed	1626	0	1626	0	1626	100	0	0	0	0
Salt marsh	333	0	0	0	0	0	100	30	72	19
Dune	1886	0	0	0	0	0	415	22	115	300
Woodland	17666	0	0	0	0	0	530	3	37	493
Total	32354	839	1626	966	3431	11	5754	16	2114	3690

XIII.E.3. Seed dispersal

An important limiting factor in the restoration of species-rich grassland is the dispersal of seeds. Seeds of species not (yet) present in a reserve or in an allotment can easily be dispersed by hay-making machinery or by grazing animals. There is only a small difference in the planning of a hay-making scheme or the transportation of sheep or cattle from floristically-rich allotments to species-poor allotments and the deliberate re-introduction of species. The re-introduction by means of sowing of rare and endangered species should at least be considered seriously, but it should be carried out under controlled conditions. Transplantation should be avoided since the establishment should be an expression of the suitability of the habitat, including the possibilities for germination.

Sampling procedures in collecting and propagation procedures for multiplication can both narrow the genetic base of a population (Frankel & Soulé, 1981). This may have consequences for the ecological adaptibility and the genetic vulnerability to pests and diseases. Genetic heterogeneity is likely to be an advantage on both grounds and should be safeguarded by representative collecting and propagation as far as the available material permits.

If the collection of seeds is difficult or very time consuming some cut material could be transported and spread, since Wells et al. (1981) found that hay obtained from floristically-rich meadows in the U.K. contained seeds of dozens of species.

XIII.E.4. Management by authorities in charge of the management or by farmers?

In the United Kingdom, Germany and France rough grazing of grassland has always been a traditional agricultural practice. Some semi-natural grassland in nature reserves is still being grazed, but with the management of the sward as the primary objective and not maximum productivity. In many situations this grazing is the best way of maintaining attractive amenity grasslands because it more or less involves a continuation of the earlier agricultural practices. But there can be many problems, such as the provision and maintenance of fencing and water, and concern for the well-being of the stock. In these circumstances the cutting of swards offers a more readily controllable alternative management (Green, 1980). The authorities in charge of the management are apparently aiming to keep the management of nature reserves in their own hands without cooperating with farmers who might be interested in the 'by-products' of nature management.

In the first half of this century the organizations for nature conservation in the Netherlands used to have farmer tenants, who hired the land and worked on it, thus continuing the former agricultural practices. The authorities in charge received money for the management and the reserves were well kept. This arrangement became more and more difficult to enforce, because the modern farmers wanted to use up-to-date agricultural methods and were, therefore, no longer interested in using the land. Consequently from the 1950's and 1960's onwards the manager had to manage the nature reserves himself using employed labourers (Westhoff, 1971). Because of the high costs involved they tried to interest the farmers again in the exploitation of their grassland in return for little or no money, and in the 1970's and 1980's an increasing area of nature reserves was once again cut by farmers or grazed by their livestock. Even without the application of fertilizers and in spite of the timing of hay-making these farmers are interested in so-called 'wild hay' as a cheap supplement to the ordinary forage.

The question arises as to what extent the interests of the farmers and the authorities in charge of the management can overlap. The supposed successional sere on wet to very wet substrates *Rumici-Alopecuretum* via *Holcus lanatus* communities and *Calthion palustris* to *Junco-Molinion* (Everts et al., 1984) is taken as an example in order to illustrate this point. The nature conservation interest of these plant communities increases in this sequence. Some agricultural characteristics of the plant communities have been summarized in Table XIII.4. It is obvious that the agricultural qualities of the plant communities decline during succession. Korevaar (1986) demonstrated that the forage quality decreases if the first cut is made later in the season (Table XIII.5). He also pointed out that the quality of the forage is, even with an application of 80kgN · ha^{-1} · yr^{-1} (Table XIII.6), lower than the 900 VEM · kgdm^{-1} standard for 5000 kg milk · yr^{-1}. The normal application of 200 kg N · ha^{-1} · yr^{-1} can result in an energy value of 1010 VEM · kgdm^{-1} (Boxem, 1973). A classification of grass species according to the digestibility of organic matter suggests that less digestible species appear in the course of succession (Korevaar, 1986). The sequence of the species is, however, not the same during the whole season since it depends on the age of the grasses. It is not known how the herbs should be ranked. The experiments which involved an application of 80 kg N · ha^{-1} · yr^{-1} artificial fertilizer and later cutting resulted in a decline of *Alopecurus geniculatus* and an increase of *Holcus lanatus* and *Rumex acetosa.* This initial stage of a *Holcus lanatus* community renders a forage with a net energy value of 875 VEM (net energy for milk production · kgdm^{-1}). Cattle with a present-day normal milk yield need, however, forage with a net

Table XIII.4. The plant community in the supposed successional series (Everts et al. 1984) and their date of first cutting, fertilizer application, yield, degree of agricultural value according to the botanical composition and the energy value of the crop. (1) Boxem (1973), (2) Korevaar (1986), (3) Klapp (1965), (4) Pannekoek (1982), (5) Oomes et al. (1980).

Grassland vegetation	Date of first cut	Fertilizer application kg N · ha^{-1} · yr^{-1}	Yield ton · ha^{-1} · yr^{-1}	Degree of agricultural value according to De Vries et al. (1942)	Energy value VEM · kgdm^{-1} (net energy for milk production)
Rumici-Alopecuretum	May/June	200	9.7–12.6 (1)	8.3	1010 (2)
Molinio-Arrhenatheretea	June/July	100	8.7–10.8 (2)	5.3	805–885 (2)
Calthion palustris	July/August	light	3.5– 4.5 (3)	4.5	?
Junco-Molinion	August/September	0	1.9– 4.9 (4,5)	1.8	?

Table XIII.5. The effect of date of first cutting on the yield and some quality parameters of forage with no nitrogen application but sufficient phosphate and potassium (after Korevaar, 1986).

Date of first cut	Yield kgdm · ha^{-1}	Crude protein g · kgdm^{-1}	Crude fibre g · kgdm^{-1}	Organic matter digestion %	Digestible crude protein g · kgdm^{-1}	Energy value (net energy for milk production) VEM · kgdm^{-1}
6 May	580	196	173	80.3	142	985
20 May	1490	155	226	76.5	102	913
3 June	2790	126	265	71.1	79	834
17 June	4060	103	289	65.6	62	761
1 July	4560	103	279	64.5	55	750
15 July	4780	99	275	62.9	50	731

Table XIII.6. The effect of application of nitrogen fertilizer with sufficient phosphate and potassium on some quality parameters of forage (after Korevaar, 1986).

Fertilizer application kg N · ha^{-1} · yr^{-1}	Crude protein g · kgdm^{-1}	Crude fibre g · kgdm^{-1}	Organic matter digestion %	Digestible crude protein g · kgdm^{-1}	Energy value (net energy for milk production) VEM · kgdm^{-1}
0	194	232	70.1	121	815
40	201	238	70.1	130	819
80	224	235	70.7	150	838

energy value of 900 VEM and some concentrates. The shortage cannot be replenished unlimitedly with concentrates since it will result in an unbalanced ratio of roughage and concentrates (Korevaar, 1986).

It can be concluded that even an initial stage of a *Holcus lanatus* community is not feasible on a present-day normal farm. The Dutch government would like to interest farmers in nature management and tries, therefore, to come to an agreement on management practices, compensating with money for postponing the first cutting date, leading to a decrease of the yield. Korevaar (1986) has said that such agreements can result in the establishment of an initial stage of a *Holcus lanatus* community. Although the hatching success of meadow birds and hence the ornithological value can increase by a delay in the first cutting date (Beintema & Müskens, 1987), the botanical value of such grasslands is, however, still poor. The roughage of grasslands with a higher botanical value is only usable for young stock and dry cows. Their possible use in agriculture is, therefore, important for nature conservation.

XIII.F. Gaps in the available knowledge

Many of the gaps in the available knowledge on nature management practices were revealed in the previous sections. These gaps are summarized in this section. It should, however, be kept in mind that this list is not exhaustive.

Gaps or preferred topics for research, with respect to the grazing management of the vegetation, are listed below. Some of these have been recorded by Thalen et al. (1987) and others come from discussions of the group of Dutch grazing ecologists (BON).

A. Refinements of the grazing models and parameters included:

- Incorporation of a herbaceous component in the structural types of woodland, i.e. simulation of the undergrowth in forests.
- A further subdivision of the structural classes in terms of vegetation types.
- A further subdivision of the herbaceous vegetation structural classes with respect to phenological stages.
- Incorporation of forage quality parameters.
- Quantification of decomposition and herbage losses by trampling.
- Effects of grazing on the availability of nutrients.
- Quantification of the rate of regrowth after grazing.
- A simulation of the spatial distribution of the macro-structural classes.
- A simulation study of the development of the vegetation in relation to the performance of the grazer in a large ecosystem.
- A simulation study of the effects of one or more types of herbivores on a large area with widely diverging habitats from wet to dry, from nutrient-rich to nutrient-poor and from grassland to woodland.
- A simulation approach to the effects of grazing on the maintenance or restoration of rare and endangered vegetation types and plant species.
- Studies of the position of young stock and dry cows in agriculture aimed at an assessment of their possible use for nature mangement.

B. Field experiments:

- A comparison of grazing impact in nature reserves of different size to work out the size-effects.
- A comparison of permanent grazing and seasonal grazing by different types of herbivores.
- The effect of changing stocking densities in the course of a year.
- A comparison of the effects of different stocking densities including seasonal as well as continuous grazing.
- A study of the effects of low stocking densities in a very large heterogeneous area, including different landscape types.
- Simultaneous grazing by more than one type of large herbivore.

Gaps or preferred topics for research with respect to the cutting regimes are listed below:

- The effects of sod removal on the successional series in valley grasslands.
- The possibilities of 'fixing' intermediate stages in the successional series.
- The effects of grazing the aftermath in meadows.

- A comparison of the performance of small populations of rare and endangered species in cut and grazed areas.
- The possibilities of reducing the soil fertility of former arable fields by growing a crop of cereals, sugarbeets or potatoes before the start of a hay-making regime.
- The correlations between fluctuations in plant populations and weather conditions.
- A modeling approach for predicting the effects of different management regimes on the species composition with special attention for rare and endangered plants species.
- A modeling approach for predicting the effects of different cutting regimes on the production and forage quality with respect to include such practices in the regular farming.

A few gaps in knowledge or preferred topics for research with respect to seed dispersal and re-introduction can finally be mentioned:

- What distances can be covered by seeds of rare or endangered species?
- What are the possibilities of re-introducing rare or endangered species up to and including their establishment, i.e. the seeding stage?
- What genetic differences exist in the constitution of seeds from populations which are far away from each other and what is the ecological significance of such differences with respect to re-introduction?
- What genetic differences exist in the constitution of seeds collected from small and larger populations and what is the ecological significance of such differences?

Besides all the afore-mentioned gaps in botanical knowledge it should be kept in mind that far less knowledge exists on the effects of nature management practices on the fauna.

Summary

Introduction

In Western Europe and, in particular, in the Netherlands nature management can be seen as a feed-back against the degradation of flora and vegetation. It can be subdivided in management for maintenance and management for restoration. Maintenance management aims at the preservation of already existing nature conservation interests. It is carried out in many countries. Restoration management aims at the increase of the nature conservation interest of abandoned areas or areas under intensive agricultural practices. This type of management was mainly restricted to the Netherlands at the start of this study in 1971. The results of restoration management were, however, hardly systematically documented. Moreover, no comparative field experiments were carried out. It seems, therefore, that at the present time, in the absence of the necessary knowledge and conceptual framework, restoration management is mainly based on trial and error, precedent or intuition, rather than rational decisions. In the 1970's and the 1980's comparative experiments were also started in Germany, Switzerland, Denmark and the United Kingdom, but most experiments are carried out in the Netherlands. Restoration management is not yet carried out at the present time outside the Netherlands.

The aim of this study is to contribute to the ecological knowledge of nature management thus providing a sounder basis for the activities of the authorities in charge of the management. In the EEC this knowledge is urgently needed. In those countries there is, as a result of the agricultural over-production, a growing interest in various types of 'set aside' schemes for several millions of hectares of marginal land.

The main items discussed in this study are:

(i) Land use dynamics, or what is the transition from agricultural practices to nature management?

(ii) Vegetation dynamics, or does restoration management create plant communities which become richer in species and does it create more plant communities than were present at the start of the management (these criteria were, namely, used as the main parameters on nature conservation interest)?

(iii) Ecosystem dynamics, or what are the causes of observed changes in the plant communities, namely, nutrient removal, increasing soil salinity or changes in the structure of the sward?

(iv) Population dynamics, or if plant communities become richer in species where do the new species come from?

(v) Management, or how is nature management applied at the present time?

In order to provide answers for these questions detailed comparative studies were carried out on sites with different treatments, namely, grazed and cut at different intervals, removal of sods and ploughing. The field data were collected in the more or less natural landscape of the formerly abandoned salt marsh of the island of Schiermonnikoog, and in the Drentsche A river area in the semi-natural landscape of the formerly fertilized valley grasslands and on the more or less cultural landscape of the grasslands reclaimed from former heathland.

Land use dynamics

It was supposed in literature that the high number of species and plant communities of the old agricultural landscapes were related to a high degree of abiotic variation and/or the differences in intensity of human interference and their local constancy. It appeared from this study on land use dynamics that the human impact on the scale of individual allotments has not been as constant as was always supposed. The numbers of plant species and plant communities are, however, high in such allotments provided the hydrological conditions are not disturbed and the level of nutrients is low. The conclusion is, therefore, that a high number of species and plant communities is primarily due to abiotic variation and is secondary to patterns of human interference. As a consequence, restoration management can be particularly successful in environments with (a spatial variation in) suited abiotic conditions, namely undisturbed hydrological conditions and a low soil fertility.

Dynamics in the composition of the vegetation

At the start of restoration management, the initial vegetation of the formerly reclaimed and fertilized grassland in the Drentsche A area mainly consisted of basal communities of the *Molinio-Arrhenatheretea* with elements of the *Lolio-Potentillion anserinae* and the *Caricion curto-nigrae*. A number of species indicating nutrient-rich soil conditions decreased in most management treatments, probably due to the cessation of fertilizer applications viz. *Lolium perenne, Phleum pratense, Poa pratensis, Elymus repens, Alopecurus geniculatus*. Species considered to indicate intermediate and nutrient-poor soil conditions increased viz. *Agrostis capillaris*. In one case the increase was earlier and greater in grazed (3 sheep · ha^{-1}) than in cut plots viz. *Hypochaeris radicata*. Considerable numbers of species considered to indicate nutrient-poor soil conditions viz. *Erica tetralix, Calluna vulgaris* and *Juncus squarrosus* which appeared in the area in which sods were removed. These species emerged to a lesser degree in the ploughed area where nutrient-poor sand from the C-horizon was brought to the surface.

Both hay-making and grazing resulted in an increase or maintenance of the number of species, with the exception of the wet and lightly grazed plots in which the number of species decreased. Whether the number of species and the equitability were higher in the grazed than in the cut areas in the dry grassland depended on the local grazing intensity.

The adjacent heathland area, which was also fenced in, became overgrown with grassland species. Even species considered to indicate nutrient-rich soil conditions appeared viz. *Poa trivialis* and *Poa pratensis* in the heavily grazed areas. In the lighter grazed areas heathland species became overgrown by *Molinia coerulea*. This cannot only be ascribed to grazing, since it also occurred in the abandoned heathland area. It was probably due to changes in the hydrological conditions outside the study area and 'acid rain'. The abandoned heathland changed into woodland (*Betula pubescens*).

In the upper and middle course of the valley grasslands the standard hay-making regime, namely in July, was carried out. On dry, sandy soils stands with *Poo-Lolietum* elements can develop into stands with *Lolio-Cynosuretum* elements. From stands with *Lolio-Cynosuretum* elements succession towards stands with *Festuco-Cynosuretum* elements was observed ten years after the cessation of fertilizer application. Already existing stands with *Festuco-Cynosuretum* elements did not, however, change during ten years of observation. On wet, peaty soils only was the succession from stands with *Lolio-Cynosuretum* elements to stands with *Calthion palustris* and *Caricion curto-nigrae* elements observed during a period of ten years. The community of *Carex acutiformis* (*Calthion palustris*) with *Glyceria maxima* was succeeded by the variant with *Crepis paludosa*. The number of species considered to indicate nutrient-poor soil conditions decreased in some stands despite continual hay-making without fertilizer applications. This was probably due to superficial drainage. The number of species increased in most dry areas, but only in a minority of the wet areas. In some cases the number of plant communities increased, so it seemed that already existing abiotic differences were reflected again in the featuring of different plant communities.

The comparison of different cutting regimes in the upper course of the valley grasslands revealed an increase in the number of species in the treatments July and September hay-making, September hay-making, July hay-making and burning. It tended to decrease with abandoning and September mulching every second year. In the more productive lower course area, however, the number of species decreased in most management regimes with the exception of September hay-making every year and the July hay-making regimes in which cases it maintained itself. It only tended to increase in the July and September hay-making regime. The overall decrease of species of the *Calthion palustris* and the increase of species considered to indicate nutrient-rich soil conditions can probably be related to changes in the hydrological conditions outside the study area.

Only a few plant communities exist on an abandoned salt marsh. The *Halimionetum portulacoidis* is found on the lower salt marsh, the *Artemisietum maritimae* somewhat higher, whereas the remainder of the salt marsh is covered by a facies of *Festuca rubra* and particularly *Elymus pycnanthus*. More vegetation types were found after 15 years of grazing (1.6 cattle $\cdot$ ha^{-1}) of this salt marsh, namely, the communities of *Puccinellia maritima, Artemisia maritima, Juncus gerardii, Juncus maritimus, Elymus pycnanthus, Festuca rubra* and *Armeria maritima*. Hay-making resulted in a dominance of *Festuca rubra* in most plant communities.

The present number of species in the grazed salt marsh was reached five years after the resumption of grazing. It initially increased quicker in the mown than in the grazed plots but dropped later below that of the level of the grazed plots. Only the *Puccinellia maritima* community showed a decrease in the number of species in the grazed area due to the destruction of the turf on the soft soil. It took over five years before the number of species in a recent exclosure dropped below that of the continuously grazed communities. The equitability, however, decreased more quickly.

Dynamics in the structure of the vegetation

Utilization patterns have developed in the grazed areas. The utilization factor of cattle i.e. the herbage utilization (foraging and trampling) expressed as a proportion of the net herbage accumulation was, on average, 0.53 on the salt marsh. It varied from 0.37 in the *Festuca rubra* to 0.62 in the *Juncus maritimus* community. A vegetational pattern has developed on the salt marsh resulting in extensive heavily grazed areas and smaller lightly grazed patches of the *Juncus maritimus* community. These patches range from nearly one ha to small isolated clumps of *Juncus maritimus* plants.

The utilization factor of sheep in the grassland area (formerly reclaimed from heathland) varied from 0.09 in the *Agrostis capillaris* to 0.30 in the *Holcus lanatus* and 0.89 in the *Leontodon autumnalis* community. The *Juncus effusus* community was hardly touched at all. These communities cover relatively large areas and the grazing pattern is, therefore, called a macro-pattern. On a more detailed scale a mosaic of taller turfs, ranging from 0.5 ha to several square metres, interspersed with shorter turf, emerged in the area with relatively high utilization factors. This mosaic is referred to as a micro-pattern. It revealed a pulsating character, namely, the lightly grazed patches spread during the spring and early summer because the herbage accumulation was higher than the amount utilized, but the fringes were removed again by the sheep during the late summer and autumn when herbage utilization exceeded the accumulation. It is obvious that these patterns cannot develop under hay-making regimes.

The heavily grazed areas have a higher forage quality than the lightly grazed patches. The herbivores maintain these sites by their feeding strategy. Heavy grazing results in a sward structure without standing dead material or accumulated litter and a high leaf: stem ratio. This implies a high protein content of the forage and a high digestibility, irrespective of its species composition. Hence such a site attracts the animals again.

Ecosystem dynamics

In the salt marsh as well as in the inland plant communities the above-ground standing crop showed strong year-to-year fluctuations as a result of varying weather conditions. After the cessation of the application of fertilizers a decrease of the biomass was expected. The sharp decline of the root biomass in most management practices was remarkable, particularly when compared with the unchanged or slowly declining above-ground standing crop. The relationship between the annual maximum above-ground standing crop, including litter, and the number of species deviated from the optimum curve that was expected from literature. This might be due to the different densities of the sward. More species are found in a high but open sward than in a low sward with much litter, although their above-ground biomass is similar.

Changes in the species composition as a result of the management regimes seem not to be related to soil chemical changes in both the salt marsh and inland grasslands.

Under a grazing regime, species of the lower salt marsh appeared higher on the salt marsh than in the abandoned area. The soil salinity had, however, not increased. Grazing in the abandoned area increased the number of species by removing the standing crop and accumulated litter. Upper salt marsh and dune species cannot spread down towards the lower salt marsh because of their low salt tolerance. Lower salt marsh species can, however, invade the upper salt marsh in grazed areas since they grow very well in less saline environments provided the light is not intercepted by the canopy. Lower salt marsh species extended into the cut areas to a lesser degree, probably since *Festuca rubra* has built a dense turf.

The nutrient status of the soil is expected to decline after the application of fertilizers is ceased, particularly, when nutrients are removed with the hay crop. The removal of nutrients was very limited, compared to the mineral pool in the soil. The above-ground standing crop decreased only little. Moreover, the strong fluctuations in the above-ground standing crop suggest great influences of weather conditions on the mineralization rates of dead plant material and litter and this may more strongly affect the nutrient status of the site than the removal of herbage. Since dung and urine return to the soil the mineral removal by grazing is even smaller than resulting from hay-making. Nevertheless, hay-making or grazing without fertilizer application showed a slow increase in the number of species and a quicker increase in species considered to indicate nutrient-poor soil conditions. These responses appear mainly due to changes in the structure of the sward as a result of timing and frequency of cutting or grazing of the above-ground standing crop and hardly due to impoverishment of the soil by nutrient removal. This seems to be a long-term process.

Population dynamics

Most species that have disappeared from the established vegetation will also vanish from the seed bank since many of the species which are found in grasslands have a transient or only a short living persistent seed bank. It seems that a persistent seed bank is particularly important for the maintenance of populations after bad seed setting or when the vegetation is cut or grazed too early. A permanent seed bank, however, can play a role in the regeneration of populations after their earlier disappearance from the established vegetation. Since in our areas the permanent seed bank mainly contains pioneers and weeds and no species characteristic of semi-natural grasslands, the latter group of species has to re-establish itself from populations from elsewhere whether by wind dispersal, grazing animals or hay-making machinery.

The re-establishment of species becomes more and more hampered by the fragmentation of the semi-natural landscapes, which considerably increases the distance from neighbouring populations. The deliberate re-introduction of species was, therefore, studied.

Two limiting factors are identified for such an establishment. The first one is the poor dispersal mechanism of many species and the second one is the lack of gaps. Species with large seeds can grow through the established surrounding vegetation, but small-seeded species need gaps for their establishment. The occurrence of gaps is strongly determined by the management practices i.e. the timing and frequency of hay-making or the intensity of grazing.

Nature management

Cutting and grazing regimes can result in the restoration of plant communities with nature conservation interest. It should be kept in mind that botanical interests might impinge on zoological interests. Far less knowledge exists, however, on the effects of restoration management practices on the fauna.

The process of restoration may be very slow, or sometimes not be achieved at all, for a variety of reasons. The accumulation of nutrient-rich humus can have been so strong under former agricultural practices that the nutrient pool can only be effectively depleted by sod removal. Nutrient removal can be delayed by the present day mineral input from precipitation which may compensate for 50% or more of the annual nutrient removal by hay-making. Superficial drainage can result in increased mineralization of organic material in the top soil and hence counteract nutrient removal. The groundwater table also influences the vegetation. Even if it remains constant a diminished seepage flow may be compensated for by rainwater or inundation water from a nearby brook. As a result the groundwater quality and hence the vegetation will change considerably. Thus the effects of nature management practices inside a reserve can be seriously counteracted by influences from outside the reserve.

References

Adam, P. (1978). Geographical variation in British salt marsh vegetation. J. Ecol. 66: 339–366.

Adam, P. (1981). The vegetation of British salt marshes. New Phytol. 88: 143–196.

Agnew, A.D.Q. & Flux, J.E.C. (1970). Plant dispersal by hares (Lepus capensis) in Kenya. Ecology 51: 735–737.

Alberda, T. (1974). Productivity measurements of some natural grasslands under different grazing pressures by rabbits. Neth. Contr. Internat. Biol. Progr. Final Report 1966–1971, pp 24–25.

Albert, P. (1982). Halophyten. In: Kinzel, H. (Ed.). Pflanzenökologie und Mineralstoffwechsel, pp 33–215. Ulmer, Stuttgart.

Allden, W.G. & Whittaker, I.A.McD. (1970). The determinants of herbage intake by sheep: the interrelationship of factors influencing herbage intake and availability. Aust. J. Agric. Res. 21: 755–766.

Allersma, G.J.R. (1977). Beweiding op de Oosterkwelder van Schiermonnikoog. Internal Report Dept. of Plant Ecology, Univ. Groningen/Research Institute for Nature Management, Leersum (mimeo).

Al-Mufti, M.M., Sydes, C.L., Furness, S.B., Grime, J.P. & Band, S.R. (1977). A quantitative analysis of shoot phenology and dominance in herbaceous vegetation. J. Ecol. 65: 759–791.

Alstorphius Grevelink, P.W. (1844). Statistiek van de provincie Drenthe. Schierbeek, Groningen.

Altena, H.J. (1983). Doorzaaien als middel om een soortenrijker grasland te krijgen. Internal Report 49, Centre for Agrobiological Research, Wageningen (mimeo).

Andic, C. (1971). Die Zeitlichen Veränderungen einiger Wachstumsfaktoren und die Ertragsbildung von Pflanzenbeständen verschiedener Grünlandpflanzengesellschaften. Thesis, Univ. Kiel.

Anon. (1943). Nature conservation and nature reserves. Report. Br. Ecol. Soc.

Armstrong, W., Wright, E.J., Lythe, S. & Gaynard, T.J. (1985). Plant zonation and the effects of the spring-neap tidal cycle on soil aeration in a Humber salt marsh. J. Ecol. 73: 323–339.

Arnold, G.W. (1964). Factors within plant associations affecting the behaviour and performance of grazing animals. In: Crisp, D.J. (Ed.). Grazing in terrestrial and marine environments, pp 133–154. Blackwell, Oxford.

Austin, M.P. (1982). Use of a relative physiological performance value in the prediction of performance in multispecies mixtures from monoculture performance. J. Ecol. 70: 559–570.

Bailey, R.E. & Putman, R.J. (1981). Estimation of fallow deer (Dama dama) populations from faecal accumulation. J. Appl. Ecol. 18: 697–702.

Bakker, D. & Fresco, L.F.M. (1969). De flora van Schiermonnikoog. Waddenbulletin 4: 8–12.

Bakker, J.P. (1978). Changes in a salt-marsh vegetation as a result of grazing and mowing – a five-year study of permanent plots. Vegetatio 38: 77–87.

Bakker, J.P. (1983). Use and management of salt marshes on sand-dune islands. In: Dijkema, K.S. & Wolff, W.J. (Eds.). Flora and vegetation of the Wadden Sea islands and coastal areas. Report 9 of the Wadden Sea Working Group, pp 290–302. Balkema, Rotterdam.

Bakker, J.P. (1984). Effecten van begrazing op de vegetatie van de Oosterkwelder op Schiermonnikoog. De Levende Natuur 85: 41–46.

Bakker, J.P. (1985). The impact of grazing on plant communities, plant populations and soil conditions on salt marshes. Vegetatio 62: 391–398.

Bakker, J.P. (1987). Restoration of species-rich grassland after a period of fertilizer application. In: Van Andel, J., Bakker, J.P. & Snaydon, R.W. (Eds.). Disturbance in grasslands, pp 185–200, Junk, Dordrecht.

Bakker, J.P. & Allersma, G. (1979). Beweiding op de Oosterkwelder van Schiermonnikoog. Waddenbulletin 14: 146–156.

Bakker, J.P., Dekker, M. & De Vries, Y. (1980). The effect of different management practices on a grassland community and the resulting fate of seedlings. Acta Bot. Neerl. 29: 469–482.

Bakker, J.P. & Ruyter, J.C. (1981). Effects of five years of grazing on a salt marsh vegetation. Vegetatio 44: 84–100.

Bakker, J.P. & De Vries, Y. (1982). Enige indrukken van beekdalvegetaties in Noordoost-Polen. Natura 79: 381–393.

Bakker, J.P., De Bie, S., Dallinga, J.H., Tjaden, P. & De Vries, Y. (1983). Sheep-grazing as a management tool for heathland conservation and regeneration in the Netherlands. J. Appl. Ecol. 20: 541–560.

Bakker, J.P., De Leeuw, J. & Van Wieren, S.E. (1984). Micropatterns in grassland vegetation created and sustained by sheep-grazing. Vegetatio 55: 153–161.

Bakker, J.P. & De Vries, Y. (1985 A). The results of different cutting regimes in grassland taken out of the agricultural system. In: Schreiber, K.F. (Ed.). Sukzession auf Grünlandbrachen, pp 51–57. Münstersche Geogr. Arb. 20.

Bakker, J.P. & De Vries, Y. (1985B). Ueber die Wiederherstellung artenreicher Wiesengesellschaften unter verschiedenen Mahdsystemen in den Niederlanden. Nat. Landschaft 60: 292–296.

Bakker, J.P., Dijkstra, M. & Russchen, P.T. (1985). Dispersal, germination and early establishment of halophytes and glycophytes on a grazed and abandoned salt marsh gradient. New Phytol. 101: 291–308.

Bakker, J.P., Brouwer, C., Van Den Hof, L. & Jansen, A. (1987). Vegetational succession, management and hydrology in a brookland (The Netherlands). Acta Bot. Neerl. 36: 39–58.

Bakker, T.W.M., Klijn, J.A. & Van Zadelhoff, F.J. (1979). Duinen en duinvalleien. Een landschapsecologische studie

van het Nederlandse duingebied. Pudoc, Wageningen.
Balsberg, A. (1982). Plant biomass, primary production and litter disappearance in a Filipendula ulmaria meadow ecosystem, and the effects of Cadmium. Oikos 38: 72–90.
Baradziej, E. (1978). The effect of sheep folding on some chemical properties of soil in the Hieracio-Nardetum strictae association in the Gorce Mountains (Western Carpathians). Proc. Botan. Cracow 6: 135–150.
Bardet, A.C., Kooi, P.B., Waterbolk, H.T. & Wieringa, J. (1983). Peelo, historisch geografisch and archeologisch onderzoek naar de ouderdom van een Drents dorp. Meded. K. Ned. Akad. Wet. afd. Letterkunde, N.R. 46: 1–25.
Barkmann, J.J., Doing, H. & Segal, S. (1964). Kritische Bemerkungen und Vorschläge zur quantitativen Vegetationsanalyse. Acta Bot. Neerl. 13: 394–419.
Barkmann, J.J. & Stoutjesdijk, Ph. (1987). Microklimaat, vegetatie en fauna. Pudoc, Wageningen.
Beddows, A.R. (1961). Biological flora of the British Isles: Holcus lanatus. J. Ecol. 49: 421–430.
Beeftink, W.G. (1965). De zoutvegetatie van Z.W.-Nederland, beschouwd in Europees verband. Meded. Landbouwhogeschool Wageningen 65: 1–167.
Beeftink, W.G. (1977 A). The coastal salt marshes of western and northern Europe: an ecological and phytosociological approach. In: Chapman, V.J. (Ed.). Wet coastal ecosystems, pp 109–155. Elsevier, Amsterdam.
Beeftink, W.G. (1977 B). Salt marshes. In: Barnes, R.S.K. (Ed.). The coastline, pp 93–121. Wiley, London.
Beeftink, W.G., Daane, M.C., De Munck, W. & Nieuwenhuize, J. (1978 A). Vegetation dynamics in salt marshes as a consequence of some traditional cultural practices of man. Phytocoenosis 7: 279–298.
Beeftink, W.G., Daane, M.C., De Munck, W. & Nieuwenhuize, J. (1978 B). Aspects of population dynamics in Halimione portulacoides communities. Vegetatio 36: 31–43.
Behre, K.E. (1976). Beginn und Form der Plaggenwirtschaft in Nortwestdeutschland nach Pollenanalytischen Untersuchungen in Ostfriesland. Neue Ausgrabungen und Forschungen in Niedersachsen 10: 197–224.
Behre, K.E. (1979). Zur Rekonstruktion ehemaliger Pflanzengesellschaften an der Deutschen Nordseeküste. In: Wilmanns, O. & Tüxen, R. (Eds.). Werden und Vergehen von Pflanzengesellschaften, pp 181–214. Cramer, Vaduz.
Behre, K.E. (1985). Die ursprungliche Vegetation in den deutschen Marschgebieten und deren Veränderung durch prähistorische Besiedlung und Meeresspiegelbewegungen. Verh. Gesellsch. für Oekologie 13: 85–96.
Beintema, A.J. & Müskens, G.J.D.M. (1987). Nesting success of birds breeding in a Dutch agricultural grassland. J. Appl. Ecol. 24: 743–758.
Berendse, F. (1986). Stikstof mineralisatie en strooiselproduktie in voedselarme oecosystemen. Vakbl. Biol. 66: 430–433.
Berendse, F. & Kwant, R. (1985). Een wiskundig model voor de heide. Tijdschr. K. Ned. Heide Mij. 78: 371–375.
Bertness, M.D. & Ellison, A.M. (1987). Determinants of pattern in a New England salt marsh plant community. Ecol. Monogr. 57: 129–147.
Bieleman, J. (1985). De Landbouw in de periode 1600–1850. In: Heringa, J., Blok, D.P., Buist, M.G. & Waterbolk, H.T. (Eds.). Geschiedenis van Drenthe, pp 327–372. Boom, Meppel.
Bieleman, J. (1987). Boeren op het Drentse zand 1600–1910. Een nieuwe visie op de oude landbouw. Hes, Utrecht.
Bijhouwer, J.T.P. (1971). Het Nederlandse landschap. Kosmos, Amsterdam.
Bink, F.A. & Van Der Made, J.G. (1986). Dagvlinders en grote herbivoren. De Levende Natuur 87: 168–175.
Bird, E.C.F. (1969). Coasts. M.I.T. Press, Cambridge.
Black, J.L. & Kenney, P.A. (1984). Factors affecting diet selection by sheep. II. Height and density of pasture. Aust. J. Agric. Res. 35: 565–578.
Bleuten, W. (1971). Een geomorphologische studie van het eiland Schiermonnikoog. Internal Report Dept. of Fysical Geography, Univ. Utrecht/Research Institute for Nature Management, Leersum (mimeo).
Blom, C.W.P.M. (1977). Effects of trampling and soil compaction on the occurrence of some Plantago species in coastal sand dunes. II. Trampling and seedling establishment. Oecol. Plant. 12: 363–381.
Bobbink, R. & Willems, J.H. (1987). Increasing dominance of Brachypodium pinnatum in chalk grasslands: a threat to a species-rich ecosystem. Biol. Conserv. 40: 301–314.
Böcker, R., Kowarik, I. & Bornkamm, R. (1983). Untersuchungen zur Anwendung der Zeigerwerte nach Ellenberg. Verh. Gesellsch. für Oekologie 11: 35–56.
Boedeltje, G. & Bakker, J.P. (1980). Vegetation, soil, hydrology and management in a Drenthian brookland (The Netherlands). Acta Bot. Neerl. 29: 509–522.
Boeker, P. (1959). Samenauslauf aus Mist und Erde von Triebwegen und Ruheplätze. Z. Acker Pflanzen B. 108: 77–92.
Boersma, R.A. (1988). Een nieuwe typologie voor kweldervegetaties. Internal Report Dept. of Plant Ecology, Univ. Groningen/Research Institute for Nature Management, Leersum (mimeo).
Bokdam, J. (1987). Foerageergedrag van jongvee in het Junner Koeland in relatie tot het voedselaanbod. In: De Bie, S., Joenje, W. & Van Wieren, S.E. (Eds.). Begrazing in de natuur, pp 165–186. Pudoc, Wageningen.
Bokdam, J., Gleichman, M. & Batterink, M. (1986). Vier jaar begrazing met runderen op de Wolfhezerheide. Huid en Haar 5: 186–196.
Booy, A.H. (1979). Met namen in Börck. Landinrichtingsdienst, Utrecht.
Booy, A.H. (1986). IJzeroer in Drenthe. Ontstaan, voorkomen en winning. Historisch Jaarboek Drenthe 103: 66–88.
Both, J.G. & Van Wirdum, G. (1981). Waterhuishouding, bodem en vegetatie van enkele Gelderse natuurgebieden. Report 81/18. Research Institute for Nature Management, Leersum.
Bottema, S. (1987). De invloed van de vegetatie op de fauna in Nederland gedurende het laat Quartair. Nederlands Bouwtijdschr. 59: 287–294.
Bouwer, K. (1985). De ontwikkeling van het cultuurlandschap.

In: Heringa, J., Blok, D.P., Buist, M.G. & Waterbolk, H.T. (Eds.). Geschiedenis van Drenthe, pp 91–140. Boom, Meppel.
Boxem, T. (1973). Stikstofbemesting en bruto opbrengst van grasland. Stikstof 73: 536–546.
Braaksma, A. (1984). De invloed van verschillende beheersmaatregelen op de kweldervegetatie van Schiermonnikoog. Internal Report Dept. of Plant Ecology, Univ. Groningen (mimeo).
Bradshaw, A.D. (1965). Evolutionary significance of phenotypic plasticity in plants. Adv. Genet. 13: 115–155.
Bradshaw, A.D. (1980). Mineral nutrition. In: Rorison, I.H. & Hunt, R. (Eds.). Amenity grassland: an ecological perspective, pp 101–118. Wiley, London.
Brandsma, O. (1984). Micropatronen in de door schapen beweide graslandvegetatie van het Westerholt. I. Het beschrijven en vastleggen van het mogelijk pulserend systeem van het micropatroon. Internal Report Dept. of Plant Ecology, Univ. Groningen/Research Institute for Nature Management, Leersum (mimeo).
Brandsma, O.H. & Bakker, J.P. (1984). Het Westerholt VI. De zaadvoorraad in relatie tot de samenstelling van de graslandvegetatie. De Levende Natuur 86: 176–180.
Braun-Blanquet, J. & Pavillard, J. (1930). Vocabulair der Plantensociologie. Versl. en Meded. Inst. voor Vegetatieonderzoek voor Nederland 1: 1–22.
Breymeyer, A. (1978). Analysis of the trophic structure of some grassland ecosystems. Pol. Ecol. Stud. 4: 55–128.
Briemle, G. (1985). Vegetations- und Standortentwicklung auf Niedermoor unter den Einfluss verschiedener Pflegemassnahmen – Erste Tendenzen nach fünf Versuchsjahren. Telma 15: 197–221.
Brongers, J.A. (1976). Air photography and celtic field research in the Netherlands. Thesis, Univ. Groningen.
Buchwald, K. (1984). Zum Schutz des Gesellschaftsinventars vorindustriell geprägter Kulturlandschaften in Industriestaaten – Fallstudie Naturschutzgebiet Lüneburger Heide. Phytocoenologia 12: 395–432.
Bülow-Olsen, A. (1980 A). Nutrient cycling in grassland dominated by Deschampsia flexuosa and grazed by nursing cows. Agro-Ecosystems 6: 209–220.
Bülow-Olsen, A. (1980 B). Changes in the species composition in an area dominated by Deschampsia flexuosa as a result of cattle grazing. Biol. Conserv. 18: 257–270.
Burny, J. (1986). Het landgebruik in en rond de vallei van de Zwarte Beek te Koersel (Limburgse Kempen) in het begin van de twintigste eeuw. Het Oude Land van Loon 41: 79–111.
Burrichter, E., Pott, R., Raus, T. & Wittig, R. (1980). Die Hudelandschaft 'Borkener Paradies' im Emstal bei Meppen. Abhandlungen aus dem Landesmuseum für Naturkunde zu Münster in Westfalen 42.
Buth, G.J.C., Verdonschot, P.F.M. & De Wolf, L. (1982). Decomposition of three halophytes in different habitats of an Eastern Scheldt salt march. Hydrobiol. Bull. 16: 103–112.
Buth, G.J.C. & Voesenek, L.A.C.J. (1987). Decomposition of standing and fallen litter of halophytes in a Dutch salt marsh. In: Huiskes, A.H.L., Blom, C.W.P.M. & Rozema, J. (Eds.). Vegetation between land and sea, pp 146–162. Junk, Dordrecht.
Buttenschøn, J. & Buttenschøn, R.M. (1978). The effect of browsing by cattle and sheep on trees and bushes. Nat. Jutl. 20: 79–94.
Buttenschøn, J. & Buttenschøn, R.M. (1982 A). Grazing experiments with cattle and sheep on nutrient poor, acidic grassland and heath. I. Vegetation development. Nat. Jutl. 21: 1–18.
Buttenschøn, J. & Buttenschøn, R.M. (1982 B). Grazing experiments with cattle and sheep on nutrient poor, acidic grassland and heath. II. Grazing impact. Nat. Jutl. 21: 19–27.
Buytendorp, B., Gorzeman, J., Morel, T. & Willems, J. (1983). Maaibeheer en bovengronds verloop van de nutrienten in een kalkgrasland. Publ. Natuurhist. Genootsch. Limburg 33: 33.

Campino Johnson, I. (1978). Einfluss der Nutzungsintensität auf Kompartimente von Grünland Oekosystemen. Thesis, Univ. Lahn-Giessen.
Cancrinus, S. (1950). Verdwijnende cultuurwaarden III. De vloei- en stroomlanden. Nieuwe Drentsche Volksalmanak 68: 163–167.
Casparie, W.A. (1982). The neolithic wooden trackway XXI in the raised bog at Nieuw-Dordrecht (The Netherlands). Palaeohistoria 24: 115–164.
Casparie, W.A. & Groenman-van Waateringe, W. (1980). Palynological analysis of dutch barrows. Palaeohistoria 22: 7–65.
Centraal Bureau voor de Statistiek (1983). Algemene Milieustatistiek 1979–1982. Staatsuitgeverij, Den Haag.
Chacon, E. & Stobbs, T.H. (1976). Influence of progressive defoliation of a grass sward on the eating behaviour of cattle. Aust. J. Agric. Res. 27: 709–727.
Champness, S.S. & Morris, K. (1948). The population of buried viable seeds in relation to contrasting pasture and soil types. J. Ecol. 36: 149–173.
Chapin, F.S. (1980). The mineral nutrition of wild plants. Annu. Rev. Ecol. Syst. 11: 233–281.
Charles, W.N., Mc Cowan, D. & East, K. (1977). Selection of upland swards by Red Deer (Cervus elaphus L.) on Rhum. J. Appl. Ecol. 14: 55–64.
Chippindale, H.G. & Milton, W.E. (1934). On the viable seeds present in the soil beneath pastures. J. Ecol. 22: 508–531.
Christiansen, W. (1937). Beobachtungen an Dauerquadraten auf der Lotseninsel Schleimünde. Schr. Naturwiss. Ver. Schleswig-Holstein 12: 69–84.
Clarkson, D.T. & Hanson, J.B. (1980). The mineral nutrition of higher plants. Annu. Rev. Plant Physiol. 31: 239–298.
Clason, A.T. (1980 A). Jager, visser, veehouder, vogellijmer. In: Chamalaun, M. & Waterbolk, H.T. (Eds.). Voltooid verleden tijd?, pp 131–146. Intermediair, Amsterdam.
Clason, A.T. (1980 B). Zeldzame huisdierrassen. Thieme, Zutphen.
Clason, A.T. (1986). Knagers, wroeters en gravers van vroeger en nu. Huid en Haar 5: 134–140.

Cook, C.W., Harris, L.E. & Young, M.C. (1967). Botanical and nutritive content of diets of cattle and sheep under single and common use on mountain range. J. Anim. Sci. 26: 1169–1174.
Crawley, M.J. (1983). Herbivory, the dynamics of animal-plant interactions. Blackwell, Oxford.
Czerwinsky, Z., Jakubczyk, H., Tatur, A. & Traczyk, T. (1974). Analysis of a sheep pasture ecosystem in the Pieniny Mountains (The Carpenthians). VII. The effect of penning-up sheep on soil, microflora and vegetation. Ekol. Pol 22: 547–558.

Daines, T. (1980). The use of grazing patterns in the management of Dohne Sourveld. Proc. Grassl. Soc. South Afr. 15: 185–188.
Dallinga, J.H. & Tjaden, P. (1981). Vijf jaar maaien en beweiden op het Westerholt. Internal Report Dept. of Plant Ecology, Univ. Groningen/Research Institute for Nature Management, Leersum (mimeo).
Dallinga, J.H., Tjaden, P. & Bakker, J.P. (1982 A). Het Westerholt IV. Veranderingen in de grasland vegetatie bij beweiden en hooien. De Levende Natuur 84: 77–86.
Dallinga, J.H., Tjaden, P. & Bakker, J.P. (1982 B). Het Westerholt V. Veranderingen in de heidevegetatie bij beweiden. De Levende Natuur 84: 113–120.
Davies, W. (1925). The relative palatability of pasture plants. J.Min. Agric. 32: 106–116.
Davies, M.S. & Snaydon, R.W. (1976). Rapid population differentiation in a mosaic environment. III. Measures of selection pressure. Heredity 36: 59–66.
Davy, A.J. & Taylor, K. (1974). Seasonal patterns of nitrogen availability in contrasting soils in the Chiltern hills. J. Ecol. 62: 793–807.
De Bakker, H. & Schelling, J. (1966). Systeem voor bodemclassificatie van Nederland. De hogere niveaus. Pudoc, Wageningen.
De Bie, S. (1976). Onderzoek naar beweiding als natuurtechnische beheersmaatregel op het Westerholt (Drentsche A). Internal Report Dept. of Plant Ecology, Univ. Groningen/Research Institute for Nature Management, Leersum (mimeo).
De Boer, J.M.U. (1974). Onderzoek naar de verdeling van de mest en de invloed van bemesting op de Oosterkwelder van Schiermonnikoog. Internal Report Dept. of Plant Ecology, Univ. Groningen/Research Institute for Nature Management, Leersum (Mimeo).
De Both, M.T.J., De Muinck Keizer, M., Wouda, A., Bakker, B. & Drenth, W. (1983). Seizoensfluctuaties in de bodem. Een studie naar de veranderingen in bodemeigenschappen gedurende het groeiseizoen 1981 in het beekdal van de Drentse Aa. Internal Report Dept. of Plant Ecology, Univ. Groningen (mimeo).
De Bruijn, O. (1980). The Carex-flora of varied landscapes in the Netherlands, an example of decreasing ecological diversity. Acta Bot. Neerl. 29: 359–376.
Decleer, K. (1987). Het belang van ongewervelden in het natuurbeheer. Internal Report Dept. of Animal Ecology, Univ. Gent (mimeo).
De Dainville, F.S.J. (1964). Le langage de géographes. Picard, Paris.
De Gans, W. (1980). Outlines of the late Quaternary history of the Drentsche Aa valley. Geol. Mijnbouw 59: 353–362.
De Gans, W. (1981). The Drentsche A valley system. Thesis, Univ. Amsterdam.
De Haan, H., Ybema, R. & Reitsma, D.T. (1983). Engelsmanplaat. Geschiedenis van en gebeurtenissen rond een zandbank. Stichting 't Fiskerhúske, Moddergat.
De Hullu, E. (1985). The influence of sward density on the population dynamics of Rhinanthus angustifolius in a grassland succession. Acta Bot. Neerl. 34: 23–32.
Deinum, B. (1976). Kwaliteit van ruwvoer. Syllabus, Univ. Wageningen (mimeo).
Deinum, B. & Van Soest, P.J. (1969). Prediction of forage digestibility from some laboratory procedures. Neth. J. Agr. Sci. 17: 119–127.
De Jong, J. (1975). Pollen analysis and radiocarbon dating of a peat deposit from the island of Schiermonnikoog (The Netherlands). Ber. Rijksdienst Oudheidkundig Bodemonderzoek 25: 183–186.
De Leeuw, J. & Bakker, J.P. (1986). Sheep-grazing with different foraging efficiencies in a Dutch mixed grassland. J. Appl. Ecol. 23: 781–793.
Den Hartog, C. (1952). Plantensociologische waarnemingen op Schiermonnikoog. Kruipnieuws 14: 2–24.
De Roo, H.C. (1953). Enkele bodemkundige aantekeningen over de Drentse essen. Boor en Spade 6: 59–76.
De Smidt, J.T. (1975). Nederlandse heidevegetaties. Thesis, Univ. Utrecht.
De Smidt, J.T. (1979). Origin and destruction of northwest european heath vegetation. In: Wilmanns, O. & Tüxen, R. (Eds.). Werden und Vergehen von Pflanzengesellschaften, pp 411–435. Cramer, Vaduz.
De Vries, D.M., 't Hart, M.L. & Kruijne, A.A. (1942). Een waardering van grasland op grond van de plantkundige samenstelling. Landbouwkd. Tijdschr. 54: 245–265.
De Vries, J. (1964). Etymologisch woordenboek. Utrecht.
De Vries, V. (1961). Vegetatiestudie op de Westpunt van Vlieland. Thesis, Univ. Amsterdam.
Dickinson, N.M. (1983). Decomposition of grass litter in a successional grassland. Pedobiologia 25: 117–126.
Dickinson, N.M. (1984). Seasonal dynamics and compartmentation of nutrients in a grassland meadow in lowland England. J. Appl. Ecol. 21: 695–701.
Dickinson, N.M. & Polwart, A. (1982). The effect of mowing regime on an amenity grassland ecosystem: above- and below-ground components. J. Appl. Ecol. 19: 569–577.
Dierschke, H. (1974). Saumgesellschaften im Vegetations- und Standortgefälle an Waldrändern. Scr. Geobot. 6: 1–246.
Dierschke, H. (1984). Natürlichkeitsgrade von Pflanzengesellschaften under besonderer Berücksichtigung der Vegetation Mitteleuropas. Phytocoenologia 12: 173–184.
Dierschke, H. (1985). Experimentelle Untersuchungen zur Bestandesdynamik von Kalkmagerrasen (Mesobromion) I. Vegetationsentwicklung auf Dauerflächen 1972–1984. In:

Schreiber, K.F. (Ed.). Sukzession auf Grünlandbrachen, pp 9–24. Münstersche Geogr. Arb. 20.
Dijkema, K.S. (1975). Vegetatie en beheer van kwelders en landaanwinningswerken aan de Waddenzeekust van Noord-Groningen. Meded. Werkgroep Waddengebied 2: 1–49.
Dijkema, K.S. (1980). Towards a vegetation and landscape map of the Danish, German and Dutch Waddensea Islands and mainland coastal areas. Acta Bot. Neerl. 29: 523–531.
Dijkema, K.S. (1983). The salt marsh vegetation of the mainland coast, estuaries and halligen. In: Dijkema, K.S. & Wolff, W.J. (Eds.). Flora and vegetation of the Waddensea islands and coastal areas. Report 9 of the Waddensea Working Group, pp 185–220. Balkema, Rotterdam.
Dijkema, K.S. (1984). Salt marshes in Europe. Council of Europe, Strasbourg.
Dijkstra, M. (1984). Verder onderzoek van enkele mogelijke oorzaken van uitbreiding van lage kweldersoorten onder beweiding met jongvee op de Oosterkwelder. Internal Report Dept. of Plant Ecology, Univ. Groningen (mimeo).
Dirven, J.G.P. & Wind, K. (1980). De beworteling van enige graslandsoorten bij verschillende bodemvruchtbaarheid. Stikstof 95/96: 354–359.
Dirven, J.G.P. & Wind, K. (1982). De invloed van bemesting op de beworteling van verschillende grassoorten en -rassen. Meded. Dept Landbouwplantenteelt Graslandkd. 61.
Don, R., Van Hoof, W.C.M., Jansen, S.R.J. & Jonker, A.J.P. (1980). Patroon, proces en dynamiek van vegetatie en milieu op de Boschplaat van Terschelling. Internal Report Dept. of Geobotany, Univ. Nijmegen (mimeo).
Dore, W.G. & Raymond, L.C. (1942). Pasture studies XXIV. Viable seeds in pasture soil and manure. Sci. Agri. 23: 69–79.
Draaisma, A. (1958). Het productieniveau onderzoek. I. Teelt en bemesting op bouwland in de praktijk. Versl. Landbouwkd. Onderz. 64.9: 116–120.
Drent, R.H. & Prins, H.H.T. (1987). The herbivore as a prisoner of its food supply. In: Van Andel, J., Bakker, J.P. & Snaydon, R.W. (Eds.). Disturbance in grasslands, pp 131–147. Junk, Dordrecht.
Duffey, E., Moris, M.G., Sheail, J., Ward, L.K., Wells, D.A. & Wells, T.C.E. (1974). Grassland ecology and wildlife management. Chapman & Hall, London.
During, H.J., Schenkeveld, A.J., Verkaar, H.J. & Willems, J.H. (1985). Demography of short-lived forbs in chalk grassland in relation to vegetation structure. In: White, J. (Ed.). The population structure of vegetation, pp 341–369. Junk, Dordrecht.
During, H.J. & Willems, J.H. (1986). The impoverishment of the bryophyte and lichen flora of the Dutch chalk grasslands in the thirty years 1953–1983. Biol. Conserv. 36: 143–158.
Durwen, K.J. (1982). Zur Nutzung von Zeigerwerten und artspecifische Merkmalen der Gefässpflanzen Mitteleuropas für Zwecke der Landschaftsökologie und -Planung mit Hilfe der E.D.V. Internal Report Dept. Landschaftsökologie, Univ. Münster (mimeo).

Ebbinge, B., Canters, K. & Drent, R. (1975). Foraging routines and estimated daily food intake in Barnacle Geese wintering in the northern Netherlands. Wildfowl 26: 5–19.
Ebel, F. & Rauschert, S. (1982). Die Bedeutung der Botanischen Gärten für die Erhaltung gefährdeter und vom Aussterben bedrohter heimischer Pflanzenarten. Arch. Naturschutz Landschaftsforsch. 22: 187–199.
Edelman, C.H. (1974). Harm Tiesing over landbouw en volksleven in Drenthe I, II. Van Gorcum, Assen.
Edelman-Vlam, R.W. (1969). Veldnamen in de gemeente Rolde. Meded. van het Inst. voor Naamkunde, Leuven en de Commissie voor Naamkunde en Nederzettingsgeschiedenis, Amsterdam 1: 138–156.
Edwards, P.J. & Hollis, S. (1982). The distribution of excreta on New Forest grassland used by cattle, ponies and deer. J. Appl. Ecol. 19: 953–964.
EEC Committee (1979). Umweltfolgen der Brachlegung von Kulturflächen. Mitt. über Landwirtschaft 62. EGKS, Brussel.
Egloff, T. (1986). Auswirkungen und Beseitigung von Düngungseinflüsse auf Streuweisen. Veröff. Geobot. Inst. Tech. Hochsch. Stift. Rübel Zür. 89: 1–183.
Eilers, H.P. (1979). Production ecology in an Oregon coastal salt marsh. Estuarine Coastal Mar. Sci. 8: 399–410.
Ekstam, U. & Sjögren, E. (1973). Studies of past and present change in deciduous forest vegetation on Oeland. Zoon Suppl. 1: 123–135.
Elberse, W.Th., Van Den Bergh, J.P. & Dirven, J.G.P. (1983). Effects of use and mineral supply on the botanical composition and yield of old grassland on heavy clay soil. Neth. J. Agric. Sci. 31: 62–88.
Ellenberg, H. (1953). Physiologisches und ökologisches Verhalten derselben Pflanzenarten. Ber. Deutsch. Bot. Geselsch. 65: 351–362.
Ellenberg, H. (1954). Steppenweide und Waldheide. Erdkunde 8: 188–194.
Ellenberg, H. (1978). Vegetation Mitteleuropas mit den Alpen. Ulmer, Stuttgart.
Ellenberg, H. (1979). Zeigerwerte der Gefässpflanzen Mitteleuropas. Scr. Geobot. 9: 1–122.
Ernst, A. (1976). Tien jaar Stroomdallandschap Drentsche A. Nat. Landschap 30: 13–17.
Ernst, W.H.O. (1978). Discrepancy between ecological and physiological optima of plant species. A re-interpretation. Oecol. Plant. 13: 175–188.
Ernst, W.H.O. (1982). Schwermetallpflanzen. In: Kinzel, H. (Ed.). Pflanzenökologie und Minerallstoffwechsel, pp 472–506. Ulmer, Stuttgart.
Esselink, P., Schotel, J. & Van Gils, H. (1989). Nitrogen and phosphorus limited primary production of annuals. I.T.C. publication (in press).
Everitt, B.S. (1974). Cluster analysis. Heinemann, London.
Everts, F.H., Grootjans, A.P. & De Vries N.P.J. (1980). De vegetatie van de madelanden in het stroomdal van de Drentse Aa. Internal Report Dept. of Plant Ecology, Univ. Groningen Research Institute for Nature Management, Leersum (mimeo).
Everts, F.H., Grootjans, A.P. & De Vries, N.P.J. (1984). Ve-

getatiekartering van de Drentse Aa. Internal Report State Forestry Commission, Assen/Dept. of Plant Ecology, Univ. Groningen (mimeo).

Fastabend, H. & Von Raupach, F. (1961). Zur Kenntnis der Plaggenböden in Nordwestdeutschland. Geol. Jahrb. 78: 139–172.

Fenner, M. (1978). A comparison of the abilities of colonizers and close-turf species to establish from seed in artificial swards. J. Ecol. 66: 953–963.

Fenner, M. (1985). Seed ecology. Chapman & Hall, London.

Firbas, F. (1949). Waldgeschichte Mitteleuropas. I. Fischer, Jena.

Fliervoet, L.M. (1984). Canopy structure of Dutch grasslands. Thesis, Univ. Nijmegen.

Floate, M.J.S. (1970). Mineralization of nitrogen and phosphorus from organic materials and animal origin and its significance in the nutrient cycle in grazed uplands and hill soils. J.Br.Grassl. Soc. 25: 295–302.

Floate, M.J.S., Eadie, J., Black, J.S. & Nicholson, I.A. (1973). The improvement of Nardus dominant hill pasture by grazing control and fertilizer treatment and its economic assessment. Colloquium Proc. 3: 33–39.

Frankel, O.M. & Soulé (1981). Conservation and evolution. Cambridge University Press, Cambridge.

Frame, J. & Hunt, I.V. (1971). The effect of cutting and grazing-systems on herbage production from grass swards. J. Br. Grassl. Soc. 26: 171–173.

Fransen, J. (1987). Slechte vooruitzichten voor Europa door de verzuring. Nat. Milieu 11: 4–8.

Fresco, L.F.M. (1971). Compound analysis: a preliminary report as a new numerical approach in phytosociology. Acta Bot. Neerl. 20: 589–599.

Fresco, L.F.M., Hermans, E. & Van Der Lans, W. (1984). Soil phosphorus as a factor controlling vegetation composition in moist hay fields: a statistical evaluation of some methods for the determination of available phosphate. Plant Soil 78: 259–269.

Fresco, L.F.M., Van Laarhoven, H.P.M., Loonen, M.J.J.E. & Moesker, T. (1987). Ecological modeling of short-term plant community dynamics under grazing with and without disturbance. In: Van Andel, J., Bakker, J.P. & Snaydon, R.W. (Eds.). Disturbance in grasslands, pp 149–165. Junk, Dordrecht.

Gabrielson, I.N. (1947). Management of nature reserves on the basis of modern scientific knowledge. Proc. and Papers 6th Technological Meeting I.U.C.N., pp 27–35, London.

Galle, F. & Bontemps, J. (1975). Vegetatiekundig en bodemkundig onderzoek in een transect in de beweide enclave 'de Groede' op de Boschplaat, Terschelling. Internal Report Dept. of Geobotany, Univ. Nijmegen (mimeo).

Garming, R.W. (1955). Het carspel Emmen omstreeks 1650. Nieuwe Drentse Volksalmanak 73: 149–178.

Gauch, H.G. (1982). Multivariate analysis in community ecology. Cambridge University Press, Cambridge.

Gerth, H. (1978). Wirkungen einiger Landschaftspflegeverfahren auf die Pflanzenbestände und Möglichkeiten der Schafweide auf feuchten Grünlandbrachen. Thesis, Univ. Kiel.

Gibson, C.W.D., Watt, T.A. & Brown, V.K. (1987). The use of sheep grazing to recreate species-rich grassland from abandoned arable land. Biol. Conserv. 42: 165–183.

Gillham, M.E. (1955). Ecology of the Pembrokeshire Islands. III. The effect of grazing on the vegetation. J. Ecol. 43: 172–206.

Gillham, M.E. (1957 A). Vegetation of the Exe estuary in relation to water salinity. J. Ecol. 45: 735–756.

Gillham, M.E. (1957 B). Coastal vegetation of Mull and Iona in relation to salinity and soil reaction. J. Ecol. 45: 757–778.

Gimingham, C.H. (1972). Ecology of heathlands. Chapman & Hall, London.

Gisi, U. & Oertli, J.J. (1981). Oekologische Entwicklung in Brachland verglichen mit Kulturwiesen. II. Veränderungen in ober- und unterirdischer Pflanzenmasse. Oecol. Plant. 2: 79–86.

Glässer, E. (1967). Zur Entstehungsgeschichte der bäuerlichen Kulturlandschaft. Geogr. Rundschau 19: 383–389.

Glitz, D. (1980). Erfahrungen mit der Ausbringung von Wildpflanzen im Hamburger Raum. In: Ausbringung von Wildpflanzen, pp 56–71. Tagungsbericht 5/80 der Akademie für Naturschutz und Landschaftspflege, Laufen-Salzach.

Goedewagen, M.A.J. & Schuurman, J.J. (1950). Wortelproductie van bouw- en grasland als bron van organische stof. Landbouwkd. Tijdschr. 62: 469–482.

Goldberg, D.E. & Werner, P. (1983). The effects of size opening in vegetation litter and litter cover on seedling establishment of golden rods (Solidago spp.). Oecologia Berl. 60: 149–155.

Graham, D. & Petterson, B.D. (1982). Responses of plants to low, non freezing temperatures: proteins, metabolism and acclimatization. Annu. Rev. Plant Physiol. 33: 347–372.

Grant, S.A., Suckling, D.E., Smith, H.K., Torvell, L., Forbes, T.D.A. & Hodgson, J. (1985). Comparative studies of diet selection by sheep and cattle: the hill grasslands. J. Appl. Ecol. 73: 987–1004.

Grant, S.A., Torvell, L., Smith, H.K., Suckling, D.E., Forbes, T.D.A. & Hodgson, J. (1987). Comparative studies of diet selection by sheep and cattle: blanket bog and heather moor. J. Appl. Ecol. 75: 947–960.

Gray, A.J. & Scott, R. (1977). Biological flora of the British Isles. Puccinellia maritima. J. Ecol. 65: 699–716.

Gray, A.J., Parsell, R.J. & Scott, R. (1979). The genetic structure of plant populations in relation to the development of salt marshes. In: Jefferies, R.L. & Davy, A.J. (Eds.). Ecological processes in coastal environments., pp 43–64. Blackwell, Oxford.

Green, B.H. (1972). The relevance of seral eutrophication and plant competition to the management of successional communities. Biol. Conserv. 4: 378–384.

Green, B.H. (1980). Management of extensive amenity grassland by mowing. In: Rorison, I.H. & Hunt, R. (Eds.). Amen-

ity grassland, pp 155–161. Wiley, Chichester.
Green, B.H. (1981). A policy on introductions in Britain. In: Synge, H. (Ed.). The Biological Aspects of rare plant conservation, pp 403–412. Wiley, Chichester.
Green, B.H. (1983). The management of herbaceous vegetation for Wildlife Conservation. In: Way, J.M. (Ed.). Management of vegetation, pp 99–116. British Crop Protection Counsil, Croydon.
Greig, J. (1984). The palaeoecology of some British hay meadow types. In: Van Zeist, W. & Casparie, W.A. (Eds.). Plants and ancient man, studies in palaeoethnobotany, pp 213–226. Balkema, Rotterdam.
Grime, J.P. (1973). Control of species diversity in herbaceous vegetation. J. Environ. Manage. 1: 151–167.
Grime, J.P. (1979). Plant strategies and vegetation processes. Wiley, Chichester.
Grime, J.P. (1981). An ecological approach to management. In: Rorison, I.H. & Hunt, R. (Eds.). Amenity grassland, pp 13–55. Wiley, Chichester.
Grime, J.P., Mason, G., Curtis, A.V., Rodman, J., Band, S.R., Mowforth, M.A.G., Neal, A.M. & Shaw, S. (1981). A comparative study of germination characteristics in a local flora. J. Ecol. 69: 1017–1059.
Groenendijk, A.M. (1984). Primary production of four dominant salt marsh angiosperms in the S.W. Netherlands. Vegetatio 57: 143–152.
Groenendijk, A.M. (1987). Ecological consequences of a stormsurge barrier in the Oosterschelde salt marshes. Thesis, Univ. Utrecht.
Groenman-Van Waateringe, W. (1979). The origin of crop weed communities composed of summer annuals. Vegetatio 41: 57–59.
Groenman-Van Waateringe, W. (1983). The early agricultural utilization of the Irish landscape: the last word on the elm decline? In: Reeves-Smyth, T. & Hamond, F. (Eds.). Landscape archeology in Ireland, pp 217–232. B.A.R. 116, Oxford.
Groenman-Van Waateringe, W. (1986). Grazing possibilities in the Neolithic of the Netherlands based on palynological data. In: Behre, K.E. (Ed.). Anthropogenic indicators in pollendiagrams, pp 187–202. Balkema, Rotterdam.
Grootjans, A.P. (1980). Distribution of plant communities along rivulets in relation to hydrology and management. In: Wilmanns, O. & Tüxen, R. (Eds.). Epharmonie, pp 143–170. Cramer, Vaduz.
Grootjans, A.P. (1985). Changes of groundwater regime in wet meadows. Thesis, Univ. Groningen.
Grootjans, A.P. & Ten Klooster, W.P. (1980). Changes of groundwater regime in wet meadows. Acta Bot. Neerl. 29: 541–554.
Grootjans, A.P., Schipper, P.C. & Van Der Windt, H.J. (1985). Influence of drainage on N-mineralization and vegetation response in wet meadows. I. Calthion palustris stands. Oecol. Plant. 6: 403–417.
Grootjans, A.P., Schipper, P.C. & Van Der Windt, H.J. (1986). Influence of drainage on N-mineralization and vegetation response in wet meadows. II. Cirsio-Molinietum stands. Oecol. Plant. 7: 3–14.
Gross, K.L. (1984). Effects of seed size and growth form on seedling establishment of six monocarpic perennial plants. J. Ecol. 72: 369–387.
Gross, K.L. & Werner, P. (1982). Colonizing abilities of 'biennial' plant species in relation to ground cover: implications for their distributions in a successional sere. Ecology 63: 921–931.
Grubb, P.J. (1977). The maintenance of species-richness in plant communities: the importance of the regeneration niche. Biol. Rev. 52: 107–145.
Grubb, P.J. (1984). Some growth points in investigative plant ecology. In: Cooley, J.H. & Golley, F.B. (Eds.). Trends in ecological research in the 1980's, pp 51–74. Plenum, New York.
Grubb, P.J. (1985). Plant populations in relation to habitat, disturbance and competition: problems of generalization. In: White, J. (Ed.). The Population Structure of Vegetation, pp 595–621. Junk, Dordrecht.
Gupta, P.L. & Rorison, I.H. (1975). Seasonal differences in the availability of nutrients down a podzolic profile. J. Ecol. 63: 521–534.

Haarnagel, W. (1979). Die Grabung der Feddersen Wierde. Feddersen Wierde 2: 1–364.
Hacker, J.B. & Minson, D.J. (1981). The digestibility of plant parts. Herbage Abstr. 51: 459–482.
Handke, K. & Schreiber, K.F. (1985). Faunistisch-ökologische Untersuchungen auf unterschiedlich gepflegten Parzellen einer Brachfläche im Taubergebiet. In: Schreiber, K.F. (Ed.). Sukzession auf Grünlandbrachen, pp 155–186. Münstersche Geogr. Arb. 20.
Hansen, D. (1982). Entwicklung und Beeinflussung der Netto primär produktion auf Vorlandflächen und im Vogelschutzgebiet Hauke-Haien-Koog. Schriftenreihe Inst. f. Wasserwirtschaft und Landschaftökologie 1: 1–273. Christian-Albrechts-Univ., Kiel.
Hansson, A.C. & Andren, O. (1986). Below-ground plant production in a perennial grass ley (Festuca pratensis) assessed with different methods. J. Appl. Ecol. 23: 657–666.
Härdtle, W. (1984). Vegetationskundliche Untersuchungen in Salzwiezen der ostholsteinischen Ostsee Küste. Mitt. Arbeitsgem. Geobot. Schleswig-Holstein u. Hamburg 34: 1–142.
Harper, J.L. (1977). Population biology of plants. Academic Press, London.
Harrison, A.F. (1978). Phosphorus cycles of forest and upland grassland ecosystem and some effects of land management practices. In: Potter, R. & Fitzsimmon, D.W. (Eds.). Phosphorus and the environment: its chemistry and biochemistry, pp 175–199. Elsevier, Amsterdam.
Harsema, O.H. (1974). Archeologisch onderzoek op het Hijkerveld gem. Beilen. Voorlopig bericht van de campagnes 1969 en 1970. Nieuwe Drentse Volksalmanak 91: 161–168.
Harsema, O.H. (1979). Noordbarge. Bull. K. Ned. Oudheidkd. Bond 75: 145–153.

Harsema, O.H. (1980). Het Drents Plateau. De latere prehistorische bewoning. In: Chamalaun, M. & Waterbolk, H.T. (Eds.). Voltooid verleden tijd?, pp 83–102. Intermediair, Amsterdam.

Harsema, O.H. (1984). Boeren in Neolithicum, Bronstijd en IJzertijd. In: Abrahamse, J. Bottema, S., Buruma, M.H., De Vroome, H.W. & Waterbolk, H.T. (Eds.). Het Drentse Landschap, pp 32–39. Walburg Pers, Zutphen.

Heady, H.F. (1975). Rangeland management. Mc Graw-Hill, New York.

Heard, C.A.H. (1980). Interrelationships between the factors influencing the intake and digestibility of forage by ruminants. Proc. Grassl. Soc. South Afr. 15: 123–128.

Heijink, J. (1974). Heideontwikkeling op voormalige cultuurgronden. Report 1162 Stichting voor Bodemkartering, Wageningen.

Heil, G.W., Werger, M.J.A., De Mol, W., Van Dam, D. & Heijne, B. (1988). Capture of atmospheric ammonium by grassland canopies. Science 239: 764–765.

Heimans, J. (1954). L'accessibilité, term nouveau en phytogéographie. Vegetatio 5/6: 142–146.

Heinemeyer, H.D. (1982). Vergelijkend onderzoek naar beweide en uit beweiding genomen vegetaties op de Groede, Terschelling. Internal Report Hugo de Vries Laboratory, Univ. Amsterdam/Research Institute for Nature Management, Leersum (mimeo).

Hendriks, A. (1984). Het effect van de structuur van de vegetatie op kieming en vestiging bij 15 kunstmatig geïntroduceerde plantensoorten op het Westerholt. Internal Report Dept. of Plant Ecology, Univ. Groningen/Research Institute for Nature Management, Leersum (mimeo).

Hendrikse, K. & Wuring, G. (1981). De relatie biomassa-soorts diversiteit van graslandvegetaties in de Lauwerszeepolder en het stroomdal van de Drentsche A. Internal Report Dept. of Plantecology, Univ. Groningen (mimeo).

Heringa, J. (1977). De bevolking van Drenthe in 1630. Nieuwe Drentse Volksalmanak 94: 49–61.

Heringa, J. (1982). Drense historische studiën V. De buurschap en haar marke. Provinciaal Bestuur Drenthe, Assen.

Hilder, E.J. (1966). Distribution of excreta by sheep at pasture. Proc. Xth Int. Grassl. Congress Helsinki: 977–984.

Hillegers, H.P.M. (1984). Begrazing met Mergellandschapen in Zuid Limburg. De Levende Natuur 85: 178–184.

Hillegers, H.P.M. (1985). Exozoöchoor transport van diasporen door Mergellandschapen. Natuurhist. Maandbl. 74: 54–56.

Hodgson, J. (1979). Nomenclature and definitions in grazing studies. Grass Forage Sci. 14: 11–18.

Hodgson, J. (1981). Sward measurement handbook. British Grassland Society, Maidenhead.

Hodgson, J. & Grant, S.A. (1981). Grazing animals and forage resources in the hills and uplands. In: Frame, J. (Ed.). The effective use of forage and animal resources in the hills and uplands, pp 41–57, Symp. Br. Grassl. Soc. no. 12.

Hofman, R. & Griffioen, C. (1980). Het zaadkapitaal onder enkele graslanden in Zd. Limburg (Gerendal). Internal Report Dept. of Vegetation Science and Plant Ecology, Univ. Utrecht (mimeo).

Hoogveld, J. & Muller, H.J.M. (1985). Zaadkapitaal en kieming en vestiging van ingezaaid zaad in vijf graslanden bij het Anloër Diepje. Internal Report Dept. of Plant Ecology, Univ. Groningen (mimeo).

Hopkinson, C.S., Gosselink, J.G. & Parrondo, R.T. (1978). Above-ground production of seven salt march species in coastal Louisiana. Ecology 59: 760–769.

Howe, C.D. & Chancellor, R.J. (1983). Factors affecting the viable seed content of soils beneath lowland pastures. J. Appl. Ecol. 20: 915–922.

Hülbusch, K.H. (1986). Eine pflanzensoziologische 'Spurensicherung' zur Geschichte eines 'Stücks Landschaft'. Landschaft und Stadt 18: 60–72.

Hunter, R.F. (1954). The grazing of hill pasture sward types. J. Br. Grassl. Soc. 9: 195–208.

Hunter, R.F. (1962). Hill sheep and their pasture: a study of sheep grazing in southeast Scotland. J. Ecol. 50: 651-680.

Huston, M. (1979). A general hypothesis of species diversity. Am. Nat. 13: 81–101.

Hutchinson, K.J. & King, K.L. (1980). Management impacts on structure and function of sown grasslands. In: Breymeyer, A.I. & Van Dyne, G.M. (Eds.). Grasslands, system analysis and man, pp 823–852. Cambridge University Press, Cambridge.

Ijjas, I. (1969). Effect of compactness and initial moisture content of the soil on the process of capillary rise. In: Rijtema, P.E. & Wassink, H. (Eds.). Water in the unsaturated zone, pp 547–559. Internat. Ass. Sci. Hydrol., Leuven.

Irmler, U. & Heydemann, B. (1985). Der Einfluss der Rinder Beweidung auf die Struktur der Salzwiezen-Biozönose. Verh. Gesellsch. Oekologie 13: 71–76.

Irmler, U. & Heydemann, B. (1986). Die ökologische Problematik der Beweidung von Salzwiesen an der Niedersächsischen Küste- am Beispiel der Leybucht. Naturschutz Landschaftspflege Niedersachsen Beiheft 15.

Isbary, G. (1936). Das Inselgebiet von Ameland bis Rottemeroog. Morphologische und hydrographische Beiträge zur Entwicklungsgeschichte der friesische Inseln. Thesis, Univ. Hamburg.

Iversen, J. (1949). The influence of prehistoric man on vegetation. Dan. Geol. Unders. IV. Raekke 3.

Iversen, J. (1967). Naturens udvikling siden sidste istid. In: Nørrevang, A. & Meyer, T.J. (Eds.). Danmarks Natur, pp 345–445. Politikens Forlag, Copenhagen.

Iversen, J. (1969). Retrogressive development of a forest ecosystem demonstrated by pollen diagrams from fossil mor. Oikos (Suppl.) 12: 35–49.

Ivlev, V.S. (1961). Experimental ecology of the feeding of fishes. Yale University Press, New Haven.

Jacobs, J. (1974). Quantitative measurement of food selection. A modification of the forage ratio and Ivlev's Electivity Index. Oecologia Berl. 14: 413–417.

Janssen, C.R. (1972). The palaeoecology of plant communities in the Dommel valley, North Brabant, Netherlands. J. Ecol. 60: 411–437.

Janssen, C.R. (1979). Pollenassoziation als Ausdruck der Vegetation. In: Willmanns, O. & Tüxen, R. (Eds.). Werden und Vergehen von Pflanzengesellschaften, pp 253–261. Cramer, Vaduz.

Jelgersma, S., De Jong, J., Zagwijn, W.H., Regteren Altena, J.F. (1970). The coastal dunes of the Western Netherlands; geology, vegetational history and archeology. Meded. Rijks Geol. Dienst Nieuwe Serie no. 21.

Jensen, A. (1978). Skallingen. In: Jensen, A. & Ovesen, C.H. (Eds.). Drift og pleje af våde områder i de nordiske lande, pp 164–172. Report Botanical Institute, Univ. Aarhus.

Jensen, A. (1985). The effect of cattle and sheep grazing on salt-marsh vegetation at Skallingen, Denmark. Vegetatio 60: 37–48.

Jensen, A., Blicher-Mathiesen, G. & Skovhus, K. (1987). Kunstgødning af marsk-og strandeng. Indflydelse på udbytte og kvalitet af planteproduktionen i naturlige plantesamlund. Marginaljorder og miljøinteresser Tekniker Rapport 30.

Jerling, L. (1982). Population dynamics of a perennial herb (Plantago maritima) along a distributional gradient- a demographic study. Thesis, Univ. Stockholm.

Job, D.A. & Taylor, J.A. (1978). The production, utilization and management of upland grazings on Plynlimon, Wales. J. Biogeogr. 5: 173–191.

Joenje, W. (1978). Plant colonization and succession on embanked sandflats; a case study in the Lauwerszeepolder. Thesis, Univ. Groningen.

Joenje, W. (1985). The significance of grazing in the primary vegetation succession on embanked sandflats. Vegetatio 62: 399–406.

Joenje, W. & Thalen, D.C.P. (1968). Het groene strand van Schiermonnikoog. De Levende Natuur 71: 97–107.

Joenje, W. & Wolff, W.J. (1979). Functional aspects of salt marshes in the Wadden Sea area. In: Wolff, W.J. (Ed.). Flora and vegetation of the Wadden Sea. Report 3 of the Wadden Sea Working Group, pp 161–170. Balkema, Rotterdam.

Kapfer, A. (1987). Untersuchungen zur Renaturierung ehemaliger Streuwiesen im Südwestdeutschen Alpenvorland- Erste Tendenzen nach 4 Versuchsjahren. In: Schubert, R. & Hilbig, W. (Eds.). Erfassung und Bewertung Anthropogener Vegetationsveränderungen, pp 179–215. Wissenschaftliche Beiträge Martin Luther Univ., Halle-Wittenberg.

Kapfer, A. & Pfadenhauer, J. (1986). Vegetationskundliche Untersuchungen zur Pflege von Pfeiffengrass-Streuwiesen. Nat. Landschaft 61: 428–432.

Kasteleijn, H.W. (1985). Beweiding op de Oosterkwelder van Schiermonnikoog. Internal Report Dept. of Plant Ecology, Univ. Groningen/Research Institute for Nature Management, Leersum (mimeo).

Kauppi, M. (1967). Ueber den Einfluss der Beweidung auf die Vegetation der Uferwiesen an der Bucht Liminganlahti im Nordteil des Bottnischen Meerbusens. Aquilo Ser. Bot. 6: 347–369.

Keefe, C.W. (1972). Marsh production: a summary of the literature. Contrib. Mar. Sci. 16: 164–181.

Kempf, N., Lamp, J. & Prokosch, P. (1987). Salzwiezen: geformt von Küstenschutz, Landwirtschaft oder Natur? Tagungsbericht I der Umweltstiftung WWF-Deutschland, Husum.

Kenney, P.A. & Black, J.L. (1984). Factors affecting diet selection by sheep. I. Potential intake rate and acceptability of feed. Aust. J. Agric. Res. 35: 551–563.

Kershaw, K.A. (1958). An investigation of the structure of a grassland community. I. The pattern of Agrostis tenuis. J. Ecol. 46: 571–592.

Kettner, P. (1972). Primary production of salt marsh communities on the Island of Terschelling in the Netherlands. Thesis, Univ. Utrecht.

Keuning, H.J. & Naarding, J. (1963). Het Esdorp. Bijdragen en Meded. Naamkundecommissie K. Ned. Akad. Wet. 20.

Kinzel, H. (1982). Pflanzenökologie und Mineralstoffwechsel. Ulmer, Stuttgart.

Kirby, C.J. & Gosselink, J.G. (1976). Primary production in Louisiana Gulf coast Spartina alterniflora marsh. Ecology 57: 1052–1059.

Kirkwood, R.C. (1964). A study of some factors causing mat formation of reseeded hill pasture swards. J. Br. Grassl. Soc. 19: 387–395.

Klaasen, H.P. (1961). Het Heggenlandschap in Europa. Nat. Landschap 15: 80–90.

Klapp, E. (1965). Grünlandvegetation und ihre Standort. Parey, Berlin.

Klapp, E. (1971). Wiesen und Weiden. Parey, Berlin.

Klötzli, F. (1965). Qualität der Rehäsung. Veröff. Geobot. Inst. Eidg. Tech. Hochsch. Stift. Rübel Zür. 38: 1–186.

Kmoch, H.G. (1952). Ueber den Umfang und einige Gesetzmässigkeiten der Wurzelmassenbildung unter Grasnarben. Z. Acker Pflanzenb. 95: 363–380.

Knörzer, K.H. (1971). Urgeschichtler Unkräuter im Rheinland; ein Beitrag zur Entstehungsgeschichte der Segetalgesellschaften. Vegetatio 23: 89–111.

Knörzer, K.H. (1975). Entstehung und Entwicklung der Grünlandvegetation im Rheinland. Decheniana 127: 195–214.

Kocks, G.H. (1979). Gemeente Vries, het land van de Zweedse Kornoelje. Knoop, Haren.

Koeman, C. (1970). Een Franse topografische kaart van Drenthe uit de jaren 1811–1813. Nieuwe Drentse Volksalmanak 88: 89–101.

Kogel, T.J. & De Jong, D.J. (1983). Vegetatiekartering van de schorren in de Oosterschelde en het Krammer-Volkerak, 1978. Nota 80.20. Rijkswaterstaat, Deltadienst, Middelburg.

Kooi, P.B. (1979). Pre-Roman urnfields in the North of the Netherlands. Thesis, Univ. Groningen.

Koopmans, J. (1960). Het productie niveau onderzoek. II. De bemesting van grasland in de praktijk. Versl. Landbouwkd. Onderz. 66.5: 35–39.

Kopecký, K. & Hejný, S. (1974). A new approach to the classification of anthropogenic plant communities. Vegetatio 29: 17–20.

Kopecký, K. & Hejný, S. (1978). Die Anwendung einer 'deduktiven Methode syntaxonomischer Klassifikation' bei der Bearbeitung der strassenbegleitenden Pflanzengesellschaften Nordostböhmens. Vegetatio 36: 43–51.

Körner-Grohne, U. (1967). Geobotanische Untersuchungen auf der Feddersen Wierde. Feddersen Wierde 1: 1–358.

Korevaar, H. (1986). Produktie en voederwaarde van gras bij gebruiks- en bemestingsbeperkingen voor natuurbeheer. Thesis, Univ. Wageningen.

Kottmann, H., Schnöppe, W., Willers, T. & Wittig, R. (1985). Heath conservation by sheep grazing: a cost-benefit analysis. Biol. Conserv. 31: 67–74.

Koutstaal, B.P., Markusse, M.M. & De Munck, W. (1987). Aspects of seed dispersal by tidal movements. In: Huiskes, A.H.L., Blom, C.W.P.M. & Rozema, J. (Eds.). Vegetation between land and sea, pp 226–233, Junk, Dordrecht.

Kruijne, A.A., De Vries, D.M. & Mooi, H. (1967). Bijdrage tot de oecologie van Nederlandse graslandplanten. Versl. Landbouwkd. Onderz. 696. Pudoc, Wageningen.

Krüsi, B. (1981). Phenological methods in permanent plot research. Veröff. Geobot. Inst. Erdg. Tech. Hochsch. Stift. Rübel Zür. 75: 1–115.

Kubitzki, K. (1961). Zur Synchronisierung der nordwesteuropäischen Pollendiagramme. Flora 150: 43–72.

Küchler, A.W. (1967). Vegetation mapping. Ronald Press Company, New York.

Kuiper, E., De Graaf, A. & Vos, H. (1981). Ervaringen met begrazing bij de Vereniging tot Behoud van Natuurmonumenten in Nederland. Internal Report Vereniging tot Behoud van Natuurmonumenten in Nederland, 's Graveland (mimeo).

Lambert, A.M. (1971). The making of the Dutch landscape. Seminar, London.

Lange, E. (1976). Zur Entwicklung der natürlichen und anthropogenen Vegetation in frühgeschichtlicher Zeit. Feddes Repert. 87: 5–30.

Langlands, K.P. & Bennett, I.L. (1973). Stocking intensity and pastoral production. II. Herbage intake of Merino sheep grazed at different stocking rates. J. Agric. Sci. Camb. 81: 205–209.

Lantinga, E. (1985). Productivity of grasslands under continuous and rotational grazing. Thesis, Univ. Wageningen.

Laws, R.M. (1970). Elephants as agents of habitat and landscape change in East Africa. Oikos 21: 1–15.

Leege, T.A., Herman, D.J. & Zamora, B. (1981). Effects of cattle grazing on mountain meadows in Idaho. J. Range Manage. 34: 324–328.

Leertouwer, J. (1966). Onderzoek naar macro- en microgradienten in pH and kalkgehalte in relatie met de vegetatie op Schiermonnikoog. Internal Report Dept. of Plant Ecology, Univ. Groningen (mimeo).

Levitt, J. (1972). Responses of plants to environmental stresses. Academic Press, New York.

Linkola, K. (1930). Ueber das Vorkommen von Samenkeimlingen bei Pollakanthen in den natürlichen Pflanzengesellschaften. Ann. Zool. Soc. Zool. Bot. Fenn. Vanamo 2: 150–169.

Lippe, E., De Smidt, J.T. & Glenn-Lewin, D.C. (1985). Markov models and succession: a test from a heathland in the Netherlands. J. Ecol. 73: 775–791.

Little, D.A. (1982). Utilization of minerals. In: Hacker, J.B. (Ed.). Nutritional limits to animal production from pastures, pp 259–283. C.A.B. Farnham-Royal, United Kingdom.

Lodge, R.W. (1959). Cynosurus cristatus. Biological flora of the British Isles. J. Ecol. 47: 511–518.

Looijen, R.C. (1983). Grazing pressure and vegetation dynamics at the Oosterkwelder salt marsh on Schiermonnikoog. Acta Bot. Neerl. 33: 360–361.

Looijen, J., Looijen, R. & Zomer, H. (1984). Tien jaar begrazing op de Oosterkwelder van Schiermonnikoog. Internal Report Dept. of Plant Ecology, Univ. Groningen/Research Institute for Nature Management, Leersum (mimeo).

Looijen, R.C. & Bakker, J.P. (1987). Utilization of different salt marsh plant communities by cattle and geese. In: Huiskes, A.H.L., Blom, C.W.P.M. & Rozema, J. (Eds.). Vegetation between land and sea, pp 52–64. Junk, Dordrecht.

Londo, G. (1971). Patroon en proces in duinvalleivegetaties langs een gegraven meer in de Kennemerduinen. Thesis, Univ. Nijmegen.

Londo, G. (1974). Successive mapping of dune slack vegetation. Vegetatio 29: 51–61.

Londo, G. (1976). The decimal scale for relevés of permanent quadrats. Vegetatio 33: 61–64.

Londo, G. (1982 A). Natuurbouw, een middel om meer natuur te krijgen. Tijdschr. K. Ned. Heide Mij. 93: 273–278.

Londo, G. (1982 B). Kleinschalige natuurbouw proeftuin Scherpenzeel. Plantesoorten 1967–1976. Report 82/16. Research Institute for Nature Management, Leersum.

Londo, G. (1984). Zijn uitplanten en uitzaaien zinvolle maatregelen bij het natuurbeheer? De Levende Natuur 85: 130–135.

Londo, G. (1988). Nederlandse freatofyten. Pudoc, Wageningen.

Loopstra, I.L. & Van Der Maarel, E. (1984). Toetsing van de ecologische soortengroepen in de Nederlands flora aan het systeem van indicatiewaarden volgens Ellenberg. Report 381 Rijksinstituut voor onderzoek in de bos- en landschapsbouw 'De Dorschkamp', Wageningen.

Ludlow, M.M., Stobbs, T.H., Davis, R. & Charles-Edwards, D.A. (1982). Effect of sward structure of two tropical grasses with contrasting canopies on light distribution, net photosynthesis and size of site harvested by grazing cattle. Aust. J. Agric. Res. 33: 187–201.

Mansholt, S.L. (1969). Landbouwstructuurbeleid in de EEG. Landbouwkd. Tijdschr. 81: 208–213.

Marrs, R.H. (1985). Techniques for reducing soil fertility for nature conservation purposes: a review in relation to research at Roper's Heath, Suffolk, England. Biol. Conserv. 34: 307–332.

Mc Arthur, R.H. & Wilson, E.O. (1967). The theory of island

biogeography. Princeton University Press, Princeton.
Mc Naughton, S.J. (1979). Grazing as an optimization process: grass ungulate relationships in the Serengeti. Am. Nat 113: 691–703.
Mc Naughton, S.J. (1983). Serengeti grassland ecology: the role of composite environmental factors and contingency in community organization. Ecol. Monogr. 53: 291–320.
Mc Naughton, S.J. (1984). Grazing lawns: animals in herds, plant form and coevolution. Am. Nat. 124: 863–886.
Melman, T.C.P., Clausman, P.H.M.A. & Udo De Haes, H.A. (1985). Voedselrijkdom-indicatie van graslanden. Meded. 19 Centrum voor Milieukunde, Univ. Leiden.
Mennema, J. & Weeda, E.J. (1983). Vascular plants. In: Dijkema, K.S. & Wolff, W.J. (Eds.). Flora and vegetation of the Wadden Sea islands and coastal areas. Report 9 of the Wadden Sea Working Group, pp 38–50. Balkema, Rotterdam.
Miedema, M. (1983). Vijfentwintig eeuwen bewoning in het terpen gebied ten NW van Groningen. Thesis, Univ. Amsterdam.
Miles, J. (1974). Effects of experimental interference with stand structure on establishment of seedlings in Callunetum. J. Ecol. 62: 675–687.
Miles, J. (1985). The pedogenic effects of different species and vegetation types and the implications of succession. J. Soil Sci. 36: 571–584.
Miles, J. (1987). Soil variation caused by plants: a mechanism of floristic change in grassland? In: Van Andel, J. Bakker, J.P. & Snaydon, R.W. (Eds.). Disturbance in grasslands, pp 37–49. Junk, Dordrecht.
Milner, C. & Gwynne, D. (1974). The Soay sheep and their food supply. In: Jewell, P.A., Milner, C. & Morton Boyd, J. (Eds.). Island survivors: the ecology of the Soay Sheep of St. Kilda, pp 273–325. Athlone Press, London.
Milner, C. & Hughes, R.E. (1968). Methods for the measurement of the primary production of grassland. IBP Handbook No. 6. Blackwell, Oxford.
Milton, W.E.J. (1939). The occurrence of buried viable seeds in soils at different elevations and on a salt marsh. J. Ecol. 27: 149–159.
Milton, W.E.J. (1953). The palatability of herbage on undeveloped grasslands in West Central Wales. Emp. J. Exp. Agric. 21: 116–122.
Montag, A. (1980). Erfahrungen mit der Ausbringung von Pflanzenarten in Niedersachsen. In: Ausbringung von Wildpflanzen, pp 51–55. Tagungsbericht 5/80 der Akademie für Naturschutz und Landschaftspflege, Laufen-Salzach.
Moore, I. (1966). Grass and grasslands. Collins, London.
Morris, M.G. (1978). Grassland management and invertebrate animals – a selective review. Sci. Proc. R. Dublin Soc. Series A 6: 247–257.
Morris, M.G. (1979). Responses of grassland invertebrates to management by cutting. II. Heteroptera. J. Appl. Ecol. 16: 417–432.
Morris, M.G. (1981). Responses of grasland invertebrates by cutting. IV. Positive responses of Auchenorhyncha. J. Appl. Ecol. 18: 763–771.
Müller, P. (1955). Verbreitungsbiologie der Blütenpflanzen. Veröff. Geobot. Inst. Eidg. Tech. Hochsch. Stift. Rübel Zür. 30: 1–152.
Murphy, J. & Riley, J.P. (1962). A modified single solution method for the determination of phosphate in natural waters. Ann. Chim. 27: 31–36.
Mølgaard, P. (1977). Comparative effect of grass on establishment and performance of Taraxacum officinale. Oikos 29: 376–382.

Nagy, J.G. & Tengerdy, R.P. (1968). Anti-bacterial action of essential oils of Artemisia as an ecological factor. Appl. Microbiol. 16: 441–444.
Naveh, Z. & Whittaker, R.H. (1980). Structural and floristic diversity of shrublands and woodlands in northern Israel and other Mediterranean areas. Vegetatio 41: 171–190.
Nicholson, I.A., Paterson, I.S. & Currie, A. (1970). A study of vegetational dynamics: selection by sheep and cattle in Nardus pasture. In: Watson, A. (Ed.). Animal populations in relation to their food resources. Symp. British Ecological Society no. 10, pp 129–143. Blackwell, Oxford.
Niemann, E. (1963). Beziehungen zwischen Vegetation und Grundwasser. Arch. Naturschutz Landschaftsforsch. 3: 3–36.
Niemann, E. (1973). Grundwasser und Vegetationsgefüge. Nova Acta Leopold. 6 Band 38: 1–147.
Nilsson, J. (1970). Notes on the biomass and productivity of below-ground organs of a South-Swedish hay-meadow. Bot. Not. 123: 183–194.
Norder, J. & Ruyter, J.C. (1977). Vijf jaar beweiding op de Oosterkwelder, Schiermonnikoog. Internal Report Dept. of Plant Ecology, Univ. Groningen/Research Institute for Nature Management, Leersum (mimeo).
Norman, M.J.T. (1957). The influence of various grazing treatments upon the botanical composition of a downland permanent pasture. J. Br. Grassl. Soc. 12: 246–256.

Oberdorfer, E. (1983). Süddeutsche Pflanzengesellschaften. III. Fischer, Jena.
Odum, E.P. (1959). Fundamentals of ecology. Saunders, Philadelphia.
Oomes, M.J.M., Korevaar, H. & Altena, H.J. (1980). Productie en botanische samenstelling van extensief gebruikt grasland. Internal Report Centre for Agrobiological Research, Wageningen (mimeo).
Oomes, M.J.M. & Mooi, H. (1981). The effect of cutting and fertilizing on the floristic composition and production of an Arrhenatherion elatioris grassland. Vegetatio 47: 233–239.
Oomes, M.J.M. & Mooi, H. (1985). The effect of management on succession and production of formerly agricultural grassland after stopping fertilization. In: Schreiber, K.F. (Ed.). Sukzession auf Grünlandbrachen, pp 59–67. Münstersche Geogr. Arb. 20.
Oosterveld, P. (1975). Beheer en ontwikkeling van natuurreservaten door begrazing. Nat. Landschap 29: 161–171.
Oosterveld, P. (1976). Integratie van voormalige landbouwgronden door middel van een extensief graasbeheer met IJs-

landse ponies in de Baronie Cranendonck. Contactblad voor Oecologen 12: 99–109.

Oosterveld, P. (1979). Maaien, grazen of stuiven. Duin 2: 3–8.

Oosterveld, P. (1983). Eight years of monitoring rabbits and vegetation development on abandoned arable fields grazed by ponies. Acta Zool. Fenn. 174: 71–74.

Opperman, R. (1987). Tierökologische Untersuchungen zum Biotopmanagement in Feuchtwiesen. Ergebnisse einer Feldstudie an Schmetterlingen und Heuschrecken in Württenbergischen Alpenvorland. Nat. Landschaft 62: 235–241.

Osbourn, D.F. (1980). The feeding value of grass and grass products. In: Holmes, W. (Ed.). Grass, its production and utilization, pp 70–123. Blackwell, Oxford.

Otto, H.A. (1973). Het gedrag van runderen op de Groede, Terschelling, met de nadruk op graasaspecten. Internal Report Dept. of Nature Conservation, Univ. Wageningen/Research Institute for Nature Management, Leersum (mimeo).

Packham, J.R. & Liddle, M.J. (1970). The Cefni salt marsh, Anglesey, and its recent development. Field Stud. 3: 331–356.

Pannekoek, G. (1982). Onderzoek naar de veranderingen in de vegetatie en bodemvruchtbaarheid van dertien blauwgraslandpercelen. Internal Report Centre for Agrobiological Research, Wageningen (mimeo).

Pape, J.C. (1970). Plaggen soils in the Netherlands. Geoderma 4: 229–255.

Parsons, A.J., Colett, B. & Lewis, J. (1983). Changes in the structure and physiology of a perennial ryegrass sward when released from a continuous stocking management: implications for the use of exclusion cages in continuously stocked swards. Grass Forage Sci. 39: 1–9.

Parsons, A.J., Leafe, E.L., Colett, B., Penning, P.D. & Lewis, J. (1984). The physiology of grass production under grazing. II. Photosynthesis, crop growth and animal intake of continuously grazed swards. J. Appl. Ecol. 20: 127–139.

Peet, R.K., Glenn-Lewin, D.C. & Walker Wolf, J. (1983). Prediction of man's impact on plant species diversity. In: Holzner, W., Werger, M.J.A. & Ikusima, I. (Eds.). Man's impact on vegetation, pp 41–54. Junk, The Hague.

Pegtel, D.M. (1976). On the ecology of two varieties of Sonchus arvensis L. Thesis, Univ. Groningen.

Pegtel, D.M. (1987). Soil fertility and the composition of seminatural grassland. In: Van Andel, J., Bakker, J.P. & Snaydon, R.W. (Eds.). Disturbance in grasslands, pp 37–50. Junk, Dordrecht.

Penman, H.L. (1948). Natural evaporation from open water, bare soil and grass. Proc. R. Soc. Acad. 193: 120–145.

Pennington, W. (1969). The history of British vegetation. English University Press, London.

Perkins, D.F. (1978). The distribution and transfer of energy and nutrients in the Agrostis-Festuca grassland ecosystem. In: Heal, O.W. & Perkins, D.F. (Eds.). Production ecology of British Moors and Montane Grasslands, pp 375–396. Springer, Berlin.

Perkins, D.F., Jones, V., Millar, R.O. & Neep, P. (1978). Primary production, mineral nutrients and litter decomposition in the grassland ecosystem. In: Heal, O.W. & Perkins D.F. (Eds.). Production ecology of British Moors and Montane grasslands, pp 304–331. Springer, Berlin.

Persson, S. (1981). Ecological indicator values as an aid in the interpretation of ordination diagramms. J. Ecol. 69: 71–84.

Persson, S. (1984). Vegetation development after the exclusion of grazing cattle in a meadow area in the south of Sweden. Vegetatio 55: 65–92.

Pfadenhauer, J. & Maas, D. (1987). Samenpotential in Niedermoorböden des Alpenvorlandes bei Grünlandnutzung unterschiedlicher Intensität. Flora 179: 85–97.

Pfadenhauer, J., Kapfer, A. & Maas, D. (1987). Renaturierung von Futterwiesen auf Niedermoortorf durch Aushagerung. Nat. Landschaft 62: 430–434.

Pieters, J.H. (1971). Een bodemvruchtbaarheidsonderzoek op grasland in de Friese Wouden. Report 12 Institute for Soilfertility, Haren.

Pigott, C.D. (1956). The vegetation of upper Teesdale in the North Pennines. J. Ecol. 44: 545–586.

Plewczynska-Kuras, U. (1974). Analysis of a sheep pasture ecosystem in the Pieniny mountains (The Carpathians). IV. Biomass of the upper and underground parts of plants and of organic detritus. Ekol. Pol. 22: 517–526.

Poissonet, P., Romane, F., Thiault, M. & Trabaud, L. (1978). Evolution d'une garrigue de Quercus coccifera soumise a divers traitements: quelques résultats des cinq premières années. Vegetatio 38: 135–142.

Postma, H. (1982). Hydrography of the Wadden Sea: movements and properties of water and particulate matter. Report 2 of the Wadden Sea Working Group. Balkema, Rotterdam.

Pott, R. (1986). Der pollenanalytische Nachweis extensiver Waldbewirtschaftungen in den Haubergen des Siegerlandes. In: Behre, K.E. (Ed.). Anthropogenic indicators in pollendiagrams, pp 125–134. Balkema, Rotterdam.

Pratt, R.M., Putman, R.J., Ekins, J.R. & Edwards, P.J. (1986). Use of habitat by free-ranging cattle and ponies in the New Forest, southern England. J. Appl. Ecol. 23: 539–557.

Prentice, I.C., Van Tongeren, O. & De Smidt, J.T. (1987). Simulation of heathland vegetation dynamics. J. Ecol. 75: 203–219.

Prillevitz, F.C. (1969). Gronden uit de cultuur. Landbouwkd. Tijdschr. 81: 311–319.

Prins, F.W. (1974). Beweiding op de Oosterkwelder, Schiermonnikoog. Internal Report Dept. of Plant Ecology, Univ. Groningen/Research Institute for Nature management, Leersum (mimeo).

Prins, W.H. (1983). Limits to nitrogen fertilizer on grassland. Thesis, Univ. Wageningen.

Prins, H. & Ydenberg, R. (1978). Grazing and manipulation of food resources by barnacle geese (Branta leucopsis) and brent geese (Branta bernicla) wintering on Schiermonnikoog. Internal Report Dept. of Zoology, Univ. Groningen (mimeo).

Prins, H.H.Th., Ydenberg, R.C. & Drent, R.H. (1980). The interaction of Brent Geese (Branta bernicla) and sea plaintain (Plantago maritima) during spring staging: field observations and experiments. Acta Bot. Neerl. 29: 585–596.

Proefstation voor de Akker- en Weidebouw (1965). Een graslandvegetatiekartering van het ruilverkavelingsgebied 'Vries'. Report 186, Wageningen (mimeo).

Proefstation voor de Akker- en Weidebouw (1966). Een graslandvegetatiekartering van het ruilverkavelingsgebied 'Anlo'. Report 211, Wageningen (mimeo).

Putman, R.J. (1986). Grazing in temperate ecosystems. Large herbivores and the ecology of the New Forest. Croom Helm, London.

Pyke, G.H. (1984). Optimal foraging theory, a critical review. Annu. Rev. Ecol. Syst. 15: 523–575.

Pyke, G.H., Pulliman, H.R. & Charnov, E.L. (1977). Optimal foraging: a selective review of theory and tests. Q. Rev. Biol. 52: 137–154.

Quarrie, S.A. (1981). Genetic variability and inheritability of drought-induced abscisic acid accumulation in spring wheat. Plant Cell Environ. 4: 147–151.

Rahmann, M., Rahmann, H., Kempf, N., Hoffmann, B. & Gloger (1987). Auswirkungen unterschiedlicher landwirtschaftlicher Nutzung auf die Flora und Fauna der Salzwiesen an der ostfriesischen Wattenmeerküste. Senckenb. Marit. 19: 163–193.

Ranwell, D.S. (1968). Coastal marshes in perspective. Univ. Strathclyde Regional Studies Group Bull. 9: 1–26.

Ratcliffe, D.A. (1984). Post-medieval and recent changes in British vegetation: the culmination of human influence. New Phytol. 98: 73–100.

Rawes, M. & Heal, O.W. (1978). The blanket bog a part of Pennine Moorland. In: Heal, O.W. & Perkins, D.F. (Eds.). Production ecology of British Moors and Montane grasslands, pp 224–243. Springer, Berlin.

Rawes, M. & Welch, D. (1969). Upland productivity of vegetation and sheep at Moor House National Nature Reserve, Westmoreland, England. Oikos Suppl. 11: 9–72.

Rawes, M. & Welch, D. (1972). Trials to recreate floristically-rich vegetation by plant introduction in the Northern Pennines, England. Biol. Conserv. 4: 135–140.

Regnéll, G. (1980). A numerical study of successions in an abandoned, damp calcareous meadow in S. Sweden. Vegetatio 43: 123–130.

Reitsma, D.T. & Bakker, J.P. (1986). Landgebruik op Schiermonnikoog, vroeger en nu. Waddenbulletin 21: 94–100.

Renes, J. (1981). Typologieën en bewonings- en perceelsvormen. Een overzicht van bestaand werk in Nederland en buurlanden. Pudoc, Wageningen.

Research Institute for Nature Management (1979). Natuurbeheer in Nederland; Levensgemeenschappen. Pudoc, Wageningen.

Rinsema, W.T. (1979). Bemesting en meststoffen. Educaboek, Culemborg.

Roberts, H.A. (1981). Seed banks in soils. Adv. Appl. Biol. 6: 1–56.

Roozen, A.J.M. & Westhoff, V. (1985). A study on long-term salt marsh succession using permanent plots. Vegetatio 61: 23–32.

Rorison, I.H. (1980). The current challenge for research and development. In: Rorison, I.H. & Hunt, R. (Eds.). Amenity grassland, an ecological perspective, pp 3–10. Wiley, Chichester.

Rosén, E. (1973). Sheep grazing and changes of vegetation on the limestone heath of Oeland. Zoon Suppl. 1: 137–151.

Rosén, E. (1982). Vegetation development and sheep grazing in limestone grassland of south Oeland, Sweden. Acta Phytogeogr. Suec. 72: 1–104.

Rosenthal, G. (1984). Vegetationskundliche und ökologische Untersuchungen zur Sukzession in feuchtem Grünland. Internal Report Dept. of Geobotany, Univ. Göttingen (mimeo).

Rosenthal, G., Müller, J. & Cordes, H. (1985). Vegetations- und standortliche Untersuchungen zur Sukzession auf feuchtem Grünland. Verh. Gesellsch. für Oekologie 13: 235–242.

Ross, M.A. & Harper, J.L. (1972). Occupation of biological space during seedling establishment. J. Ecol. 60: 77–88.

Rowell, T.A., Guarino, L. & Harvey, H.J. (1985). The experimental management of vegetation at Wicken Fen, Cambridgeshire. J. Appl. Ecol. 22: 217–227.

Rozema, J. (1976). Vegetatie-zonering op de strandvlakte van Schiermonnikoog. Waddenbulletin 11: 144–148.

Rozema, J., Bijwaard, P., Prast, G. & Broekman, R. (1985). Ecophysiological adaptations of coastal halophytes from foredunes and salt marshes. Vegetatio 62: 499–521.

Runge, F. (1980). Die Pflanzengesellschaften Mitteleuropas. Aschendorf, Münster.

Runge, F. (1985). 21-, 10-, und 8-jährige Dauerquadratuntersuchungen in aufgelassener Grünlandereien. In: Schreiber, K.F. (Ed.). Sukzession auf Grünlandbrachen, pp 45–49. Münstersche Geogr. Arb. 20.

Sarukhán, J.S. & Harper, J.L (1973). Studies on plant demography: Ranunculus repens, R. bulbosus and R. acris. I. Population flux and survivorship. J. Ecol. 61: 675–716.

Scarnecchia, D.L., Nastis, A.S. & Malechek, J.C. (1985). Effects of forage availability on grazing behaviour of heifers. J. Range Manage. 38: 177–180.

Schat, H. (1983). Germination ecology of some dune slack pioneers. Acta Bot. Neerl. 32: 203–212.

Scheltens, J. (1905). Het hout en de bosschen van ons gewest in den germaanschen voortijd. Nieuwe Drentse Volksalmanak 23: 208–226.

Schenkeveld, A.J. & Verkaar, H.J. (1984). On the ecology of short-lived forbs in chalk grasslands: Distribution of germinative seeds and its significance for seedling emergence. J. Biogeogr. 11: 251–260.

Schiefer, J. (1981). Bracheversuche in Baden-Württenberg. Vegetations- und Standortsentwicklung auf 16 verschiedenen Versuchsflächen mit unterschiedlichen Behandlungen. Beih. Veröff. Naturschutz Landschaftsplege, Baden Württ. 22: 1–325.

Schiefer, J. (1982). Kontroliertes Brennen als Landschaftspflegemassnahme? Einige Erfahrungen mit dem kontrollierten Brennen auf Brache-Versuchsflächen. Nat. Landschaft 57: 264–268.

Schiefer, J. (1983). Möglichkeiten der Aushagerung von nährstoffreichen Grünlandflächen. Veröff. Naturschutz Landschaftsflege, Baden Württ. 57/58: 33–62.

Schimmel, H.J.W. (1955). De Drentse beken en beekdalen en hun betekenis voor natuurwetenschap en landschapsschoon. Internal Report State Forestry Commission, Utrecht (mimeo).

Schlichting, F, & Blume, H.P. (1966). Bodenkundliches Praktikum. Parey, Berlin.

Schmeisky, H. (1974). Vegetationskundliche und Oekologische Untersuchungen in Strandrasen des Graswarders vor Heiligenhafen, Ostsee. Thesis, Univ. Göttingen.

Schmeisky, H. (1977). Der Einfluss von Weidetieren auf Salspflanzengesellschaften an der Ostsee. In: Tüxen, R. (Ed.). Vegetation und Fauna, pp 481–498. Cramer, Vaduz.

Schmeisky, H. (1978). Die Vegetationsentwicklung in Dauerquadraten auf dem Graswarder (Heiligenhafen/Ostsee) nach Beeindigung der Beweidung. In: Schmeisky, H. (Ed.). Mitteilungen aus dem Ergänzungsstudium ökologische Umweltsicherung 4/78, pp 71–85. Gesamthochschule, Kassel.

Schmidt, W. (1981). Ungestörte und gelenkte Sukzession auf Brachäckern. Scr. Geobot. 15: 1–199.

Schmidt, W. (1984). Der Einfluss des Mulchens auf die Entwicklung von Ackerbrachen. Ergebnisse aus 15 jährigen Dauerflächenbeobachtungen. Nat. Landschaft 59: 47–55.

Schmidt, W. (1985). Mahd ohne Düngung- Vegetationskundliche und ökologische Ergebnisse aus Dauerflächeuntersuchungen zur Pflege von Brachfläche. In: Schreiber, K.F. (Ed.). Sukzession auf Grünlandbrachen, pp 81–99. Münstersche Geogr. Arb. 20.

Schreiber, K.F. (1980). Brachfläche in der Kulturlandschaft. In: Oekologische Probleme in Agrarlandschaften, pp 61–93. Univ. Hohenheim.

Schreiber, K.F. & Schiefer, J. (1985). Vegetations- und Stoffdynamik in Grünlandbrachen – 10 Jahre Bracheversuche in Baden Württenberg. In: Schreiber, K.F. (Ed.). Sukzession auf Grünlandbrachen, pp 111–153. Münstersche Geogr. Arb. 20.

Schultz, W. (1987). Einfluss der Beweidung von Salzwiesen auf die Vogelfauna. In: Kempf, N., Lamp, J. & Prokosch, P. (Eds.). Salzwiezen: Geformt von Küstenschutz, Landwirtschaft oder Natur?, pp 255–270. Tagungsbericht I der Umweltstiftung WWF-Deutschland.

Schulz, H. (1974). Langjährige Untersuchungen über den Einfluss von Vielschnitt und Beweidung auf Ertrag und Pflanzenbestand einer Dauergrünlandfläche. Z. Acker PflanzenB. 140: 144–156.

Schutter, A., Ter Heerdt, G.N.J. & Bakker, J.P. (1987). Het Westerholt VIII. Veranderingen in de heidevegetatie na 10 jaar beweiden. De Levende Natuur 88: 72–77.

Schuurman, J.J. & Goedewaagen, M.A.J. (1965). Methods for the examination of root system and roots. Pudoc, Wageningen.

Schwaar, J. (1976). Paläogeobotanische Untersuchungen im Belmer Bruch bei Osnabrück. Abh. Naturwiss. Ver. Bremen 38: 207–257.

Schwaar, J. (1980). Möglichkeiten der Arterhaltung durch Neu- und Wiederansiedlung. In: Ausbringung von Wildflanzen, pp 30–40. Tagungsbericht 5/80 der Akademie für Naturschutz und Landschaftspflege, Laufen-Salzach.

Schwaar, J. (1985). Natürliche Vegetationsentwicklung auf Brachland und Wiedereinbürgerung bedrohter Pflanzensippen. Drosera 1: 35–48.

Schwab, R.F. & Schwab-Vos, T. (1974). Beweiding op de Oosterkwelder, Schiermonnikoog. Internal Report Dept. of Plant Ecology, Univ. Groningen/Research Institute for Nature Management, Leersum (mimeo).

Selle, W. (1941). Die Bestockungsanteil der Buche, Hainbuche, Eiche und Birke in Nordwestdeutschland auf Grund von Pollenanalytischen Untersuchungen. Zeitschr. Forst- und Jagdwesen 73: 65–108.

Short, J. (1986). The effect of pasture availability on food intake, species selection and grazing behaviour of kangoroos. J. Appl. Ecol. 23: 559–571.

Siepel, H., Van De Bund, C.F., Van Wingerden, W.K.R.E., Bink, F.A., Bongers, W., Mabelis, A.A., Roelofsen, G.J. Meijer, J. & Den Boer, M.H. (1987). Beheer van graslanden in relatie tot de ongewervelde fauna: ontwikkeling van een monitorsysteem. Report 87/29. Research Institute for Nature Management, Leersum.

Siira, J. (1970). Studies on the ecology of the seashore meadows of the Bothnian Bay with special reference to the Liminka area. Aquilo Ser. Bot. 9: 1–109.

Silvertown, J. (1980). The dynamics of a grassland ecosystem: botanical equilibrium in the Park Grass Experiment. J. Appl. Ecol. 19: 491–504.

Silvertown, J.W. & Dickie, J.B. (1981). Seedling survivorship in natural populations of nine perennial chalk grassland plants. New Phytol. 88: 555–558.

Sissingh, H.A. (1969). Die Lösung der Bodenphosphorsäure bei wässiger Extraktion in Verbindung mit der Entwicklung einer neuen P-Wasser Methode. Sonderheft 2. Z. Landwirtschaftwiss. Forsch. 23/11: 110–120.

Slicher Van Bath, B.H. (1945). Mensch en Land in de Middeleeuwen. Thesis, Univ. Amsterdam.

Slicher Van Bath, B.H. (1963). De agrarische geschiedenis van West-Europa. Spectrum, Utrecht.

Slobodkin, L.B. & Saunders, H.L. (1969). On the contribution of environmental predictability to species diversity. In: Woodwell, G.M. & Smith, H.H. (Eds.). Diversity and stability in ecological systems, pp 82–95. Brookhaven National Laboratory, New York.

Smalley, A.E. (1959). The growth cycle of Spartina and its relation to the insect populations in the marsh. Proc. Salt Marsh Conf. March 1958, Sapelo Island Georgia.

Smit, G., Timmer, V.H.M. & Wieringa, J. (1977). Aspecten van de historische geografie van Langelo. Driemaandelijkse Bladen 29: 129–150.

Snaydon, R.W. (1980). Ecological aspects of management – a

perspective. In: Rorison, I.H. & Hunt, R. (Eds.). Amenity grassland, pp 219–231. Wiley, Chichester.
Snaydon, R.W. & Davies, M.S. (1972). Rapid population differentiation in a mosaic environment. II Morphological variation in Anthoxanthum odoratum. Evolution 26: 390–405.
Snaydon, R.W. & Davies, M.S. (1976). Rapid population differentiation in a mosaic environment. IV. Populations of Anthoxanthum odoratum at sharp boundaries. Heredity 37: 9–25.
Snaydon, R.W. & Howe, C.D. (1986). Root and shoot competition between established ryegrass and invading grass seedlings. J. Appl. Ecol. 23: 677–674.
Snow, A.A. & Vince, S.W. (1984). Plant zonation in an Alaskan salt marsh. II An experimental study of the role of edaphic conditions. J. Ecol. 72: 669–684.
Spedding, C.R.W. (1971). Grassland ecology. Oxford University Press, Oxford.
Steen, E. (1985). Root and rhizome dynamics in a perennial grass crop during an annual growth cycle. Swed. J. Agric. Res. 15: 25–30.
Sterk, A.A., Groenhart, M.C. & Mooren, J.F.A. (1983). Aspects of the ecology of some microspecies of Taraxacum in the Netherlands. Acta Bot. Neerl. 32: 385–415.
Stichting voor Bodemkartering/Provinciale Planologische Dienst Drenthe (1987). Op weg met het verleden. Cultuurhistorische kartering van Drenthe. Provinciaal Bestuur van Drenthe, Assen.
Stobbs, T.H. (1973 A). The effect of plant structure on the intake of tropical pastures. I Variation in the bite size of grazing cattle. Aust. J. Agric. Res. 24: 809–819.
Stobbs, T.H. (1973 B). The effect of plant structure on the intake of tropical pastures. II Differences in sward structure, nutritive value and bite size of animals grazing Setaria anceps and Chloris gayana at various stages of growth. Aust. J. Agric. Res. 24: 821–829.
Stöcklin, J. & Gisi, U. (1985). Bildung und Abbau der Streu in bewirtschafteten und brachliegenden Mähwiesen. In: Schreiber, K.F. (Ed.). Sukzession auf Grünlandbrachen, pp 101–109. Münstersche Geogr. Arb. 20.
Stoddart, L.A., Smith, A.D. & Box, T.W. (1975). Range management. Mc. Graw-Hill, New York.
Streefkerk, J. & Van Hoorn, D. (1985). Hydrologisch onderzoek in het stroomdal van de Drentsche A. Internal Report State Forestry Commission, Utrecht (mimeo).
Stuart Chapin III, F. (1980). The mineral nutrition of wild plants. Annu. Rev. Ecol. Syst. 11: 233–260.
Succow, M. (1971). Die Talmoore der nordostdeutschen Flachlandes, ein Beitrag zur Charakterisierung der Moortyps 'Niedermoor'. Arch. Naturschutz Landschaftsforsch. 11: 133–168.
Sykora, K.V. (1982). Syntaxonomy and synecology of the Lolio-Potentillion Tüxen 1947 in the Netherlands. Acta Bot. Neerl. 31: 65–95.
Sykora, K.V. (1983). The Lolio-Potentillion anserinae Tüxen 1947 in the northern part of the Atlantic domain. Thesis, Univ. Nijmegen.
Taylor, A.A., De-Felice, J. & Havill, D.C. (1982). Seasonal variation in nitrogen availability and utilization in an acidic and calcareous soil. New Phytol. 92: 141–152.
Ter Borg, S.J. (1972). Variability of Rhinanthus serotinus in relation to the environment. Thesis, Univ. Groningen.
Ter Borg, S.J. (1979). Some topics in plant population biology. In: Werger, M.J.A. (Ed.). The study of vegetation, pp 11–55. Junk, The Hague.
Ter Borg, S.J. (1985). Population biology and habitat relations of some hemiparasitic Scrophulariaceae. In: White, J. (Ed.). The population structure of vegetation, pp 463–487. Junk, Dordrecht.
Ter Heerdt, G.N.J. & Schutter, A. (1984). Resultaten van 10 jaar maaien en beweiden op het Westerholt. Internal Report Dept. of Plant Ecology, Univ. Groningen (mimeo).
Ter Heerdt, G.N.J., Schutter, A. & Bakker, J.P. (1986). Het Westerholt VII. Veranderingen in de graslandvegetatie na 10 jaar beweiden. De Levende Natuur 87: 145–149.
Thalen, D.C.P. (1971). Variation in some salt marsh and dune vegetations in the Netherlands with special reference to gradient situations. Acta Bot. Neerl. 20: 327–342.
Thalen, D.C.P. (1981). Grote grazers en snelle snoeiers. In: Verslag van het Veluwe-Symposium, pp 21–39. Gelderse Milieufederatie, Arnhem.
Thalen, D.C.P. (1984 A). Begrazingsbeheer en begrazingsonderzoek: een ontwikkelingsschets. De Levende Natuur 85: 35–40.
Thalen, D.C.P. (1984 B). Large animals as tools in the conservation of diverse habitats. Acta Zool. Fenn. 172: 159–163.
Thalen, D.C.P. (1987). Begrazing in een Nederlands perspectief. In: De Bie, S., Joenje, W. & Van Wieren, S.E. (Eds.). Begrazing in de natuur, pp 3–14. Pudoc, Wageningen.
Thalen, D.C.P., Poorter, H., Lotz, L.A. & Oosterveld, P. (1987). Modelling the structural changes in vegetation under different grazing regimes. In: Van Andel, J., Bakker, J.P. & Snaydon, R.W. (Eds.). Disturbance in grasslands, pp 167–183. Junk, Dordrecht.
't Hart, M.L. (1943). Beweidingsvraagstukken. Landbouwkd. Tijdschr. 55: 728–737.
Thompson, K. & Grime, J.P. (1979). Seasonal variation in the seed banks of herbaceous species in ten contrasting grassland habitats. J. Ecol. 67: 893–921.
Thompson, K. & Grime, J.P. (1983). A comparative study of germination responses to diurnally-fluctuating temperatures. J. Appl. Ecol. 20: 141–156.
Tidelski, F. (1933). Zur Waldgeschichte der Schleswig-Holsteinischen Geest. Schr. Naturwiss. Ver. Schleswig-Holstein 20: 1–19.
't Mannetje, L. (1978). Measuring quantity of grassland vegetation. In: 't Mannetje (Ed.). Measurement of grassland vegetation and animal production, pp 63–95. Commonwealth Agricultural Bureaux, Hurley.
Troels-Smith, J. (1960). Ivy, Mistletoe and Elm, climate indicators, fodder plants. Dan. Geol. Unders. IV. Raekke 4: 1–32.
Turner, R.E. (1976). Geographic variations in salt marsh mac-

rophyte production: a review. Contrib. Mar. Sci. 20: 47–68.

Tutin, T.G., Heywood, V.H., Burges, N.A., Moore, D.M., Valentine, D.H., Walters, S.M. & Webb, D.A. (1964–1980). Flora Europaea I-V. Cambridge University Press, Cambridge.

Tüxen, R. (1974). Die Haselüner Kuhweide. Die Pflanzengesellschaften einer mittelalterlichen Gemeindeweide. Mit. flor.-soz. Arbeitsgem. N.F. 17: 69–102.

Tyler, G. (1969). Regional aspects of baltic shore-meadow vegetation. Vegetatio 19: 60–86.

Tyler, G. (1971). Distribution and turnover of organic matter and minerals in a shore meadow ecosystem. Studies in the ecology of Baltic sea-shore meadows IV. Oikos 22: 265–291.

Tyler, C. (1984). Calcareous fens in South Sweden. Previous use, effects of management and management recommendations. Biol. Conserv. 30: 69–89.

Van Altena, S.C. & Minderhoud, J.W. (1972). Keimfähige Samen von Gräsern und Kräutern in der Narbenschicht der Niederländische Weiden. Z. Acker Pflanzen B. 136: 95–109.

Van Andel, J. & Van Den Bergh, J.P. (1987). Disturbance of grasslands – Outline of the theme. In: Van Andel, J., Bakker, J.P. & Snaydon, R.W. (Eds.). Disturbance in grasslands, pp 3–13, Junk, Dordrecht.

Van Ast, A., Ter Borg, S.J. & Van Groenendael, J.M. (1987). De oorzaak van de achteruitgang van biggekruid in onze bermen. De Levende Natuur 88: 88–93.

Van Bemmel, A.C.V. (1963). Niederländische Heideschafe. Z. Saugetierkd. 28: 248–255.

Van Berkel, B. & Schoemaker, R. (1985). Micro-patronen in de door schapen beweide grasland vegetatie van het Westerholt. II. Mogelijke verklaringen voor het in stand houden van het micro-patroon op basis van verschillen in productiviteit, structuur en kwaliteit van de vegetatie. Internal Report Dept. of Plant Ecology, Univ. Groningen/Research Institute for Nature Management, Leersum (mimeo).

Van Breemen, A.A.M. & Van Leeuwen, B.H. (1983). The seed bank of three short-lived monocarpic species, Cirsium vulgare, Echium vulgare and Cynoglossum officinale. Acta Bot. Neerl. 32: 245.

Van De Laar, J. & Slim, P.A. (1982). Veranderingen op verlaten landbouwgronden in de Baronie Cranendonck met behulp van begrazing door IJslandse pony's. Brabants Landschap 58: 25–38.

Van Den Berg, B.M., Fresco, L.F.M. & Bakker, J.P. (1985). Vegetation succession approached with a non-Markovian transitition model. Acta Bot. Neerl. 34: 427–428.

Van Den Bergh, J.P. (1979). Changes in the composition of mixed populations of grassland species. In: M.J.A. Werger (Ed.). The study of vegetation, pp 57–80. Junk, The Hague.

Van Der Duim, M. (1973). Beweiding op de Oosterkwelder, Schiermonnikoog. Internal Report Dept. of Plant Ecology, Univ. Groningen/Research Institute for Nature Management, Leersum (mimeo).

Van Der Hammen, Th. (1951). Late glacial flora and periglacial phenomena in the Netherlands. Leidse Geol. Meded. 17: 171–183.

Van Der Linden, J.A. (1973). Topografische en militaire kaart van het Koninkrijk der Nederlanden. Fibula-Van Dishoeck, Bussum.

Van Der Maarel, E. (1971). Plant species diversity in relation to management. In: Duffey, E. & Watt, A.S. (Eds.). The scientific management of animal and plant communities for conservation, pp 45–63. Blackwell, Oxford.

Van Der Marel, E. (1980). Vegetation development in a former orchard under different treatments. A preliminary report. Vegetatio 43: 95–102.

Van Der Meijden, R., Weeda, E.J., Adema, F.A.C.B. & De Jonckheere, G.J. (1983 A). Flora van Nederland. Wolters-Noordhoff, Groningen.

Van Der Meijden, R., Arnols, E.J.M., Adema, F., Weeda, E.J. & Plate, C.L. (1983 B). Standaardlijst van de Nederlandse flora 1983. Rijksherbarium, Leiden.

Van Der Vegte, F.W. (1978). Population differentiation and germination ecology in Stellaria media. Oecologia Berl. 37: 231–245.

Van Der Waals, J.D. & Butler, J.J. (1974). Bargeroosterveld. Reallexikon der germanischen Altertumskunde 2: 54–58.

Van Der Woude, J.D. (1983). A late medieval clearing at the edge of the Dinkel valley (eastern Netherlands). Acta Bot. Neerl. 32: 313–321.

Van De Veen, H.E. (1985). Natuurontwikkelingsbeleid en bosbegrazing. Landschap 2: 14–28.

Van De Waerdt, J. (1982). Enquète betreffende natuurtechnische begrazing in Nederland. Deelrapport Friesland, Groningen, Drenthe en Overijssel. Internal Report Research Institute for Nature Management, Leersum (mimeo).

Van Dijk, A.J. & Bakker, J.P. (1980). Beweiding en broedvogels op de Oosterkwelder van Schiermonnikoog. Waddenbulletin 15: 134–140.

Van Duuren, L., Bakker J.P. & Fresco, L.F.M. (1981). From intensively agricultural practices to hay-making without fertilization. Vegetatio 47: 241–258.

Van Dyne, G.M., Brockington, N.R., Szocs, Z., Duek, J. & Ribic, C.A. (1980). Large herbivore subsysytem. In: Breymeyer, A.I. & Van Dyne, G.M. (Eds.). Grasslands, system analysis and man, pp 260–537. Cambridge University Press, Cambridge.

Van Es, W.A. (1967). Wijster, a native village beyond the Imperial frontier. Thesis, Univ. Groningen.

Van Es, A.J.H. (1978). Feed evaluation for ruminants. I. The system in use from May 1977 onwards in the Netherlands. Livest. Prod. Sci. 5: 331–345.

Van Es, A.J.H. (1979). Fysiologische aspecten van plantenvezel in de diervoeding. In: Ruwe celstof of voedingsvezel? De rol van plantecelwandbestanddelen in de voeding van dier en mens, pp 12–21. Symposium, Univ. Wageningen.

Van Gijn, A.L. & Waterbolk, H.T. (1984). The colonization of the salt marshes of Friesland and Groningen: the possibility of a transhumant prelude. Palaeohistoria 26: 101–122.

Van Gool, C.R. & Van Der Hoog, C.A. (1986). De zaadvoorraad in grasland. Internal Report Dept. of Agricultural Plantbreeding and Grassland science, Univ. Wageningen (mimeo).

Van Groenendael, J.M. (1985). Floravervalsing, de mening van een oecoloog. De Levende Natuur 86: 138–142.
Van Hecke, P., Impens, I. & Behaeghe, T.J. (1981). Temporal variation of species composition and species diversity in permanent grassland plots with different fertilizer treatments. Vegetatio 47: 221–232.
Van Heuveln, B. (1980). Vegetation and soil in a Drenthian brook valley (The Netherlands). Acta Bot. Neerl. 29: 555–564.
Van Leeuwen, C.G. (1966). A relation theoretical approach to pattern and process in vegetation. Wentia 15: 25–46.
Van Leeuwen, C.G. (1973). Oecologie en natuurtechniek. Nat. Landschap 27: 57–67.
Van Leeuwen, C.G., Londo, G. & Van Wijngaarden, A. (1971). Verslag van een studiereis naar Gotland en Oeland in 1971. Internal Report Research Institute for Nature Management, Leersum (mimeo).
Van Noordwijk-Puijk, K, Beektink, W.G. & Hogeweg, P. (1979). Vegetation development on salt marsh-flats after disappearance of the tidal factor. Vegetatio 29: 1–13.
Van Soest, P.J. (1975). Voluntary intake in relation to chemical composition and digestibility. J. Anim. Sci. 24: 834–844.
Van Soest, P.J. (1982). Nutritional ecology of the ruminant. O. & B Books, Corvallis.
Van Tooren, B.F., During, H.J. & Lensink, M.J.G. (1987). The influence of the bryophyte layer on the microclimate in chalk grasslands. Abstr. Bot. 9: 219–230.
Van Wieren, S.E. (1977). Het Westerholt I. Geologie, ontginning en bodem. De Levende Natuur 80: 279–286.
Van Wieren, S.E. (1985). Het jaar van de Schotse Hooglander. Huid en Haar 4: 19–30.
Van Wieren, S.E. (1987). Grote herbivoren in het natuurbeheer. In: De Bie, S., Joenje, W. & Van Wieren, S.E. (Eds.). Begrazing in de natuur, pp 200–228. Pudoc, Wageningen.
Van Zeist, C. (1978). De invloed van het maaitijdstip op de vegetatieve en generatieve vermeerdering van graslandplanten, en op de samenstelling van de vegetatie. Internal Report Centre for Agrobiological Research, Wageningen (mimeo).
Van Zeist, W. (1959). Studies on the Post-Boreal vegetational history of South-eastern Drenthe (Netherlands). Acta Bot. Neerl. 8: 156–184.
Van Zeist, W. (1974). Palaeobotanical studies of settlement sites in the coastal area of the Netherlands. Palaeohistoria 16: 223–371.
Van Zeist, W. (1980). Prehistorische cultuurplanten. In: Chamalaun, M. & Waterbolk, H.T. (Eds.). Voltooid verleden tijd?, pp 147–165. Intermediair, Amsterdam.
Van Zeist, W., Van Hoorn, T.C., Bottema, S. & Woldring, H. (1976). An agricultural experiment in the unprotected salt marsh. Palaeohistoria 18: 11–153.
Van Zeist, W. & Palfenier-Vegter, R.M. (1979). Agriculture in medieval Gasselte. Palaeohistoria 21: 267–299.
Van Zeist, W., De Roller, G.J., Palfenier-Vegter, R.M., Harsema, O.H. & During, H. (1986). Plant remains from medieval sites in Drenthe, The Netherlands. Helinium 26: 226–274.
Veenstra, H.J. (1980). Introduction to the geomorphology of the Wadden Sea area. In: Dijkema, K.S., Reineck, H.E., Wolff, W.J. (Eds.). Geomorphology of the Wadden Sea area. Report 1 of the Wadden Sea Working Group, pp 8–19. Balkema, Rotterdam.
Veldman, D. (1983). De invloed van verschillende beheersmaatregelen op de grasland- en heidevegetatie van het Westerholt. Internal Report Dept. of Plant Ecology, Univ. Groningen/Research Institute for Nature Management, Leersum (mimeo).
Vera, F.W.M. (1986). Grote plantenetende zoogdieren. Huid en Haar 5: 214–228.
Verkaar, H.J., Schenkeveld, A.J. & Van De Klashorst, M.P. (1983). The ecology of short-lived forbs in chalk grasslands: dispersal of seeds. New Phytol. 95: 335–344.
Vermeer, J.G. & Berendse, F. (1983). The relationship between nutrient availability, shoot biomass and species richness in grassland and wetland communities. Vegetatio 53: 121–126.
Vermeulen, F.H.B. (1954). Invloed van de afstand tot de boerderij op de vruchtbaarheid van graslanden. Landbouwvoorlichting 11: 424–426.
Vestergaard, P. (1978). Graesningens betydning for vegetationen på øst-danske strandenge. In: Jensen, A. & Ovesen, C.H. (Eds.). Drift og pleje af våde områder a de nordiske lande, pp 144–155. Report Botanical Institute, Univ. Aarhus.
Vestergaard, P. (1985). Effects of mowing on the composition of Baltic salt-meadow communities. Vegetatio 62: 383–390.
Vink, A.P.A. (1975). Landuse in advancing agriculture. Springer, Berlin.
Von Glahn, H. (1987). Zur Bestimmung der in Norddeutschland vorkommenden Quecken (Arten, Unterarten und Bastarde der Gattung Agropyron s.l.) nach vegetativen Merkmalen under besonderer Berücksichtigung der Küsten region. Drosera 12: 1–27.
Von Thünen, J.H. (1842). Der isolierten Staat im Beziehung auf Landwirtschaft und Nationalökonomie. Fischer, Stuttgart.
Vreugdenhil, A. & Van Wieren, S.E. (1976). De veranderingen in de vegetatie na drie jaar maaien en beweiden op het Westerholt. Internal Report Dept. of Plant Ecology, Univ. Groningen/Research Institute for Nature Management, Leersum (mimeo).
Vyvey, Q. (1983). Study of the seed bank in Schoeno-Juncetum subnodulosi and in Molinietum caeruleae. Acta Bot. Neerl. 32: 246–247.

Waisel, Y. (1972). Biology of Halophytes. Academic Press, New York.
Wallentius, H.G. (1973). Above ground primary production of a Juncetum gerardii on a Baltic sea-shore meadow. Oikos 24: 200–219.
Wams, T.J. & Wemelsfelder, F. (1984). Grondwaterkwaliteit en vegetatie op Schiermonnikoog. Internal Report Dept. of Plant Ecology, Univ. Groningen (mimeo).
Warwick, S.I. & Briggs, D. (1979). The genecology of lawn weeds. III. Cultivation experiments with Achillea millefolium, Bellis perennis, Plantago major and Prunella vulgaris

collected from lawns and contrasting grassland habitats. New Phytol. 83: 509–536.
Waterbolk, H.T. (1954). De praehistorische mens en zijn milieu. Thesis, Univ. Groningen.
Waterbolk, H.T. (1964). The bronzen age settlement of Elp. Helinium 4: 97–131.
Waterbolk, H.T. (1973 A). De oorspronkelijke structuur der Drentse zanddorpen. In: Archeologie en Historie (Festschrift Brunsting), pp 429–439. Fibula – Van Dishoek, Bussum.
Waterbolk, H.T. (1973 B). Odoorn im frühen Mittelalter. Bericht der Grabung 1966. Neue Ausgrabungen und Forschungen in Niedersachsen 8: 25–89.
Waterbolk, H.T. (1976). Oude bewoning in het Waddengebied. In: Abrahamse, J., Joenje, W. & Van Leeuwen-Seelt, N. (Eds.). Waddenzee, pp 210–221. Landelijke Vereniging tot Behoud van de Waddenzee, Harlingen.
Waterbolk, H.T. (1977). Walled enclosures of the Iron Age in the north of the Netherlands. Palaeohistoria 19: 97–172.
Waterbolk, H.T. (1979). Siedlungskontinuität im Küstengebiet der Nordsee zwischen Rhein und Elbe. Probleme der Küstenforschung im südlichen Nordseegebiet 13: 1–21
Waterbolk, H.T. (1982). Mobilität von Dorf, Ackerflur und Gräberfeld in Drenthe seit der Latènezeit. Offa 39: 97–137.
Waterbolk, H.T. (1984). Gebruik van het landschap in de Romeinse tijd en de Vroege Middeleeuwen. In: Abrahamse, J., Bottema, S., Buruma, M.H. De Vroome, H.W. & Waterbolk, H.T. (Eds.). Het Drentse landschap, pp 40–47. Walburg Pers. Zutphen.
Waterbolk, H.T. (1985). Archeologie. In: Heringa, J., Blok, D.P., Buist, M.G. & Waterbolk, H.T. (Eds.). Geschiedenis van Drenthe, pp 15–90. Boom, Meppel.
Waterbolk, H.T. & Harsema, O.H. (1979). Medieval farmsteads in Gasselte (Province of Drenthe). Palaeohistoria 21: 227–265.
Wathern, P. & Gilbert, O.L. (1978). Artificial diversification of grassland with native herbs. J. Environ. Manage. 7: 29–42.
Watkinson, A.R. & Davy, A.J. (1985). Population biology of salt marsh and sand dune annuals. Vegetatio 62: 487–497.
Weaver, S.E. & Cavers, P.B. (1979). The effects of date of emergence order on seedling survival rates in Rumex crispus and R. obtusifolius. Can. J. Bot. 57: 730–738.
Weber, H.E. (1986). Waldumwandlung durch Beweidung in Niederschasen. Nat. Landschaft 61: 330–333.
Weeda, E.J. & Mennema, J. (1983). Phytogeographical aspects. In: Dijkema, K.S. & Wolff, W.J. (Eds.). Flora and vegetation of the Wadden Sea islands and coastal areas. Report 9 of the Wadden Sea Working Group, pp 221–242. Balkema, Rotterdam.
Welch, D. (1984). Studies in the grazing of heather moorland in North-east Scotland. I. Site description and patterns of utilization. J. Appl. Ecol. 21: 179–195.
Welch, D. (1985). Studies in the grazing of heather moorland in North-east Scotland. IV. Seed dispersal and plant establishment in dung. J. Appl. Ecol. 22: 461–472.
Welch, D. (1986). Studies in the grazing of heather moorland in North-east Scotland. V. Trends in Nardus stricta and other unpalatable graminoids. J. Appl. Ecol. 23: 1047–1058.
Welch, D. & Rawes, M. (1965). The herbage production of some Pennine grasslands. Oikos 16: 39–47.
Wells, T.C.E. (1971). A comparison of the effects of sheep grazing and mechanical cutting on the structure and botanical composition of chalk grassland. In: Duffey, E. & Watt, A.S. (Eds.). The scientific management of animal and plant communities for conservation, pp 497–515. Blackwell, Oxford.
Wells, T.C.E. (1980). Management options for lowland grassland. In: Rorison, I.H. & Hunt, R. (Eds.). Amenity grassland. An ecological perspective, pp 175–195. Wiley, Chichester.
Wells, T.C.E. (1983). The creation of species-rich grasslands. In: Warren, A. & Goldsmith, F.B. (Eds.). Conservation in perspective, pp 215–232. Wiley, Chichester.
Wells, T.C.E., Sheail, J., Ball, D.F. & Ward, L.K. (1976). Ecological studies on the Porton Ranges: relationships between vegetation, soils and land-use history. J. Ecol. 64: 589–626.
Wells, T., Bell, S. & Frost, A. (1981). Creating attractive grasslands using native plant species. Nature Conservancy Council, Shrewsbury.
Werger, M.J.A. (1983). Wurzel/Spross-Verhältnis als Merkmal der Pflanzenstrategie. In: Symposium Wurzelökologie und ihre Nutzanwendung, pp 323–334. Bundesanstalt Gumpenstein, Irdning.
Werger, M.J.A., Smeets, P.J.A.M., Helsper, H.P.G. & Westhoff, V. (1978). Oekologie der subalpinen Vegetation des Lauschbachtales. Tirol. Verh. Zool.-Bot. Gesellsch. Wien 116/117: 111–125.
Wesson, G. & Wareing, P.F. (1967). Light requirements of buried seeds. Nature 213: 600–601.
Westhoff, V. (1945). Biologische problemen der natuurbescherming. Inleiding gehouden voor het N.J.N. congres op 15-8-1945 te Drachten, 13 pp. (mimeo).
Westhoff, V. (1947). The vegetation of dunes and salt marshes on the Dutch islands of Terschelling, Vlieland and Texel. Thesis, Univ. Utrecht.
Westhoff, V. (1949). Schaakspel met de natuur. Nat. Landschap 3: 54–62.
Westhoff, V. (1971). The dynamic structure of plant communities in relation to the objects of conservation. In: Duffey, E. & Watt, A.S. (Eds.). The scientific management of animal and plant communities for conservation, pp 3–14. Blackwell, Oxford.
Westhoff, V. (1983). Man's attitude towards vegetation. In: Holzner, W., Werger, M.J.A. & Ikusima, I. (Eds.). Man's impact on vegetation, pp 7–24. Junk, The Hague.
Westhoff, V. (1985). Nature management in coastal areas of Western Europe. Vegetatio 62: 523–532.
Westhoff, V. (1987). Salt marsh communities of three West Frisian islands, with some notes on their long-term succession during half a century. In: Huiskes, A.H.L., Blom, C.W.P.M. & Rozema, J. (Eds.). Vegetation between land and sea, pp 16–40. Junk, Dordrecht.
Westhoff, V. & Van Dijk, J. (1952). Experimenteel successie-

onderzoek in natuurreservaten, in het bijzonder in het Korenburgerveen bij Winterswijk. De Levende Natuur 55: 5–16.
Westhoff, V. & Den Held, A.J. (1969). Plantengemeenschappen in Nederland. Thieme, Zutphen.
Westhoff, V., Bakker, P.A., Van Leeuwen, C.G., Van der Voo, E. & Westra, R. (1970). Wilde Planten I. Natuurmonumenten, Amsterdam.
Westhoff, V., Bakker, P.A., Van Leeuwen, C.G., Van der Voo, E.E. & Zonneveld, J.S. (1973). Wilde Planten III. Natuurmonumenten, Amsterdam.
Westhoff, V. & Sykora, K.V. (1979). A study on the influence of desalination on the Juncetum gerardii. Acta Bot. Neerl. 28: 505–512.
Westhoff, V., Schamineé, J. Sykora, K.V. (1983). Aufzeichnungen zur Vegetation der Schwedischen Inseln Oeland, Gotland and Stora Karlsö. Tuexenia 3: 179–198.
Wetzel, D. (1966). Pflanzenbestand, Nutzung und Futterwert auf Grünland des norddeutschen Küstenvorlands. Kali Briefe 4: 1–6.
Wheeler, B.D. (1983). Vegetation, nutrients and agricultural land use in a north Buckinghamshire valley fen. J. Ecol. 71: 529–544.
Wheeler, B.D. & Giller, K.E. (1982). Species richness of herbaceous fen vegetation in broadland, Norfolk, in relation to the quantity of above-ground plant material. J. Ecol. 70: 179–200.
White, D.J.B. (1961). Some observations on the vegetation of Blakeney Point, Norfolk, following the disappearance of the rabbits in 1954. J. Ecol. 49: 113–118.
Whittaker, R.H. (1965). Dominance and diversity in land plant communities. Science 147: 250–260.
Whittaker, R.H. (1972). Evolution and measurement of species diversity. Taxon 21: 213–251.
Whittaker, R.H. & Gauch, H.G. (1973). Evaluation of ordination techniques. In: Whittaker, R.H. (Ed.). Handbook of Vegetation Science. Part V. Ordination and classification of communities, pp 289–321. Junk, The Hague.
Wielgolaski, F.E. (1975). Comparison of plant structure on grazed and ungrazed tundra meadows. In: Wielgolaski, F.E. (Ed.). Fennoscandian tundra ecosystems. I. Plants and micro-organisms, pp 86–93. Springer, Berlin.
Wielgolaski, F.E. (1976). The effect of herbage intake by sheep on primary production, ratios top-root dead-live aboveground parts (Hardangervidda, Norway). Pol. Ecol. Stud. 2: 67–76.
Wieniger, F.M. (1936). Nachtrag zur Schnellstickstoffbestimmungsmethode mit Silen als Beschleuniger. Wochenschr. für die Brauerei 53: 251–252.
Wieringa, J. (1968). Drentse veldnamen. Boor en Spade 16: 110–118.
Wieringa, J. (1979). Veldnamen. In: Kocks, G.H. (Ed.). Gemeente Vries, het land van de Zweedse kornoelje, pp 141–153. Knoop, Haren.
Wild, A. (1979). Physiologie der Photosynthese höherer Pflanzen. Die Anpassung an die Lichtbedingungen. Ber. Dtsch. Bot. Ges. 92: 341–364.
Wilkinson, S.R. & Lowry, R.W. (1973). Cycling of mineral nutrients in pasture ecosystems. In: Butler, G.W. & Bailey, R.W. (Eds.). Chemistry and Biochemistry of Herbage, pp 247–315. Academic Press, London.
Willems, J.H. (1980). Observations on northwest European limestone grassland communities. Proc. K. Ned. Akad. Wet. Ser. C 83: 279–306.
Willems, J.H. (1983 A). Species composition and above ground phytomass in chalk grassland with different management. Vegetatio 52: 171–180.
Willems, J.H. (1983 B). The seed bank as a part of the vegetation. Acta Bot. Neerl. 32: 243.
Willems, J.H. (1985). Growth form spectra and species diversity in permanent grassland plots with different management. In: Schreiber, K.F. (Ed.). Sukzession auf Grünlandbrachen, pp 35–43. Münstersche Geogr. Arb. 20.
Williams, E.D. (1978). The botanical composition of the Park Grass plots at Rothamsted 1856–1976. Rothamsted Experimental Station, Harpenden.
Williams, E.D. (1984). Changes during 3 years in the size and composition of the seed bank beneath a long-term pasture as influenced by defoliation and fertilizer regime. J. Appl. Ecol. 21: 603–615.
Willms, W., Bailey, A.W. & Mc Lean, A. (1980). Effect of burning or clipping Agropyron spicatum in the autumn on the spring foraging behaviour of mule deer and cattle. J. Appl. Ecol. 17: 69–84.
Wind, K. (1980). Botanische samenstelling van grasland bij extensivering van het gebruik. Meded. Dept. Landbouwplantenteelt Graslandkd. 52.
Witschi, P.A. & Michalk, D.L. (1979). The effect of sheep treading and grazing on pasture and soil characteristics of irrigated annual pastures. Aust. J. Agric. Res. 30: 741–750.
Wittig, R. (1979). Geschichte, Verbreitung und Funktion der Westfälischen Wallhecken. Natur- und Landschaftskunde Westfalen 15: 1–9.

Zagwijn, W.H. (1974). Hoe de duinen zijn ontstaan. In: Croin Michielsen, N. (Ed.). Meyendel, duin water, leven, pp 46–50. Van Hoeve, Den Haag.
Zeevalking, H.J. & Fresco, L.F.M. (1977). Rabbit grazing and species diversity in a dune area. Vegetatio 35: 193–196.
Zoller, H., Bischof, N., Erhardt, A. & Kienzle, U. (1984). Biocoenosen von Grenzertragsflächen und Brachland in den Berggebieten der Schweiz, Hinweize zur Sukzession zum Naturschutzwert und zur Pflege. Phytocoenologia 12: 373–394.

Subject index

GPSR Compliance
The European Union's (EU) General Product Safety Regulation (GPSR) is a set of rules that requires consumer products to be safe and our obligations to ensure this.

If you have any concerns about our products, you can contact us on

ProductSafety@springernature.com

In case Publisher is established outside the EU, the EU authorized representative is:

Springer Nature Customer Service Center GmbH
Europaplatz 3
69115 Heidelberg, Germany

www.ingramcontent.com/pod-product-compliance
Ingram Content Group UK Ltd.
Pitfield, Milton Keynes, MK11 3LW, UK
UKHW061515220826
474687UK00005B/80
* 9 7 8 0 7 9 2 3 0 0 6 8 7 *